Severe Convective Storms and Tornadoes

Observations and Dynamics

Howard B. Bluestein

Severe Convective Storms and Tornadoes

Observations and Dynamics

Springer

Published in association with
Praxis Publishing
Chichester, UK

PRAXIS

Professor Howard B. Bluestein
School of Meteorology
University of Oklahoma
Norman
Oklahoma
U.S.A.

SPRINGER–PRAXIS BOOKS IN ENVIRONMENTAL SCIENCES

ISBN 978-3-642-05380-1 ISBN 978-3-642-05381-8 (eBook)
DOI 10.1007/978-3-642-05381-8
Springer Heidelberg New York Dordrecht London

Library of Congress Control Number: 2012943848

Cover design: Jim Wilkie
Project management: OPS Ltd., Gt. Yarmouth, Norfolk, U.K.

Printed on acid-free paper

Springer is part of Springer Science+Business Media (www.springer.com)

Contents

To the memory of my late mother and father,
to my wife Kathleen, and
to whoever put the bop in the bop shoo-bop.
If severe convection be the food of love, play on

Preface

During the World's Exposition in Chicago in 1893, Frederick Turner Jackson delivered a talk entitled "The Significance of the Frontier in American History". As the frontier of new land to be explored was coming to an end, he raised the question of how American society might change in response to its end. The severe convection "frontier" in the U. S. is steadily disappearing now because—with the advent of the Internet, cellphone technology, and cable television channels devoted exclusively to the weather—severe convective storms and tornadoes are being observed and documented all the time, even in remote places, by both meteorologists and non-meteorologists alike, and made available for mass viewing. Observing a tornado used to be a very rare occurrence.

While the observational severe convection frontier is disappearing, the knowledge frontier is still with us, as is the beauty of severe convective phenonmena. I wrote this book in response to a need for updated material for a graduate course on convective clouds and storms, with an emphasis on severe convective storms and tornadoes, that I have taught at the University of Oklahoma roughly once every other year for the past three decades. It has become very difficult for students to learn just from my class lectures and journal articles covering more than three decades. This course has evolved considerably, especially in the last decade and a half, with the advent of mobile Doppler radars and more sophisticated numerical models. It is hoped that this text will be useful not only to students, but also as a reference for researchers and forecasters.

The contents of this book are heavily influenced by an introductory course on convection taught by Prof. Norm Phillips at MIT in 1970, but not fully appreciated by the author then, and by the American Meteorological Society's (AMS) 1963 monograph on *Severe Local Storms*, edited by Dave Atlas, which contains seminal contributions by the editor, Ted Fujita, Chester Newton, and Frank Ludlam, among others. While there have been more recent contributions such as the latest, updated AMS monograph on severe local storms, which contains disparate contributions from many authors, Kerry Emanuel's textbook on many types of convection (but without an emphasis on severe convection or tornadoes), and Bob Houze's textbook on clouds (which covers a very broad range of topics), I felt compelled to produce a work from my own perspective as an avid observationalist and participant in over three decades of storm chasing, mainly with mobile instruments. This text should be considered a work in progress; since the pace of research in severe convective storms and tornadoes is rapid, I encourage the student and other readers to keep abreast of more recent journal articles. Despite this book's assured obsolescence in a relatively short time, I hope that most of the core dynamical issues addressed herein will be "current" for much longer.

No attempt has been made to be all inclusive; some topics have been ignored altogether and the student/reader must look elsewhere for detailed treatments on, for example, moist thermodynamics, cloud and precipitation microphysics, numerical modeling techniques for convective clouds, data assimilation techniques for cloud models, objective analysis of data, lightning and other electrical phenomena, radar meteorology, and shallow convection. By doing so it is hoped that the topics discussed herein will be adequate for a one-semester course. Students can take more specialized courses on the topics not covered in detail or ignored altogether. It is also recognized that there may be some overlap between the topics covered in this text and some topics covered in mesoscale meteorology courses (e.g., density currents and gravity waves may also be considered purely mesoscale phenomena and not exclusively associated with convection). Density currents are most frequently driven by water phase changes in convective clouds, so they are detailed here; gravity waves, on the other hand, frequently occur in the absence of convection, so we do not detail their dynamics.

To a better understanding of the wind and rain and hail ...

Howie "Cb" Bluestein
Norman, OK and Boulder, CO
December 2011

Acknowledgments

As the Beatles once sang, "I get high with a little help from friends." So be it for this book. Many colleagues, friends, and students contributed in various ways to this text. At the University of Oklahoma (OU) I am indebted to the chairs of the School of Meteorology the late Rex Inman, and Jeff Kimpel, Claude Duchon, Bill Beasley, Fred Carr, and Dave Parsons and to OU colleagues Kelvin Droegemeier, Pete Lamb, and OU president David L. Boren for their support. Much of my research funding has come from the National Science Foundation with the aid of the late Ron Taylor, and Steve Nelson and Brad Smull. I thank the reviewers of research proposals and manuscripts who have anonymously supported and assisted my work and that of my students. Colleagues and friends Rich Rotunno and Morris Weisman at NCAR, and John Brown at NOAA, in Boulder, have provided many years of stimulating scientific and recreational interaction and assistance with various aspects of this work. Lance Bosart at the University of Albany/SUNY has for years been an enabler of weather addicts such as me while at the same time holding me accountable to the advancement of science. The late Bob Burpee (formerly at the Hurricane Research Division and National Hurricane Center in Miami) was always supportive of my convective interests. I am particularly thankful for the many graduate and undergraduate students at OU over the years, too numerous to list in entirety, who have chased storms with me both formally as part of their courses of research and informally just for the fun of it. Teaching has always been a two-way street and the students have willingly shared their enthusiasm for meteorological phenomena of the buoyant, rotational, and violent kind. Graduate students Jana Houser, Jeff Snyder, Mike French, Robin Tanamachi, Vivek Mahale, Chris Weiss, David Dowell, and Matthew Kramar were "instrumental" in conducting our field experiments using mobile Doppler radars. Rodger Brown at NSSL, Keith Browning (ret.), Dave Atlas (ret.), and Jerry Brotzge at OU assisted with fact checking and other matters. I also thank collaborators Al Bedard (formerly at NOAA WPL), the late Wes Unruh (formerly at LANL), Andy Pazmany (formerly at the University of Massachusetts and currently at ProSensing), the late Bob McIntosh (formerly at U. Mass.), Steve Frasier (U. Mass.), Bob Bluth (NPS), Ivan PopStefanija (ProSensing), and others for support with in situ instruments and radars over the years. The late Joanne Simpson at EML in Miami was kind enough to take me, as a graduate student, on a flight into and out of big thunderstorms over central Florida during her randomized cloud-seeding experiment. Dave Jorgensen (NSSL) and Roger Wakimoto (formerly at UCLA and currently at NCAR) made it possible for me to participate in data collection on the NOAA P-3 and ELDORA aircraft, respectively. Joe Golden (formerly at NOAA) kindly allowed me to fly around waterspouts with him. OU colleagues Brian Fiedler and Al Shapiro have generously provided time to discuss various scientific problems. I appreciate Kerry Emanuel's (MIT) support over the years, especially during an extended visit to MIT, and for the example he has set for research excellence. His textbook on convection and the work and textbook on clouds by Bob Houze (University of Washington) and Bob's research have represented standards to which I have aspired. Much inspiration has come from the work of the late Ted Fujita (formerly at the University of Chicago), and Don Burgess, Bob Davies-Jones and his colleagues at NSSL. The late Pauline Austin and Speed Geotis (formerly at MIT) freely exchanged ideas about convection when I was in graduate school; Speed entertained me during a stint on the R.V. *Gillis* in the Intertropical Convergence Zone in the western north Atlantic during GATE in the late summer of 1974, when I operated the onboard radar. Norm Phillips' "notes" at MIT also played an important role in this textbook. I thank Ed Kessler, the first director of NSSL, for suggesting that I move to Norman to study severe convective storms and tornadoes and my advisor at MIT, the late Fred Sanders for providing the mentoring necessary to set me on my convective way. Finally, I thank Clive Horwood and Romy Blott at Praxis Publishing and Robert Doe at Springer Publishing for their support, Jim Wilkie for designing the cover, and Neil Shuttlewood for his expert copyediting.

Figures

Abbreviations and acronyms

ARPS	Advanced Regional Prediction System
BAMEX	Bow Echo and MCV EXperiment
BWER	Bounded Weak Echo Region
CAPE	Convective Available Potential Energy
CAPS	Center for Analysis and Prediction of Storms
CCL	Convective Condensation Level
CCOPE	Cooperative COnvective Precipitation Experiment
CDI	Cloud-base Detrainment Instability
CIN	Convective INhibition
CINDE	Convective INitiation and Downburst Experiment
CISK	Conditional Instability of the Second Kind
COHMEX	COoperative Huntsville Meteorological EXperiment
COMET	Cooperative Program for Operational Meteorology, Education, and Training
COPS	Cooperative Oklahoma Profiler Study
CSU	Colorado State University
DCZ	Deep Convergence Zone
DOW	Doppler On Wheels
DPE	Dynamic Pipe Effect
DRC	Descending Reflectivity Core
EF	Enhanced Fujita
EL	Equilibrium Level
ELDORA	ELectra DOppler RAdar
FAA	Federal Aviation Administration
FAST	Fore–Aft Scanning Technique
FFD	Forward Flank Downdraft
GATE	GARP (Global Atmospheric Research Program) Atlantic Tropical Experiment

GBVTD	Ground Based Velocity Track Display
HCR	Horizontal Convective Roll
HP	High Precipitation
IPV	Isentropic Potential Vorticity
JAWS	Joint Airport Weather Studies
JDOP	Joint Doppler Operational Project
LANL	Los Alamos National Laboratory
LCL	Lifting Condensation Level
LES	Large Eddy Simulation
LFC	Level of Free Convection
LLJ	Low Level Jet
LM	Left Moving
LP	Low Precipitation
M-CLASS	Mobile Cross chain LORAN Atmospheric Sounding System
MAUL	Moist Absolutely Unstable Layer
MCS	Mesoscale Convective System
MCV	Mesoscale Convective Vortex
MIST	MIcroburst and Severe Thunderstorm Project
MIT	Massachusetts Institute of Technology
MLCAPE	Mixed Layer or Mean Layer CAPE
MTT	Morton, Taylor, and Turner (1956)
MUCAPE	Most Unstable CAPE
MWR-05XP	Meteorological Weather Radar-2005, X-band, Phased Array
NCAR	National Center for Atmospheric Research
NEXRAD	NEXt generation RADar
NHRE	National Hail Research Experiment
NIMROD	Northern Illinois Meteorological Research on Downbursts
NOAA	National Oceanic and Atmospheric Administration
NOCM	Non Occluding Cyclic Mesocyclogenesis
NSSL	National Severe Storms Laboratory
NSSP	National Severe Storm Project
NST	Non Supercell Tornado
NWS	National Weather Service
OCM	Occluding Cyclic Mesocyclogenesis
OU	Oklahoma University
PGF	Pressure Gradient Force
PROFS	Program for Regional Observing and Forecasting Services
QLCS	Quasi-Linear (Mesoscale) Convective System
RAMS	Regional Atmospheric Modeling System
RASS	radio acoustic sounding system
RaXPol	Rapid-scan (mechanically scanning, not electronically scanning) polarimetric, X-band, Doppler radar
RFD	Rear Flank Downdraft
RFGF	Rear Flank Gust Front
RKW	Rotunno, Klemp, and Weisman

RM	Right Moving
RMW	Radius of Maximum Wind
ROTATE	Radar Observations of Tornadoes And Thunderstorms Experiment
RPV	Remotely Piloted Vehicle
SBCAPE	Surface Based CAPE
SESAME	Severe Environmental Storms And Mesoscale Experiment
SMART-R	Shared Mobile Atmospheric Research and Teaching Radar
SPC	Storm Prediction Center
SREH	Storm Relative Environmental Helicity
STEPS	Severe Thunderstorm Electrification and Precipitation Study
TDWR	Terminal Doppler Weather Radar
TIV	Tornado Intercept Vehicle
TOTO	TOtable Tornado Observatory
TRAP	Tornado Research Airplane Project
TVS	Tornado Vortex Signature
TWISTEX	Tactical Weather Instrumented Sampling in/near Tornadoes EXperiment
TWOLF	Truck-Mounted Wind Observing Lidar Facility
UAS	Unmanned Aerial System
UAV	Unmanned Aerial Vehicle
VAD	Velocity Azimuth Display
VORTEX	Verification of the Origins of Rotation in Tornadoes EXperiment
VSI	Value of a Statistical Injury
VSL	Value of a Statistical Life
WEC	Weak Echo column
WEH	Weak Echo "eye" or Hole
WER	Weak Echo Region
WPL	Wave Propagation Laboratory
WRF	Weather Research and Forecasting Model

1

Introduction

"... oh now feel it comin' back again
like a rollin' thunder chasing the wind
forces pullin' from the center of the earth again
I can feel it."
Lyrics from *Lighting Crashes* by Live

1.1 BASIC DEFINITION OF SEVERE CONVECTIVE STORMS AND SCOPE OF THE MATERIAL

Severe convective storms worldwide inflict damage to property and crops, disrupt air, sea, and ground travel and outdoor activity, and, in the most extreme cases, inflict injuries and even death. While the adjective "severe" generally refers to weather phenomena that produce damage, what is damaging to one type of structure may not be damaging to another, owing to differences in the integrity of construction and the nature of the underlying surface. In the U. S., "severe" weather associated with *local* storms (as opposed to storms that are much larger in scale such as extratropical and tropical cyclones) is defined more precisely by the Storm Prediction Center (SPC) of the National Weather Service (NWS) as having one or more of the following: tornadoes, winds equal to or in excess of $25.8\,\mathrm{m\,s^{-1}}$ (58 mph), or hail 2.5 cm (1 inch) or greater in diameter, regardless of whether or not there is actual damage; it is noted that prior to January 5, 2010 the minimum hail size criterion was only 1.9 cm (3/4 inch).

It is perhaps a shortcoming of the U. S. definition of severe weather that flooding and lightning are not included, even though each of these also may be responsible for damage, injuries, and death. To maintain a manageable focus, however, *this textbook discusses only the physics of the airflow and cloud and precipitation distribution (with little regard for cloud particle type or precipitation type) in severe convective storms.* The reader is directed elsewhere for detailed discussions of cloud microphysics and precipitation formation, including the formation of large hail (e.g., Knight and Knight, 2001), the hydrological consequences of excessive rainfall (i.e., flooding), and cloud electrification and its consequences (e.g., Williams, 2001). Forecasting techniques using numerical models initialized by observational data are also not covered in much detail, in part because at the time of this writing there is a flurry of activity using data assimilation techniques that is in a state of rapid flux and, consequently, attempts to detail them might not be useful, since the art and science of data assimilation are changing so rapidly.

The purpose of this textbook is to summarize what we have learned in approximately the last half-century about the kinematics and dynamics and, to a lesser extent, the thermodynamics of severe convective storms. I do not use the term "thunderstorm", because it is possible that a severe convective storm does not produce lightning and I would not want to exclude this class of storms from discussion. In addition, while the adjective "convective" simply denotes the movement of air in general, we generally use the adjective "convective" to denote small-scale movements of air in deep cumulus clouds or cumulonimbus clouds.

Advances in observing systems, particularly in radars, and advances in computer technology and numerical modeling techniques have stimulated and made possible fruitful studies of the structure and dynamics of severe convective storms. Through the analysis of observational data (from both quantitative measurements and from visual observations) and the results of controlled numerical experiments, the fundamental processes responsible for determining the convective storm type and the severe weather associated with each type of convective storm have been identified.

After a brief history in this chapter of the major field programs and numerical simulation experiments aimed at understanding the physical processes responsible for severe convective storms is given, the dynamical and, to a much lesser extent, the thermodynamic frameworks used to diagnose the behavior of severe convective storms are discussed mathematically and explained physically in Chapter 2. Thermodynamics is given short shrift because the details are mostly important for numerical modelers and numerical modeling is not a major focus of this book. Students and readers are referred elsewhere (e.g., Emanuel, 1994) and to many journal articles (see the reference lists for specific works) for further discussions on thermodynamics. Also, it is assumed that the reader has some knowledge of radar meteorology. Some additional information, however, is embedded within the main body of the text on the maturing area of polarimetric radar technology and its applications to severe convective storm studies.

The author believes that students will gain an increased appreciation for the theory after they have become aware of some of the major problems and solutions to them that have been grappled with and proposed by scientists, engineers, computer scientists, and amateur meteorologists and have become more acquainted with the actors involved in the scenes of the theater of severe storm meteorology.

To the latter end, in this chapter the reader is referred to Figures 1.1–1.10, which are a non-exhaustive mosaic of some of the major players in severe storm research in the past six decades and some of the instrumentation used in field experiments.

Modern observing systems and some of the techniques used to study severe convective storms using them are described briefly in many of the chapters, but students and readers are referred elsewhere for more details. The following chapter (Chapter 2) describes the basic physics and thermodynamics of convection, including the older plume and thermal models along with a brief introduction to the classic linear analysis of Rayleigh–Bénard convection. The dynamics and behavior of the two basic types of convective building blocks, "ordinary cells" and "supercells" are detailed in Chapters 3 and 4, respectively. The former includes discussion of density–current dynamics, downdrafts, and multicell behavior. Chapter 5 is a treatment of the dynamics of larger conglomerates of convective cells, "mesoscale convective systems (MCSs)", Tornadoes, which can form in either type of convective cell and in some MCSs, are then discussed in Chapter 6. There is a brief discussion of short-range forecasting, the possible effects of climate change on severe convective storm occurrence, and topics for future research in Chapter 7.

1.2 A BRIEF HISTORY OF SEVERE STORM FIELD PROGRAMS AND NUMERICAL MODELING EFFORTS

1.2.1 Field programs and instrument development

The seminal field program for the study of severe convective storms was the *Thunderstorm Project*, conducted near Orlando, Florida in the summer of 1946 and near Wilmington, Ohio in the summer of 1947. An objective of this field program was to learn enough to be able to reduce the likelihood of thunderstorm-related airplane crashes. In situ measurements were made in and around thunderstorms in aircraft and at the surface. Rawinsonde measurements were made profiling the vertical thermodynamic and wind structure of the troposphere, and radar reflectivity data were collected, enabling the identification of the entire life cycle of convective storms. Most field programs since then have been focused on the study of convective storms, severe or otherwise, and have been variations on the original theme of the Thunderstorm Project. The University of Chicago–based leaders of the Thunderstorm Project, Horace Byers and his graduate student Roscoe Braham, named the precipitation region associated with a discrete updraft in a convective storm a "cell", in analogy to a biological cell, and identified three distinct stages in the life cycle of a thunderstorm cell. Braham's M.Sc. thesis was incorporated into a classic and much referenced (even up to the current time) volume attributed to both Byers and Braham and published in 1949.

Figure 1.1. Chester Newton (1997; Val Morin, Quebec, Canada); Ted Fujita (1960s; courtesy of the University of Chicago News Office); Keith Browning (2003; Reading, U. K.); Yoshi Ogura (2008; Tokyo, Japan); Norm Phillips (2004; Seattle, WA); Neil Ward with his vortex simulator (circa late 1960s; courtesy of History of Science Collections, University of Oklahoma Libraries); Ralph Donaldson (1999; Framingham, MA); Dave Atlas (2001; Florida Keys) and Tom Seliga, who did pioneering work with polarimetric radar (2001; Florida Keys); Doug Lilly (2004; Seattle, WA). Photographs by the author except where otherwise indicated.

Figure 1.2. Ed Kessler (1998; Oklahoma City); Rodger Brown (2009; Norman, OK); Les Lemon (2010; Norman, OK); Bob Davies-Jones (1979; Norman, OK; 2008); Ed Brandes (courtesy of Ed Brandes; 2005); Joe Golden (1993; aboard a NOAA helicopter, near Key West, FL); Chuck Doswell, III (1998; Oklahoma City); John Snow (1998; Oklahoma City). Photographs by the author except where otherwise indicated.

Figure 1.3. Bob Wilhelmson (2010); Joe Klemp (2010; Boulder, CO); Bob Schlesinger (2008); Rich Rotunno (1980, 2004); Dave Hoadley (2008; Denver, CO); Morris Weisman (1984, 2007; Boulder, CO). Photographs by the author except where otherwise indicated.

Figure 1.4. Peter Ray (c. 1980–1990; at the NSSL Doppler radar); Bob Houze (2006, Atlanta, GA); Kerry Emanuel (1999, Cambridge, MA); Dick Johnson (2012, New Orleans, LA); Brad Smull (2010; Norman, OK); Steve Lewellen (courtesy of Dave Lewellen); Tim Marshall (2011; Pittsburgh, PA); Don Burgess (2009; Norman, OK); Kelvin Droegemeier (2010). Photographs by the author except where otherwise indicated.

Figure 1.5. Joanne Simpson (2004, Seattle, WA); Gerry Heymsfield (2009, Williamsburg, VA); Tim Samaras (2012, Denver, CO); Dick Doviak (2004, Norman, OK); Al Bedard deploying TOTO (Totable Tornado Observatory) (1980; eastern Colorado); Bill McCaul, Glen Lesins, Rich Fulton, and Mohan Ramamurthy about to release a portable radiosonde (1984); Roscoe Braham (2006, Atlanta, GA); Wes Unruh and the first portable Doppler radar at the LANL (1987; Los Alamos, NM); Steve Gaddy, Greg Martin, and Sam Contorno probing a tornado with the LANL radar (1989; near Hodges, TX); the author probing a tornado with the LANL radar (1991; near Enid, OK; courtesy of Herb Stein); the author and the University of Massachusetts mobile W-band Doppler radar (1995; courtesy of Herb Stein); the University of Massachusetts mobile W-band Doppler radar probing a tornado (1999; near Stockton, KS). Photographs by the author except where otherwise indicated.

Figure 1.6. Dusan Zrnic (1994; Mt. Tamalpais State Park, CA); Roger Wakimoto (2004; Seattle, WA); Andy Pazmany and the University of Massachusetts mobile W-band Doppler radar (1993; Norman, OK); Dave Jorgensen aboard the NOAA P-3 (2005, while flying though category 5 Hurricane Rita; courtesy Dave Jorgensen; photo by Frank Marks); NOAA P-3 with airborne Doppler radar in the tail section (1995, during VORTEX); ELDORA (Electra Doppler Radar) aircraft (1995, during VORTEX; Oklahoma City); Jerry Straka (courtesy of Jerry Straka). Photographs by the author except where otherwise indicated.

Figure 1.7. Lou Wicker (2009; Norman, OK); Josh Wurman and the first Doppler on Wheels (1996; Norman, OK); Ivan PopStefanija, Paul Buczynski, and Bob Bluth at the MWR-05XP (2007; Norman, OK); Dave Lewellen (2012; New Orleans, LA); Josh Wurman during VORTEX2 (2009); Steve Frasier (2008, Savannah, GA); Erik Rasmussen and Al Moller (2009; Norman, OK). Photographs by the author except where otherwise indicated.

Figure 1.8. Yvette Richardson and Paul Markowski (2009; in the field during VORTEX2); David Dowell (2009; in the Field Coordinator Truck during VORTEX2); Chris Weiss (2009, during VORTEX2); Kevin Knupp (2012, New Orleans, LA); Ming Xue (courtesy Ming Xue); Matt Parker (2012; New Orleans, LA); Robin Tanamachi (2012; New Orleans, LA); Brian Fiedler (2011; Norman, OK). Photographs by the author except where otherwise indicated.

Figure 1.9. Jeff Trapp (courtesy of Jeff Trapp); University of Massachusetts mobile, polarimetric, X-band Doppler radar (U. Mass. X-Pol) probing a supercell in the Oklahoma Panhandle (2008); MWR-05XP (Meteorological Weather Radar-2005, X-band, Phased-array) (2007; Texas Panhandle); DOW6 (Doppler on Wheels) (2009; during VORTEX2); Rapid-DOW (2010; during VORTEX2); two SMART-R (Shared Mobile Atmospheric Research and Teaching Radar) C-band mobile Doppler radars (2009; during VORTEX2); NOAA mobile, X-band, polarimetric Doppler radar (NOXP) with Chris Schwarz and Don Burgess (2009; Norman, OK); Texas Tech Ka-band mobile Doppler radar probing a supercell (2010; during VORTEX2; western Kansas); NOAA Field Coordination Vehicle (2009; Norman, OK). Photographs by the author except where otherwise indicated.

Figure 1.10. Pennsylvania State/NSSL Mobile Mesonet Vehicles (2010; during VORTEX2); Center for Severe Weather Research Mobile Mesonet Vehicle (2010; during VORTEX2); StickNet Portable Weather Stations (2010; during VORTEX2); Texas Tech Tornado Pods (2010; during VORTEX2); Brian Argrow (left) and University of Colorado unmanned aircraft system (2010; Denver, CO); the TIV (Tornado Intercept Vehicle) (2009); University of Alabama at Huntsville (UAH) MAX (Mobile Alabama X-band) Doppler radar (2009; Boise City, OK); UAH MIPS (Mobile Integrated Profiling System) (2009; Boise City, OK); Greg Forbes (2010; at the Weather Channel, Atlanta, GA); RaXPol (Rapid-scan X-band polarimetric) mobile Doppler radar probing a supercell in Oklahoma (2011). Photographs by the author except where otherwise indicated.

An impetus for the study of severe convective storms in particular was an unlikely set of events that occurred in central Oklahoma on March 20 and March 25, 1948. After a tornado on March 20 struck Tinker Air Force Base, near Oklahoma City, causing about $10 million in damage, similar meteorological conditions became evident five days later. Major Ernest Fawbush and Capt. Robert Miller at Tinker Air Force Base issued a non-public tornado warning; a tornado did in fact strike again, causing about $6 million in damage. This successful tornado warning may have been the result of beginner's luck, but it was arguably a tipping point in the history of severe storm forecasting. The U.S. Weather Bureau in Kansas City took interest in the forecasting techniques developed by Fawbush and Miller, and convened a meeting between the Air Force personnel and forecasters from Washington, D.C. Thus began formal efforts to forecast severe local storms. Earlier efforts were subdued because it was thought that tornado forecasts would panic the general populace.

In 1953 a hook echo was serendipitously observed in a tornadic storm in Champaign, IL and the relationship between the location of a tornado and this radar signature was noted. The characteristic shape of the radar echo of the storm would be recognized years into the following decade as a special class of convective storm. In addition, there was a chance occurrence of a tornado near a radar site near Waco, Texas also in 1953.

The beginning of the systematic study of severe convective storms thus began in response to both practical needs and serendipity. The radar, perfected during World War II, and the rawinsonde, which became part of the operational network also around World War II, made such subsequent studies possible.

Another important part of observational field experiment history was the separate efforts of the first amateur storm-chasers Roger Jensen and David Hoadley in the mid-1950s. The central U. S., where severe storms occur most often, was a near ideal place to chase storms by automobile, given the relative flatness of the landscape and long view to the horizon, the availability of a relatively dense road network, the relative frequency of severe storms during the spring, and the ease in crossing political boundaries (states). Storm-chasers (though not termed "storm-chasers" until the early 1970s) provided some of the first systematic visual observations, though serendipitously obtained photographs of severe storms and tornadoes were also correlated with radar imagery by non-chaser pioneers such as Ted Fujita. Fujita's landmark analysis of photographs taken of a tornadic storm in and near Fargo, North Dakota on June 20, 1957 led to nomenclature such as the "wall cloud" and the "tail cloud". Fujita, who had analyzed debris patterns created by the atomic bomb dropped on Hiroshima at the end of World War II, had been lured to the University of Chicago by Horace Byers to do meteorological research.

Other significant visual media collected by non-professional meteorologists was a movie of a tornado in Dallas, Texas on April 2, 1957, in which flying debris was clearly seen. The meteorologist Walter Hoecker photogrammetrically analyzed the debris motion, thereby obtaining the most reliable estimates to date of wind speeds in tornadoes. So, non-meteorologists chased storms but the fruits of their labors remained mainly unknown to professional meteorologists, while meteorologists made use of serendipitously obtained photographs and movies taken by non-meteorologists and serendipitously obtained radar measurements from fixed site radars.

There were some organized efforts, however, in the 1950s to study some aspects of severe storms. The pilots known as the "Rough Riders" penetrated squall lines in the central U. S. during the Tornado Research Airplane Project (TRAP). This project evolved into the National Severe Storm Project (NSSP) in 1961. An interesting finding from this project was that air tended to flow around storms, as if storms were obstacles to the flow.

Neil Ward, who was responsible later for refining the tornado chamber, was probably the first professional meteorologist to storm-chase and combine visual observations with other measurements. He observed a tornadic storm near Geary, Oklahoma on May 4, 1961.[1] The following year the type of storm Ward observed was named a "supercell" by Keith Browning. Browning had been at Imperial College working with Frank Ludlam and then at the Air Force Cambridge Research Laboratory (AFCRL) radar in Sudbury, Massachusetts, under Dave Atlas. Ward subsequently worked in Norman, Oklahoma at the National Severe Storms Laboratory (NSSL), which formed as a merger in 1964 between the Weather Radar Laboratory in Norman and the NSSP under the guidance of Ed Kessler, who had done seminal work early in his career as a student at MIT on precipitation and cloud kinematics and some radar studies. The same year Ward went storm chasing, Fred Bates from St. Louis University in 1961 documented a tornado he observed from an aircraft and noted how the tornado was far removed from the precipitation in its parent storm.

Small field programs whose main purpose was to collect hailstones were conducted in 1964 based at the University of Wyoming in Laramie, in 1965 and 1966 in South Dakota and northeastern Colorado as part of Project Hailswath, and at NSSL in 1966. There were larger hail field efforts in northeastern Colorado organized by NCAR (National Center for Atmospheric Research), the University of Wyoming, and the Desert Research Institute at the University of Nevada during the National Hail Research Experiment (NHRE) from 1972 to 1974. Other programs included experiments in Switzerland, South Africa, and Alberta (Canada), as well as in Bulgaria. While the aforementioned field programs were focused on hail production, much was learned also about the parent storms producing the hail, which in many instances were supercells.

In the late 1960s and early 1970s, waterspouts in the Florida Keys were studied by making airborne measurements. Airborne measurements made from aircraft have been important over the years (Figure 1.11). Joe Golden from NOAA described the waterspout life cycle based on many visual observations from aircraft; Verne Leverson, Pete Sinclair from Colorado State, and Golden made in situ measurements from aircraft; and Ron Schwiesow from NOAA and colleagues made remote wind measurements with a Doppler lidar (without the capability of range discrimination) from aircraft. It was around this time that meteorological Doppler radar was being developed, especially at NSSL, NCAR, the Air Force Cambridge Research Laboratory, and Cornell Aeronautical Labora-

[1] According to Ward, "... transportation and communications were furnished by the Oklahoma Highway Patrol."

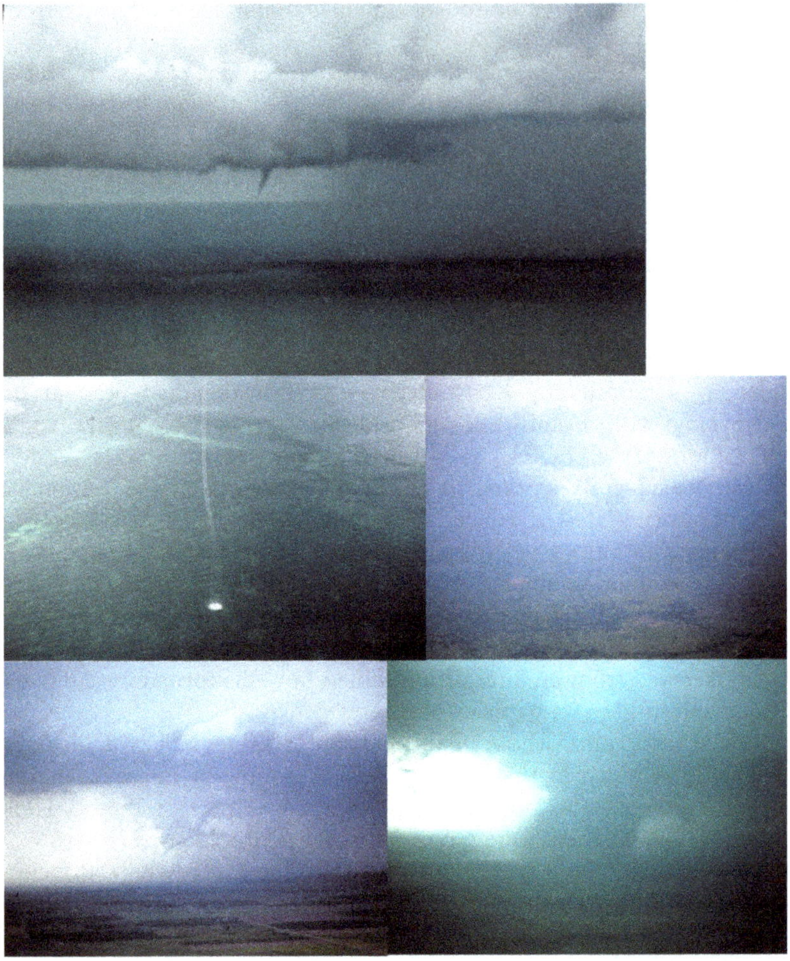

Figure 1.11. Waterspouts and tornadoes as seen from airborne platforms. (Top) Waterspout in the Florida Keys from a private aircraft on September 3, 1971; (middle row, left to right) waterspout in the Florida Keys from a NOAA helicopter on August 24, 1993, tornado from the NOAA P-3 on May 29, 1994; (bottom row) tornado over eastern Colorado on June 8, 1994 from the NOAA P-3, tornado over the Texas Panhandle on June 8, 1995 from ELDORA. (photographs by the author)

tory, that NSSL began conducting a Tornado Intercept Project, spurred on in part by Bruce Morgan, an engineer from Notre Dame University, who had proposed driving an instrumented, armored vehicle into tornadoes and making in situ measurements. (This idea was not realized until the early 2000s when IMAX photographer Sean Casey developed the Tornado Intercept Vehicle or TIV). NSSL and the University of Oklahoma (OU), under the direction of Bob Davies-Jones, conducted storm intercept experiments in Oklahoma between 1975 and 1986. Alan Moller and Chuck Doswell, students at OU during the early 1970s, combined many of their photographs to devise a visual model of tornadic supercells, which is still used to this day. Some of the participants in the early NSSL/OU storm intercept field activities who at the time of this writing are still active in tornado research include the author and undergraduate students Erik Rasmussen and Lou Wicker, who accompanied the author on early storm chases.

On May 24, 1973 visual, storm intercept observations were combined with Doppler radar measurements from a fixed site research Doppler radar at NSSL to study the complete life cycle of a tornado that struck Union City, Oklahoma. The data set collected was the first comprehensive combined Doppler radar and visual documentation of a supercell tornado. Rodger Brown, Don Burgess, Les Lemon, and Bob Davies-Jones were prominent members of the team that studied this tornado. Ralph Donaldson at the Air Force Cambridge Research Laboratory in Massachusetts provided some of the early ground-laying work for the identification of the mesocyclone in supercells. The addition of a second Doppler radar west-northwest of NSSL made it possible for dual-Doppler analysis of the wind field when storms passed through the small NSSL dual-Doppler network. An outline of dual-Doppler and other analysis techniques is given in the Appendix.

Under the direction of Peter Ray at NSSL, a number of excellent cases of tornado formation in supercells were analyzed when the parent supercells traversed the NSSL dual-Doppler network, and were studied in the late 1970s and early 1980s. In addition, Doppler spectra in tornadoes, from which the maximum wind speed could be estimated, were collected by the first NSSL Doppler radar and analyzed by Dusan Zrnic, Dick Doviak, and colleagues at NSSL. Doppler radar data were also collected in a wide spectrum of storm types during NHRE (National Hail Research Experiment), conducted in northeastern Colorado during the summers of 1972–1974 and 1976, mainly to test the usefulness of seeding clouds to reduce damaging hail.

The main objective of JDOP (Joint Doppler Operational Project) conducted first in 1978 (and in succeeding years) at NSSL was to provide verification of the signatures detected on Doppler radar. JDOP was influential in the development of the future national Doppler radar system (NEXRAD) implemented in the U. S. more than a decade later. SESAME (Severe Environmental Storms and Mesoscale Experiment) was conducted during the spring of 1979 in the Southern Plains. Several large tornadoes that occurred during SESAME were studied extensively. Storm chasing also took hold in the late 1970s and early 1980s at Texas Tech University in Lubbock with Erik Rasmussen and Tim Marshall and subsequently elsewhere. Storm intercept field programs were carried out in eastern Colorado in the mid-1980s by NOAA's PROFS (Program for Regional Observing and Forecasting Services) and by NCAR. Non-supercell thunderstorms and microbursts were studied extensively in the Denver area, especially during JAWS (Joint Airport Weather Studies) in 1982. In 1981 CCOPE (Cooperative Convective Precipitation Experiment) was conducted in southeastern Montana using three fixed site Doppler radars and aircraft; High Plains supercells were studied and penetrated by an armored aircraft (T-28) which made in situ measurements of updrafts and downdrafts and thermodynamic measurements.

Efforts to make in situ ground-based measurements during storm chases began in the early 1980s at NSSL and OU. TOTO (Totable Tornado Observatory) was designed by Al Bedard at NOAA in Boulder, Colorado and first used by the author at OU in 1981. After 1984, NSSL continued to use TOTO through 1985. In situ thermodynamic and wind measurements inside tornadoes were attempted, but were only minimally successful. In 1986 Fred Brock and colleagues at OU developed smaller instrument packages named "Turtles." These packages were much easier to deploy than TOTO and increased the chances of a direct hit by a tornado. They were used with varying amounts of success in 1988, 1989, 1991, and 1993.

Efforts were also made to make measurements from more conventional airborne platforms. Stirling Colgate at the Los Alamos National Laboratory (LANL), carrying through on an idea that originated at Purdue University in the late 1960s by Ernie Agee and colleagues in the early 1980s, fired instrumented lightweight rockets from an airplane into tornadoes; however, because the rockets were required by federal regulations to be lightweight and, therefore, were rather fragile, they were not successful. Fred Bates at St. Louis University in 1963 first proposed that an unmanned aerial vehicle (UAV) be used to make in situ measurements near tornadoes. Almost 25 years later, Karl Bergey at OU and his students developed two RPVs (remotely piloted vehicles) capable of taking video as they flew by storms. Meteorological instruments, however, were never installed. A UAS (unmanned aerial system), another name for an RPV, was used with instruments by a group from the Universities of Colorado and Nebraska in 2010 during VORTEX2 (an experiment to be discussed subsequently).

In 1984 the author first released portable radiosondes into and around severe convective storms. Subsequent mobile sounding systems were used during COHMEX (Cooperative Huntsville Meteorological Experiment) in Alabama in 1986 and during CINDE (Convective Initiation and Downburst Experiment) in 1987 near Denver.

While airborne Doppler lidar measurements were first made in waterspouts during the previous decade, the first measurements using a *pulsed* Doppler lidar system, one in which range information is available, were made in 1981 in Oklahoma along gust fronts and around cumulus towers by the author and his student Bill McCaul using a NASA Marshall Space Flight Center airborne system. This lidar system scanned at an angle in the direction of the flight track and, alternately, at the same angle in a direction opposite to the flight track, so that in space dual-Doppler analysis is possible if one neglects the difference in time or the observations (Figure 1.12). Such a dual-Doppler technique is referred to as pseudo-dual-Doppler analysis. (One may also scan a storm from a fixed site radar at two different times as the storm propagates by. If the storm undergoes little if any evolution during the time period between scans, dual-Doppler analysis may be possible if the difference in viewing angle is appropriate. Such an analysis is also referred to as "pseudo-dual-Doppler analysis".)

The first airborne Doppler radar measurements in severe convective storms were made by NSSL (at X-band) under the guidance of Peter Ray and subsequently Dave Jorgensen. Data in supercells were collected successfully for the first time in the Southern Plains during COPS-91 (Cooperative Oklahoma Profiler

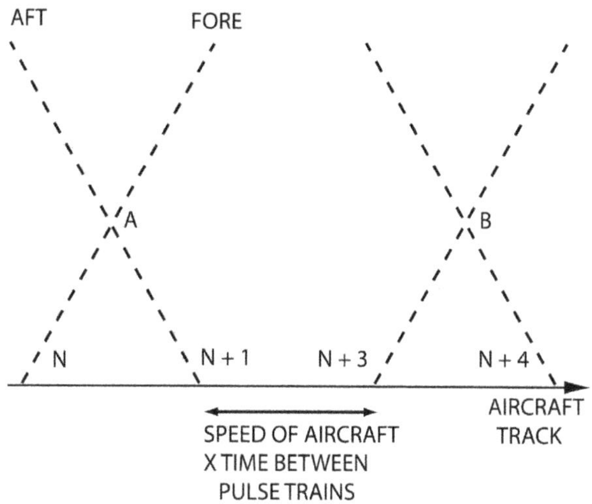

Figure 1.12. Illustration of pseudo dual-Doppler analysis technique as wind vectors are resolved at fixed points in space from intersecting oblique beams, but at slightly different times. In this example, the fore and aft beams of pulse trains in the fore and aft directions at an angle from the flight track are noted at points A and B. The spacing between the N and N + 1, etc. pulse trains is given by the product of the speed of the aircraft with the time between the transmission of the pulse trains (adapted from McCaul *et al.*, 1987).

Studies) using a NOAA P-3 aircraft and radar system that had been used for a number of years in hurricanes. Early on, two aircraft, each with its own radar, were used and each aircraft flew nearly perpendicular flight legs; later, a scanning technique similar to the one used with NASA Doppler lidar was implemented and called FAST (fore–aft scanning technique).

The first ground-based, portable, X-band Doppler radar was built at LANL and converted for meteorological use by Wes Unruh and colleagues at LANL. First used by the author in 1987, a number of tornadoes were probed at close range between 1987 and 1995 and maximum wind speeds were estimated from Doppler velocity spectra. A high-frequency (W-band, 95 MHz), very narrow beam (0.6°), truck-mounted, scanning, mobile Doppler radar was built at the University of Massachusetts at Amherst by a group headed by Bob McIntosh; the group included Andy Pazmany, who led the field effort. Having undergone a number of upgrades (the antenna was replaced in 1999 with a dish having a half-power beam-width of only 0.18°), the fine-resolution radar has been used since 1993; however, high-quality data in tornadoes were not collected until 1999, for the very first time, during a tornado outbreak in central Oklahoma on May 3.

During the first year of VORTEX (Verification of the Origins of Rotation in Tornadoes Experiment) in 1994, airborne Doppler radar measurements by the NOAA P-3 system near Newcastle, Texas were collected for the first time in which the entire life history of a tornado was documented. During the second year of VORTEX in 1995, a number of excellent data sets documenting the life histories of tornadoes in Kansas and Texas were collected using the new airborne Doppler radar system ELDORA (Electra Doppler Radar) aboard an NCAR P-3 aircraft, under the guidance of Roger Wakimoto from UCLA. In 1994 the FAST scanning technique included scans not only in the plane of the track of the aircraft, but also at other elevation angles, as each fore and aft scan sequence included a rotation about the axis of the flight track (Figure A.3).

VORTEX also marked the first use of a mobile mesonet, whose instrumented vehicles were used to make in situ thermodynamic and wind measurements in strategically selected locations within supercells. The mobile mesonet was developed by Jerry Straka at OU and by Erik Rasmussen and Sherman Fredrickson at NSSL. Mobile soundings were obtained using a system recently developed at NCAR in Boulder, Colorado, which made use of radio navigation aids (M-CLASS: Mobile Cross chain LORAN Atmospheric Sounding System) to locate the sonde packages (these instruments have since been superseded by sondes that use GPS navigation information). VORTEX was an important milestone in that it was coordinated, making use of a number of different observing platforms, some old and new. The second year of VORTEX was significant also because Josh Wurman at OU, with collaboration among OU, NSSL, and NCAR (including Jerry Straka, Erik Rasmussen, Mitch Randall, and Allen Zahrai), built a truck-mounted, pulsed Doppler radar system named the Doppler on Wheels (DOW). The DOW made the first successful volumetric maps of the wind field in tornadoes, at X-band, in 1995, during VORTEX. Finally, an instrument package (a later generation version of the Turtle) from Bill Winn at New Mexico Tech was placed directly in the path of a tornado and successful thermodynamic measurements were made.

After VORTEX, smaller field programs were conducted on an almost annual basis in the late 1990s and afterwards using the DOW and subsequent versions of the DOW during ROTATE (Radar Observations of Tornadoes and Thunderstorms Experiment). Sporadic, unnamed, small field programs were conducted collaboratively by the University of Massachusetts and the University of Oklahoma with the U. Mass. W-band radar and the U. Mass. X-band radar. The latter was first used in 2001 without Doppler or polarimetric capabilities, but in 2002 was expanded to include them. Two C-band mobile Doppler radars (SMART-R, Shared Mobile Atmospheric Research and Teaching Radar) developed at Texas A&M University, the University of Oklahoma, NSSL, and Texas Tech University were used in Oklahoma in 2004 to obtain dual-Doppler coverage of a tornadic supercell. The Rapid (X-band) DOW made its debut in 2005 and the Naval Postgraduate School MWR-05XP (Meteorological Weather Radar-2005, X-band, Phased Array) made its debut in 2007. Both radars scan electronically in elevation but mechanically in azimuth. The latter also includes limited electronic scanning in azimuth to keep the beam nearly stationary in space and frequency hopping to increase the number of independent samples.

Since VORTEX, many single and some dual-Doppler data sets have been collected of supercells and tornadoes. Of these, the most significant were the first mobile dual-Doppler data set collected in a tornadic supercell in eastern

Oklahoma in 1997, single-Doppler coverage of a tornadic supercell in Spencer, South Dakota in 1998, single-Doppler coverage (at X-band and W-band) of large tornadoes in central Oklahoma on May 3, 1999 during a major tornado outbreak and of tornadoes in Kansas and Oklahoma in 2004, rapid scan X-band measurements covering the entire life history of a tornado in a supercell in Kansas in 2008, and the complete life history of a tornado in southeastern Wyoming on June 5, 2009, during the first year of VORTEX2. This VORTEX2 tornado had dual-Doppler collection with mobile radars and volumetric, single-Doppler coverage with rapid update times.

Tim Samaras, an engineer, has made some remarkable in situ measurements in tornadoes during TWISTEX, with a Turtle-like instrument, especially in 2003, when he also recorded some breathtaking video as the tornado passed overhead. In addition, during VORTEX2, in situ mobile mesonet observations were made along with mobile rawinsonde observations and other probes (such as the Sticknets and Tornado Pods), other successors to the Turtles of the late 1980s. The modern version of the instrumented tank proposed back in the early 1970s was realized by the TIV (Tornado Intercept Vehicle). This instrument was highly publicized by a Discovery Channel television series and has made some in situ measurements in tornadoes. During the second year of VORTEX2, in 2010, Texas Tech successfully obtained Ka-band Doppler radar measurements in tornadoes; the Ka-band radar is a compromise between the high spatial resolution, narrow-beam W-band radar, which has a severely restricted range when there is attenuation by intervening heavy precipitation, and the less high spatial resolution of the less narrow beam X-band radar which, while still adversely affected by attenuation from heavy precipitation, is not as severely range limited.

The year *after* year 2 of VORTEX2, 2011, ironically was very unusual in terms of the number of severe weather events in the U. S. A record number of tornadoes struck the U. S. in the spring, inflicting devastating damage in cities such as Joplin, MO (May 22), Raleigh, NC (April 16), and Tuscaloosa, Alabama (April 27), while also striking the airport in St. Louis, Missouri (April 22) and portions of Springfield, Massachusetts (June 1). A tornado even hit Auckland, New Zealand on May 3. Despite our increased knowledge about tornadoes and severe convective storms and increases in our skill in predicting and monitoring them, over 500 people lost their lives in the U. S. as a result of tornadoes, more than any year since 1953. The first rapid-scan (mechanically scanning, not electronically scanning) polarimetric, X-band, Doppler radar, RaXpol, was used by the author and his graduate students for the first time and scanned much of the life history of an EF-5 tornado from relatively close range, in central Oklahoma on May 24.

While VORTEX2, ROTATE, and the U. Mass./OU collaborative field programs were designed mainly to study tornadoes, STEPS (Severe Thunderstorm Electrification and Precipitation Study) in 2000, centered in northwest Kansas, was focused on precipitation and lightning in supercells and BAMEX (Bow Echo and MCV Experiment) in the central U. S. in 2003 was focused on severe weather–producing features found in mesoscale convective systems.

1.2.2 Numerical model simulation experiments

Numerical modeling of severe convection and convective systems began in the mid-1970s, when computing power became adequate for simulating simple, two-dimensional features using models. The foundation for the modeling of convective storms had been laid in the 1960s and early 1970s by Yoshi Ogura at the University of Illinois at Champaign-Urbana and Norm Phillips and Jule Charney at MIT, Carl Hane at Florida State University, and later NSSL did some early non-hydrostatic two-dimensional squall line work. Robert Schlesinger at the University of Wisconsin Madison did pioneering work with a simplified model that excluded the existence of sound waves while retaining the compressibility of air and, in the late 1970s, produced successful three-dimensional simulations of supercells. The presence of sound waves due to the compressibility of air created problems because sound waves are of such a high frequency that numerical integration through each sound wave would require very short time steps and therefore be computationally prohibitive. Bob Wilhelmson at the University of Illinois at Champaign-Urbana and Joe Klemp at NCAR devised a method of numerical integration to include the full compressibility of air. Their model became known as the Klemp–Wilhelmson model and was used for almost two decades at NCAR, before being superseded by the WRF (Weather Research and Forecasting Model) at NCAR, which today is probably the most widely used community model in the U. S. Interestingly, Joe Klemp began his research after getting training in chemical engineering—not in atmospheric science. His ability to conduct research with a technical background other than meteorology should be an inspiration to those students wishing to undertake research in severe convective storms and tornadoes, and have backgrounds in physics, applied mathematics, and engineering disciplines.

Other models that have been widely used include the ARPS (Advanced Regional Prediction System) at CAPS (Center for Analysis and Prediction of Storms) at OU under the direction of Ming Xue; RAMS (Regional Atmospheric Modeling System) at CSU (Colorado State University) under the direction of Bill Cotton; and other models at the University of Wisconsin Madison and Pennsylvania State University. Results of significant studies from other less-used models devised by Lou Wicker at NSSL and George Bryan at NCAR, among others, also appear in the literature. Horizontal resolution is commonly employed down to 250 m and sometimes down to 100 m. It is possible to use coarser resolution to model the parent storm and with a nested grid use finer resolution to model sub-storm features like tornadoes. When doing so, one must match boundary conditions carefully. Computer time, storage, and speed are the current limiting factors in using models with very fine grid spaceing. When complex cloud and precipitation microphysics are employed the speed and storage requirements must be significantly increased.

Some LES (large-eddy simulation) models[2] have been used to study tornadoes as isolated vortices making contact with the ground. With a horizontal resolution

[2] The shortest scales of motion are filtered out and parameterized.

of 1 m, it is not yet possible to include the entire parent storm at that fine a resolution. Dave and Steve Lewellen and collaborators at West Virginia University have been pioneers in this arena. Rich Rotunno at NCAR, building on earlier work by Roger Smith, Lance Leslie, Bruce Morton, A. Szillinsky, Lou Berkofsky, F. Wippermann, Francis Harlow, and Leland Stein, simulated tornado-like vortices in the late 1970s and did some seminal nested grid simulations of tornado-like vortices in supercells in the mid-1980s. The way subgrid-scale parameterizations are handled is very important.

More recently, it has been possible to use radar observations to fit numerical models to the real atmosphere state and make predictions, a vision first put forward by Doug Lilly at OU in the 1980s and implemented with work by Kelvin Droegemeier, Ming Xue, and his colleagues at OU/CAPS. Since model equations are highly nonlinear, the forecasts produced by numerically integrating the equations in time are highly sensitive to both the initial conditions and to the type of model physics used to represent microphysical and turbulent processes. Ensemble forecasts are being produced at CAPS that make use of numerical integrations under many different slightly perturbed initial conditions and subject to different parameterizations of microphysics and subgrid-scale turbulence. The mean of all these integrations is thought to be a good guess at what will happen, while the spread from all the different integrations is thought to be a good measure of the forecast uncertainty.

1.3 METHODS TO BE EMPLOYED

In this text I will emphasize a number of aspects of severe convective storms and tornadoes: recent (beginning in the mid–20th century) historical context, basic descriptions of convective phenomena including photographs and radar imagery (with a bias toward the U. S., where data are most easily available to the author), Doppler radar wind analyses, polarimetric data, in situ measurements, idealized analytical modeling, similarity theory, linear stability analyses, and three-dimensional, non-hydrostatic cloud modeling results will all be considered. In particular, analysis of real data and model data, employing Lagrangian parcel analysis of vorticity and circulation, and analyses of horizontal vorticity to represent vertical circulations in convective storms will be discussed.

Historical context is important because it gives credit to the pioneers who led us to where we are today and to understand why certain avenues of research have been pursued. Again, the reader is reminded of the non-exhaustive group of old and recent photographs of some of the major figures and instrumentation involved in research in severe convective storms and tornadoes in Figures 1.1–1.10. In recent years we have been able to obtain unprecedented high spatial and temporal resolution observations and high spatial resolution numerical simulations of tornadoes and convective storms. Understanding the basic dynamics of the observed phenomena, however, requires our using multiple approaches from our "bag of tricks", which includes idealized modeling (cutting to the basic mechanisms, even if the models themselves are not completely realistic), linear stability analyses (even if the mechanisms are ultimately nonlinear), and horizontal and vertical vorticity and circulation analyses, in addition to accepting the vastly more realistic, but more complicated results of three-dimensional numerical simulations. Merely blindly accepting the black box results from models will not do: the student and researcher must simplify and verify the results from the fully nonlinear, three-dimensional simulations in order to acquire a basic understanding of the physics involved. I intend to attempt to do so in what follows, even though the modern student may be put off by the use of seemingly "old-fashioned" analytical techniques. In my opinion, a well-educated student should be able to analyze a problem using multiple methods to check his/her understanding for consistency.

A short, separate chapter on forecasting is included at the end of this text. However, suggested applications for forecasting techniques are implicitly and explicitly in some instances scattered throughout the text. It is hoped that forecasters in different parts of the world can adapt various aspects of the observations and theory presented to problems specific to their areas.

1.4 GENERAL MONOGRAPHS AND BOOKS

Atlas, D. (Ed.) (1963) *Severe Local Storms*, AMS Monograph 5, American Meteorological Society, Boston, pp. 1–128, 141–156, 177–223.

Atlas, D. (Ed.) (1990) *Radar in Meteorology: Battan Memorial and 40th Anniversary Radar Meteorology Conference*, American Meteorological Society, Boston, Chapters 14–16, 24.

Bluestein, H. B. (1999) *Tornado Alley: Monster Storms of the Great Plains*, Oxford University Press, New York, 180 pp.

Byers, H. R. and R. R. Braham, Jr. (1949) *The Thunderstorm*, U.S. Government Printing Office, Washington, D.C., 287 pp.

Church, C. *et al.* (Eds.) (1990) *The Tornado: Its Structure, Dynamics, Prediction, and Hazards*, AGU Monograph 79, American Geophysical Union, Washington, D.C., 637 pp.

Doswell, C. (Ed.) (2001) *Severe Convective Storms*, AMS Monograph 28, American Meteorological Society, Boston, Chapters 1–5, 7, 9, and 11.

Emanuel, K. A. (1994) *Atmospheric Convection*, Oxford University Press, New York, Chapters 1–6, 9–11.

Fujita, T. T. (1992) *The Mystery of Severe Storms*, Wind Research Laboratory, Department of Geophysical Sciences, The University of Chicago, 298 pp.

Houze, R. A., Jr. (1993) *Cloud Dynamics*, International Geophys. Series 53, San Diego, CA, Chapters 7–9.

Kessler, E. (Ed.) (1986). *Thunderstorm Morphology and Dynamics*, University of Oklahoma Press, Norman, OK, Chapters 1, 4, 5, 7, 9, 10, and 15.

Ray, P. (Ed.) (1986) *Mesoscale Meteorology and Forecasting*, American Meteorological Society, Boston, Chapters 15–18, pp. 331–436.

Wakimoto R. M. and Srivastava, R. C. (Eds.) (2001) *Radar and Atmospheric Science: A Collection of Essays in Honor of David Atlas*, AMS Monograph 30, American Meteorological Society, Boston, Chapters 4, 5.

1.5 REFERENCES AND BIBLIOGRAPHY

Bluestein, H. B. (1999) A history of storm-intercept field programs. *Wea. Forecasting*, **14**, 558–577.

Browning, K. A. (1962) Cellular structure of convective storms. *Meteor. Mag.*, **91**, 342–350.

Browning, K. A. and R. J. Donaldson, Jr. (1963) Airflow and structure of a tornadic storm. *J. Atmos. Sci.*, **20**, 533–545.

Davis, C. *et al.* (2004) The Bow Echo and MCV Experiment: Observations and opportunities. *Bull. Amer. Meteor. Soc.*, **85**, 1075–1093.

Fankhauser, J. C. (1971) Thunderstorm–environment interactions determined from aircraft and radar observations. *Mon. Wea. Rev.*, **99**, 171–192.

Galway, J. G. (1971) Early severe thunderstorm forecasting and research by the United States Weather Bureau. *Wea. Forecasting*, **7**, 564–587.

Hildebrand, P. H., W.-C. Lee, C. A. Walther, C. Frush, M. Randall, E. Loew, R. Neitzel, R. Parsons, J. Testud, F. Baudin, and A. LeCorec (1996) The ELDORA/ASSTRAIA airborne Doppler weather radar: High-resolution observations from TOGA COARE. *Bull. Amer. Meteor. Soc.*, **77**, 214–232.

Johns, R. H. and C. A. Doswell, III (1992) Severe local storms forecasting. *Wea. Forecasting*, **7**, 588–612.

Knight, C. A. and N. C. Knight (2001) *Hailstorms*, AMS Monograph 28, American Meteorological Society, Boston, pp. 223–248.

Rasmussen, E. N., J. M. Straka, R. Davies-Jones, C. A. Doswell, III, F. H. Carr, M. D. Eilts, and D. R. MacGorman (1994) Verification of the Origins of Rotation in Tornadoes Experiment: VORTEX. *Bull. Amer. Meteor. Soc.*, **75**, 995–1006.

Schaefer, J. T. (1986) Severe thunderstorm forecasting: A historical perspective. *Wea. Forecasting*, **1**, 164–189.

Ward, N. B. (1961) Radar and surface observations of tornadoes of May 4, 1961. *Proceedings Ninth Weather Radar Conference, Kansas City*, American Meteorological Society, Boston, pp. 175–180.

Williams, E. R. (2001) *The Electrification of Severe Storms*, AMS Monograph 28, American Meteorological Society, Boston, pp. 527–561.

2

The basic equations

"A cloud is made of billows upon billows upon billows that look like clouds. As you come closer to a cloud you don't get something smooth, but irregularities at a smaller scale."

Benoit Mandelbrot

"There are no rules of architecture for a castle in the clouds."

G.K. Chesterton—*The Everlasting Man*

The three fundamental physical laws that are used to diagnose the physical processes in severe convective storms are as follows: Newton's equation of motion applied to a fluid (called the "Navier–Stokes equations" after their formulators or, more simply, the "equations of motion"), the first law of thermodynamics applied to a fluid, and a statement of the continuity (conservation) of mass, including dry air, moist air, and the various forms of water substance (see Bluestein, 1992, 1993; Holton and Hakim, 2012; or other elementary textbooks for basic treatments).

2.1 THE EQUATIONS OF MOTION

2.1.1 The horizontal equation of motion

The equations of motion are usually separated into a horizontal component and a vertical component, each of which is slightly different. The inviscid (i.e., frictionless) horizontal equation of motion (its "primitive" form contains no simplifications or, more precisely, has the smallest possible number of simplifications) in vector form is as follows:

$$D\mathbf{v}_h/Dt = \partial\mathbf{v}_h/\partial t + \mathbf{v}\cdot\nabla\mathbf{v}_h = -1/\rho\nabla p \tag{2.1}$$

where D/Dt indicates a derivative following air parcel motion; $\mathbf{v}$ is the three-dimensional wind velocity; $\mathbf{v}_h$ is the horizontal component of the wind velocity; ρ is the density of moist air, which includes water vapor and other water substances such as liquid water and ice; p is pressure; and t is time. An air parcel is a tiny blob of air, a collection of air molecules. This air parcel is carried along with the three-dimensional wind and always contains the same collection of air molecules (i.e., the same collection of material). Unlike a mass in rigid body mechanics, however, the parcel may change shape.

The acceleration induced by the Coriolis force $\sim fU$, where $f = 2\Omega \sin \phi$, U is the horizontal wind speed, Ω is the rotation rate of the Earth about its axis, and ϕ is the latitude. It follows that accelerations induced by the Coriolis force are significant for time scales $\geq 1/f$. The Coriolis force is therefore not included unless the time scale of the phenomenon considered is at least 6 h. For a typical horizontal wind speed $\sim 10\,\mathrm{m\,s^{-1}}$ at midlatitudes, where $f \sim 10^{-4}\,\mathrm{s^{-1}}$, the acceleration due to the Coriolis force is $\sim 10^{-3}\,\mathrm{m\,s^{-2}}$, while the acceleration following air parcel motion for convection is $\sim 10\,\mathrm{m\,s^{-1}}/10\,\mathrm{min} \sim 10\,\mathrm{m\,s^{-1}}/10^3\,\mathrm{s} = 10^{-2}\,\mathrm{m\,s^{-2}}$, which is an order of magnitude greater. Thus, for ordinary convective storms, the Earth's rotation plays no direct role, while for long-lived, mesoscale convective systems Earth's rotation must be taken into account, but is a modifying—not necessarily a fundamental—factor in nearly all cases.

Molecular and turbulent friction are $\sim \mu\, \partial^2 U/\partial z^2$, where μ is the molecular/eddy coefficient of viscosity; μ for molecular viscosity is small and the turbulent term is significant typically only in the lowest kilometer or so, where $\partial^2 U/\partial z^2$ is relatively large. For the sake of *simplicity* and in an effort to focus on the smallest number of important physical processes possible, turbulent and molecular friction are not included here or in most subsequent equations except much later when tornadoes are discussed; it is thus assumed that all the variables are time and space averaged for the scales of motion we are considering and that subgrid-scale turbulence is ignored.

2.1.2 Buoyancy and the vertical equation of motion: defying gravity

The inviscid vertical equation of motion (without including the effects of Earth's rotation) is as follows:

$$Dw/Dt = -1/\rho\, \partial p/\partial z - g \tag{2.2}$$

where w is the vertical component of the wind (the vertical velocity); and g is the acceleration of gravity (which is taken to be a constant).

According to (2.2), vertical accelerations in convective clouds are simply a result of the imbalance between the vertical pressure gradient force and the force of gravity. The pressure can be decomposed into a part that is hydrostatic, which is associated with a balance between the vertical pressure gradient force and gravity, and a part that is non-hydrostatic, which is not. It is the vertical gradient of the latter that is responsible for vertical accelerations. It can be assumed that the base state of the atmosphere on the large scale is hydrostatic and that, although it may vary in space, it does so on scales much larger than those of convective clouds. Such an assumption is fine in most cases but not, for example, when a strong, surface front or sharp outflow boundary/gust front is present.

Nevertheless, let

$$p = \bar{p}(z) + p'(x, y, z, t) \tag{2.3}$$

and

$$\rho = \bar{\rho}(z) + \rho'(x, y, z, t) \tag{2.4}$$

and

$$-1/\bar{\rho}\, \partial\bar{p}/\partial z = g \tag{2.5}$$

where p is the total pressure; ρ is the total density; $\bar{p}$ and $\bar{\rho}$ are the pressure and density of the hydrostatic "base state" (the horizontally averaged values of p and ρ over a broad area); and p' and ρ' are the "perturbation" pressure and density, where in this case the perturbation quantities are at least an order of magnitude less in magnitude than the base-state quantities. The reader is reminded that this formulation requires modification when the basic state $\bar{\rho}$ varies horizontally as well (e.g., near a front or other baroclinic zone).

Because the perturbation quantities are so small, products of perturbation terms with perturbation terms are much less than products of perturbation terms with base-state terms, etc., it follows that

$$1/(\bar{\rho} + \rho') \approx 1 - (\rho'/\bar{\rho}) \tag{2.6}$$

and the vertical equation of motion (2.2) may then be rewritten as

$$Dw/Dt = -(1/\bar{\rho})\, \partial p'/\partial z - (\rho'/\bar{\rho})g \tag{2.7}$$

Archimedean buoyancy B is given by the second term on the RHS (right-hand side) of (2.7)

$$B = -(\rho'/\bar{\rho})g \tag{2.8}$$

The vertical equation of motion is similar to the horizontal equation of motion, except that there is an additional term that represents the acceleration induced by the buoyancy force. (If the atmosphere were in hydrostatic balance, then $Dw/Dt = 0$, so that $B = -1/\bar{\rho}\, \partial p'/\partial z$.) The concept of Archimedean buoyancy is nice because it is easy to visualize buoyancy in terms of a less dense object (i.e., less dense than water) released in water rising or a denser object sinking. In the atmosphere, however, the distinction between "the object" and "the water" is not always clear and the density of the surrounding air representing the water varies horizontally as well as vertically.

The effect of Archimedean buoyancy may be understood quantitatively by considering a box (of dimensions $\Delta x \times \Delta y \times \Delta z$) of fluid of density ρ_1 embedded within a fluid of density ρ_2 (Figure 2.1). If the fluid surrounding the buoyant box, the "environment", is in hydrostatic balance, then

$$-1/\rho_2\, \partial p_2/\partial z - g = 0 \tag{2.9}$$

where p_2 is the pressure in the environment. So, the vertical pressure gradient force in the environment for a volume the size of the buoyant box is the

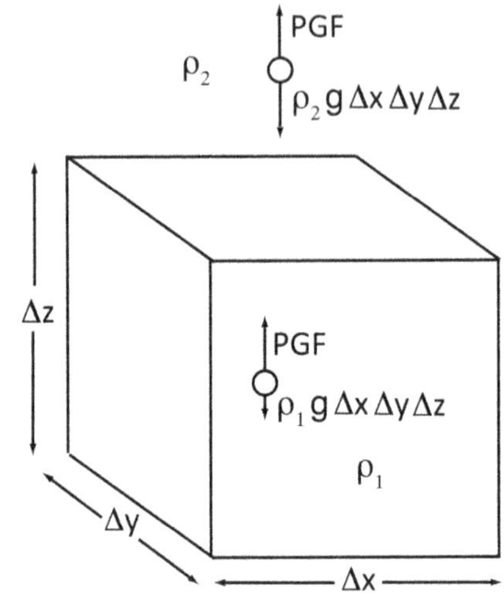

Figure 2.1. Idealized illustration of Archimedean buoyancy for a box of fluid. Outside the box the fluid is in hydrostatic balance. Since the pressure gradient force (PGF) is the same outside the box as it is inside the box, but the density inside the box ρ_1 in this case is less than the density outside the box ρ_2, there is a net upward buoyancy force. Forces inside and outside the box are indicated.

acceleration due to the vertical pressure gradient force multiplied by the mass of the fluid in the same volume as that of the box; that is, the upward force is $-1/\rho_2\, \partial p_2/\partial z(\Delta x\, \Delta y\, \Delta z)\rho_2 = -\partial p_2/\partial z(\Delta x\, \Delta y\, \Delta z)$ and the equal and opposite downward force due to gravity is $-\rho_2 g(\Delta x\, \Delta y\, \Delta z)$. If we assume that the pressure in the buoyant box is the same as that in the environment (a good assumption at the outset only), then the upward pressure gradient force experienced by the box is still $-\partial p_2/\partial z(\Delta x\, \Delta y\, \Delta z)$. However, the buoyant box's downward force due to gravity is only $-\rho_1 g(\Delta x\, \Delta y\, \Delta z)$ if, for example, ρ_1 is less than ρ_2. The box of fluid is *not* in hydrostatic balance and there is a net upward force of $g(\Delta x\, \Delta y\, \Delta z)(\rho_2 - \rho_1)$. The vertical acceleration (force/mass) experienced by the buoyant box is therefore

$$Dw/Dt = g(\rho_2 - \rho_1)/\rho_1 = B \tag{2.10}$$

Buoyancy is therefore due to the net difference in weight between the denser environment and that of the less dense box of fluid. The formulations for buoyancy in (2.8) and (2.10) are essentially identical, but the latter is exact, while the former is approximate: for the former, $B = -(\rho'/\bar{\rho})g$, where $\rho_1 = \rho_2 + \rho'$ and $\bar{\rho}$ is slightly less than ρ_2, because the horizontal average of ρ is weighted mostly by ρ_2 and much less by the less dense ρ_1; in this instance, the average value of ρ' is not zero, as it usually is taken to be, because the only perturbation is the density of the buoyant cube. In (2.10), ρ_1 appears in the denominator rather than ρ_2, reflecting the influence of the slightly lower density of the buoyant cube.

From the ideal gas law for dry air ($p = \rho R_d T$, where R_d is the gas constant for dry air) and (2.3) and (2.4), it is seen that

$$\rho'/\bar{\rho} = p'/\bar{p} - T'/\bar{T} \tag{2.11}$$

where $T = \bar{T}(z) + T'(x, y, z, t)$. Thus,

$$B = -(p'/\bar{p} - T'/\bar{T})g \tag{2.12}$$

In other words, buoyancy depends not only on the temperature excess or deficit over that of the environment, but also on the pressure excess or deficit over that of the environment. It turns out, however, that $p'/\bar{p}$ can be neglected in comparison with $T'/\bar{T}$ for substantially subsonic flow. To see this, consider the horizontal equation of motion (2.1) expressed in terms of the base state and perturbations (2.3) and (2.4)

$$D\mathbf{v}_h/Dt = \partial\mathbf{v}_h/\partial t + \mathbf{v}\boldsymbol{\cdot}\boldsymbol{\nabla}\mathbf{v}_h = -1/\bar{\rho}\,\nabla p' \tag{2.13}$$

Suppose that each component of the wind $u, v, w \sim U$, which is typically $\sim 10\,\mathrm{m\,s^{-1}}$. Then it follows from (2.13) that

$$|-1/\rho\,\nabla p'| \sim |\mathbf{v}\boldsymbol{\cdot}\boldsymbol{\nabla}\mathbf{v}_h| \tag{2.14}$$

and therefore that

$$\sim p'/\bar{\rho} \sim U^2 \tag{2.15}$$

From the ideal gas law it follows that

$$\sim p'/\bar{p} \sim U^2/(R_d\bar{T}) \approx U^2/c^2 \tag{2.16}$$

where the speed of sound c in dry air is given by

$$c = (\gamma R_d \bar{T})^{1/2} \tag{2.17}$$

and

$$\gamma = C_p/C_v = 1.4 \sim 1 \tag{2.18}$$

where C_p and C_v are the specific heat of air at constant pressure and volume, respectively. Thus, with the exception of perhaps the most intense tornadoes, $U \ll c \sim 300\,\mathrm{m\,s^{-1}}$ such that the square of the Mach number $U^2/c^2 \sim 100/(300)^2 \sim 10^{-3} \ll T'/T \sim 5/300 \sim 10^{-2}$, so that to a good approximation

$$B = (T'/\bar{T})g \tag{2.19}$$

It will be shown later that, if buoyancy is expressed in terms of potential temperature rather than temperature, we do not have to be concerned with the Mach number in the computation of buoyancy.

Suppose now that air includes both water vapor and condensate; the pressure of the air is the sum of the partial pressure of the dry air and that of the water vapor, while the condensate does not contribute to the total pressure. Suppose also that we treat the air containing both water vapor and condensate as if it were dry (i.e., with no water vapor or condensate) and note that the density is modified by the less dense water vapor and the more dense condensate, so that

$$p = \rho_d R_d T + \rho_v R_v T = (\rho_d + \rho_v) R_d T_v = (\rho_d + \rho_v + \rho_l + \rho_i) R_d T_{\mathrm{cv}} \tag{2.20}$$

where ρ_l is the density of all liquid substances (including cloud droplets, drizzle drops, and raindrops); ρ_i is the density of all ice substances (includes all ice crystals, graupel, and hailstones); T_v is the virtual temperature; and T_{cv} is the cloud virtual temperature. It follows from (2.20) and the observation that the water vapor mixing ratio $r_v \ll 1$ that

$$T_v = T[1 + (R_v/R_d)r_v]/(1 + r_v) \approx (1 + 0.609r_v)T \tag{2.21}$$

From (2.20) and (2.21) it is seen that the cloud virtual temperature

$$T_{cv} = (1 + 0.609r_v - r_l - r_i)T \tag{2.22}$$

where r_l is the liquid water mixing ratio; and r_i is the ice mixing ratio. The buoyancy for moist air containing both liquid and ice condensate is therefore

$$B = gT'(1 + 0.609r_v - r_l - r_i)/\bar{T} \tag{2.23}$$

An upward (downward) buoyancy force is thus diminished (increased) by the loading of water substance and increased (decreased) by the presence of water vapor.

It will be demonstrated later that the *total* vertical acceleration following an air parcel is given not only by Archimedean buoyancy, but also by the vertical *perturbation* pressure gradient force. This force is in part associated with air that is moved out of the way ahead of the air parcel and air moving in behind the air parcel as it is accelerated vertically by Archimedean buoyancy. This air movement is accomplished by a perturbation high-pressure area ahead of the air parcel to force air out of the way and by a perturbation low-pressure area behind the air parcel to suck air in behind it. When the pressure gradient force is directed in the opposite direction to air motion, as it is in this case, the pressure gradient force is said to be "adverse". If the air parcel is moving vertically at a constant speed and the adverse vertical perturbation pressure force exactly counteracts the buoyancy force, the flow is hydrostatic (as noted earlier). If the air parcel *accelerates* vertically, then the vertical perturbation pressure gradient does not exactly counteract the buoyancy force. It will be shown shortly how one can compute the vertical pressure gradient force and thereby determine net vertical acceleration. There is also a component to the vertical perturbation pressure gradient force associated with the wind field in the environment; this component will also be discussed later.

2.2 THERMODYNAMICS

The first law of thermodynamics for air is as follows:

$$dQ = C_v\, dT + p\, d\alpha \tag{2.24}$$

where Q is heat energy; C_v is the specific heat at constant volume of a volume of air containing a mixture of dry air, water vapor, and other water substances (liquid and frozen); and the specific volume $\alpha = 1/\rho$. Diabatic heating from changes in the phase of water substance, turbulent heat transfer from the surface, and radiation causes changes in temperature and specific volume. The dynamics of convective storms is affected most by the latent heat released or absorbed when water droplets condense from water vapor, when water droplets evaporate, when ice crystals form directly from water vapor, when water droplets freeze into ice, when ice melts into water, and when ice particles sublimate ("to err is human, to change phase, sublime"). The bulk change-in-phase processes (as opposed to changes to individual particles) are referred to as "cloud microphysics". Since cloud microphysical processes are not completely understood and not easily observable and quantified, they are parameterized in terms of quantities that can in fact be measured, such as temperature, humidity, pressure, etc.

Turbulent sensible heat transfer from a heated land surface during the day when the Sun is out or when cold air passes over a much warmer body of water is often linked to cloud formation. Radiative cooling at cloud top, heating at cloud base, or horizontal gradients in radiative heating (e.g., at cirrus anvil edges) can also play significant roles, but are not of primary dynamical importance for convective storms.

The adiabatic form of the thermodynamic equation, expressed in terms of the time rate of change of variables, is given by

$$C_p\, DT/Dt - 1/\rho\, Dp/Dt = 0 \tag{2.25}$$

In this formulation changes in temperature following air parcel motion are related only to changes in pressure following air parcel motion. The adiabatic form of the thermodynamic equation is useful for describing the thermodynamic changes associated with horizontal and vertical air motions outside of convective storms: rising air is cooled and sinking air is warmed if the stratification is stable, the amount of cooling/warming varying as the lapse rate of temperature.

It is useful to express the adiabatic form of the thermodynamic equation in terms of potential temperature

$$\theta = T(p/p_0)^{R/C_p} \tag{2.26}$$

where $D\theta/Dt = 0$ (i.e., potential temperature is conserved). When the air is saturated, θ is replaced by θ_e (equivalent potential temperature). From the definition of potential temperature, the adiabatic form of the thermodynamic equation (2.25) may be expressed as

$$1/\rho\, D\rho/Dt = \gamma\, 1/p\, Dp/Dt \tag{2.27}$$

This equation will be useful later when we analyze sound waves. In particular, the left-hand side is part of the equation of continuity discussed in the following section and allows us to relate changes in density to changes in pressure.

2.3 CONSERVATION OF MASS, AND THE BOUSSINESQ AND ANELASTIC APPROXIMATIONS

The equation of continuity, which is a statement of the conservation of mass, is as follows for a compressible atmosphere:

$$1/\rho \, D\rho/Dt = -\nabla \cdot \mathbf{v} \tag{2.28}$$

When air converges in three dimensions, some of it is compressed so that its density increases; it is also deformed such that some air is drawn in from some sides and is expelled out other sides. To see this, think of what happens when you squeeze a tube of toothpaste; if the toothpaste is compressible, then its density increases somewhat, but some of the toothpaste squirts out. The time derivative in (2.28) creates some problems in understanding convective clouds because it permits the existence of sound waves, which can propagate and therefore must be accounted for. Sound waves are not thought to be important for the dynamics of convective storms, unless perhaps you believe you can shout at a storm and change its behavior (something that storm chasers often do, by "badmouthing" a storm so as to conjure up a more intense storm) or that a convective storm can contain some aural information and can thereby communicate with you. If one integrates pressure with time, then rapid pressure fluctuations associated with sound waves overwhelm the slower pressure fluctuations associated with storm-scale processes. Just as if one integrates pressure with time for the hydrostatic primitive equations, the rapid pressure fluctuations associated with gravity waves overwhelm the slower pressure fluctuations associated with synoptic-scale, quasi-geostrophic processes. A technique must be devised for filtering out sound waves, unless it can be shown the dynamics of some parts of severe convective storms do respond to or are dependent on sound waves. The original motivation for simplifying the dynamical equations that describe convection was to eliminate sound waves.

Under certain circumstances a compressible fluid such as the atmosphere may be treated like an incompressible fluid (i.e., one in which $1/\rho \, D\rho/Dt = 0 = -\nabla \cdot \mathbf{v}$) and sound waves are not allowed. In this situation, air parcels are deformed when there is divergence or convergence in one plane, but convergence in one plane must be compensated for by divergence in the direction normal to the plane. Think of squeezing a tube of toothpaste but this time, no matter how hard you squeeze the toothpaste, it does not get any denser if the toothpaste is incompressible.

Let us now see if there are any conditions under which we can neglect the time derivative term in the compressible form of the continuity equation. From the adiabatic form of the thermodynamic equation (2.27), the equation of continuity (2.28), and the ideal gas law, it is seen that

$$-\nabla \cdot \mathbf{v} = C_v/C_p \, D(\ln p)/Dt = C_v/C_p(\partial/\partial t + \mathbf{v} \cdot \nabla_z + w \, \partial/\partial z) \ln p \tag{2.29}$$

The last term on the RHS of (2.29), with the aid of the hydrostatic approximation and the ideal gas law, may be expressed as $\partial/\partial z(\ln p) \sim g/R\bar{T} = 1/H$, where H is the scale height of the atmosphere, which is ~10 km. We would like to avoid making an assumption about the time scale, so we estimate the magnitude of the time derivative term—the first on the RHS of (2.29)—in terms of the horizontal advection term—the second on the RHS of (2.29)—as $\sim \mathbf{v}/p \cdot \nabla_z p \sim U/p|\nabla_z p|$, where U is the scale of the horizontal wind speed. We are thus assuming that the time derivative term is the same order of magnitude as the horizontal advection term, but not necessarily the same magnitude as the vertical advection term. We found earlier (2.15) that $p'/\bar{p} \sim U^2$, so that $|\mathbf{v}/p \cdot \nabla_z p| \sim U^2/R\bar{T}|\nabla_z|\mathbf{v}|| \sim U^3/R\bar{T}L$, where L is the horizontal scale. So, in summary

$$\partial/\partial t(\ln p) \sim U^3/R\bar{T}L \tag{2.30}$$

$$\nabla_z(\ln p) \sim U^3/R\bar{T}L \tag{2.31}$$

$$\partial/\partial z(\ln p) \sim 1/H \tag{2.32}$$

The LHS of (2.29) is $\sim U/L$ and the scale of vertical velocity $w \sim W$, so that the scale of (2.29) may be expressed as follows:

$$U/L \sim C_v/C_p(U^3/R\bar{T}L + U^3/R\bar{T}L + W/H) \tag{2.33}$$

From the continuity equation (2.28) it is seen that $U/L \sim W/D$, where D is the vertical scale and not necessarily the same scale as the scale height H. Substituting for W in the third term on the RHS of (2.33) and making use of the formula for the speed of sound (2.17), it follows that:

$$1 \sim U^2/c^2 + C_v/C_p\, D/H \tag{2.34}$$

where we have combined the first and second terms because they are the same. The first term on the RHS of (2.34) can therefore be neglected when $U \ll c$; that is, when wind speeds are much less than the speed of sound, which is $\sim 300\,\mathrm{m\,s^{-1}}$ (i.e., for low Mach numbers). This condition holds for most winds in convective storms $\sim 30\,\mathrm{m\,s^{-1}}$ or less, but definitely not for strong tornadoes, in which wind speeds are $\sim 100\,\mathrm{m\,s^{-1}}$. We conclude then that the time derivative term (or the horizontal advection term) may be neglected for typical non-tornadic, subsonic wind speeds, but not for strong tornadoes. If sound waves must be taken into account when considering the dynamics of strong tornadoes, then the consequences of a relatively high ratio between the wind speed and the speed of sound (Mach number) must be taken into account and the first term on the RHS of (2.34) must be included.

Under some conditions, the second term on the RHS of (2.34), which represents the vertical advection of pressure (cf. (2.29)), can also be neglected. This term may be neglected when $D \ll H$ (i.e., when the vertical scale is much less than the scale height or, in other words, when convection is "shallow").

2.3.1 The Boussinesq approximation

For shallow, substantially subsonic convection, the continuity equation may be simplified to a form equivalent to the incompressible form of the continuity equation:

$$\nabla \cdot \mathbf{v} = 0 \tag{2.35}$$

This approximation is referred to as the Boussinesq approximation, named after the French mathematician and physicist who discussed the approximation in the late 19th century.

The Boussinesq equations refer to the set of equations for which the incompressible form of the continuity equation is used or, in other words, density is considered a constant in the continuity equation *but* for which variations in density *are* included in the vertical equation of motion when coupled with gravity (Archimedean buoyancy). Thus, the effects of compressibility are retained in the vertical equation of motion, but not in the continuity equation. Strictly speaking, the Boussinesq equations describe air motions for *shallow* cumulus clouds and a *shallow* boundary layer, the region near the ground that is affected by it (typically up to ~1 km AGL, but up to as much as 3 km AGL or more in heated, arid regions). Any change in density following air parcel motion is considered negligible because the vertical variation of density over the path it has taken is negligible.

The frictionless equation of motion in vector form, subject to the Boussinessq approximation, is

$$D\mathbf{v}/\partial t = -1/\bar{\rho}\,\nabla p' + B\mathbf{k} \tag{2.36}$$

where $\bar{\rho}$ is a constant, because ρ in the denominator of the pressure gradient term can be treated as a constant in a shallow layer.

From the ideal gas law it can be shown that

$$1/\rho\, D\rho/Dt = 1/p\, Dp/Dt - 1/T\, DT/Dt \tag{2.37}$$

For adiabatic processes

$$1/p\, Dp/Dt = C_p/RT\, DT/Dt \tag{2.38}$$

Subject to the Boussinesq approximation, it follows that

$$1/\rho\, D\rho/Dt = -\nabla \cdot \mathbf{v} = 0 = 1/p\, Dp/Dt - 1/T\, DT/Dt = C_v/C_p\, 1/p\, Dp/Dt \tag{2.39}$$

So, $1/p\, Dp/Dt = 0$ and from (2.38) it follows that

$$1/T\, DT/Dt = 0 \tag{2.40}$$

If we let $T = \bar{T}(z) + T'(x, y, z, t)$, then from (2.41) we find that

$$DB/Dt + wB\, d\ln \bar{T}/dz = 0 \tag{2.41}$$

2.3.2 Anelastic approximation

For substantially subsonic convection that is "deep" (i.e., for which $D \sim H$), the equation of continuity (2.29) may be simplified as

$$\nabla \cdot \mathbf{v} = -C_v/C_p(w\,\partial/\partial z)(\ln p) \tag{2.42}$$

This approximation is referred to as the *anelastic* approximation; Yoshi Ogura and Norman Phillips first coined this term in a seminal paper published in 1962. Stigler's Law of Eponymy is evident here: according to this law, any scientific discovery named for someone was actually discovered earlier by someone else. In this case, the Australian applied mathematician and fluid dynamicist George Batchelor had come up with a nearly identical formulation back in 1953, while Jule Charney and Yoshi Ogura had used it in 1960 without naming it. (It also appears that Boussinesq's work itself was pre-dated by A. Oberbeck in 1879, though my knowledge of the German language is not adequate and I must rely on translations by others.) For adiabatic flow, it is seen from (2.39) that (2.42) may be expressed as

$$\nabla \cdot \mathbf{v} = -1/\rho\,\partial\rho/\partial z \tag{2.43}$$

For severe convective systems, which are typically deep (i.e., extend up to the tropopause and slightly beyond, ~10–15 km AGL), the anelastic continuity equation is often a better approximation. It is a simplification of (2.28) in which local time derivatives and horizontal gradients are neglected, and only the vertical variations of density are retained. The anelastic continuity equation is used in some numerical models and in some Doppler radar wind analysis schemes. To keep analyses of the dynamics of convection as simple as possible, however, the Boussinesq continuity equation is most frequently used. By ignoring the vertical effects of compressibility, the overall physics are changed only slightly and fundamental results are not altered qualitatively. The results of numerical simulations conducted with models that are fully compressible (using (2.28)) support the analysis of storm dynamics in a qualitative sense using the Boussinesq approximation.

Remember that we constrained the first and second terms in the continuity equation (2.33) to be of the same scale. From the Eulerian form of the thermodynamic equation (2.29) we estimate the time scale $\tau \sim D/W$ by assuming that the time derivative term is the same order of magnitude as the vertical advection term. For deep convection, $D \sim 10\,\text{km}$ and $W \sim 10\,\text{m s}^{-1}$, so that $\tau \sim 10^3\,\text{s}$, or about 10 min. From the vertical equation of motion (2.7), we estimate $\tau \sim W/B$, where B is the buoyancy. So, for $B \sim (10\,\text{m s}^{-2})$ (1 K/300 K), $\tau \sim 300\,\text{s} = 5\,\text{min}$. We can now compare the magnitude of the time derivative term in (2.29) with the horizontal advection term in (2.29) by assuming $\tau \sim$ 5–10 min, the reciprocal of which is approximately the Brunt–Väisälä frequency (~1/300–1/1,000 Hz), which is much lower than that of audible sound waves. So

$$\partial \ln p/\partial t \sim 1/\tau \tag{2.44}$$

and

$$\mathbf{v} \cdot \nabla_z \ln p \sim U/L \tag{2.45}$$

where L is the horizontal scale. For $\tau \sim 500$–$1{,}000\,\mathrm{s}$ and $U \sim 10\,\mathrm{m\,s^{-1}}$, $L \sim U\tau \sim 5 \times 10^3$–$10^4\,\mathrm{m} = 5$–$10\,\mathrm{km}$, which makes sense from what we know about the wind field in convective storms.

Although we reject sound waves as being dynamically unimportant, there is some evidence that convective storms, and intense vortices such as tornadoes in particular, can generate detectable sound waves in the infrasound region. Al Bedard of the former Wave Propagation Laboratory (WPL) in Boulder, Colorado did some seminal work in this area, backed by earlier work by Roy Arnold and collaborators during the 1970s and stimulated by the theoretical work of Abdul Abdullah at NCAR in the mid-1960s. More recently, David Schecter at North West Research Associates has numerically simulated infrasound in convective storms.

In any event, three-dimensional cloud models have been developed that permit sound waves and, thus, do not make use of the Boussinesq or anelastic approximation. Numerical procedures such as "time-splitting" have been developed that allow one to include the full effects of compressibility without actually representing all the terms in the model equations at the highest frequencies: relatively low-frequency processes such as advection and buoyancy are separated from relatively high-frequency, sound wave propagation processes such as the pressure gradient force and divergence, each of which is integrated using different time steps. *For diagnostic purposes, it is sufficient though to use the Boussinesq approximation to examine most dynamical effects.*

Although we have carefully determined how well the Boussinesq and anelastic approximations work and noted what their benefits and liabilities are, we must be aware that we have not yet determined if energy is conserved for the full set of equations. Since the purpose of this text is to understand dynamics—not how to construct a numerical model—we will defer discussions of energy conservation elsewhere (see the list of references).

2.3.3 Water substance

Conservation of mass is extended to include water vapor and the various forms of water substance through the following equation:

$$Dq/Dt = -\nabla\cdot(q\mathbf{v}) + q\nabla\cdot\mathbf{v} + E + S - C - D \tag{2.46}$$

where q is the specific humidity; E is the evaporation rate per unit mass of moist air; S is the sublimation rate; C is the condensation rate; D is the deposition rate; and the first two terms on the right-hand side come from the advective term. The various types of water substance can be broken down into many more categories (e.g., the deposition and sublimation rates can be specified separately for different types of ice crystals and other forms of frozen water) than those represented in (2.46), which is a highly simplified representation of what actually happens in the atmosphere. Additional relations can be specified for conversion rates from ice to liquid water (e.g., due to melting, etc.), the shedding of water on the surface of melting hailstones, etc. Our inability to make in situ measurements of cloud particles and hydrometeors everywhere simultaneously in a convective storm for its entire duration is a major obstacle in our quest to understand completely and to be able to predict the evolution of convective storms. It will be seen later that polarimetric radars offer some hope for bridging the gap in our being able to map out hydrometeors three dimensionally in a convective storm.

It is not practical to follow individual hydrometeors around in a convective storm; the spatial resolution required is much greater than the spatial resolution used in numerical cloud models. Microphysical processes are therefore parameterized, just as subgrid-scale turbulence is parameterized based on "bulk" quantities (i.e., those on larger, resolvable scales). The most accurate technique for representing the microphysical processes in (2.46) in numerical models is the "bin" (also known as "spectral") method. In the bin method, the size or mass of hydrometeors within each bin (i.e., range of diameter or mass) is apportioned according to microphysical and kinematic processes. When many bins are used, as is necessary for realistic results, a relatively long duration of computer time is needed.

To reduce computer time, bulk multi-moment parameterization schemes are often used. In this case, a specified distribution function for hydrometeor size (assumed a priori and not generated for each case, as happens with the bin method) is used. For a quantity M

$$M_k = \int_0^\infty D^k N(D)\, dD \tag{2.47}$$

where k is the moment; D is the diameter of the hydrometeor; $N(D)$ is the hydrometeor size distribution (i.e., the variation of the number of hydrometeors at each diameter D), also called "particle size distribution", such that $N(D)\, dD$ is the number per unit volume of hydrometeors having a diameter from D to $D + dD$. Based on observations of raindrops, snow crystals, and hailstone diameters (hail of course, may not be spherical; Figure 2.2), a gamma distribution is often assumed, so that

$$N(D) = N_0 D^\alpha e^{-\lambda D} \tag{2.48}$$

where N_0, α, and λ are the intercept parameter, shape parameter, and slope parameter, respectively. The limitations to assuming a gamma distribution must be acknowledged: for example, bimodal distributions of hydrometeors are not well represented by a gamma distribution. For example, when large water droplets break up into smaller ones the distribution may be bimodal.

When $k = 3, M_3$, a measure of the mass of the hydrometeors (because the cube of the diameter is both proportional to the volume—which, when multiplied by density, yields mass—and to the mass mixing ratio) is used as the only prognostic variable, in much the same way as a scheme is referred to as a "one-moment" scheme because only one moment, the third moment, is a prognostic variable. If $\alpha = 0$ and N_0 is assumed to be a constant, then only λ is allowed to vary. Different relations are used for different types of hydrometeors; the more different classes of hydrometeors included, the more like a bin scheme the parameterization becomes.

Figure 2.2. Hailstone that fell in Vivian, SD on July 23, 2010. This hailstone as of 2011 is the largest hailstone ever documented in terms of both weight (31 oz.) and circumference (19 inches). It was studied in Charlie Knight's Cold Laboratory at NCAR, where the author took this photograph. The radar echo associated with a convective storm having large hail may have a spike or flare attached to it as a result of three-body scattering (hail to ground, ground to hail, and hail back to radar).

Keeping N_0 fixed, however, results in some errors because the consequences of evaporation are not represented accurately and because processes that affect one part of the size spectrum differently from other parts are also not represented accurately: size sorting is not simulated well.

When $k = 0, M_0$, the number of hydrometeors in the volume (cf. (2.47)) is added as a second prognostic variable and the bulk parameterization scheme is known as a "two-moment" scheme. Both N_0 and λ can vary. Two-moment schemes, however, over-represent size sorting.

When $k = 6, M_6$ (cf. (2.47)), which is proportional to the radar reflectivity factor Z (i.e., proportional to the sixth power of the diameters of the hydrometeors when they are liquid and when their diameters are much shorter than the wavelength of the radar radiation, the case in the Rayleigh scattering region) is added as a third prognostic variable, the bulk parameterization scheme is then known as a "three-moment" scheme. It has been found that improvements in the realism of simulations when a one-moment scheme is changed to a two-moment scheme are greater than when a two-moment scheme is changed to a three-moment scheme.

The student and researcher are referred to Jerry Straka's 2009 book *Cloud and Precipitation Microphysics* for more details on bin and bulk microphysics schemes used in models and Pruppacher and Klett's 1998 book *Microphysics of Clouds and Precipitation* for more detailed derivations of governing equations, etc. The main point the student should come away with from this discussion of microphysical parameterization is that it is difficult to represent accurately the thermodynamic consequences of cloud and precipitation formation and dissipation. Since the amount of cloud and precipitation material affect buoyancy (cf. (2.23)), which appears in the vertical equation of motion (2.7) and therefore affects the kinematics of the convective cloud, cloud microphysics can significantly affect the dynamics and behavior of convective storms.

2.4 THE VORTICITY AND CIRCULATION EQUATIONS

The dynamics of severe convective storms are often elucidated by using "derived" forms of the equations of motion (2.13) and (2.7). By applying the curl operator to the three-dimensional equation of motion represented by the combination (2.13) and (2.7), the following time-dependent, *prognostic* vorticity equation is found:

$$D/Dt(\nabla \times \mathbf{v}) = [(\nabla \times \mathbf{v}) \cdot \nabla]\mathbf{v} - \nabla \times (1/\bar{\rho}\, \nabla p') + \nabla \times (B\mathbf{k}) \tag{2.49}$$

where $\mathbf{k}$ is a unit vector pointing upward; and $\nabla \times \mathbf{v}$ is the three-dimensional vorticity. The first term on the RHS of (2.49) is the stretching and tilting term, the second term is the solenoidal term, and the third term is the baroclinic term. In a Boussinesq atmosphere the solenoidal term vanishes because density is treated as a constant (since it comes from the pressure gradient term in the equation of motion). In (2.49), Earth's vorticity is not included, but could be added to the vertical component of vorticity to account for the behavior of convective phenomena that persist for a relatively long time (i.e., when the Coriolis force becomes significant).

The vertical component of (2.49) is

$$D\zeta/Dt = \underbrace{\partial\zeta/\partial t}_{1} + \underbrace{\mathbf{v}_h \cdot \nabla\zeta}_{2} + \underbrace{w\, \partial\zeta/\partial z}_{3} = \underbrace{-\delta\zeta}_{1} + \underbrace{\mathbf{k}\cdot(\partial\mathbf{v}/\partial z \times \nabla w)}_{2} + \underbrace{\mathbf{k}\cdot[\nabla p' \times \nabla(1/\bar{\rho})]}_{3} \tag{2.50}$$

where $\zeta = \mathbf{k} \cdot \nabla \times \mathbf{v} = \partial v/\partial x - \partial u/\partial y$ is the vertical component of vorticity; $\delta = \partial u/\partial x + \partial v/\partial y$ is horizontal divergence; x and y are coordinate axes that point to the east and north, respectively; and u and v are the components of the

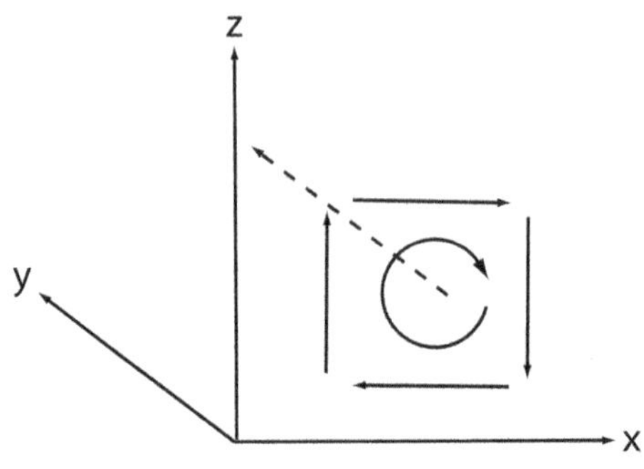

Figure 2.3. An idealized illustration of positive vorticity about the y-axis when $\partial u/\partial z > 0$ and $\partial w/\partial x < 0$. The dashed vector indicates vorticity. The curved arrow denotes the sense of rotation, while the direction of the dashed vector may be determined by curling one's fingers about the sense of the curl such that the thumb points in the direction of the vorticity vector according to the "right-hand rule".

horizontal wind in the x and y-directions, respectively. Terms 2 and 3 on the left-hand side of (2.50) (when multiplied by -1) represent horizontal and vertical advection of vorticity, while terms 1, 2, and 3 on the right-hand side represent stretching (the divergence term), tilting, and the solenoidal effect. The latter occurs when the atmosphere is baroclinic (i.e., when isobars are not parallel to lines of constant density, so that there is a nonzero horizontal gradient in the pressure gradient force). The solenoidal term (3 on the RHS), however, is zero in a Boussinesq atmosphere. The vorticity equation is in part more convenient than the equations of motion for dynamical analyses because pressure does not appear explicitly. As before, Earth's vorticity is not included, but could be added to the vertical vorticity to account for the behavior of convective phenomena that last for a relatively long period of time.

The component of (2.49) along the y axis (Figure 2.3) is

$$D/Dt(\partial u/\partial z - \partial w/\partial x) = \underbrace{(\partial u/\partial x + \partial w/\partial z)(\partial u/\partial z - \partial w/\partial x)}_{1} + \underbrace{(\partial v/\partial z\, \partial u/\partial y - \partial v/\partial x\, \partial w/\partial y)}_{2} - \underbrace{\partial B/\partial x}_{3} \quad (2.51)$$

where $\partial u/\partial z - \partial w/\partial x$ is the component of vorticity about the y axis. Term 1 represents the stretching of vorticity about the y axis, 2 represents tilting of vorticity about the z and/or x axes onto the y axis, and 3 represents baroclinic generation. The latter is created when there is a horizontal gradient of vertical acceleration due to a horizontal gradient of buoyancy (Figure 2.4). Again, Earth's vorticity is not included, but its effects could be accounted for by adding it to the vertical vorticity.

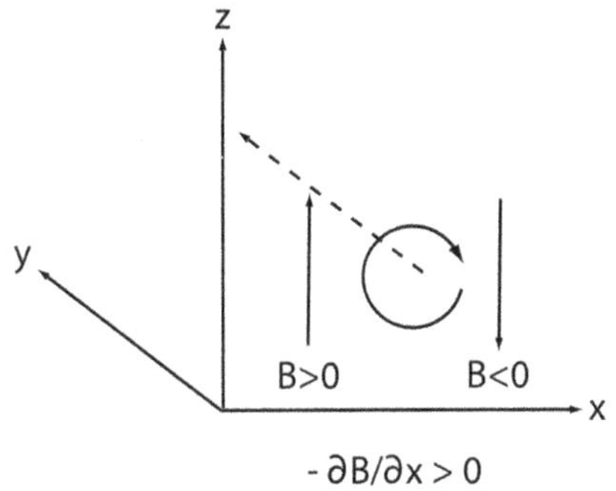

Figure 2.4. An idealized illustration of the generation of vorticity about the y-axis when buoyancy decreases in the x-direction. Imagine a paddle wheel placed in the flow; it will begin to rotate about the y-axis because the left side experiences an upward acceleration while the right side experiences a downward acceleration.

Equations (2.50) and (2.51) are often used to analyze convective storm dynamics—a similar equation for the x-component of vorticity $(\partial w/\partial y - \partial v/\partial z)$ can also be used. The vertical vorticity equation (2.50) is most useful for analyzing the formation, maintenance, and dissipation of vortices such as tornadoes, meso-cyclones, and meso-anticyclones. It is most useful when expressed in a Lagrangian framework, so that the processes contributing significantly to the vertical vorticity associated with the vortex, which is being transported along with the mean flow, can be determined. Trajectories must be calculated to follow air parcels along and one must carefully select trajectories that are representative: one must select those that begin/end in specific regions and pass through the vortex being analyzed. In some instances small differences in the beginning (or, in the case of backward trajectories, ending) positions may lead to trajectories that pass through very different locations in space.[1]

To overcome the problem of having to compute many parcel trajectories, it is useful to analyze the macroscopic measure of vorticity, circulation C, where one makes use of Stokes' theorem, so that

$$C = \oint \mathbf{v} \cdot \mathbf{dl} = \iint (\nabla \times \mathbf{v} \cdot \mathbf{n})\, dA \tag{2.52}$$

where the line integral of the component of wind velocity over the perimeter ($\mathbf{l}$ is tangent to the edge of the fluid element) of a two-dimensional fluid element is given by the component of vorticity normal to the fluid element ($\mathbf{n}$ is a unit vector pointing normal to the plane surface of A, in the direction defined by the right-hand rule) integrated over the area defined by the two-dimensional fluid element (A) (Figure 2.5). In other words, circulation is the vorticity normal to the fluid element multiplied by its area or, equivalently, vorticity is circulation per unit area.

[1] When windspeeds are high, as in a tornado, the effects of small changes in position (and time increment used to compute the trajectories) are amplified.

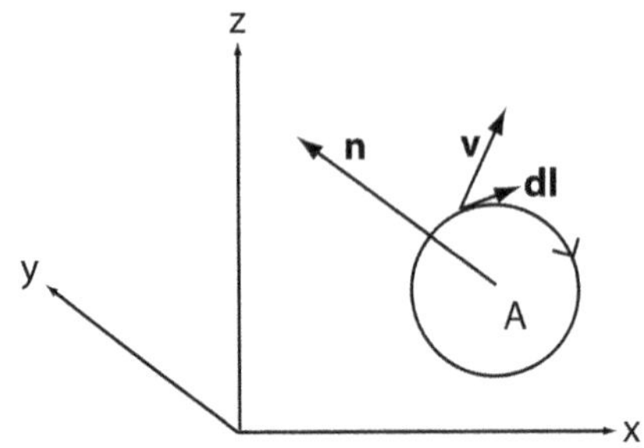

Figure 2.5. An idealized illustration of circulation computation. Area A of a plane fluid element, sense of direction of line integral of $\mathbf{v}\cdot\mathbf{dl}$ (arrow around closed curve), unit vector $\mathbf{n}$ normal to the surface of A, using the right-hand rule to determine the direction it's pointing.

According to V. Bjerknes's first circulation theorem

$$DC/Dt = \oint B\mathbf{k}\cdot\mathbf{dl} \tag{2.53}$$

which is obtained by using the definition of circulation (2.52), the three-dimensional equation of motion (2.36), and noting that

$$\oint \nabla(p/\bar{\rho})\,\mathbf{dl} = 0 \tag{2.54}$$

because $\bar{\rho}$ is treated as a constant. The same two-dimensional fluid element must be followed along in space and its change in shape must be accounted for. When the fluid element is well behaved, the fluid element can be followed. In some instances, however, the fluid element with time becomes so distorted that it is difficult to work with accurately.

From Stokes' theorem, (2.52) may be expressed as

$$DC/Dt = \iint (\nabla \times B\mathbf{k})\cdot\mathbf{dA} \tag{2.55}$$

In the absence of friction, in a Boussinesq atmosphere (which acts as if it were incompressible) without gradients in buoyancy, circulation is conserved. Also, if the material curve always lies in the x–y-plane, circulation is conserved. Circulation is generated when there are horizontal buoyancy gradients and the fluid element has a nonzero projection onto the vertical plane.

To analyze circulation dynamics, one is careful to choose a circuit for a material curve that encompasses just the vortex under consideration. If one chooses a circuit that contains two counterrotating vortices, then the total circulation may be nearly zero; if one chooses a circuit that contains less than the complete vortex, then the analysis may be in error because a significant part of the vortex has been omitted. It is an art to determine circuits to be analyzed and one must be sure that the orientation of the plane chosen is normal to the local vorticity vector. Also, one cannot trace a material circuit too far back in time because the shape of the circuit may have been extremely distorted[2] and friction may have changed the shape and/or fractured the material circuit. Friction, of course, can also reduce circulation. Also, circulation tendencies may reverse sign if the material curve is traced too far back in time.

For the special case of vertical vortices, the flux form of the Eulerian vertical vorticity equation in a frictionless, Boussinesq atmosphere may be used; it is obtained by rewriting (2.50) less the solenoidal term as

$$\partial\zeta/\partial t = \nabla_h \cdot [-\zeta \mathbf{v}_h + w(\nabla \times \mathbf{v})_h] \tag{2.56}$$

where

$$\nabla_h = \partial/\partial x\,\mathbf{i} + \partial/\partial y\,\mathbf{j} \tag{2.57}$$

$$\mathbf{v}_h = u\mathbf{i} + v\mathbf{j} \tag{2.58}$$

$$(\nabla \times \mathbf{v})_h = (\partial w/\partial y - \partial v/\partial z)\mathbf{i} + (\partial u/\partial z - \partial w/\partial x)\mathbf{j} \tag{2.59}$$

and the latter is the horizontal vorticity vector.

Equation (2.56) is integrated over an area (fluid element) in the horizontal plane so that

$$\iint \partial\zeta/\partial t\, dA = \partial/\partial t \iint \zeta\, dA = \partial C/\partial t \tag{2.60}$$

Using the divergence theorem, we find that (2.60) may be expressed as a line integral:

$$\partial C/\partial t = \oint ([-\zeta \mathbf{v}_h + w(\nabla \times \mathbf{v})_h] \cdot \mathbf{n}\, \mathbf{dl} \tag{2.61}$$

where $\mathbf{n}$ is a unit vector normal to the line composing the perimeter of the fluid element and pointing outward from it. The terms on the RHS of (2.61) represent the net horizontal flux of vertical vorticity into the area and the net vertical flux of horizontal vorticity into the area. If we include Earth's vorticity, then $-f\mathbf{v}_h$ is included inside the brackets on the RHS of (2.61). Subgrid-scale mixing may also be included.

The horizontal vorticity equation (one component of it) (2.51) is used to diagnose circulations in the vertical plane. It is very useful, for example, for understanding the dynamics of vertical circulations in two-dimensional squall lines, in circularly symmetric clouds, and across gust fronts.

2.5 THE DIVERGENCE EQUATION AND THE BUOYANCY FORCE

By applying the divergence operator to the full three-dimensional equation of motion—combination of (2.13) and (2.7)—we obtain the following equation:

$$\begin{aligned} 1/\bar{\rho}\,\nabla^2 p' = {} & -[(\partial u/\partial x)^2 + (\partial v/\partial y)^2 + (\partial w/\partial z)^2 \\ & + 2(\partial u/\partial y\, \partial v/\partial x + \partial w/\partial x\, \partial u/\partial z + \partial w/\partial y\, \partial v/\partial z)] + \partial B/\partial z \end{aligned} \tag{2.62}$$

[2] As for trajectories, when windspeeds are high, as in a tornado, the effects of small changes in position (and time increment used to compute the trajectories) are amplified.

Since the flow in a Boussinesq fluid is non-divergent, this divergence equation is time independent (*diagnostic*, as opposed to *prognostic*). It is used to compute perturbation pressure from the three-dimensional distribution of wind. While the circulations associated with convective storms are computed from the vorticity equation, the three-dimensional pressure field that is consistent with the circulations is computed from the divergence equation. It is therefore not appropriate to infer that the pressure field *causes* the wind field, but rather that it is consistent with it. We can, however, use the pressure field to compute pressure gradient forces that will change the wind field in the future. Both (2.49) and (2.62) are used in tandem to analyze storm dynamics.

2.5.1 Buoyancy-induced and dynamically induced pressure perturbations

Since the operator on p' is linear (a Laplacian operator), we can find solutions of p' for each term on the right-hand side of (2.62) and add up all solutions to find the total p'. To facilitate the physical interpretation of perturbation pressures, we let

$$p' = p'_b + p'_d \tag{2.63}$$

where p'_b is the perturbation pressure due to the buoyancy term alone; and p'_d is the perturbation pressure due to the "dynamic terms" (i.e., those involving the wind field and its variations).

Appropriate boundary conditions on p' must be selected so that (2.62) can be solved. It may assumed that, far from a region of nonzero B, $p'_b = 0$. These lateral and vertical boundaries may therefore be prescribed far from regions of significantly nonzero B. Otherwise, one may use the equations of motion (2.13) and (2.7) to compute the horizontal and vertical gradients of perturbation pressure at the boundaries in terms of the three-dimensional wind field and its spatial and temporal gradients (Neumann boundary conditions) and in terms of B. If the outer boundaries are chosen so that the atmosphere is resting and there is no buoyancy there, then the boundary conditions become $p'_d = 0$. The values of perturbation pressure are actually determined to within an arbitrary constant, but the gradients of perturbation pressure are exact; the mean value of the perturbation pressure is usually assumed to be zero.

If the fully compressible, time-dependent version of the equation of continuity (2.28) were used, then the divergence equation would contain a term involving time derivatives ($-\partial/\partial t\, \nabla \cdot \mathbf{v}$) and it would therefore no longer be diagnostic. Suppose now for the purposes of illustration that the atmosphere were fully compressible. For simplicity, let's assume that the atmosphere has no variations in the y-direction and that there is no mean (horizontal or vertical) flow. Then the linearized equation of motion (2.13) in the x-direction is

$$\partial u'/\partial t + 1/\bar{\rho}\, \partial p'/\partial x = 0 \tag{2.64}$$

where u' is the perturbation part of the zonal wind. The compressible form of the continuity equation when $v = w = 0$ (i.e, when the motion of the fluid is in the x-direction only) is

$$1/\rho \, D\rho/Dt = -\partial u/\partial x \tag{2.65}$$

The adiabatic form of the thermodynamic equation (2.27) may therefore be expressed as

$$\partial u/\partial x + 1/(\gamma p) \, Dp/Dt = 0 \tag{2.66}$$

Linearized about a resting atmosphere, (2.66) may be combined to yield the following:

$$p \, \partial u'/\partial x + 1/\gamma \, \partial p'/\partial t = 0 \tag{2.67}$$

Eliminating u' from the equation of motion (2.64) and the combined continuity and adiabatic thermodynamic equation (2.66), the following wave equation for perturbation pressure is obtained:

$$\partial^2 p'/\partial x^2 - 1/c^2 \, \partial^2 p'/\partial t^2 = 0 \tag{2.68}$$

where c is the speed of sound (2.17). When the Boussinesq version of the divergence equation (2.62) is simplified so that the mean winds in the atmosphere are calm (i.e., a "resting" atmosphere), there are no variations in y, and $B = 0$, we find that in this case the linearized version of (2.62) is

$$\partial^2 p'/\partial x^2 = 0 \tag{2.69}$$

Thus, the Boussinesq divergence equation does *not* contain the time derivative term. In nature, sound waves transmit information relating the pressure field to the wind field. By eliminating them, we in effect assume that their speed is infinite—since the speed of sound appears in the denominator of the factor multiplying the time derivative term in (2.68)—so that information linking the pressure field to the wind field is instantaneous and they are linked by a Poisson equation. (An analogous situation exists in synoptic meteorology, when the atmosphere is analyzed using quasi-geostrophic theory. Information linking the pressure field to the wind field is assumed to be instantaneous, through the geostrophic wind relation. However, in nature inertial gravity waves carry the information linking them together and it is assumed there is a Poisson equation linking them, the time derivative term being eliminated when the speed of inertial gravity waves is infinite.)

The pressure field retrieved from the divergence equation can be used to separate the effects of the vertical perturbation pressure gradient force due to dynamics (by means of terms involving the wind field) from those due to buoyancy (the vertical derivative of buoyancy) as in (2.63). The vertical equation of motion (2.7) may be expressed using (2.63) as

$$Dw/Dt = -1/\bar{\rho} \, \partial p'_d/\partial z + [(-1/\bar{\rho}) \, \partial p'_b/\partial z + B] \tag{2.70}$$

The vertical acceleration due to buoyancy is then $(-1/\bar{\rho} \, \partial p'_b/\partial z + B)$ the "effective" buoyancy—not just B alone. For example, underneath a buoyant parcel of air, $\partial B/\partial z > 0$; from (2.62) it is seen that $1/\bar{\rho} \, \nabla^2 p'_b > 0$. Within a volume, p'_b averages out to zero and therefore must be positive in some locations

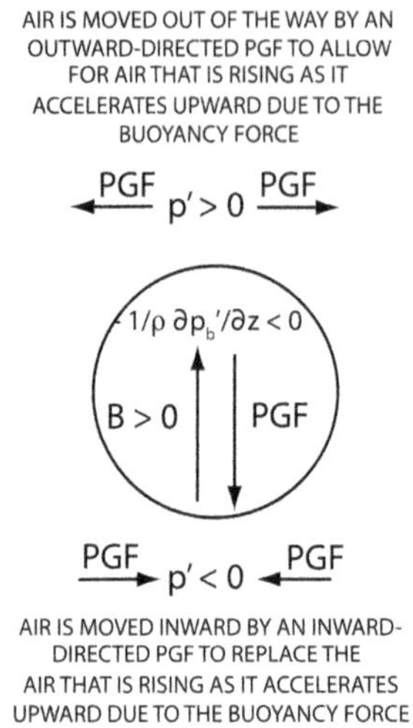

Figure 2.6. An idealized illustration of how an air parcel (circle) that is accelerated upward by positive buoyancy is opposed by a downward-directed perturbation pressure gradient force as air is forced outward just above the parcel to make way for air in the buoyant parcel that is accelerated upward and air is forced inward just below the parcel to replace the air that is accelerated upward.

and negative in others. Since maxima (minima) in p_b' tend to be correlated with positive (negative) pressure perturbations and ρ is always positive, p_b' tends to be positive when $\nabla^2 p_b' < 0$ (i.e., when there is a local maximum in p_b') and negative when $\nabla^2 p_b' > 0$ (i.e., when there is a local minimum in p_b'). In other words, p_b' is negatively correlated with $\partial B/\partial z$. We arrive then at the following conclusion: underneath a buoyant parcel of air the perturbation pressure is negative and above it the pressure is positive. Thus, there is a downward-directed perturbation pressure gradient force that acts to reduce the effects of Archimedean buoyancy (Figure 2.6). It is only when the two do not cancel each other out that there is a true buoyancy effect. When the two cancel each other out, the atmosphere is hydrostatic. When p_b' is neglected, as it is when employing the parcel method, B is in fact the entire buoyancy force. Since there *is* a counteracting perturbation pressure gradient force, any estimates of acceleration based on Archimedean buoyancy alone are an overestimate. The reader is reminded that our analysis has neglected turbulent mixing with the environment of the buoyant air parcel (the entrainment of environmental air); if mixing were taken into account, the value of B would be reduced even more.

Let us now consider what vertical circulations are produced by buoyancy gradients alone at the outer edges of the buoyant air parcel using the horizontal component (for simplicity, only in the y-direction) of the Boussinesq vorticity equation (2.61)

$$D/Dt(\partial u/\partial z - \partial w/\partial x) = -\partial B/\partial x \tag{2.71}$$

From (2.71) we see that to the left and right of a buoyant air parcel vorticity is generated in the $-y$ and $+y$-directions, respectively (Figure 2.7d–f). For a (Figure 2.7b) fixed value of B, more air must be displaced laterally outward above a flattened buoyant air parcel and more air must be displaced laterally inward below a flattened buoyant air parcel than that for a more spherical air parcel (Figure 2.7a). It follows then that the adverse pressure gradient force is more intense for the flattened air parcel. The smaller the aspect ratio (D/L, where D is the depth and L is the width) of the air parcel, the more the downward-directed pressure gradient force opposes upward Archimedean buoyancy. As D/L approaches zero, the atmosphere becomes hydrostatic. Also, when the aspect ratio is small, the effects of the gradient of B in generating vertical circulations are concentrated only along the far edges of the air parcel. When the aspect ratio approaches infinity very little air is displaced directly above and below the air parcel, very high up and very low below (Figure 2.7c). Figures 2.7d, e, and f repeat a similar analysis of the effect of the aspect ratio on vertical accelerations, but from the perspective of horizontal vorticity produced by the vertical variation of the horizontal hydrostatic pressure gradient force.

The qualitative analysis in the last paragraph is made more quantitative by considering a region of buoyancy having the following distribution in space:

$$B(x,z) = B_0 \sin(\pi z/H)\cos(\pi x/L) \tag{2.72}$$

This idealized two-dimensional distribution of buoyancy represents a maximum in the middle of the troposphere ($z = H/2$) and at $x = 0$; it is periodic in x. It is like a quasi-spherical slab of a bubble placed in the middle of the troposphere (Figure 2.8, left panel). (Not including variations in the y-direction does not change the results of our analysis qualitatively.)

The divergence equation (2.62) for Archimedean buoyancy only is

$$1/\bar{\rho}\,\nabla^2 p'_b = \partial B/\partial z \tag{2.73}$$

It follows from (2.72) and (2.73) that the following

$$p'_b/\bar{\rho} = -(B_0\pi/H)/[(\pi/L)^2 + (\pi/H)^2]\cos(\pi z/H)\cos(\pi x/L) \tag{2.74}$$

is a solution to the divergence equation (2.73) subject to boundary conditions of vanishing $p'_b/\bar{\rho}$ at $z = 0, H$ (the top and bottom of the model troposphere) and at $x = \pm L/2$, etc.

Using (2.74), we can now calculate the acceleration induced by the vertical perturbation pressure gradient force (inferred from Figure 2.8, right panel)

$$-1/\bar{\rho}\,\partial p'_b/\partial z = [B_0 \sin(\pi z/H)\cos(\pi x/L)]\{-1/[(H/L)^2 + 1]\} \tag{2.75}$$

As the aspect ratio H/L approaches zero, $-1/\bar{\rho}\,\partial p'_b/\partial z$ becomes $-B_0 \sin(\pi z/H)\cos(\pi x/L)]$, which is equal and opposite to B (2.73), and $Dw/Dt = 0$, which is the hydrostatic case. On the other hand, when H/L approaches infinity, $-1/\bar{\rho}\,\partial p'_b/\partial z$ vanishes, so that Archimedean buoyancy B contributes solely to vertical accelerations in the frictionless equation of motion (2.7). This analysis suggests that infinitely narrow convective bubbles are most efficient

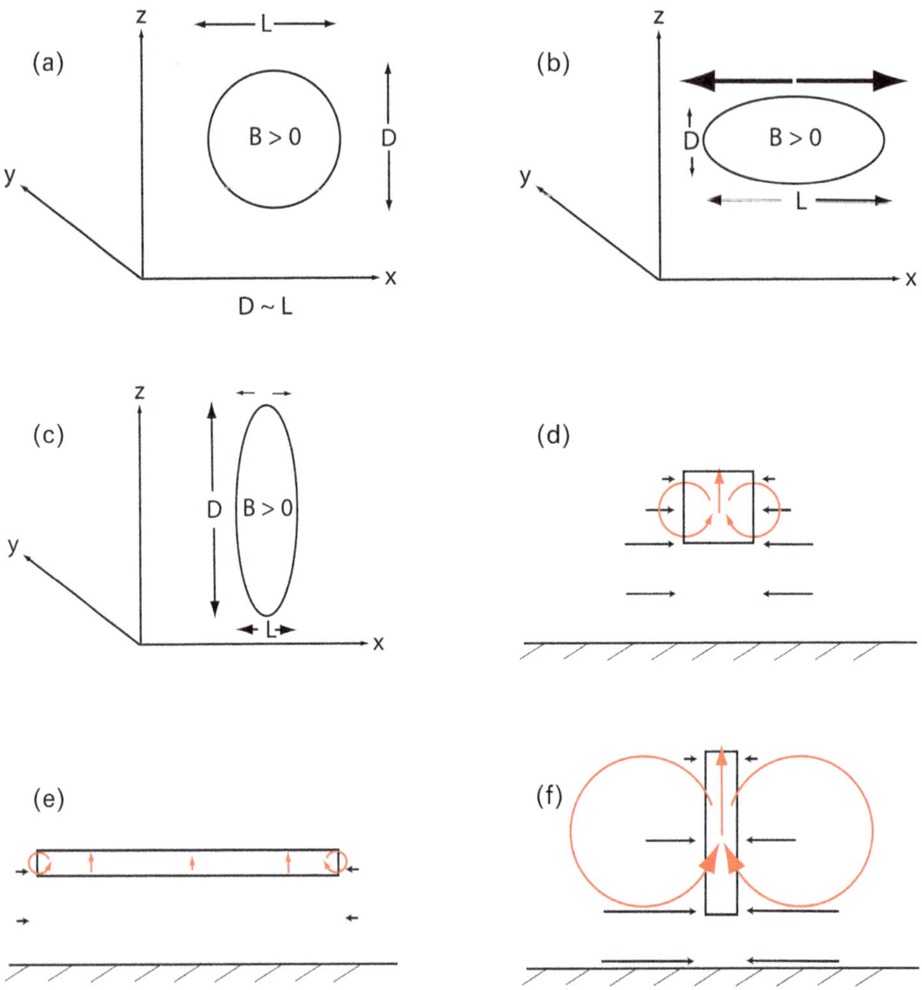

Figure 2.7. An idealized illustration of the way in which the aspect ratio (D/L) of a buoyant air parcel affects how much air must be moved *out* above it and *in* beneath it, which in turn affects the magnitude of the perturbation pressures above and below it. (a) Spherical air parcel; (b) flattened air parcel: D/L is small; (c) narrow air parcel: D/L is large. In (b) and (c) the relative amount of air that is moved out of the way above the rising air parcel is denoted by the boldness of vectors. In (b) much more air than in (c) must be moved out of the way. (d) Alternative way of illustrating how the aspect ratio affects buoyancy: vertical cross section showing projection of box enclosing air that is warmer than that of its environment. The hydrostatic horizontal pressure gradient force at left and right sides of the box are indicated by black vectors. The greatest horizontal pressure gradient forces are at the bottom of the box, where the differences between the hydrostatic pressure outside and inside the box are at their maximum. Red streamlines show how air is accelerated. (e) As in (d), but for the case when D/L is small. The accelerations are weak. (f) As in (d), but for the case when D/L is large. The accelerations are intense.

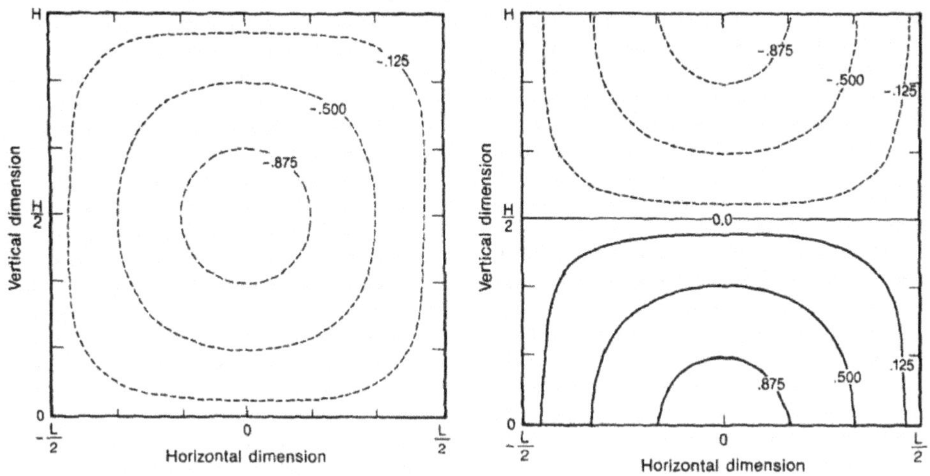

Figure 2.8. The buoyancy distribution from (2.72) (left panel) and the perturbation pressure field associated with it (right panel) (from Parsons and Kropfli, 1990), but with negative buoyancy, rather than positive buoyancy. Note how there is an upward-directed perturbation pressure gradient force ($p' < 0$ above and $p' > 0$ below) that acts opposite to the imposed negative buoyancy.

at using Archimedean buoyancy as a force. Finally, for the case when $H = L$, it is seen that the vertical acceleration due to buoyancy is cut down to one half of what it would have been in the absence of the opposing vertical perturbation pressure gradient force. The reader is reminded that this simplified analysis does *not* include the effects of turbulent entrainment of environmental air into the bubble and the effects of phase change of water substance. In addition, this analysis does not include the effects of the interaction of the updraft in the bubble with the environmental wind field, which so far has been assumed to be calm. The latter (via p'_d) will be considered later when we analyze downdrafts, microbursts, and supercell convection.

2.5.2 Retrieval of pressure and buoyancy fields from the wind field

To retrieve the thermodynamic profiles from wind observations (particularly those synthesized from Doppler radar wind data), one can use a technique pioneered by Tzvi Gal-Chen in the early 1980s at the University of Oklahoma, which is based on a divergence equation for the Boussinesq system of equations, as follows: the perturbation pressure field at each level is solved from the horizontal (two-dimensional) version of the divergence equation (2.62); the buoyancy field does not appear in the horizontal divergence equation

$$\nabla_h^2 p' = -1/\bar{\rho}[\partial/\partial x(Du/Dt) + \partial/\partial y(Dv/Dt)] \tag{2.76}$$

Since time derivatives appear, it is necessary that the Doppler radars sample the atmospheric wind field frequently enough to minimize aliasing of high-frequency fluctuations. In addition, the wind field must be low-pass filtered to reduce the effect of noise, since spatial derivatives are computed.

Perturbation pressure p' in theory may be calculated at each level by inverting the Laplacian operator, subject to appropriate boundary conditions; if the storm is localized, then we can use $p' = 0$ on the boundaries (Dirichlet boundary conditions) because the lateral boundaries are far from the storm; if they are relatively near the storm, then the horizontal pressure gradient at the lateral boundaries can be solved from the wind field using the horizontal equations of motion, such that at the lateral boundaries (Neumann boundary conditions)

$$\partial p'/\partial x = -\bar{\rho}\, Du/Dt \tag{2.77}$$

$$\partial p'/\partial y = -\bar{\rho}\, Dv/Dt \tag{2.78}$$

If there is no detectable radar echo outside the storm, however, it is not possible to use (2.77) and (2.78) because there are no Doppler wind data there.

Because there are instrument and sampling errors in the determination of the Doppler wind field, solution of the perturbation pressure field exactly using (2.76) is not possible, especially when $\partial/\partial y(Du/Dt) \neq -\partial/\partial x(Dv/Dt)$. An alternative approach is to solve for perturbation pressure using variational analysis by minimizing the cost function

$$J = \iint \{[\partial p'/\partial x - (\bar{\rho}\, Du/Dt)]^2 + [\partial p'/\partial y - (\bar{\rho}\, Dv/Dt)]^2\}\, dx\, dy \tag{2.79}$$

where the domain of the integral is over the analysis region at each level. This procedure amounts to fitting a pressure field to the wind field by minimizing the difference between the fitted pressure gradient and the equation of motion. The resulting Euler equation as shown by Gal-Chen, however, is identical to (2.76). Inverting the horizontal Laplacian operator in (2.76) effectively filters out noise as it fits the pressure field to the wind field.

The vertical variation of p' along with the wind field are used to solve for buoyancy B via the vertical equation of motion (2.70), without distinguishing between p'_d and p'_b as follows:

$$B = Dw/Dt + 1/\bar{\rho}\, \partial p'/\partial z \tag{2.80}$$

It is assumed that the perturbation pressures at each level represent true deviations from the average, an assumption that may not be entirely valid and must be checked. A measure of how well retrieved perturbation pressures may be estimated is by the "consistency check" quantity E, where

$$E = \frac{\iint \{[-1/\bar{\rho}\, \partial p'/\partial x - Du/Dt]^2 + [-1/\bar{\rho}\, \partial p'/\partial y - Dv/Dt]^2\}\, dx\, dy}{\iint \{[Du/Dt]^2 + [Dv/Dt]^2\}\, dx\, dy} \tag{2.81}$$

and the integrals are computed over the entire radar domain at each level. The values of E are then averaged over all levels for which there are Doppler radar wind analyses. When E is less than 0.35, the retrieved perturbation pressure field is deemed adequate.

Actual temperatures (T) may be estimated by using a proximity sounding representative of the storm environment, so that from (2.19)

$$T = \bar{T}(B/g) + \bar{T} \tag{2.82}$$

where $\bar{T}$ is the environmental temperature. If water vapor content is relatively high (e.g., as it is in the boundary layer) water vapor must be accounted for as in (2.23). For retrievals of temperature inside convective storms, it is necessary to include estimates of cloud material and precipitation as in (2.23). The latter may be estimated from the radar reflectivity factor and further refined from polarimetric radar variables.

In this retrieval process, we have neglected turbulent mixing. A parameterized version of mixing is sometimes included. In addition, the Coriolis force may be included if the time scale of the feature analyzed is relatively long compared with a day.

An alternative to retrieving the temperature field from the buoyancy field by first retrieving the perturbation pressure field at separate levels using the horizontal divergence equation is to take the entire set of equations (i.e., all three equations of motion, the thermodynamic equation, and the equation of continuity) and solve for buoyancy alone by eliminating u, v, w, and p'. The resulting equation is like the ones derived in the later section on Rayleigh–Bénard convection (Section 2.9), a high-order differential equation. A large set of boundary conditions must then be specified and the equation is not easy to solve.

2.5.3 Quantitative analysis of a buoyant sphere in a resting environment

We now consider a more quantitative analysis of what happens when a thermal bubble, as exemplified by a sphere of constant buoyancy, is released into a horizontally homogeneous atmosphere in which the vertical temperature profile is $\bar{T}(z)$. The casual reader is advised to skip or just skim the following analysis. For simplicity we neglect the effects of moisture and turbulent and molecular diffusion. Let us assume that initially, at $t = 0$, a sphere of radius a having a temperature excess over the environment of $T' = \Delta T = T - \bar{T}$ is released from rest (i.e., $u = v = w = 0$ at $t = 0$). The problem, simply stated, is to find the acceleration field $(D\mathbf{v}/Dt)$ at $t = 0$.

The simplified Boussinesq equations of motion are expressed as follows:

$$\partial \mathbf{v}/\partial t = -\nabla P + B\mathbf{k}_z \tag{2.83}$$

$$\nabla \cdot \mathbf{v} = 0 \tag{2.84}$$

$$\partial B/\partial t = 0 \tag{2.85}$$

where $B = (T'/\bar{T})g$ (cf. (2.19)); $P = p'/\bar{\rho}$; and $\mathbf{k}_z$ is a unit vector pointing upward. The reader is reminded that partial—not material—derivatives appear on the left-hand sides of (2.83) and (2.85) because the advection terms are zero when the

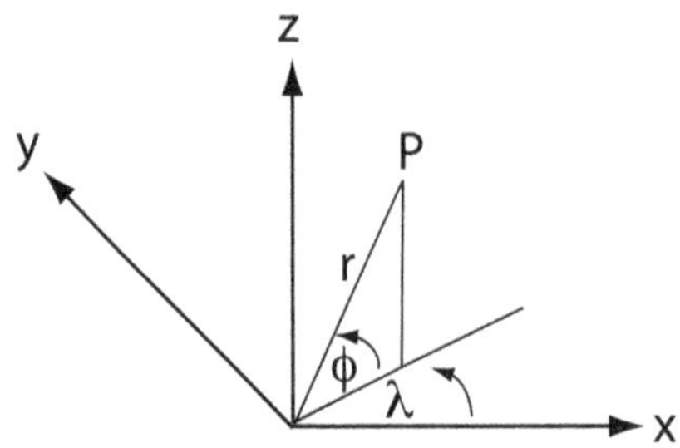

Figure 2.9. Spherical coordinates. At point P, the coordinates are given in terms of the distance r (radius) from the origin (center of the sphere), latitude (ϕ) measured from the equator, which lies in the x–y-plane, and longitude (λ) measured relative to the $+x$-axis.

atmosphere is at rest; also, the $\partial \bar{T}/\partial z$ term does not appear in the latter because at $t = 0$, $w = 0$.

In spherical coordinates (Figure 2.9), r, ϕ, and λ represent the radial distance from the center of the sphere, the latitude measured from the equator (which lies in the x–y-plane) and the longitude measured from some reference (e.g., the x-axis). The z-axis (vertical) is oriented from $\phi = -\pi/2$ to $\phi = \pi/2$. We will neglect any variations in longitude, so that $\partial/\partial\lambda = 0$. Thus, when looking down on the sphere from above there is symmetry about the z-axis. Also, we will neglect any motions in the longitudinal direction, so that $D\lambda/Dt = 0$.

The two non-trivial components of the equation of motion in spherical coordinates subject to longitudinal symmetry are as follows:

$$\partial/\partial t(r\, D\phi/Dt) = -1/r\, \partial P/\partial\phi + B(\mathbf{k}_z \cdot \mathbf{j}) \tag{2.86}$$

$$\partial/\partial t(Dr/Dt) = -\partial P/\partial r + B(\mathbf{k}_z \cdot \mathbf{k}) \tag{2.87}$$

where the former is the $\mathbf{j}$ (latitudinal) component and the latter is the $\mathbf{k}$ (radial) component, respectively, of the equation of motion.

The equation of continuity is

$$\partial/\partial\phi(r^2 \cos\phi\, D\phi/Dt) + \partial/\partial r(r^2 \cos\phi\, Dr/Dt) = 0 \tag{2.88}$$

and the thermodynamic equation is still

$$\partial B/\partial t = 0 \tag{2.89}$$

We define the radial and latitudinal components of the wind as

$$u = Dr/Dt \tag{2.90}$$

and

$$v = r\, D\phi/Dt \tag{2.91}$$

The four governing equations at $t = 0$ in four unknowns ($\partial u/\partial t$, $\partial v/\partial t$, P, and T)[3]

[3] The fourth unknown is really B. But since T is computed from B, T can also be regarded as the fourth unknown.

may now be expressed as

$$\partial v/\partial t = -1/r\, \partial P/\partial \phi + B \cos \phi \tag{2.92}$$

$$\partial u/\partial t = -\partial P/\partial r + B \sin \phi \tag{2.93}$$

$$\partial/\partial \phi(rv \cos \phi) + \partial/\partial r(r^2 u \cos \phi) = 0 \tag{2.94}$$

$$\partial B/\partial t = 0 \tag{2.95}$$

The problem is now to solve for $\partial u/\partial t$ and $\partial v/\partial t$ (at $t = 0$) in terms of the independent variables (at $t = 0$; we are not trying to determine the time-dependent behavior of the wind field beyond $t = 0$). To do so, we must eliminate P and T. Actually, since (2.95) is not obviously coupled to the other three equations, we eliminate only P.

First, we eliminate P from (2.92) and (2.93) by forming a vorticity equation (in the longitudinal direction) by multiplying (2.92) by r and then differentiating with respect to r and subtracting from this resulting equation (2.93) differentiated with respect to ϕ. Making use of

$$\partial B/\partial \phi = dB/dz\, \partial z/\partial \phi \tag{2.96}$$

and

$$\partial B/\partial r = dB/dz\, \partial z/\partial r \tag{2.97}$$

we find that

$$1/r[\partial/\partial \phi\, \partial u/\partial t - \partial/\partial r(r\, \partial v/\partial t)] = 0 \tag{2.98}$$

We now make use of the continuity equation (2.88) which is non-divergent in the r–ϕ-plane to define a streamfunction Ψ such that

$$\partial u/\partial t = -1/(r^2 \cos \phi)\, \partial/\partial \phi(\partial \Psi/\partial t) \tag{2.99}$$

and

$$\partial v/\partial t = 1/(r \cos \phi)\, \partial/\partial r(\partial \Psi/\partial t) \tag{2.100}$$

We now have eliminated two of the three variables and end up with just one equation in terms of Ψ. Since the streamfunction appears differentiated with respect to time, we define yet a new variable

$$\chi = \partial \Psi/\partial t \tag{2.101}$$

which is reminiscent of how we define the geopotential height tendency variable χ in quasi-geostrophic theory. The resulting second-order partial differential equation for χ is

$$\cos \phi\, \partial/\partial \phi(1/\cos \phi\, \partial \chi/\partial \phi) + r^2\, \partial^2 \chi/\partial r^2 = 0 \tag{2.102}$$

Once we know χ at $t = 0$, we then can use (2.99) and (2.100) to find $\partial u/\partial t$ and $\partial v/\partial t$. We need two boundary conditions to solve this equation. One is the kinematic boundary condition and the other is the dynamic boundary condition, both applied at the interface between the spherical bubble of buoyant fluid and its non-buoyant environment.

So, at $r = a$, the component of the wind normal to the sphere must be continuous across the interface (kinematic boundary condition) at $t = 0$.

$$u(r = a_-) = u(r = a_+) \tag{2.103}$$

at $t = 0$ and also at $t = t + \Delta t$. Note that $u(t = 0) = 0$.

Expanding both the left and right-hand sides of (2.103) into Taylor series expansions in time, it is seen that

$$u(r = a_-, t + \Delta t) = u(r = a_-, t = 0) + (\partial u/\partial t)_{t=0,a_-} \Delta t + \text{higher order terms} \tag{2.104}$$

and

$$u(r = a_+, t + \Delta t) = u(r = a_+, t = 0) + (\partial u/\partial t)_{t=0,a_+} \Delta t + \text{higher order terms} \tag{2.105}$$

Then, in the limit as Δt goes to zero

$$(\partial u/\partial t)_{t=0,a_-} = (\partial u/\partial t)_{t=0,a_+} \tag{2.106}$$

and so from (2.99) and (2.101) it follows that

$$(\partial \chi/\partial \phi)_{a_-} = (\partial \chi/\partial \phi)_{a_+} \tag{2.107}$$

Pressure must be continuous across the spherical bubble surface (the dynamic boundary condition) so that

$$P(r = a_-) = P(r = a_+) \tag{2.108}$$

Then the latitudinal gradient of pressure is also continuous; that is

$$(1/r\, \partial P/\partial \phi)_{a_-} = (1/r\, \partial P/\partial \phi)_{a_+} \tag{2.109}$$

It follows from the latitudinal component of the equation of motion (2.92) that

$$(\partial v/\partial t)_{a_+} - (\partial v/\partial t)_{a_-} = -(gT'/\bar{T}) \cos \phi \tag{2.110}$$

Substituting the left-hand side of (2.110) using the definition of χ—(2.100) and (2.101)—we find that

$$(\partial \chi/\partial r)_{a_-} - (\partial \chi/\partial r)_{a_+} = aB_{a_-} \cos^2 \phi \tag{2.111}$$

where $B_{a_-} = gT'/\bar{T}$. We now have two boundary conditions: one on $(\partial \chi/\partial r)_a$ and one on $(\partial \chi/\partial \phi)_a$. The solutions to (2.102) subject to the boundary conditions (2.107) and (2.111) are separable and take the form

$$\chi = R(r)\Phi(\phi) \tag{2.112}$$

Substituting (2.112) into (2.102), we find that

$$(\cos \phi/\Phi)\, d/d\phi(d\Phi/d\phi/\cos \phi) = -S \tag{2.113}$$

and

$$d^2R/dr^2 = (R/r^2)S \tag{2.114}$$

where S is the "separation constant". We now have two equations in two unknowns (R and Φ), and they are linked by the separation constant, which we will determine after applying the boundary conditions. It follows from (2.112), the boundary conditions (2.107) and (2.111), and since $R(r = a_-) = R(r = a_+)$ that

$$(d\Phi/d\phi)_{a_-}[(dR/dr)_{a_-} - (dR/dr)_{a_+}] = -2aB_{a_-} \sin\phi\cos\phi \qquad (2.115)$$

Since $[(dR/dr)_{a_-} - (dR/dr)_{a_+}]$ is a constant, we let

$$1/[(dR/dr)_{a_-} - (dR/dr)_{a_+}] \equiv A \qquad (2.116)$$

so that

$$(d\Phi/d\phi)_{a_-} = -A2aB_{a_-} \sin\phi\cos\phi \qquad (2.117)$$

Integrating (2.117) with respect to ϕ we find that

$$\Phi(\phi)_{a_-} = AaB_{a_-} \sin^2\phi + C \qquad (2.118)$$

where C is an arbitrary constant of integration. It follows from substituting (2.118) into (2.113) that $C = 0$ and $S = 2$. So

$$\chi = A'R(r)\cos^2\phi \qquad (2.119)$$

where

$$A' = AaB_{a_-} \qquad (2.120)$$

We now find $R(r)$ from the equation

$$d^2R/dr^2 - 2/r^2 R = 0 \qquad (2.121)$$

which is an "equi-dimensional" equation, for which solutions have the form

$$R = r^\alpha \qquad (2.122)$$

It follows from substituting (2.122) into (2.121) that α can be 2 or -1, so that

$$R(r) = D_1 r^{-1} + D_2 r^2 \qquad (2.123)$$

We expect R not to tend to infinity as $r \to 0$, so that $D_1 = 0$ for $r < a$. Also, as $r \to \infty$, χ should $\to 0$ (there should be no accelerations at infinity in the far environment), so that $D_2 = 0$ for $r > a$. Then

$$\chi = A'_{\text{int}} r^2 \cos^2\phi \qquad \text{for } r < a \qquad (2.124)$$

$$= A'_{\text{ext}}(1/r)\cos^2\phi \qquad \text{for } r > a \qquad (2.125)$$

where

$$A'_{\text{int}} = A'D_2 \qquad (2.126)$$

and

$$A'_{\text{ext}} = A'D_1 \qquad (2.127)$$

Applying the boundary conditions (2.107) and (2.111) to (2.124) and (2.125) at $r = a$, we find that A'_{int} and A'_{ext} are determined such that

$$\chi = (B_{a_-}/3)r^2\cos^2\phi \qquad \text{for } r < a \qquad (2.128)$$

and

$$= (B_{a_-}/3)(a^3/r)\cos^2\phi \qquad \text{for } r > a \qquad (2.129)$$

We can now calculate the local accelerations at $t = 0$ by substituting (2.128) and (2.129) into (2.99) and (2.100), using (2.101). Then

$$D\mathbf{v}/Dt = \tfrac{2}{3}B_{a\_}(\cos\phi\mathbf{j} + \sin\phi\mathbf{k}) \qquad \text{for } r < a \tag{2.130}$$

$$= \tfrac{2}{3}B_{a\_}(a/r)^3(-\tfrac{1}{2}\cos\phi\mathbf{j} + \sin\phi\mathbf{k}) \qquad \text{for } r > a \tag{2.131}$$

Since

$$\mathbf{k}_z = \mathbf{j}\cos\phi + \mathbf{k}\sin\phi \tag{2.132}$$

it follows that

$$D\mathbf{v}/Dt = \tfrac{2}{3}B_{a\_}\mathbf{k}_z \qquad \text{for } r < a \tag{2.133}$$

In other words, within the buoyant sphere the acceleration is uniform and upward; about two thirds is due to that from buoyancy alone. It follows that there must be a downward-directed pressure gradient force to counteract buoyancy and cancel the effect of one third of it. Outside the buoyant sphere

$$D\mathbf{v}/Dt = \tfrac{2}{3}B_{a\_}(a/r)^2\mathbf{k}_z \qquad \text{for } r > a, \text{ at } \phi = \pm\pi/2 \tag{2.134}$$

Thus, at $r = a$, the vertical acceleration is up along the z-axis and decays above and below the sphere. At $\phi = 0$, for $r > a$

$$D\mathbf{v}/Dt = -\tfrac{1}{3}B_{a\_}(a/r)^3\mathbf{k}_z \tag{2.135}$$

which is a downward-directed acceleration in the horizontal plane at the equator. The acceleration field is depicted in Figure 2.10. Away from the sphere, pressure gradient forces must be driving air motion because there is no buoyancy to do so. The accelerations inside and up to the edge of the buoyant bubble (at $r = a$) *are independent of the radius of the bubble* (cf. (2.130) and (2.131)). Thus, the accelerations there are a function of the buoyancy—not the size of the bubble. In other

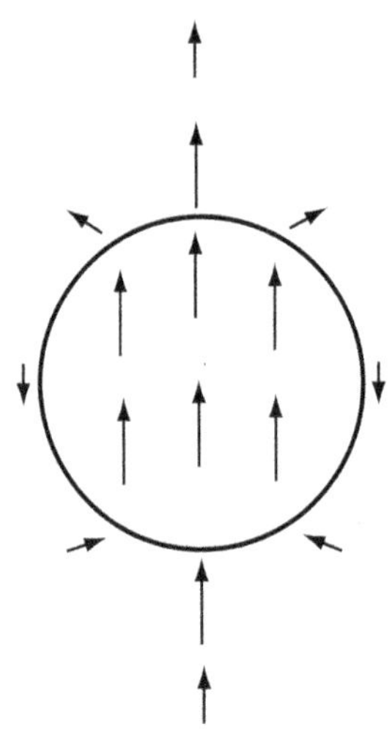

Figure 2.10. Qualitative depiction of the acceleration field (vectors) induced by a buoyant, spherical bubble (solid circle).

words, as we demonstrated in Section 2.5.1, the aspect ratio of the size of the buoyant fluid element determines how much buoyancy is counteracted by a vertical perturbation pressure gradient force: Larger spherical bubbles of equal buoyancy do not have greater accelerations inside them than smaller bubbles. It should also be noted how localized buoyancy can affect the wind field away from the spherical bubble, where there is no buoyancy. The effects of the buoyant bubble have been communicated to the outside world instantaneously, as action at a distance. Sound waves, which are not modeled by the Boussinesq equations, carry this information (i.e., that there is a buoyant sphere) at infinite speed: news travels quickly in a Boussinesq atmosphere.

To solve for the pressure field, we can take the equations of motion (2.92) and (2.93) and form a divergence equation in spherical coordinates to yield the following:

$$\nabla^2 P = 1/r^2\, \partial/\partial r(-\partial u/\partial t\, r^2 + r^2 B \sin\varphi) \\ + 1/(r^2 \sin\varphi)\, \partial/\partial\varphi(-r\, \partial v/\partial t \sin\varphi + rB \sin\varphi \cos\varphi) \qquad (2.136)$$

We can then substitute in (2.136) for $\partial u/\partial t$ and $\partial v/\partial t$ for $r < a$ and for $r > a$ using (2.130) and (2.131), and find solutions to P. It is perhaps just as useful and easier for a student to recognize that isobars are aligned perpendicular to the pressure gradient, and just evaluate $\partial P/\partial r$ and $1/r\, \partial P/\partial\varphi$ at various locations and draw isobars accordingly by looking for regions where the gradients are zero (Figure 2.11). It is noted that when doing numerical integrations in a cloud model, one cannot arbitrarily specify the initial pressure field. A diagnostic

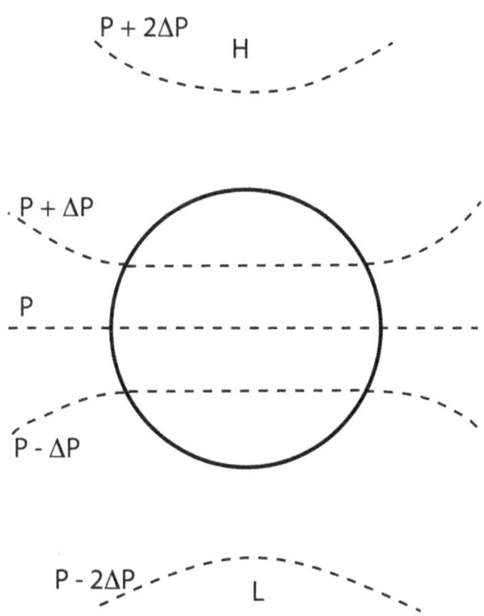

Figure 2.11. Qualitative depiction of the perturbation pressure field (isobars denoted by dashed lines) induced by a buoyant, spherical bubble (solid circle).

divergence equation like (2.136) must be solved in order to determine the initial pressure field based on the initial wind and buoyancy fields.

2.6 ERTEL'S POTENTIAL VORTICITY

Another method for analyzing convective storm dynamics involves the use of Ertel's potential vorticity Z, which is given as follows:

$$Z = 1/\rho[(\nabla \times \mathbf{v}) \cdot \nabla s] \tag{2.137}$$

where s is the specific entropy, $s = C_p \ln \theta$. (If the air is saturated and/or has water substance suspended in it, then θ must be modified.) If diabatic heating and molecular and turbulent viscosity are negligible or zero, Ertel's potential vorticity is conserved (e.g., see pp. 265–267 of Bluestein, 1992 for a derivation), so that

$$DZ/Dt = 0 = D/Dt[(\nabla \times \mathbf{v}) \cdot \nabla \theta] \tag{2.138}$$

Conservation of Ertel's potential vorticity for a fluid is like conservation of angular momentum for rigid bodies: when the gradient of potential temperature decreases (θ surfaces spread farther apart), the fluid contracts and spins up about the axis defined by the potential temperature gradient vector, and vice versa. Equation (2.138) is based on the equations of motion, the equation of continuity, and the thermodynamic equation. It can be used to estimate the future three-dimensional distribution of Z, from which, under certain conditions and using appropriate boundary conditions, it is possible to retrieve the temperature and winds fields. In severe convective storms, diabatic heating plays a prominent role, so that Z is not conserved. However, if it is assumed that the latent heat of condensation from the formation of cloud material is absorbed by the air parcel (i.e., as in the moist-adiabatic process), then θ may be replaced by θ_e, the equivalent potential temperature, which *is* conserved for moist-adiabatic processes. In addition, turbulent mixing may render Z not conserved. For some pedagogical purposes, though, we sometimes treat moist Ertel's potential vorticity as if it were conservative.

2.7 THE EXNER FUNCTION AS A VERTICAL COORDINATE, POTENTIAL TEMPERATURE AS A THERMODYNAMIC VARIABLE, AND THE PSEUDO-INCOMPRESSIBLE CONTINUITY EQUATION

Derivation of the equations of motion (2.13) and (2.7) in terms of density involves approximations because density appears in the denominator in the RHS: in particular, $1/(\bar{\rho} + \rho')$ is approximated because ρ' is much less than $\bar{\rho}$ and the $p'/\bar{p}$ term in the expression for Archimedean buoyancy B must be dealt with. It was shown earlier that the $p'/\bar{p}$ term can be neglected in comparison with $T'/\bar{T}$ for largely subsonic flow and that nonlinear perturbation terms are neglected. A commonly used alternative to pressure as a vertical coordinate is the Exner function π, which is non-dimensional:

$$\pi = (p/p_0)^{R/C_p} = T/\theta = (R\rho\theta/p_0)^{R/C_p} \tag{2.139}$$

This pressure variable is proportional to potential temperature (2.26), so that potential temperature is used rather than temperature as the temperature variable. Use of the Exner function makes mathematical analysis easier, but we think in terms of pressure which we measure with a barometer, and temperature which we measure with a thermometer; I have never heard of anyone devising an instrument to measure the Exner function, which could well be called an "exnerometer".

The equations of motion (2.1) and (2.2) and the adiabatic form of the continuity equation (2.28), with the Exner function as the pressure variable and potential temperature as the density/temperature variable, are given by

$$D\mathbf{v}_h/Dt = \partial\mathbf{v}_h/\partial t + \mathbf{v}\cdot\nabla\mathbf{v}_h = -C_p\theta\nabla_h\pi \tag{2.140}$$

and

$$Dw/Dt = -C_p\theta\,\partial\pi/\partial z - g \tag{2.141}$$

$$\nabla\cdot\mathbf{v} = (1 - 1/\kappa)D\ln\pi/Dt \tag{2.142}$$

where $\kappa = R/C_p$. Unlike the equations of motion expressed in terms of density, the equations of motion in terms of π and θ contain products of variables only; there are no variables appearing in the denominator. To derive these equations, we have made use of π and θ expressed in terms of p, and T and π, respectively; ρ does not appear because it can be expressed in terms of π and θ. The continuity equation may be derived by expressing $1/\rho\, D\rho/Dt$ as $D(\ln\rho)/Dt$ before substituting for ρ expressed in terms of π and θ. The equations of motion and the adiabatic form of the continuity equation expressed subject to the decomposition in terms of a basic state and perturbations (as in (2.3) and (2.4))

$$\pi = \bar{\pi}(z) + \pi'(x, y, z, t) \tag{2.143}$$

and

$$\theta = \bar{\theta}(z) + \theta'(x, y, z, t) \tag{2.144}$$

are as follows:

$$D\mathbf{v}_h/Dt = \partial\mathbf{v}_h/\partial t + \mathbf{v}\cdot\nabla\mathbf{v}_h = -C_p\theta\,\nabla\pi' \tag{2.145}$$

$$Dw/Dt = -C_p\theta\,\partial\pi'/\partial z + B \tag{2.146}$$ [4]

$$\nabla\cdot\mathbf{v} = -C_v/(R\bar{\pi})(w\,d\bar{\pi}/dz + D\pi'/Dt) \tag{2.147}$$

where buoyancy

$$B = g\theta'/\bar{\theta} \tag{2.148}$$

It is noted that, unlike in the derivation of B expressed in terms of T' and p' (2.11), the expression for B does not involve any assumptions about the Mach number. The base state $\bar{\pi}$ is hydrostatic, so that the hydrostatic equation is

$$C_p\theta\,\partial\bar{\pi}/\partial z = -g \tag{2.149}$$

[4] The vertical perturbation pressure gradient force and B do not correspond exactly with those in (2.7).

From the definition of potential temperature (2.26) and the ideal gas law, the first term on the RHS of (2.147) may be expressed as

$$-w/(\bar{\theta}\bar{\rho})\, d(\bar{\theta}\bar{\rho})/dz \tag{2.150}$$

We will now seek the condition under which the time derivative term can be neglected in comparison with the other terms. Let us use the following scaling:

$$u \sim U, v \sim U, w \sim W, \bar{\pi} \sim 1, \pi' \sim P^*, \bar{\theta} \sim T, t \sim \tau, x \sim y \sim L, \text{and } z \sim D \tag{2.151}$$

where $P^* \ll 1$. So the terms in the continuity equation (2.147) are approximately

$$|\nabla \cdot \mathbf{v}| \sim U/L \sim W/D \tag{2.152}$$

$$|-C_v/(R\bar{\pi})(w\, d\bar{\pi}/dz)| \sim WC_v/(RD) \sim (C_v/R)(U/L) \tag{2.153}$$

$$|-C_v/(R\bar{\pi})D\pi'/Dt| \sim C_v P^*/(R\tau) \tag{2.154}$$

The terms in the horizontal equation of motion (2.145) are scaled as follows:

$$|D\mathbf{v}_h/Dt| \sim U/\tau \tag{2.155}$$

$$|-C_p\bar{\theta}\,\nabla_h\pi'| \sim C_p TP^*/L \tag{2.156}$$

So, $U/\tau \sim C_p TP^*/L$, and therefore

$$P^* \sim UL/(C_p T\tau) \tag{2.157}$$

Now, $c^2 = (C_p/C_v)R\pi\theta \sim (C_p/C_v)RT$. Substituting for P^* in (2.154), we find that the time derivative term may be neglected in comparison with the divergence term when

$$\tau \gg L/c \tag{2.158}$$

that is, when the time scale is much longer than the time it takes a sound wave to travel the characteristic horizontal scale. So, when sound waves are filtered out and only much longer time scales are considered the adiabatic form of the continuity equation may be expressed as

$$\nabla \cdot \mathbf{v} = -w/(\bar{\theta}\bar{\rho})\, d(\bar{\theta}\bar{\rho})/dz \tag{2.159}$$

This equation is similar to the anelastic equation (2.42), except that $\bar{\theta}$ is now included and is a function of height. We did not have to make any assumptions about D/H, as we did earlier when using pressure and temperature. The anelastic equation (2.42) in effect is valid when the basic state is isentropic (i.e., is one of constant potential temperature and has a dry-adiabatic lapse rate), so that $\bar{\theta}$ disappears explicitly from (2.159). This equation (2.159) is known as the "pseudo-incompressible equation". Dale Durran first introduced it in 1989. It is valid when sound waves are filtered out *and* when $|\pi'| \ll |\pi|$. The main advantage of this approximation is that the environment may be stratified in any way and is not constrained to be dry adiabatic.

The adiabatic form of the thermodynamic equation in terms of θ is seen from (2.38) and (2.26) as

$$D\theta/Dt = 0 \tag{2.160}$$

Expanding (2.160) in its Eulerian form and using the decomposition of θ into a base state and a small perturbation (cf. (2.144)) we find that after multiplying the resulting equation by $g/\bar{\theta}$ we get the following:

$$DB/Dt + wN^2 = 0 \tag{2.161}$$

where the Brunt–Väisälä frequency squared is $N^2 = g\, d(\ln \bar{\theta})/dz$. In an adiabatically stratified atmosphere θ does not vary with height, so $N^2 = 0$ and

$$DB/Dt = 0 \tag{2.162}$$

so that buoyancy B is conserved for air parcel motion.

2.8 SIMPLE, IDEALIZED MODELS OF DRY CONVECTION: PLUMES AND BUBBLES

In the late 1940s and 1950s, realistic computer simulations of convective clouds were not yet possible with the available machines. However, simple, highly idealized models were devised, which illustrate some significant properties of simple types of cumulus convection. Much of this work was done at Cambridge and Imperial College in England and supported by laboratory experiments. Prominent contributors included Richard Scorer, Bruce Morton, G. I. Taylor, J. S. Turner, and, in the U. S., Joanne (Malkus) Simpson.

Two basic types of convection were identified: *plumes* and *thermals*. Plumes are buoyant jets in which buoyancy is supplied *steadily* from a "point" source and the buoyant region is continuous (Figure 2.12a). Examples of dry plumes include active volcanoes, power plant/industrial smoke stacks, localized forest fires, and heated mountain peaks during the daytime. Moist convection forced by continuous mesoscale lift (e.g., along an advancing front or outflow boundary) or pyrocumulus clouds forced over a forest fire are like plumes, though the source of buoyancy is not *exactly* continuous. Thermals, on the other hand, are discrete buoyant elements in which buoyancy is confined to the limited volume of the fluid (Figure 2.12b). In effect, thermals are plumes that are "turned on" for very short periods of time. So, plumes result from steady heat sources, while thermals result from heat sources turned on for very short periods of time, resulting in bursts or "puffs" or "bubbles". Thermals for moist convection are found in "multicell" convection (to be discussed in Chapter 3). *Starting plumes* are plumes that are turned on and as such are plumes with well-defined, advancing upper edges (Figure 2.12c). Starting plumes for moist convection may be similar to the beginning stages of any deep convective cloud.

Figure 2.12. Idealized representations (left) and photographs of (right) convection: (top) plume; (middle) thermal; (bottom) starting plume. (Top right) Narrow convective plume over the Continental Divide in the Indian Peaks, CO, on August 6, 2010; actually, this example might also be used as a starting plume; (middle right) conglomeration of thermals over eastern Colorado, atop a developing supercell on June 6, 1990, as viewed from an aircraft; (bottom right), pyrocumulus produced by a wildfire in central Oklahoma on September 1, 2011 with a well-defined leading cap; the bottom of the plume is being advected to the left by low-level wind (photographs by the author).

2.8.1 Similarity models of plumes and thermals

Since plumes are of fundamental importance, we will look at them in some detail and refer the reader to other texts (e.g., Emanuel, 1994 and Houze, 1993) for more complete analyses of thermals and starting plumes. We will, for relative simplicity, consider dry plumes maintained by a source of buoyancy in a homogeneous fluid. Air motion is assumed to be turbulent and independent of molecular diffusion. Suppose that F is the rate at which buoyancy is supplied by a point source. Then the flux of buoyancy

$$F \sim \text{buoyancy} \times \text{velocity} \times \text{area} \tag{2.163}$$

Since buoyancy has units of $\mathrm{m\,s^{-2}}$ (see (2.19)), velocity has units of $\mathrm{m\,s^{-1}}$, and area has units of $\mathrm{m^2}$, F must have units of $\mathrm{m^4\,s^{-3}}$. To see how the mean vertical velocity (w), the mean buoyancy (B), and the mean radius of the plume (R) vary with height, we use *similarity* theory and *dimensional analysis*. We will assume first that the mean vertical velocity and buoyancy are functions of F and z, the height above the point heat source. This assumption makes sense physically, as the stronger the buoyancy flux, the stronger the vertical velocity should be; also, the farther away one gets from the buoyancy source, the weaker the vertical velocity should be.

So, we will assume that

$$w = C_1 F^a z^b \tag{2.164}$$

where C_1 is a positive dimensionless constant. To be dimensionally correct, we know that the units of m and s must match up on the LHS and RHS of (2.164). It is necessary then that

$$4a + b = 1 \tag{2.165}$$

and

$$-3a = -1 \tag{2.166}$$

It follows that

$$w = C_1 F^{1/3} z^{-1/3} \tag{2.167}$$

which is consistent with physical intuition. Similarly, it can be shown that buoyancy

$$B = C_2 F^{2/3} z^{-5/3} \tag{2.168}$$

where C_2 is a positive constant. Equation (2.168) is also consistent with physical intuition: buoyancy decreases away from the source and is a function of buoyancy flux. Of note is the finding that buoyancy decreases with height more rapidly than vertical velocity. This finding follows even without considering the governing equations. Finally, we find that the radius of the plume

$$R = C_3 z \tag{2.169}$$

where C_3 is positive constant. In other words, the radius of the plume increases linearly with height above the source.

The vertical mass flux in the plume

$$\sim wR^2 \sim z^{-1/3}z^2 = z^{5/3} \tag{2.170}$$

Since the vertical mass flux increases with height, there must be turbulent entrainment of environmental air into the plume.

In a thermal, unlike a plume, time t is also a significant independent variable, rather than just height z. For simplicity we will assume that the environment of the thermal is neutrally stratified. We assume that the vertical velocity, buoyancy, and radius of the spherical thermal bubble are functions of the volume-integrated buoyancy of the thermal Q as

$$Q = \iiint B\,dV \tag{2.171}$$

where V is the volume and the units of Q are $m^4\,s^{-2}$. So

$$w = Q^a z^b f(r) \tag{2.172}$$

where r is the radius of the thermal bubble. To be dimensionally correct

$$4a + b = 1 \tag{2.173}$$

and

$$-2a = -1 \tag{2.174}$$

So

$$w = Q^{1/2} z^{-1} f(r) \tag{2.175}$$

Similarly, it follows that

$$B = Qz^{-3} f(r) \tag{2.176}$$

and

$$R = Cz \tag{2.177}$$

where C is a positive constant. To find the variation with time, we make use of the relation between vertical velocity and height ($w = Dz/Dt$) and (2.175). Integrating (2.175) from the ground ($z = 0$) to an arbitrary height z from $t = 0$ to an arbitrary time t, we find that

$$z^2/2 \sim Q^{1/2} f(r) t \tag{2.178}$$

so that $z \sim t^{1/2}$ and therefore $w \sim t^{-1/2}$ (and $w \sim z^{-1}$). In other words, vertical velocity decreases with time as the thermal rises. It can similarly be shown that buoyancy $B \sim t^{-3/2}$ (and $B \sim z^{-3}$) (i.e., buoyancy decreases with time as the thermal rises). It decreases because it entrains neutrally buoyant environmental air (as the volume of the thermal increases, it must be taking in outside air and mixing with it). Since the mean radius of the thermal increases linearly with height (2.177), the thermal traces out a conical cross section as it ascends, like a plume (2.169).

2.8.2 The plume dynamical model

We now return to the idealized plume model, but this time we will use the Boussinesq continuity equation, the vertical equation of motion, and the adiabatic form of the thermodynamic equation to understand its behavior more precisely. Unlike the preceding analysis where we found similarity solutions, we will now make use explicitly of governing equations. We make the additional simplifying assumptions:

a. The flow is steady state ($\partial/\partial t = 0$). (It has forever been thus. So, in the words of the former rock group The Talking Heads, "... same as it ever was ...").
b. The plume is symmetrical about its center and can therefore be described in cylindrical coordinates, where u and w are the radial and vertical components of the wind, respectively. There is therefore no azimuthal variation in any variable. The azimuthal wind component, v, is set to zero, so there is no vertical vorticity. Radial variations in variables are similar at all heights (i.e., for all values of z).
c. The radial component of radial inflow represents the entrainment of environmental air and can be parameterized as

$$u = -\alpha w \tag{2.179}$$

   where α is a positive constant (fractional) entrainment rate.
d. The radial profiles of vertical velocity and buoyancy are such that each is zero beyond the mean radius of the thermal (R) and constant within the mean radius of the thermal. Such a variation is termed a "top hat" profile (Figure 2.13), not necessarily in honor of Fred Astaire and Ginger Rogers whose 1935 movie of the same name involved a top hat (worn by the former). The top hat profile is the simplest possible profile, though a Gaussian profile may be more realistic.

We will end up with four equations in four unknowns (u, w, B, and R). First, the Boussinesq continuity equation in cylindrical coordinates is

$$1/r\,\partial/\partial r(ru) + \partial w/\partial z = 0 \tag{2.180}$$

We eliminate u as a variable by substituting the entrainment relation (2.179) into (2.180); so we now have three equations in three unknowns. After integrating the

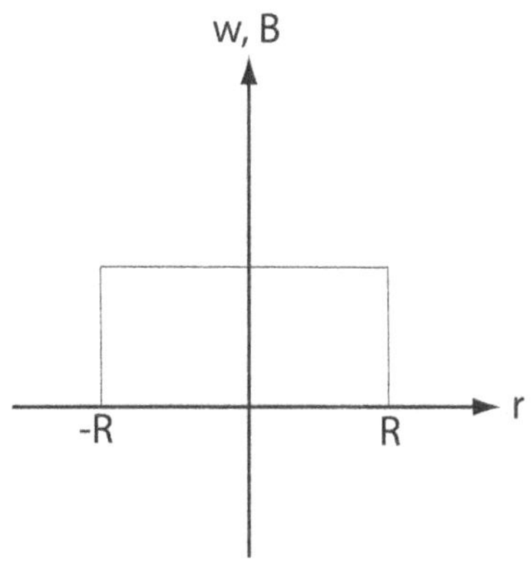

Figure 2.13. Example of a top hat profile for vertical velocity (w) and buoyancy (B).

resulting equation over a horizontal area of radius R, at arbitrary height z, we find that

$$2\pi\alpha R w = d/dz(\pi R^2 w) \tag{2.181}$$

Note that the partial derivatives become ordinary derivatives because R and w are symmetrical about $r = 0$ and radial variations are identical for each variable. This equation expresses the physical process of entrainment of mass through the (outer) boundary of the plume as being proportional to the increase with height of the vertical mass flux through the plume. The steady-state version of the inviscid vertical equation of motion (2.7) is used and the vertical perturbation pressure gradient is neglected, so that

$$u\,\partial w/\partial x + w\,\partial w/\partial z = B \tag{2.182}$$

We integrate (2.182) over a volume between two heights (z to $z + \Delta z$), making use of the Boussinesq continuity equation (2.35) and the divergence theorem to express the volume integral in terms of a surface flux. We use the boundary condition that $w = 0$ at the sides of the plume. For infinitesimally small Δz

$$d/dz(\pi R^2 w^2) = \pi R^2 B \tag{2.183}$$

The steady-state version of the adiabatic thermodynamic equation (2.161) is

$$u\,\partial B/\partial r + w\,\partial B/\partial z + wN^2 = 0 \tag{2.184}$$

It is integrated over the same volume that (2.182) was integrated and, after having used the Boussinesq continuity equation (2.35) and the divergence theorem, it is found, since (wB) at the edge of the plume vanishes, that for infinitesimally small Δz

$$d/dz(\pi R^2 wB) = -\pi R^2 wN^2 \tag{2.185}$$

Factoring out the πs, we end up with the following three, simultaneous, highly nonlinear equations in w, B, and R:

$$d/dz(R^2 w) = 2R\alpha w \tag{2.186}$$

$$d/dz(R^2 w^2) = R^2 B \tag{2.187}$$

$$d/z(R^2 wB) = -R^2 wN^2 \tag{2.188}$$

When the environment is dry adiabatic, $N^2 = 0$ and from (2.188) we find that $R^2 wB =$ constant. The reader is referred to other textbooks for exact solutions to the variables for this case and for the case of an unstable environment ($N^2 < 0$).

The much more interesting case is that of a stably stratified environment ($N^2 > 0$). In an influential paper by Morton, Taylor, and Turner in 1956 (often referred to as MTT; this might be the first case in meteorology when a triplet of authors are abbreviated as three letters; RKW, which is used in a subsequent chapter, might be the second), equations (2.186)–(2.188) were derived for the case in which the mean vertical velocity and mean buoyancy follow a more realistic

Gaussian (radial) profile. For this profile the following similar set of equations hold:

$$d/dz(R^2w) = 2R\alpha w \tag{2.189}$$

$$d/dz(R^2w^2) = 2R^2B \tag{2.190}$$

$$d/z(R^2wB) = -2R^2wN^2 \tag{2.191}$$

These equations are similar to (2.186)–(2.188); the differences are that a factor of 2 appears in the equation of motion (second equation in the set) and the thermodynamic equation (third equation in the set). The solution for these equations is sought subject to the boundary conditions that the mean vertical velocity and mean radius are zero at the ground and that the flux of buoyancy at the ground is a specified constant. The variables are first changed as follows:

$$V \equiv Rw \tag{2.192}$$

$$U \equiv R^2w \tag{2.193}$$

$$F \equiv R^2wB \tag{2.194}$$

so that the three simultaneous differential equations become

$$dU/dz = 2\alpha V \tag{2.195}$$

$$dV^2/dz = 2FU/V^2 \tag{2.196}$$

$$dF/dz = -2UN^2 \tag{2.197}$$

The boundary conditions are

$$V = U = 0 \tag{2.198}$$

at $z = 0$, and

$$F = (2/\pi)F_0 \tag{2.199}$$

at $z = 0$ and $R = 0$, where F_0 is the flux of buoyancy over the point source of heat. The salient characteristics of the solutions are as follows (Figure 2.14):

a. The mean radius of the plume first increases linearly with height, but eventually increases faster and faster with height.
b. Mean buoyancy is greatest at the ground and decreases with height until it becomes negative. It is above the height at which buoyancy drops below zero that the mean radius increases most rapidly with height.
c. A measure of mean upward vertical mass flux (R^2w) decreases with height and approaches zero rapidly near the height at which the mean radius increases towards infinity quickly.

The physical interpretation of these results is that as stable environmental air is entrained into the plume, the buoyancy of the plume decreases, vertical velocity decreases, and the plume spreads out aloft into the characteristic anvil shape. The

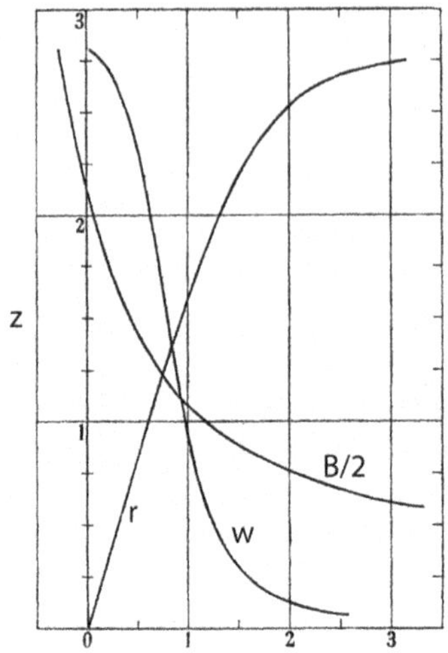

Figure 2.14. Variations of mean radius, mean buoyancy, and a measure of upward vertical mass flux as a function of height in the MTT steady-state plume in a uniform, stably stratified environment for a Gaussian profile in mean vertical velocity and mean buoyancy: r is the radius of the plume, w is vertical velocity, B is buoyancy, and z is the height, all expressed in non-dimensional units (adapted from Morton *et al.*, 1956).

top of the plume actually overshoots the equilibrium level (where the buoyancy first drops to zero). The reader is again reminded of the limitations of this simple model: not only is it steady and dry, but in addition the vertical perturbation pressure gradient is ignored, which in nature should act to reduce vertical acceleration. Finally, this model is valid for an environment in which there is no vertical shear and, in fact, no flow at all outside the plume. Yet, despite the alarming extent to which the atmosphere has been grossly oversimplified, the governing equations (2.195)–(2.197) are still highly nonlinear. Similar equations have been used to study chaos theory. The reader should be getting the feeling that, while such a simple model is so mathematically complex, an even more realistic model must be even more complex.

MTT in 1956 also derived and solved the equations for an entraining thermal in a stably stratified environment. The three governing equations, analogous to (2.189)–(2.191) for the plume model, are

$$d/dt(\tfrac{4}{3}\pi R^3) = 4\pi R^2 \alpha w \tag{2.200}$$

$$d/dt(\tfrac{4}{3}\pi R^3 w) = \tfrac{4}{3}\pi R^3 B \tag{2.201}$$

$$d/dt(\tfrac{4}{3}\pi R^3 B) = -\tfrac{4}{3}\pi R^3 W N^2 \tag{2.202}$$

For N^2 = a positive constant, the behavior of the thermal when it begins with zero radius, vertical velocity, and constant buoyancy at the outset is summarized (but not shown in a figure) as follows:

a. The radius of the thermal increases rapidly, but then increases with time much less rapidly.
b. The buoyancy of the thermal rapidly decreases as it rises and entrains stable environmental air.
c. The thermal rises rapidly at first, but then slows down and eventually reverses direction (descends) after buoyancy has decreased to zero and becomes negative. The thermal then oscillates about a level as buoyancy becomes less negative and then becomes positive again. Vertical velocity lags buoyancy during these oscillations.

Experiments have been conducted by dropping discrete masses of relatively dense fluid into a tank of less dense fluid. It was found that thermals behave like spherical vortices, as discussed mathematically by the famous fluid dynamicist Horace Lamb decades earlier. In a spherical vortex, donut-like flow is observed surrounding a thermal, similar to what we see when we watch the upper edge of a growing cumulus cloud turning inside out. A thermal may be thought of as the leading edge of a starting plume, which looks like a buoyant mass of upward-moving air turning inside out (Figure 2.10). In a convective cloud, the act of turning inside out is a result of baroclinically generated vorticity when positive buoyancy in the thermal is surrounded by an environment of neutral buoyancy. The upper edges of growing, cumuliform clouds have many small convective dimples and a cauliflower-like appearance, which suggest small-scale turbulence.

2.9 INTRODUCTION TO RAYLEIGH–BÉNARD CONVECTION

In nature, buoyancy sources are frequently not localized as in plumes and thermals, but rather are spread out over broad areas, at least in comparison with the depth of the convecting layer. Thus, the buoyancy source is not a point, but instead is distributed over a relatively wide area. The entire fluid therefore involves some overturning and convection is said to have a "global" rather than a "local" character. The classic problem of determining what motions are driven in response to the heating of an underlying surface was approached experimentally by the French physicist Henri Bénard in 1900 and solved mathematically by the English physicist and Nobel Laureate Lord Rayleigh (John William Strutt) (reported on in 1916). (Perhaps we should refer to what has been called "Rayleigh–Bénard" convection as "Ray-nard convection.")

Bénard, using oil from a sperm whale, showed that, after a certain vertical temperature gradient between two very closely spaced ($\sim$1 mm) parallel, horizontal plates is exceeded, a stationary pattern of cells of rising and sinking motion (overturning) evolves. This pattern of cells is similar in appearance to the checkerboard pattern of cumulus clouds often seen when the Sun heats up the ground or cold air flows over a warm surface (Figure 2.15) and is therefore worthy of consideration, even though the real atmosphere does not have a solid upper boundary and has much lower density than sperm whale oil, among other things. Lord Rayleigh

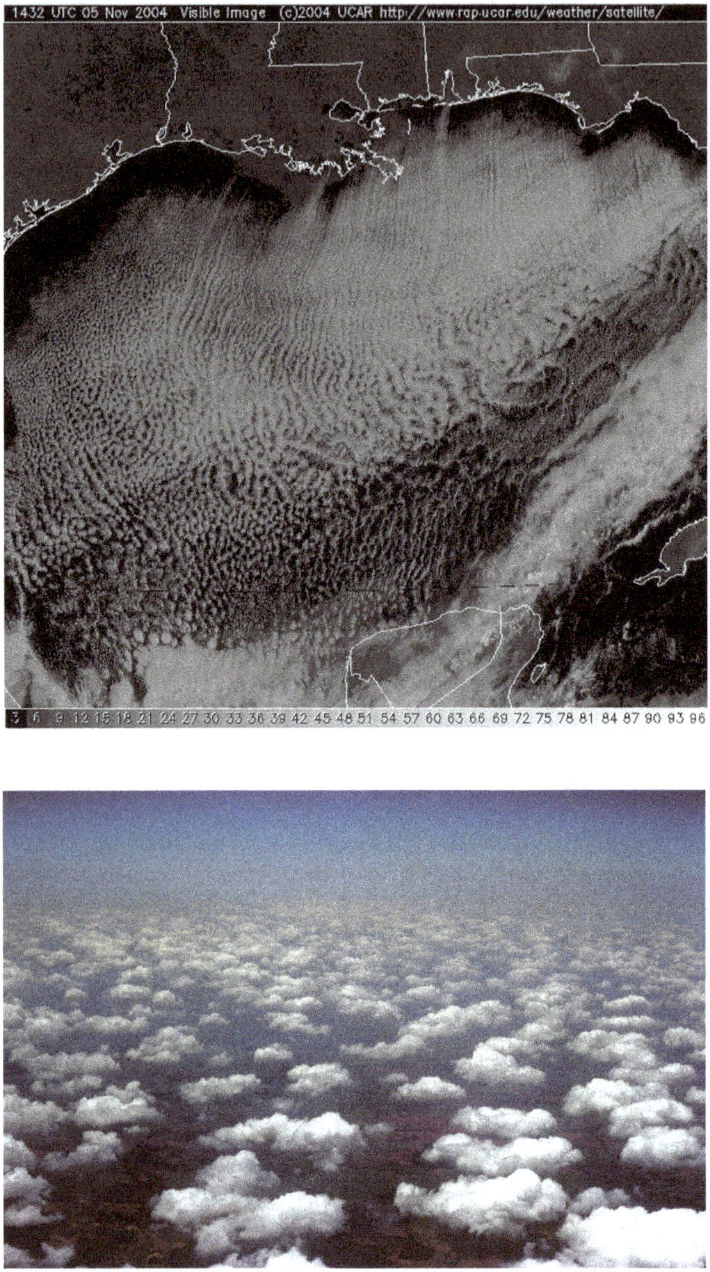

Figure 2.15. Bénard–Rayleigh-like convection represented by (a) cells of clouds over the Gulf of Mexico on November 5, 2004, when cold air from the north is flowing over the warm water surface to the south and (b) cumulus clouds over land (near the Oklahoma–Texas border on the Red River) on June 24, 1984, as viewed from an aircraft, as the surface is heated during the daytime (photograph by the author).

noted "M. Bénard does not appear to be acquainted with James Thomson's paper "On a changing tesselated structure in certain liquids" (*Proc. Glasgow Phil. Soc.*, 1881–2), where is described a like structure in much thicker layers of soapy water cooling from the surface." Alas, Rayleigh has provided more evidence for Stigler's Law of Eponymy. The essence of Lord Rayleigh's mathematical analysis is now summarized.

2.9.1 Convection in a resting atmosphere without rotation

The Boussinesq equations are used. The equations of motion are expressed in terms of density in the expression for the pressure gradient force, viscosity is now included in all three components of the equations of motion, and thermal conductivity is included in the thermodynamic equation, which is now expressed in terms of temperature, rather than potential temperature. The equations linearized about a resting basic state, in which the base-state pressure varies only with height and temperature varies linearly from the bottom plate to the top plate (separated by a distance H) (Figure 2.16) and is held fixed as a result of heating, may be expressed as follows:

$$(\partial/\partial t - \nu\nabla^2)u' = -1/\rho\, \partial p'/\partial x \tag{2.203}$$

$$(\partial/\partial t - \nu\nabla^2)v' = -1/\rho\, \partial p'/\partial y \tag{2.204}$$

$$(\partial/\partial t - \nu\nabla^2)w' = -1/\rho\, \partial p'/\partial z + B \tag{2.205}$$

$$(\partial/\partial t - \kappa\nabla^2)T' = w'\alpha' \tag{2.206}$$

$$\partial u'/\partial x + \partial v'/\partial y + \partial w'/\partial z = 0 \tag{2.207}$$

where ν is the kinematic coefficient of molecular viscosity;[5] κ is the coefficient of thermal conductivity/diffusivity; and α' is the mean temperature gradient between the two parallel plates. It should be noted that when linearizing the equations, it cannot be assumed that the perturbation wind velocities are small in magnitude compared with the basic-state wind velocities, because the basic state is resting. However, if the perturbation winds are expanded in a Taylor series expansion in time, it is seen that products of perturbation terms may be neglected in comparison with terms containing only one perturbation term for a very short period of time after the initial time. Later on, the nonlinear terms cannot be neglected. Now suppose the fluid between the two plates has the following characteristics:

$$\beta = -1/\rho\, \partial\rho/\partial T \tag{2.208}$$

where β is the coefficient of thermal expansion and the constant

$$\gamma = -\partial\rho/\partial T \tag{2.209}$$

Now

$$T' = T - \bar{T} = \int_{\bar{T}}^{T} \partial T/\partial\rho\, d\rho = -1/\gamma \int_{\rho_{\mathrm{av}}}^{\rho} d\rho = -\rho'/\gamma \tag{2.210}$$

[5] The kinematic coefficient of eddy viscosity if eddies are considered rather than molecules.

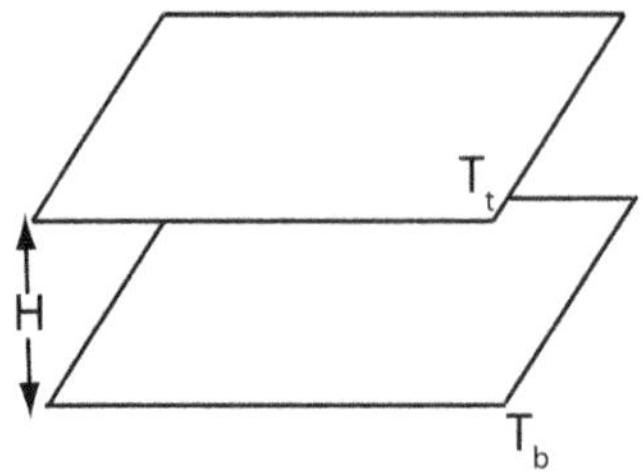

Figure 2.16. Setup for Rayleigh–Bénard convection.

It then follows that

$$B = g\beta T' \tag{2.211}$$

We define

$$\alpha = g\beta\alpha' = \text{constant} \tag{2.212}$$

(Note that this α is not the same as the entrainment rate in (2.179).) From (2.211) and (2.212), we can now express the thermodynamic equation (2.206) in terms of buoyancy, rather than temperature, as

$$(\partial/\partial t - \kappa\nabla^2)B = w'\alpha \tag{2.213}$$

So, we have five equations in five unknowns (u', v', w', p', and B). We now drop the prime notation to simplify writing of the equations, where it is understood that u, v, w, and p represent perturbation quantities.

Before we begin our analysis, we digress briefly to point out, for historical reasons, that Barry Saltzman in 1962 reported on a study of Rayleigh–Bénard convection that employed his simplifying assumption that motions are two dimensional in the x–z-plane (i.e., $\partial/\partial y = 0$: that there are convective rolls). He formed a horizontal vorticity equation from (2.203) and (2.205). The resulting vorticity equation and thermodynamic equation, which is essentially (2.51), form a system of two equations for the streamfunction in the x–z-plane (the Boussinesq continuity equation states that the flow in the x–z-plane is non-divergent, so that a streamfunction can be defined). Incidentally, Saltzman noted the similarity of his approach to that of Joanne Malkus (Simpson) and G. Witt a few years earlier. Ed Lorenz, in his famous and influential 1963 paper "Deterministic nonperiodic flow" in which he first described the essence of chaos theory, used these equations as the basis for his study. So, it turns out that the equations that Rayleigh used to describe Bénard convection were the prototypical nonlinear equations that were so influential in the argument for limits to the predictability of weather systems.

Instead of forming a vorticity equation as did Saltzman *et al.*, we now eliminate p from (2.203) and (2.204) to form a horizontal divergence equation and, using the equation of continuity, get the following:

$$(\partial/\partial t - \nu\nabla^2)\,\partial^2 w/\partial z^2 = 1/\rho\,\partial/\partial z(\partial^2 p/\partial x^2 + \partial^2 p/\partial y^2) \tag{2.214}$$

Thus, both u and v have been eliminated. We now differentiate the vertical equation of motion twice with respect to x, and also twice with respect to y.

Adding the resulting two equations to (2.214), we see that p is eliminated and the following equation is the result:

$$(\partial/\partial t - \nu\nabla^2)\nabla^2 w = (\partial^2/\partial x^2 + \partial^2/\partial y^2)B \tag{2.215}$$

We now seek to eliminate B so that we can get one equation in terms of one dependent variable, namely w. To do so, we apply the operator $(\partial/\partial t - \kappa\nabla^2)$ to (2.215) and apply the operator $(\partial^2/\partial x^2 + \partial^2/\partial y^2)$ to the thermodynamic equation (2.213) and then subtract the latter from the former to get

$$(\partial/\partial t - \kappa\nabla^2)(\partial/\partial t - \nu\nabla^2)\nabla^2 w = \alpha(\partial^2/\partial x^2 + \partial^2/\partial y^2)w \tag{2.216}$$

This is a sixth-order partial differential equation in terms of the vertical velocity w. Three physical constants/parameters that describe the characteristics of the problem are κ, ν, and α. The reader is reminded that the former two quantify the molecular diffusion of momentum and heat, while the latter quantifies the temperature gradient between the two plates and the characteristics of the fluid between them.

Before we continue on in the solution of this equation, it is useful from the standpoint of simplification to recombine these three parameters into two. The procedure follows that of Willem Malkus and George Veronis in 1958: we rewrite the equations by scaling each spatial variable by H as follows:

$$x^* = Hx, \quad y^* = Hy, \quad \text{and} \quad z^* = Hz \tag{2.217}$$

H is the typical space scale in units of space (e.g., meters) and it is assumed that the horizontal and space scales are probably similar (isotropic); in addition, H is the spacing between the two plates. The variables expressed without an asterisk are non-dimensional and $\sim O(1)$. The time scale τ is estimated from the equations of motion (2.203)–(2.205), assuming that the inertial acceleration is the same order of magnitude as the acceleration due to viscosity:

$$1/\tau \sim \nu/H^2 \tag{2.218}$$

so that

$$t^* = H^2/\nu \tag{2.219}$$

The resulting equation, which is of sixth order in w, is as follows:

$$(\sigma\,\partial/\partial t - \nabla^2)(\partial/\partial t - \nabla^2)\nabla^2 w = \mathrm{Ra}(\partial^2/\partial x^2 + \partial^2/\partial y^2)w \tag{2.220}$$

where the Prandtl (pronounced without the "t") number

$$\sigma = \nu/\kappa \tag{2.221}$$

and the Rayleigh number

$$\mathrm{Ra} = \alpha H^4/(\nu\kappa) = \beta g\alpha' H^4/(\nu\kappa) \tag{2.222}$$

The Rayleigh number varies monotonically and linearly as the temperature gradient between the two plates (α'), monotonically, and very rapidly as the space scale, but inversely proportional to both the kinematic coefficient of molecular viscosity and the coefficient of thermal conductivity/diffusivity. The physical meaning of the Rayleigh number is that when buoyancy is the same order of magnitude as the viscous term in the vertical equation of motion (i.e., the flow is laminar), it represents the ratio of the convective vertical flux of buoyancy to the molecular vertical flux of buoyancy. When the buoyancy is the same order of magnitude as the inertial acceleration the flow is turbulent. For both laminar and turbulent flows, the Rayleigh number varies monotonically as the Reynolds number and the Prandtl number. The Reynolds number expresses the relative importance of the inertial acceleration to the viscous acceleration, and is an indicator of the relative amount of turbulent flow to laminar flow.[6] The Rayleigh and Prandtl numbers determine the type of ensuing motions. We are not going to find the actual exact solutions to (2.220), but instead just seek out the conditions under which we have growing solutions (i.e., solutions that are unstable). To do so, we look for solutions of the form

$$w = W(z)\,\mathrm{Re}\left[\int_{-\infty}^{\infty}\int_{-\infty}^{\infty} e^{\omega t}\, e^{i(k_x + k_y)}\, dk_x\, dk_y\right] \tag{2.223}$$

where the complex growth rate $\omega = \omega_r + i\omega_i$ and k_x and k_y are horizontal wave numbers in the x and y-directions, where $k^2 = k_x^2 + k_y^2$. This form of solution is a Fourier series for the horizontal part of the solution; the vertical and horizontal parts of the solution are separable. We now substitute (2.223) into (2.220) and get the following "characteristic value equation":

$$[\sigma(\omega_r + i\omega_i) + k^2 - d^2/dz^2][\omega_r + i\omega_i + k^2 - d^2/dz^2](k^2 - d^2/dz^2)W = \mathrm{Ra}\, k^2 W \tag{2.224}$$

This is a sixth-order ordinary differential equation for W with respect to z; six boundary conditions are needed to solve for W. Solutions may be found for certain combinations of ω_r, ω_i, k, σ, and Ra.

The kinematic boundary condition is applied at the top plate and the bottom plate, which means that

$$W = 0 \quad \text{at } z = 0 \text{ and } 1 \tag{2.225}$$

This boundary condition means that there is no flow ($w = 0$) of fluid from or into each plate. We have a choice of two commonly used boundary conditions on the horizontal flow at the top and bottom plates: "free slip" (zero stress in which there is no friction layer; free slip is perhaps better than "pink slip" or "just *fire* the problem!") or "no slip". Let us consider the simpler one (simpler in terms of the resulting mathematics), the free-slip boundary condition, in which there is no stress at the plates, so that the fluid is allowed to move by the stationary plate. In the real atmosphere, however, there does tend to be some stress (vertical shear) at the ground. Since the stress is proportional to the vertical gradient in the flow, the vertical derivatives of the flow are constrained to be zero, so that

$$\partial u/\partial z = \partial v/\partial z = 0 \quad \text{at } z = 0 \text{ and } 1 \tag{2.226}$$

From the Boussinesq continuity equation (2.35) we know that

$$\partial u/\partial x + \partial v/\partial y = -\partial w/\partial z \tag{2.227}$$

[6] In this interpretation of the Rayleigh number, $T_b - T_t = T'$.

Differentiating (2.227) with respect to z, we find that

$$\partial/\partial x(\partial u/\partial z) + \partial/\partial y(\partial v/\partial z) = -\partial^2 w/\partial z^2 \tag{2.228}$$

Then using (2.226) we find that the LHS of (2.228) must be zero at $z = 0$ and 1, so that

$$d^2 W/dz^2 = 0 \quad \text{at } z = 0 \text{ and } 1 \tag{2.229}$$

Four boundary conditions down and two to go. The temperature of the bottom plate and that of the top plate are held fixed, so that B is a constant on each plate. If this is the case, then the plates must be perfectly conducting. So

$$\nabla_h B = 0 \quad \text{at } z = 0 \text{ and } 1 \tag{2.230}$$

We recall (2.215) and find that with (2.230)

$$(\partial/\partial t - \nu\nabla^2)\nabla^2 w = (\partial^2/\partial x^2 + \partial^2/\partial y^2)B = 0 \quad \text{at } z = 0 \text{ and } 1 \tag{2.231}$$

so that

$$d^4 W/dz^4 = 0 \quad \text{at } z = 0 \text{ and } 1 \tag{2.232}$$

The six boundary conditions for the free-slip, perfectly conducting case are that W as well as the second and fourth derivatives of W with respect to z all vanish at $z = 0$ and 1 at the top and bottom plates. A solution that satisfies the boundary conditions is the following:

$$W(z) = \sum_{n=1}^{\infty} A_n \sin(n\pi z) \tag{2.233}$$

Substituting (2.233) into (2.224) we find that

$$[\sigma(\omega_r + i\omega_i) + k^2 + n^2\pi^2][\omega_r + i\omega_i + k^2 + n^2\pi^2](k^2 + n^2\pi^2) = \text{Ra}\, k^2 \tag{2.234}$$

Since Ra is a real number, the imaginary part of the LHS must be zero. It follows that

$$\omega_i[2\sigma\omega_r + (k^2 + n^2\pi^2)(1 + \sigma)] = 0 \tag{2.235}$$

This equation is satisfied if $\omega_i = 0$, in which case there cannot be oscillating solutions and ω_r could have any value, growing or decaying (i.e., unstable or stable), or if

$$2\sigma\omega_r + (k^2 + n^2\pi^2)(1 + \sigma) = 0 \tag{2.236}$$

In the latter case, ω_i could be anything, but

$$\omega_r = -(k^2 + n^2\pi^2)(1 + \sigma)/(2\sigma) < 0 \tag{2.237}$$

because k^2, n^2, π^2, and σ are >0. So, in this case solutions could be oscillating, but stable (decaying with time): unstable, oscillating solutions are not possible. We now return to (2.234), but this time we equate the real part of the LHS to the real part of the RHS. It follows that

$$(\sigma\omega_r + k^2 + n^2\pi^2)(\omega_r + k^2 + n^2\pi^2) - \sigma\omega_i^2 = \text{Ra}\, k^2/(k^2 + n^2\pi^2) \tag{2.238}$$

We now have two simultaneous equations in ω_r and ω_i ((2.235) and (2.238)). If $\omega_i \neq 0$, then we eliminate ω_r by solving (2.236) for ω_r and substituting it into

(2.238) to get the following equation for ω_i:

$$\text{Ra}\, k^2/(k^2+n^2\pi^2) = -\{\sigma\omega_i^2 + (k^2+n^2\pi^2)^2[(\sigma-1)^2/4\sigma]\} \qquad (2.239)$$

The LHS of (2.239) is always >0. However, the RHS is always <0. Therefore ω_i must be zero (i.e., if the Rayleigh number is greater than zero), and if the bottom plate is maintained at a warmer temperature than the top plate, small-amplitude oscillations are not possible. So, let us return to (2.238) and find out what values of ω_r are allowed when $\omega_i = 0$. We find that

$$[\sigma\omega_r + k^2 + n^2\pi^2][\omega_r + k^2 + n^2\pi^2](k^2+n^2\pi^2) = \text{Ra}\, k^2 \qquad (2.240)$$

It appears from (2.240) that ω_r can be either >0 or <0 (i.e., that there could be instability or stability), depending on what the values of Ra, n, k, and σ are. For "marginal" stability, $\omega_r = 0$, so that

$$\text{Ra} = (k^2+n^2\pi^2)^3/k^2 \qquad (2.241)$$

What is the smallest value of the Rayleigh number and what is the value of k at this smallest value for $n = 1$, which is the shortest vertical mode possible? If within the expression for the Rayleigh number, only the horizontal temperature gradient between the plates varies, then by cranking up the temperature gradient we finally arrive at the "critical" Rayleigh number. (The $n = 0$ case is "degenerate", so that $w = 0$ everywhere and is therefore not considered.) This minimum value of the Rayleigh number is called the "critical" value of the Rayleigh number (Ra_c) and is found by differentiating (2.241) with respect to k and setting it to zero to find that the critical horizontal wave number, which is

$$k_c^2 = \pi^2/2 \qquad (2.242)$$

The critical value of the Rayleigh number is found by substituting (2.242) into (2.241), so that

$$\text{Ra}_c = (27/4)\pi^4 \approx 657.5 \qquad (2.243)$$

When the critical Rayleigh number is exceeded, $\omega_r > 0$ and $\omega_i = 0$, so that "stationary overturning" is possible. When the temperature gradient is turned up so that the critical Rayleigh number is exceeded, the first mode to go unstable (Ra_c increases with increasing n; cf. (2.241)) is that described by (2.242), or since

$$k_c = 2\pi/L_c = \pi/\sqrt{2} \qquad (2.244)$$

where L_c is the critical horizontal wavelength. Thus

$$L_c = 2\sqrt{2} \qquad (2.245)$$

Since $x^* = Hx$ and $y^* = Hy$ [cf. (2.217)],

$$L_c = 2\sqrt{2}H \qquad (2.246)$$

which is a nice, simple result. Eventually, of course, as the size of the perturbations grows, linear results no longer hold and matters become more complicated. However, theory does dictate what the spacing of the first-appearing cells must be.

Now, (2.246) tells us only what the critical horizontal wavelength is, but does not tell us anything about the geometry of the cells. The simplest cells would be equilateral triangles, more complicated cells would be squares, and the most complicated would be pentagons, etc.; the problem is like that of filling a bathroom floor with equilateral tiles. If j is the number of sides of a regular polygon, then the angle at each vertex of the polygon is $\pi(1-2/j)$. If the polygons fit together *evenly* as in a jigsaw puzzle, then the angle must divide into 2π an *integral* number of times, so that

$$2/(1-2/j)=m \tag{2.247}$$

where m is an integer. From (2.247), we note that

$$m=2j/(j-2)=2+4/(j-2) \tag{2.248}$$

For $j=3$, 4, and 6, m is an integer (5 does not work); if $j-2>4$ (i.e., if $j>6$), m cannot be an integer. So, equilateral triangles, squares, and hexagons are the only possible patterns of cells.

The free-slip boundary conditions at the plates we considered were easy mathematically to deal with. The analysis with no-slip boundary conditions is more complicated because the relatively simple sinusoidal variation in z for W (cf. (2.233)) does not satisfy the boundary conditions.[7] Much work has been done solving the Rayleigh–Bénard problem using other boundary conditions, and the reader is referred to Emanuel (1994) and Chandrasekhar (1961) for the gruesome details, which are in many instances rather complicated. It has been found that the critical Rayleigh number and the critical horizontal wave number (and the critical horizontal wavelength) vary for the different cases, so boundary conditions do indeed matter. The case of mixed boundary conditions (i.e., of no slip at the bottom, but free slip at the top) perhaps most closely mimic what happens in the real atmosphere, where there is indeed a rigid surface and the top of the heated boundary layer is not a like a rigid plate.

In the case when cold air flows over a relatively warm water surface and is heated from below, then the lower boundary condition must include the interaction of the air with the water, which is certainly different from what happens over a rigid land surface. Of course, flow over non-level ground also adds a complication.

Yet another variation on the problem is that of what happens when vertical heat flux is held fixed at the boundaries, rather than the temperature. In this case, the thermal conductivity of the plates is relatively low and mechanically induced fluxes of heat and momentum from the surface dominate near the surface. The boundary condition involving B is then that $\partial B/\partial z=0$ at $z=0$ and 1, rather than $\nabla_h B=0$. While variations on the problem are interesting in their own right, they will not be considered further in this text, mainly because the theories, which are complicated mathematically, are still for ultra-simple cases: they are linear, and moisture and cloud microphysics are not even considered. So, we may consider 16 possible combinations of boundary conditions: free slip or no slip at each boundary and perfectly conducting or insulating, yielding free slip at both the top and bottom, no slip at the top and bottom, free slip at the top and no slip at the

[7] When the no slip boundary condition is imposed, the first derivative of W with respect to height vanishes at the upper and lower boundaries.

bottom, no slip at the top and free slip at the bottom, and perfectly conducting or insulating for each of the former eight boundary conditions.

2.9.2 Convection in a resting atmosphere with rotation

Just for illustrative purposes, we add rotation to the problem and consider only how it qualitatively changes the results. To do this, we simply add the acceleration due to the Coriolis force in (2.203) and (2.204). The new set of horizontal equations of motion are:

$$(\partial/\partial t - \nu\nabla^2)u = -1/\rho\,\partial p/\partial x + 2\Omega \tag{2.249}$$

$$(\partial/\partial t - \nu\nabla^2)v = -1/\rho\,\partial p/\partial y - 2\Omega \tag{2.250}$$

where Ω is the local rotation rate (the Coriolis force due to the rotation of the spherical Earth is not considered here). To get an equation in terms of w alone, we first form a vertical vorticity equation from (2.249) and (2.250) and make use of the continuity equation (2.207) to eliminate $\partial u/\partial x + \partial v/\partial y$; we also express the vertical vorticity as ζ. In addition, we form a horizontal divergence equation from (2.203) and (2.204) and again use (2.207) to eliminate $\partial u/\partial x + \partial v/\partial y$; once more, we express the vertical vorticity as ζ. We then eliminate p from the vorticity and divergence equations to obtain the following:

$$(\partial/\partial t - \nu\nabla^2)\nabla^2 w = -2\Omega\,\partial\zeta/\partial z + (\partial^2/\partial x^2 + \partial^2/\partial y^2)B \tag{2.251}$$

We now eliminate B from (2.251) and the thermodynamic equation (2.213) to get

$$\begin{aligned}(\partial/\partial t - \kappa\nabla^2)(\partial/\partial t - \nu\nabla^2)^2\nabla^2 w &= \alpha(\partial/\partial t - \nu\nabla^2)(\partial^2/\partial x^2 + \partial^2/\partial y^2)w \\ &\quad - 2\Omega(\partial/\partial t - \kappa\nabla^2)(\partial/\partial t - \nu\nabla^2)\,\partial\zeta/\partial z\end{aligned} \tag{2.252}$$

Finally, we eliminate vorticity from (2.252) by differentiating the vorticity equation with respect to z (not shown):

$$(\partial/\partial t - \nu\nabla^2)\,\partial\zeta/\partial z = 2\Omega\,\partial^2 w/\partial z^2 \tag{2.253}$$

The final form of the equation for w is as follows:

$$\begin{aligned}(\partial/\partial t - \kappa\nabla^2)(\partial/\partial t - \nu\nabla^2)^2\nabla^2 w &= \alpha(\partial/\partial t - \nu\nabla^2)(\partial^2/\partial x^2 + \partial^2/\partial y^2)w \\ &\quad - (2\Omega)^2(\partial/\partial t - \kappa\nabla^2)\,\partial^2 w/\partial z^2\end{aligned} \tag{2.254}$$

This equation is of order eight, two up from what it had been before rotation was added (2.220). Although the equation has a very high order, at least it is still linear. As earlier, we scale the independent variables x, y, and z, by H and time by H^2/ν to arrive at the following:

$$\begin{aligned}(\sigma\,\partial/\partial t - \nabla^2)(\partial/\partial t - \nabla^2)^2\nabla^2 w &= \mathrm{Ra}(\partial/\partial t - \nabla^2)(\partial^2/\partial x^2 + \partial^2/\partial y^2)w \\ &\quad - T_0^2(\sigma\,\partial/\partial t - \nabla^2)\,\partial^2 w/\partial z^2\end{aligned} \tag{2.255}$$

where T_0, the Taylor number, is

$$T_0 = (2\Omega)^2 H^4/\nu^2 \tag{2.256}$$

So,

$$\sqrt{T_0} = 2\Omega U/(\nu U/H^2) \tag{2.257}$$

In other words, the square root of the Taylor number represents the relative importance of the acceleration due to the Coriolis force to viscous acceleration.

As in the case without rotation, we seek solutions of the form

$$w = W(z)\,\mathrm{Re}\left[\int_{-\infty}^{\infty}\int_{-\infty}^{\infty} e^{\omega t}\, e^{i(k_x+k_y)}\, dk_x\, dk_y\right] \tag{2.258}$$

The characteristic value equation for the Rayleigh problem including rotation is

$$\begin{aligned}[]&[\sigma(\omega_r + i\omega_i) + k^2 - d^2/dz^2][\omega_r + i\omega_i + k^2 - d^2/dz^2]^2(k^2 - d^2/dz^2)W \\ &\quad = \mathrm{Ra}[(\omega_r + i\omega_i) + k^2 - d^2/dz^2]k^2 W + T_0[(\omega_r + i\omega_i) + k^2 - d^2/dz^2]\, d^2W/dz^2\end{aligned} \tag{2.259}$$

As in the case without rotation, the simplest boundary conditions mathematically are the free-slip boundary conditions. Six of the eight boundary conditions for the free-slip, perfectly conducting case are, as for the non-rotating case, that W as well as the second and fourth derivatives of W with respect to z all vanish at $z = 0$ and 1, at the top and bottom plates. We need two additional boundary conditions.

To get them, let us consider the characteristic value equation (2.259) evaluated when ω_i and ω_r are each zero (i.e., at the onset of instability when there are no oscillations):

$$(k^2 - d^2/dz^2)[(k^2 - d^2/dz^2)^3 W - \mathrm{Ra}_c\, k^2 W - T_0\, d^2W/dz^2] = 0 \tag{2.260}$$

Recall that at the onset of instability $\mathrm{Ra} = \mathrm{Ra}_c$, the critical Rayleigh number. We integrate (2.260) twice with respect to z and find that

$$[(k^2 - d^2/dz^2)^3 - \mathrm{Ra}_c\, k^2 - T_0\, d^2/dz^2]W + F = 0 \tag{2.261}$$

where $(k^2 - d^2/dz^2)F = 0$. At $z = 0$ and 1, $W = 0$ and $d^2W/dz^2 = 0$, so $F = 0$. The remaining two boundary conditions are therefore that

$$F = 0 \quad \text{at } z = 0 \text{ and } 1 \tag{2.262}$$

which means that

$$[(k^2 - d^2/dz^2)^3 - \mathrm{Ra}_c\, k^2 - T_0\, d^2/dz^2]W = 0 \tag{2.263}$$

at $z = 0$ and 1.

For the free-slip boundary conditions, then, as for the non-rotating case

$$W(z) = \sum_{n=1}^{\infty} A_n \sin(n\pi z) \tag{2.264}$$

Substituting (2.264) into the reduced version of the characteristic value equation (2.263), we find that

$$(k^2 + n^2\pi^2)^3 + T_0(n\pi)^2 - \mathrm{Ra}_c\, k^2 = 0 \tag{2.265}$$

The smallest Rayleigh number possible, the critical Rayleigh number (Ra_c), is found for the $n = 1$ mode; that is

$$\mathrm{Ra}_c = [(k^2 + \pi^2)^3 + T_0\pi^2]/k^2 \tag{2.266}$$

Recall that without rotation (2.241) for $n = 1$

$$\mathrm{Ra}_c = (k^2 + \pi^2)^3/k^2 \tag{2.267}$$

It follows that since $T_0 = (2\Omega)^2H^4/\nu^2 > 0$, the critical Rayleigh number is *increased* by rotation, so that *the effect of rotation is stabilizing*. The reader is left to ponder whether convective storms that rotate (e.g., supercells, to be discussed later) are more stable than those that do not and whether or not that rotation is therefore responsible for their relative longevity.

The wavenumber at the minimum Rayleigh number, the critical Rayleigh number, for the $n = 1$ mode, is found by differentiating (2.266) with respect to k and setting the resultant equation to zero

$$\partial\,\mathrm{Ra}_c/\partial k = 0 = [(k^2 + \pi^2)/\pi^2]^2(2k^2/\pi^2 - 1) - T_0/\pi^4 \tag{2.268}$$

We let

$$r = k^2/\pi^2 \tag{2.269}$$

so that (2.268) can be written as

$$(r + 1)^2(2r - 1) = T_0/\pi^4 \tag{2.270}$$

The LHS of (2.270) is $2r^3$+lower order terms. So, as the RHS of (2.270) $T_0/\pi^4 \to \infty$, $2r^3 \to T_0/\pi^4$. Then the first mode to become unstable for large Taylor number, the critical mode is r_c:

$$\lim_{T_0\to\infty} r_c = (T_0/2\pi^4)^{1/3} \tag{2.271}$$

From the definition of the Rayleigh number (2.222) it follows that

$$\alpha_c = \mathrm{Ra}_c\,\nu\kappa/H^4 \tag{2.272}$$

Meanwhile, from (2.266), (2.271), and (2.269), it follows that for the $n = 1$ mode when $T_0 \to \infty$,

$$\alpha_c = \left[\tfrac{3}{2}(2\pi^4)^{1/3}T_0^{2/3}\nu\kappa\right]/H^4 \tag{2.273}$$

where it is recognized that $T_0 \times \pi^2$ and $k^6 \gg (\pi^2)^3$ in (2.266) when $T_0 \to \infty$. Substituting for the definition of T_0 (2.256) in (2.273), it follows that

$$\alpha_c = \tfrac{3}{2}(2\pi^4)^{1/3}(2\Omega/H)^{4/3}\nu^{-1/3}\kappa \tag{2.274}$$

Recall that α represents the temperature gradient between the lower and upper plate. Then for a very large Taylor number, when the coefficient of viscosity increases, the critical temperature gradient decreases. *In other words, viscosity is destabilizing when rotation is very important.* This situation is in sharp contrast to the non-rotating case, when viscosity is stabilizing (cf. (2.222)), in that increasing the viscosity decreases the Rayleigh number, which reduces it from its critical value.

The student is advised to look elsewhere (Emanuel's text and the original journal articles) for more detailed and thorough analyses of what happens when there are free-slip boundary conditions and the analyses of what happens when there are no-slip boundary conditions, etc. These treatments are more complicated and not worth the time spent on them for the purposes of this text.

2.9.3 Convection in a linearly sheared atmosphere without rotation

We now consider the case of Rayleigh–Bénard convection when there is no background rotation, but there is vertical shear in the basic state. This vertical shear may be due to the thermal wind or due to surface friction or to both. Suppose, for simplicity, that vertical shear is constant and in the x-direction only, so that

$$u = (U_0/H)z + u' \tag{2.275}$$

The unscaled equations, with the prime notation dropped for the perturbation quantities, are as follows:

$$[\partial/\partial t + (U_0/H)z\,\partial/\partial x - \nu\nabla^2]u = -1/\rho\,\partial p/\partial x \tag{2.276}$$

$$[\partial/\partial t + (U_0/H)z\,\partial/\partial x - \nu\nabla^2]v = -1/\rho\,\partial p/\partial y \tag{2.277}$$

$$[\partial/\partial t + (U_0/H)z\,\partial/\partial x - \nu\nabla^2]w = -1/\rho\,\partial p/\partial z + B \tag{2.278}$$

$$[\partial/\partial t + (U_0/H)z\,\partial/\partial x - \kappa\nabla^2]B = \alpha w \tag{2.279}$$

$$\partial u/\partial x + \partial v/\partial y + \partial w/\partial z = 0 \tag{2.280}$$

We scale each space variable as in (2.217) and time as in (2.219); wind variables are scaled as the space scale divided by the time scale (ν/H). Pressure is scaled such that the pressure gradient term is on the order of the inertial acceleration. Buoyancy is scaled using its definition (2.211) and multiplying by the dimensionless Prandtl number (2.221). The scaled equations are given by

$$(\partial/\partial t + \mathrm{Re}\,z\,\partial/\partial x - \nabla^2)u = -\partial p/\partial x - \mathrm{Re}\,w \tag{2.281}$$

$$(\partial/\partial t + \mathrm{Re}\,z\,\partial/\partial x - \nabla^2)v = -\partial p/\partial y \tag{2.282}$$

$$(\partial/\partial t + \mathrm{Re}\,z\,\partial/\partial x - \nabla^2)w = -\partial p/\partial z + \mathrm{Ra}\,B \tag{2.283}$$

$$\sigma(\partial/\partial t + \mathrm{Re}\,z\,\partial/\partial x - \nabla^2)B = w \tag{2.284}$$

$$\partial u/\partial x + \partial v/\partial y + \partial w/\partial z = 0 \tag{2.285}$$

where $\mathrm{Re} = U_0 H/\nu$ and all the dependent variables are of order one.

There have been numerous (mathematically complicated) analytical studies of the stability of the above equations. Asai's seminal work around 1970 demonstrated that when the mean state is weakly stable or relatively weakly statically unstable (e.g., when strongly heated from below) the most unstable modes (when the vertical shear is strong) take the form of rolls (in nature frequently termed *horizontal convective rolls*) oriented *perpendicular* to the mean vertical shear vector. When the mean state is very statically unstable (e.g., when very strongly heated from below), the most unstable modes take the form of rolls oriented *along* the mean vertical shear vector and the growth rates are identical to those when there is no vertical shear. Consider the above equations (2.281)–(2.285) for the case when $\partial/\partial x = 0$ (i.e., when rolls are oriented along the vertical shear vector). Then the equations (2.282)–(2.285) do not include u and only equation (2.281) still contains u. The effects of vertical shear are contained in (2.281) only in the term $\text{Re}\, w$, which represents the advection of vertical shear by the perturbation vertical velocity. The variables v, w, p, and B may be solved by considering (2.282)–(2.285) only. Thus, once w and p are solved for, u may then be calculated from (2.281); *the mean shear therefore has no effect on v and w, which make up the roll circulations along the vertical shear vector.*

Asai in the early 1970s extended his analyses to include basic states in which the vertical shear vector turns with height. Wan Shu Wu, Doug Lilly, and Bob Kerr in the early 1990s numerically examined the effects of some idealized basic states in which the vertical shear vector is unidirectional and in which the vertical shear vector turns with height. The results of the latter were that the disturbance was most likely to be helical when the shear vector curved with height. The student is referred to the original journal articles for the details.

In the real atmosphere, we frequently observe cloud streets over land during the day (Figure 2.17), which we interpret as visual manifestations of boundary

Figure 2.17. Cloud streets over land (over southern Illinois), as viewed from an aircraft, during the daytime hours on July 2, 1986 (photograph by the author).

layer roll circulations in a moist atmosphere, called "horizontal convective rolls" (HCRs). We also see them over the ocean. Observed linear patterns of seagulls soaring have been also been interpreted as evidence of roll circulations. Rolls have been inferred from measurements made aboard aircraft, in situ instruments mounted on towers, by Doppler radars and lidars (Figure 2.18), when the underlying land is being heated (rolls have also been measured when the stratification is stable; these rolls owe their existence to mechanisms involving dynamic instabilities). Insects and aerosols are accumulated in regions of convergence underneath the rising branch of rolls, so they become detectable by radar in "clear air". It is thought that it is likely that all rolls in nature may owe part of their existence to thermal instability and part to dynamic instability. Thus, the theories of Rayleigh–Bénard convection in a sheared environment may have some qualitative relevance to convection in the real atmosphere. Such roll circulations when tilted by gradients in vertical velocity (e.g., along the edge of updrafts and downdrafts) may be sources of vertical vorticity along boundaries and in some convective clouds initiated along the boundaries (Figure 2.19), and when they intersect boundaries along which rolls of another nature (e.g., frontogenetical or solenoidal circulations) are present they may influence where moist convection does or does not get triggered (Figure 2.20).

2.10 RESPONSE OF A BOUSSINESQ ATMOSPHERE TO HEAT SOURCES

In convective storms, latent heat is released by water molecules when condensation occurs and the heat is absorbed by the atmosphere. Latent heat is absorbed by water molecules and the atmosphere is cooled as precipitation evaporates, sublimates, or frozen precipitation melts. When latent heat is exchanged, buoyancy changes. In addition, when water substances change form, the loading term in (2.23) changes as the mixing ratios of various water substances change. Precipitation falling to the ground is an irreversible process: precipitation does not rise up out of the ground to become cloud droplets again (unless someone puts a powerful vacuum cleaner to the ground ...)! Furthermore, latent heat release is a highly nonlinear process: it is completely off until phase changes occur, which do so suddenly. The consequences of cooling will be discussed in the next chapter. Here, we look briefly at how the atmosphere responds to a sudden turning-on of the precipitation process and the accompanying sudden latent heat release.

Analysis of the effects of sudden onset of a heat source was carried out by Chris Bretherton and Piotr Smolarkiewicz in a 1989 journal article and reproduced in Emanuel's text. The atmosphere is assumed to be hydrostatic for both simplicity and to isolate the effects of sudden heating; it is also assumed to be resting, inviscid, and non-rotating. The equations of motion (cf. (2.203)–(2.205)), the thermodynamic equation (cf. (2.161)), and the continuity equation (2.207) are

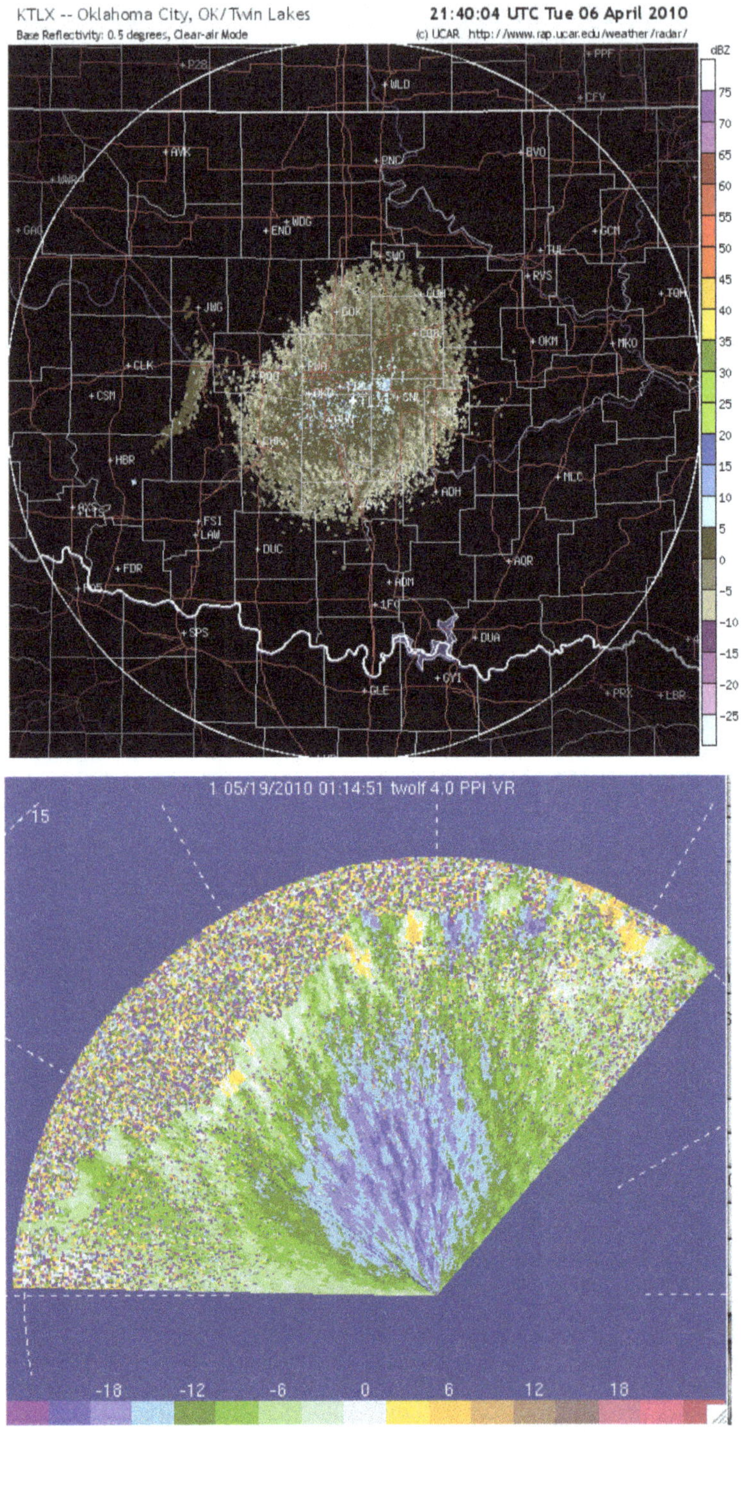
KTLX -- Oklahoma City, OK/ Twin Lakes
21:40:04 UTC Tue 06 April 2010
Base Reflectivity: 0.5 degrees, Clear-air Mode
(c) UCAR http://www.rap.ucar.edu/weather/radar/
dBZ
75
70
65
60
55
50
45
40
35
30
25
20
15
10
5
0
-5
-10
-15
-20
-25
1 05/19/2010 01:14:51 twolf 4.0 PPI VR
15
-18
-12
-6
0
6
12
18

◀ **Figure 2.18.** Evidence of clear-air boundary rolls: (top) as depicted by WSR-88D radar reflectivity at Oklahoma City, OK on April 6, 2010. The bands of clear air reflectivity are oriented in the north-northwest to south-southeast direction; the bands of enhanced clear air reflectivity are regions of enhanced insect concentration where the convergence at the bottom of the ascending branch of a roll is located. (Bottom) As depicted by a mobile Doppler lidar (TWOLF) in the Texas Panhandle on May 18, 2010 during VORTEX2. Doppler velocity in $m\,s^{-1}$; streaks in Doppler velocity are suggestive of rolls oriented in the north-northwest to south-southeast direction. The rolls are spaced several hundred meters apart. Streaks in Doppler velocity are indicative of where vertical transports of momentum increase the wind speed as a result of vertical shear in the environment (courtesy of Jana Houser).

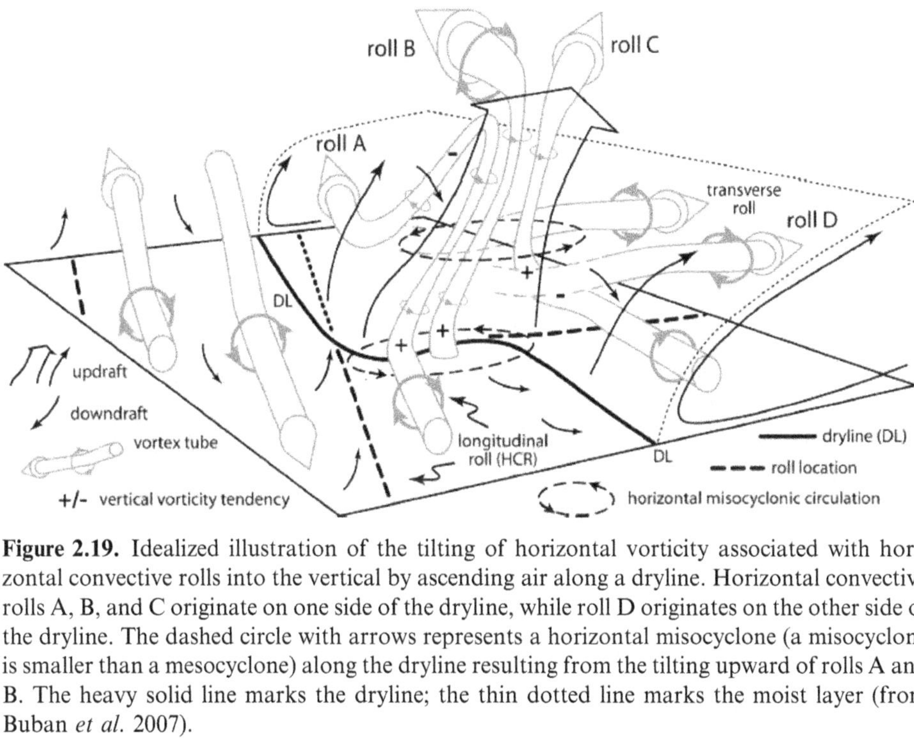

Figure 2.19. Idealized illustration of the tilting of horizontal vorticity associated with horizontal convective rolls into the vertical by ascending air along a dryline. Horizontal convective rolls A, B, and C originate on one side of the dryline, while roll D originates on the other side of the dryline. The dashed circle with arrows represents a horizontal misocyclone (a misocyclone is smaller than a mesocyclone) along the dryline resulting from the tilting upward of rolls A and B. The heavy solid line marks the dryline; the thin dotted line marks the moist layer (from Buban *et al.* 2007).

as follows:

$$\partial u'/\partial t = -1/\rho\,\partial p'/\partial x \tag{2.286}$$

$$\partial v'/\partial t = -1/\rho\,\partial p'/\partial y \tag{2.287}$$

$$0 = B - 1/\rho\,\partial p'/\partial z \tag{2.288}$$

$$\partial B/\partial t + N^2 w' = Q \tag{2.289}$$

$$\partial u'/\partial x + \partial v'/\partial y + \partial w'/\partial z = 0 \tag{2.290}$$

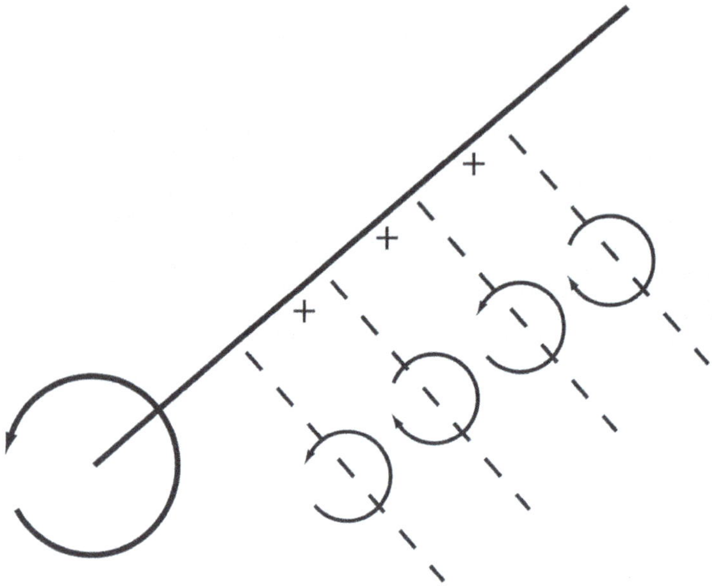

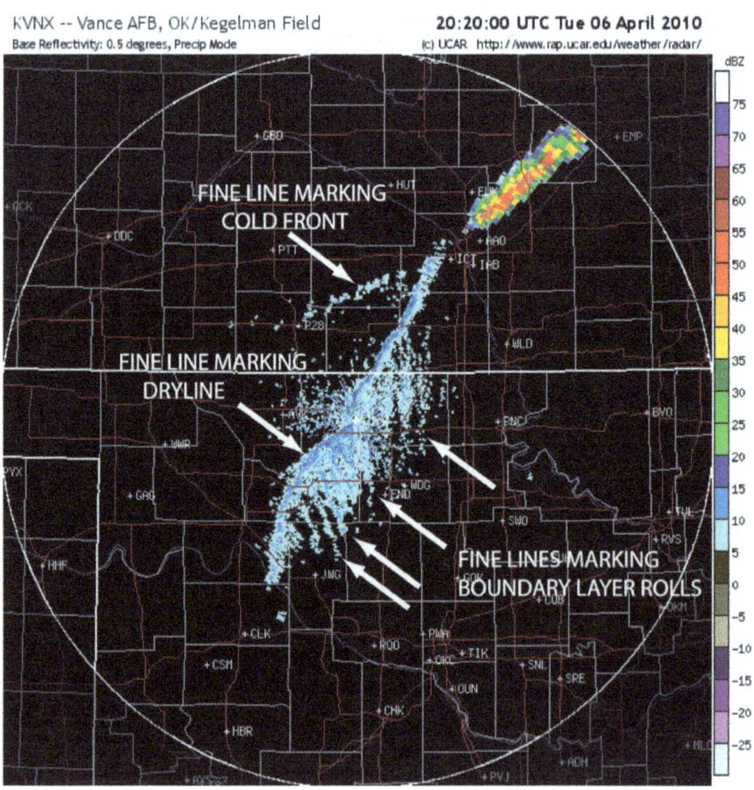
KVNX -- Vance AFB, OK/Kegelman Field
20:20:00 UTC Tue 06 April 2010
Base Reflectivity: 0.5 degrees, Precip Mode
(c) UCAR http://www.rap.ucar.edu/weather/radar/
dBZ
FINE LINE MARKING
COLD FRONT
FINE LINE MARKING
DRYLINE
FINE LINES MARKING
BOUNDARY LAYER ROLLS

◀ **Figure 2.20.** Illustration of where convection might be preferentially initiated along a surface boundary. (Top) Surface boundary denoted by solid line and sense of the vertical circulation along the boundary denoted by the large curved arrow; axes of boundary layer rolls indicated by the dashed lines, sense of vertical circulations associated with the rolls denoted by the smaller curved arrows, and places where upward motion along the surface boundary is greatest are indicated by plus signs; (bottom) radar reflectivity from the WSR-88D radar near Enid, OK on April 6, 2010 showing boundary layer roll fine lines (where convergence and upward motion are greatest) intersecting a dryline radar fine line (where convergence and upward motion are greatest).

where ρ is the mean density; Q is the heat source; and the Brunt–Väisälä frequency (N) is assumed to be constant for simplicity. All independent variables (u', v', w', and p') are eliminated except for buoyancy B, so that

$$\partial^2/\partial t^2\, \partial^2 B/\partial z^2 + N^2(\partial^2/\partial x^2 + \partial^2/\partial y^2)B = \partial/\partial t\, \partial^2 Q/\partial z^2 \tag{2.291}$$

We impose the following idealized diabatic heating function to represent condensation heating that is local, is turned on suddenly, and has a maximum at mid-levels in the atmosphere (the condensation rate varies as vertical velocity, which is greatest aloft, but also as the absolute amount of available water vapor, which is greatest at low levels), but is zero at the ground and at the tropopause ($z = D$):

$$Q = Q_0\, \delta(x) \sin(mz) H(t) \tag{2.292}$$

where $\delta(x)$ is the Dirac delta function (infinite at $x = 0$, of infinitesimal width, but whose integral is unity); $H(t)$ is a unit step function (0 before $t = 0$, and 1 at and after $t = 0$); and m is an integral multiple of π/D, so that Q is zero at $z = 0$ and D. (Students of electrical engineering make good use of the following properties of δ and H: the integral of the Dirac delta function up to an arbitrary time t is the unit step function and the derivative of the unit step function is the Dirac delta function.) This heating function, for simplicity, is a slab; there are no variations in y. It is highly idealized because in real life we cannot simply impose a heating function that is flow independent. Heat sources are intimately related to the flow.

Solutions are of the form

$$B(x, z, t) = b(x, t) \sin mz \tag{2.293}$$

so that substituting (2.293) and (2.292) into (2.291) we get the following equation for b:

$$\partial^2 b/\partial t^2 - N^2/m^2\, \partial^2 b/\partial x^2 = Q_0\, \delta(x)\, \delta(t) \tag{2.294}$$

The solution to (2.294) is

$$b = Q_0/c_m H(c_m t - |x|) \tag{2.295}$$

which is a jump in buoyancy traveling in both the $\pm x$-directions at the speed c_m, the phase speed of internal gravity waves; that is

$$c_m = N/m \tag{2.296}$$

The perturbation wind field responds as divergence/convergence supporting the vertical motions induced by buoyancy.

When rotation is included in (2.286) and (2.287) and a steady (rather than sudden) heat source is applied, buoyancy and wind field perturbations decay as the Rossby radius of deformation, as is usually demonstrated in synoptic meteorology courses.

2.11 SIMILARITY OF FLUID DYNAMICS EQUATIONS TO ELECTROMAGNETIC EQUATIONS

Since the equations that describe the dynamics of electric and magnetic fields involve the curl and divergence of both fields, it has been suggested that fluid dynamical equations, which involve the curl and divergence of the wind field as well as their relation to other quantities, are similar in some sense to the equations given by the theory of electric and magnetic fields. Owing to this similarity, Haralambos Marmanis at Brown University has proposed that some insight into the nonlinear behavior of fluids may be realized by carrying over results from electromagnetic theory to fluids.

For example, since the time rate of change of the magnetic field is related to the curl of the electric field (from Maxwell's equations) and the three-dimensional vorticity vector is related to the curl of the wind field, insights from magnetic induction may be carried over to the "induction" of the vorticity field by the wind field: the vorticity vector is analogous to the time rate of change of the magnetic field. Since the divergence of the magnetic field is zero and the divergence of the curl of the wind field (vorticity) is zero, the magnetic field is analogous to vorticity. Greg Tripoli at the University of Wisconsin Madison and Marcus Büker at Western Illinois University have proposed that studies be undertaken to explain the behavior of vorticity in some convective storms using this electromagnetic analogy. Studies in what is named "metafluid dynamics", at the time of this writing, are in their infancy and it remains to be seen if this new arena of study will bear fruit in our understanding of convective storm dynamics.

2.12 GENERAL MONOGRAPHS AND BOOKS

The reader is referred to p. 24 for a list of relevant general monographs and books.

2.13 REFERENCES AND BIBLIOGRAPHY

Abdullah, A. J. (1966) The "musical" sound emitted by a tornado. *Mon. Wea. Rev.*, **94**, 213–220.

Arnold, R. T., H. E. Bass, and L. N. Bolen (1976) Acoustic spectral analysis of three tornadoes. *J. Acoust. Soc. Amer.*, **60**, 584–593.

Asai, T. (1964) Cumulus convection in the atmosphere with vertical wind shear: Numerical experiment. *J. Meteor. Soc. Japan*, **42**, 245–259.

Asai, T. (1970a) Three-dimensional features of thermal convection in a plane Couette flow. *J. Meteor. Soc. Japan*, **48**, 18–29.

Asai, T. (1970b) Stability of a plane parallel flow with variable vertical shear and unstable stratification. *J. Meteor. Soc. Japan*, **48**, 129–138.

Batchelor, G. K. (1953) The condition for dynamical similarity of motions of a frictionless perfect-gas atmosphere. *Quart. J. Roy. Meteor. Soc.*, **79**, 224–235.

Batchelor, G. (1954) Heat convection and buoyancy effects in fluids. *Quart. J. Roy. Meteor. Soc.*, **80**, 339–358.

Batchelor, G. K. (1970) *An Introduction to Fluid Dynamics*, Cambridge University Press, Cambridge, U. K., 615 pp.

Bedard, A. J., Jr. (2005) Low-frequency atmospheric acoustic energy associated with vortices produced by thunderstorms. *Mon. Wea. Rev.*, **133**, 241–263.

Bénard, H. (1900) Les tourbillons cellulaires dans une nappe liquide. *Revue générale des Sciences pures et appliqués*, **11**, 1261–1271, 1309–1328 [in French].

Bluestein, H. B. (1992) *Synoptic-Dynamic Meteorology in Midlatitudes, Volume I: Principles of Kinematics and Dynamics*, Oxford University Press, New York, 431 pp.

Bluestein, H. B. (1993) *Synoptic-Dynamic Meteorology in Midlatitudes, Volume II: Observations and Theory of Weather Systems*, Oxford University Press, New York, 608 pp.

Boussinesq, J. (1903) *Théorie analytique de la chaleur*, Gauthier-Villars, Paris, Vol. 2, p. 172 [in French].

Bretherton, C. S. and P. K. Smolarkiewicz (1989) Gravity waves, compensating subsidence and detrainment around cumulus clouds. *J. Atmos. Sci.*, **46**, 740–759.

Buban, M. S., C. L. Ziegler, E. N. Rasmussen, and Y. P. Richardson (2007) The dryline on 22 May 2002 during IHOP: Ground-radar and in situ data analyses of the dryline and boundary layer evolution. *Mon. Wea. Rev.*, **135**, 2473–2505.

Chandrasekhar, S. (1961) *Hydrodynamics and Hydromagnetic Stability*, Oxford University Press, New York, 654 pp.

Charney, J. G. and Y. Ogura (1960) A numerical model of thermal convection in the atmosphere. *Proceedings International Symposium on Numerical Weather Prediction*, Japan Meteorological Society, Tokyo, pp. 431–452.

Doswell, C. A., III and P. M. Markowski (2004) Is buoyancy a relative quantity? *Mon. Wea. Rev.*, **132**, 855–863.

Emanuel, K. A. (1994) *Atmospheric Convection*, Oxford University Press, New York, Chapters 1–3.

Ferrier, B. S. (1994) A two-moment multiple-phase four-class bulk ice scheme, Part I: Description. *J. Atmos. Sci.*, **51**, 249–280.

Gal-Chen, T. (1978) A method for the intialization of the anelastic equations: Implications for matching models with observations. *Mon. Wea. Rev.*, **106**, 587–606.

Hane, C. E. and P. S. Ray (1985) Pressure and buoyancy fields derived from Doppler radar data in a tornadic thunderstorm. *J. Atmos. Sci.*, **42**, 18–35.

Hane, C. E., R. B. Wilhelmson, and T. Gal-Chen (1981) Retrieval of thermodynamic variables within deep convective clouds: Experiments in three dimensions. *Mon. Wea. Rev.*, **109**, 564–576.

Holton, J. R. and G. J. Hakim (2012) *An Introduction to Dynamic Meteorology*, Fifth Edition. Elsevier Academic Press, Burlington, MA, 552 pp.

Houze, R. A., Jr. (1993) *Cloud Dynamics*, Academic Press, San Diego, CA, 573 pp.

Kessler, E. (1969) *On the Distribution and Continuity of Water Substance in Atmospheric Circulations*, AMS Monograph 10(32), American Meteorological Society, Boston, 84 pp.

Lorenz, E. N. (1963) Deterministic nonperiodic flow. *J. Atmos. Sci.*, **20**, 130–141.

Malkus, J. S. and R. S. Scorer (1955) The erosion of cumulus towers. *J. Meteor.*, **12**, 43–57.

Malkus, J. S. and G. Witt (1959) The evolution of a convective element: A numerical calculation. *The Atmosphere and the Sea in Motion*, Rockefeller Institute Press, New York, pp. 425–439.

Malkus, W. V. R. and G. Veronis (1958) Finite amplitude cellular convection. *J. Fluid Mech.*, **4**, 225–260.

Marmanis, H. (1998) Analogy between the Navier–Stokes equations and Maxwell's equations: Application to turbulence. *Physics of Fluids*, **10**, 1428–1437.

Milbrandt, J. A. and M. K. Yau (2005) A multimoment bulk microphysics parameterization, Part II: A proposed three-moment closure and scheme description. *J. Atmos. Sci.*, **62**, 3065–3081.

Morton, B. R., G. I. Taylor, and J. S. Turner (1956) Turbulent gravitational convection from maintained and instantaneous sources. *Proc. Roy. Soc. A*, **234**, 1–23.

Oberbeck, A. (1879) Ueber die Wärmeleitung der Flüssigkeiten bei Berücksichtigung der Strömungen infolge von Temperaturdifferenzen. *Annalen der Physik*, **243**, 271–292 [in German].

Ogura, Y. and N. A. Phillips (1962) Scale analysis of deep and shallow convection in the atmosphere. *J. Atmos. Sci.*, **19**, 173–179.

Parsons, D. B. and R. A. Kropfli (1990) Dynamics and fine structure of a microburst. *J. Atmos. Sci.*, **47**, 1674–1692.

Phillips, N. A. (1970) 19.61 Course Notes on Convection (unpublished), MIT Department of Meteorology, Cambridge, MA.

Pruppacher, H. R. and J. D. Klett (1997) *Microphysics of Clouds and Precipitation*, Second Edition, Kluwer Academic Publishers, Dordrecht, The Netherlands, 954 pp.

Rayleigh, J. W. S. (Lord) (1916) On convective currents in a horizontal layer of fluid when the higher temperature is on the underside. *Phil. Mag.*, **32**, 529–546.

Ryzhkov, A., M. Pinsky, A. Pokrovsky, and A. Khain (2011) Polarimetric radar observation operator for a cloud model with spectral microphysics. *J. Clim. Appl. Meteor. Climatol.*, **50**, 873–894.

Saltzman, B. (1962) Finite amplitude free convection as an initial value problem, I. *J. Atmos. Sci.*, **19**, 329–341.

Schecter, D. A. and M. E. Nicholls (2010) Numerical simulation of infrasound generated by severe storms, 13A.2 (no reprint available). *25th Conference on Severe Local Storms*, American Meteorological Society, Denver, CO.

Schecter, D. A., M. E. Nicholls, J. Persing, A. J. Bedard Jr., and R. A. Pielke, Sr. (2008) Infrasound emitted by tornado-like vortices: Basic theory and a numerical comparison to the acoustic radiation of a single-cell thunderstorm. *J. Atmos. Sci.*, **65**, 685–713.

Scorer, R. S. (1957) Experiments on convection of isolated masses of buoyant fluid. *J. Fluid Mech.*, **2**, 583–594.

Squires, P. and J. Turner (1962) An entraining jet model for cumulo-nimbus updraughts. *Tellus*, **14**, 422–434.

Stommel, H. (1947) Entrainment of air into a cumulus cloud. *J. Meteor.*, **4**, 91–94.

Straka, J. M. (2009) *Cloud and Precipitation Microphysics*, Cambridge University Press, Cambridge, U. K., 406 pp.

Straka, J. M., D. S. Zrnic, and A. V. Ryzhkov (2000) Bulk hydrometeor classification and quantification using polarimetric radar data: Synthesis of relations. *J. Appl. Meteor.*, **39**, 1341–1372.

Turner, J. S. (1962) The "starting plume" in neutral surroundings. *J. Fluid Mech.*, **13**, 356–368.

Walko, R. L., W. R. Cotton, M. P. Meyers, and J. Y. Harrington (1995) New RAMS cloud microphysics, Part I: The one-moment scheme. *J. Atmos. Sci.*, **38**, 29–62.

Wilson, J. W. and D. Reum (1988) The flare echo: Reflectivity and velocity signature. *J. Atmos. Oceanic Technol.*, **5**, 197–205.

Woodward, B. (1959) The motion in and around isolated thermals. *Quart. J. Roy. Meteor. Soc.*, **85**, 144–151.

Wu, W.-S., D. K. Lilly, and R. M. Kerr (1992) Helicity and thermal convection with shear. *J. Atmos. Sci.*, **49**, 1800–1809.

Zrnic, D. S. (1987) Three-body scattering produces precipitation signature of special diagnostic value. *Radio Sci.*, **22**, 76–86.

3

Ordinary-cell convective storms

"I am the daughter of Earth and Water
And the nursling of the Sky;
  I pass through the pores of the ocean and shores
I change, but I cannot die.
For after the rain with never a strain
The pavilion of Heaven is Gale,
  And the winds and sunbeams with their convex gleams
Build up the blue dome of air,
While I gently laugh at my own cenotaph,
And out of the caverns of rain,
Like a child from the womb, like a ghost from the tomb
I arise and unbuild again."
  Percy Bysshe Shelley—*The Cloud*

In the previous chapter, we considered the dynamics of (mostly) dry convection and found that buoyancy and dynamic pressure gradient forces are of fundamental importance. In addition, vertical shear was shown to be able to organize shallow convection into lines under some circumstances. We considered the inclusion of water substance only in how it affects buoyancy (i.e., adding or increasing liquid water reduces buoyancy and adding or increasing water vapor reduces buoyancy). We will now consider the effects of water substance on convection more completely. We refer to "moist convection" as small-scale convective currents consisting of air parcels that contain water substance in potentially all of its phases.

The dynamics of moist convection is even more complicated than the already complicated dynamics of dry convection. Not only must we be concerned with buoyancy, the turbulent entrainment of environmental air, and the effects of airflow on vertical pressure gradient forces, but we must also contend with the formation of cloud particles and precipitation, which are highly nonlinear processes: If air is unsaturated initially and is cooled adiabatically, saturation is reached suddenly and latent heat is "turned on" like a switch. In other words, the properties of the air change dramatically after only a small change in the environmental temperature or pressure or both. From the standpoint of latent heat release from condensation, there is *no* heating and then suddenly there *is* heating when air becomes saturated. As shown at the end of the last chapter, heating triggers gravity waves, which "communicate" the effects of heating away from its source. If that was not complicated enough, the production of cloud droplets and ice crystals, and the conversion of cloud droplets and ice crystals into precipitation is itself highly complicated, being dependent to a great extent not only on the environmental thermodynamic conditions, but also on cloud microphysics, the dynamics and thermodynamics of individual cloud particles themselves, an important topic not treated in this text. Changes in water phase such as ice to liquid (melting) or liquid water to vapor (evaporation) cause cooling rather than heating.

In this chapter we will discuss what happens when a convective cloud grows enough that precipitation forms and falls out of its parent cloud (i.e., that the cloud grows into a cumulonimbus—remember, "nimbus" refers to precipitation). To simplify matters, we will first neglect the contributions from the terms involving the wind field in (2.62), so that dynamic perturbation pressure gradient forces ($\nabla p'_d$) are ignored for simplicity, for the time being. An important consequence of precipitation is that a downdraft can form and that air may be evaporatively cooled if it is unsaturated and precipitation falls into it. Much of this chapter will be concerned with how downdrafts and "cold pools" are formed and how they affect the subsequent evolution of convective storms.

3.1 OBSERVATIONS AND DYNAMICS

3.1.1 Conditional instability and the initiation of deep convection

The boundary layer is the region of the atmosphere where the properties of the air are significantly influenced by the ground surface (e.g., through frictional drag at the ground reducing the air speed aloft, upward turbulent heat transport from a warmer surface, or evaporation of water from the surface into the air above). Unsaturated boundary layer air can become saturated when water is evaporated into it or, more commonly, when it is adiabatically cooled. Such cooling may occur when the air is lifted to saturation, and the altitude at which this happens is called the "lifting condensation level" (LCL). Lift may be dynamic lift along a mesoscale boundary, quasi-geostrophic lift, or forced upslope (or a combination of all processes). Alternatively, saturation may occur if the air is heated (by turbulent transport of heat from below by a *deus ex machina* in the form of small-scale eddies) so that the lapse rate becomes superadiabatic and dry convection mixes moist air upward to the saturation level, which is called the "convective condensa-

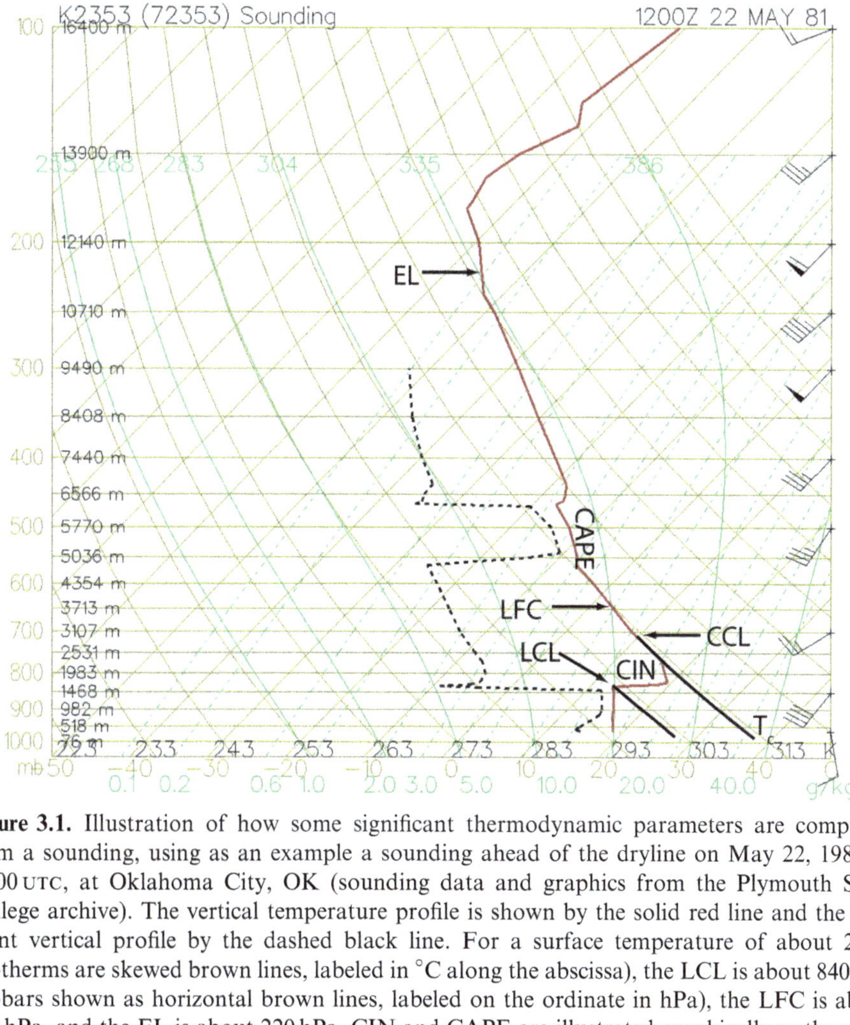

Figure 3.1. Illustration of how some significant thermodynamic parameters are computed from a sounding, using as an example a sounding ahead of the dryline on May 22, 1981 at 12:00 UTC, at Oklahoma City, OK (sounding data and graphics from the Plymouth State College archive). The vertical temperature profile is shown by the solid red line and the dew point vertical profile by the dashed black line. For a surface temperature of about 27°C (isotherms are skewed brown lines, labeled in °C along the abscissa), the LCL is about 840 hPa (isobars shown as horizontal brown lines, labeled on the ordinate in hPa), the LFC is about 650 hPa, and the EL is about 220 hPa. CIN and CAPE are illustrated graphically as the areas between the temperature and the moist adiabatic for the equivalent potential temperature of 335 K (curved green line). The convective temperature (T_c) is about 38°C for a mixed layer water vapor mixing ratio (dashed green lines) of about 11 g kg^{-1}; the CCL is about 700 hPa. Later, on this day, there were tornadic supercells to the west.

tion level" (CCL). When air reaches its LCL or CCL, cloud particles may form (we will ignore the complexities of microphysics in this discussion).

The amount of potential energy that must be overcome to lift air to its LCL is called "convective inhibition" (CIN) (Figure 3.1) and the level at which the air first becomes buoyant is called the "level of free convection" (LFC) (Figure 3.1).

If cloud particles do form, latent heat is released and, when the environmental lapse rate is greater than that of the moist adiabatic, the temperature excess of the air in the cloud (parcel) over its environment (and consequently its buoyancy) increases with height (Figure 3.1). We will consider for simplicity what happens when the aspect ratio of the cloud parcel is large (i.e., the parcel is narrow), so that we can neglect the counteracting downward perturbation pressure gradient force due to buoyancy (cf. (2.70) and (2.75)) and, consequently, buoyancy is the only vertical force.

When the layer of air in which the environmental lapse rate is greater than moist adiabatic the atmosphere is unstable because a parcel's upward vertical motion increases with time/height if it is given a nudge upward. The atmosphere is said to be in a state of "conditional instability": the air continues to accelerate upward in the absence of any more lifting if given an initial push upward. If the air were not saturated, it would not be unstable, whence the adjective "conditional" is used. If the air were unsaturated, it would cool when lifted at the more rapid dry adiabatic lapse rate and the air would be stable. The same analysis and results apply if the air parcel is pushed downward: if the lapse rate in the environment is greater than moist adiabatic, downward motion increases with time/decreasing height if it is given a nudge downward.

While it is a challenge to forecast exactly where and when the LCL or CCL may be reached and convection initiated, some convective storms are triggered in the exact same place with regularity for part of the year; some of these storms have been given names. The onset of these storms is connected to topography/orography and related solenoidal vertical circulations. The most famous is "Hector", which is a complex of storms that form over the Tiwi Islands off the coast of northern Australia. Owing to a high tropopause and strong updrafts, Hector is one of Earth's tallest convective storms, extending up to 20 km. In the case of Hector, it is the sea breeze circulation from several coasts that is in part responsible for its intensity and regularity.

The boundary layer is not the only source of moist air that may be lifted to its LCL or LFC. Air that originates above the boundary layer may also be lifted to its LCL and beyond to its LFC. Such convection that develops is referred to as "elevated" convection (Figure 3.2). Elevated convection may develop as a result of quasi-geostrophic, synoptic-scale ascent in a conditionally unstable environment, or as a result of mesoscale ascent in a conditionally unstable environment. Prominent cases of the latter include lift at and just behind a frontal zone that tilts with height toward the cold air (Figure 3.3), lift over an outflow boundary (Figure 3.3), and orographic lift (Figure 3.3) as air is forced upslope when there is enough kinetic energy for the air to be lifted a sufficient distance to reach the LFC in the face of initial stability. Lift may also occur along the rising branch of boundary layer rolls or the rising branch of coherent gravity waves.

3.1.2 Entrainment and convective initiation

In the real (non-idealized) world, as air rises environmental air is "entrained" through turbulent eddies into the cloud, from both the sides and/or the top and/or bottom (Figure 3.4): try to visualize the leading edge and the sides of a cumuliform cloud curling around and trapping pockets of unsaturated air inside the main body of the cloud. In 1947, Henry Stommel published a classic paper in which he suggested that entrainment into clouds occurs and described a method for calculating its effects. If the air outside the buoyant cloud is cooler or drier than it is inside the cloud (by definition the air outside a cloud must be drier), then the cloud's buoyancy is reduced.

Consider the vertical mass flux in a steady-state plume:

$$M = w\pi R^2 \tag{3.1}$$

where R is the radius of the plume; and w is the mean updraft velocity. It is seen from (2.189) that the fractional change of mass flux with height is

$$(1/M)\, dM/dz = [1/(wR^2)]2R\alpha w = 2\alpha/R \tag{3.2}$$

So, the entrainment rate, which is defined by the fractional rate of change of vertical mass flux ($1/M\, DM/Dt$), is inversely proportional to the width of the buoyant air parcel in a steady-state plume.

In an entraining thermal bubble, entrainment is given by the fractional rate of increase in volume V as the thermal bubble rises. So, from (2.200) it follows that

$$1/M\, dM/dz = (1/w)(1/V)\, DV/Dt = (1/w)(4\pi R^2 \alpha w)/(\tfrac{4}{3}\pi R^3) = 3\alpha/R \tag{3.3}$$

This entrainment rate for a thermal is also inversely proportional to the width of the buoyant air parcel. Wide air parcels are protected from the potentially deleterious effects of a cooler environment because mixing at the outer edges is felt less and less the farther inward you go; narrow parcels are not protected, because mixing at the outer edges easily reaches the center of the parcel. At the top of the cloud, where vertical velocity decreases to zero, the parcel spreads out into an anvil (Figure 3.5) as an artificial wall of sorts (a wall of non-buoyant or negatively buoyant air) prevents any air from rising any further, so that air is forced to spread laterally as a consequence of continuity. This wall may be at the tropopause, or below at stable layers.

The entrainment rate ($1/M\, dM/dz$) is a function of vertical velocity (cf. (3.1)). For the plume and thermal models, however, vertical velocity does not explicitly appear in the expression for the entrainment rate because the vertical derivative of the mass flux is proportional to vertical velocity. In general (i.e., not in idealized, steady-state plumes or thermals), however, the faster an air parcel moves upward, the less time it has to be diluted by environmental air. So, the lapse rate of the cloud-bearing layer can have an effect on whether or not a cloud can be initiated: if the rate of increase in buoyancy as an air parcel ascends upward is greater than the rate of reduction in buoyancy owing to the entrainment of environmental air, then the cloud has a good chance of reaching upper levels in the troposphere if there are no intervening stable layers. On the other hand, if the rate of increase in buoyancy is less than the reduction of buoyancy owing to entrainment, then the cloud will more likely not reach the upper troposphere if there are no intervening stable layers. The process of formation of a convective cloud such that it is able to

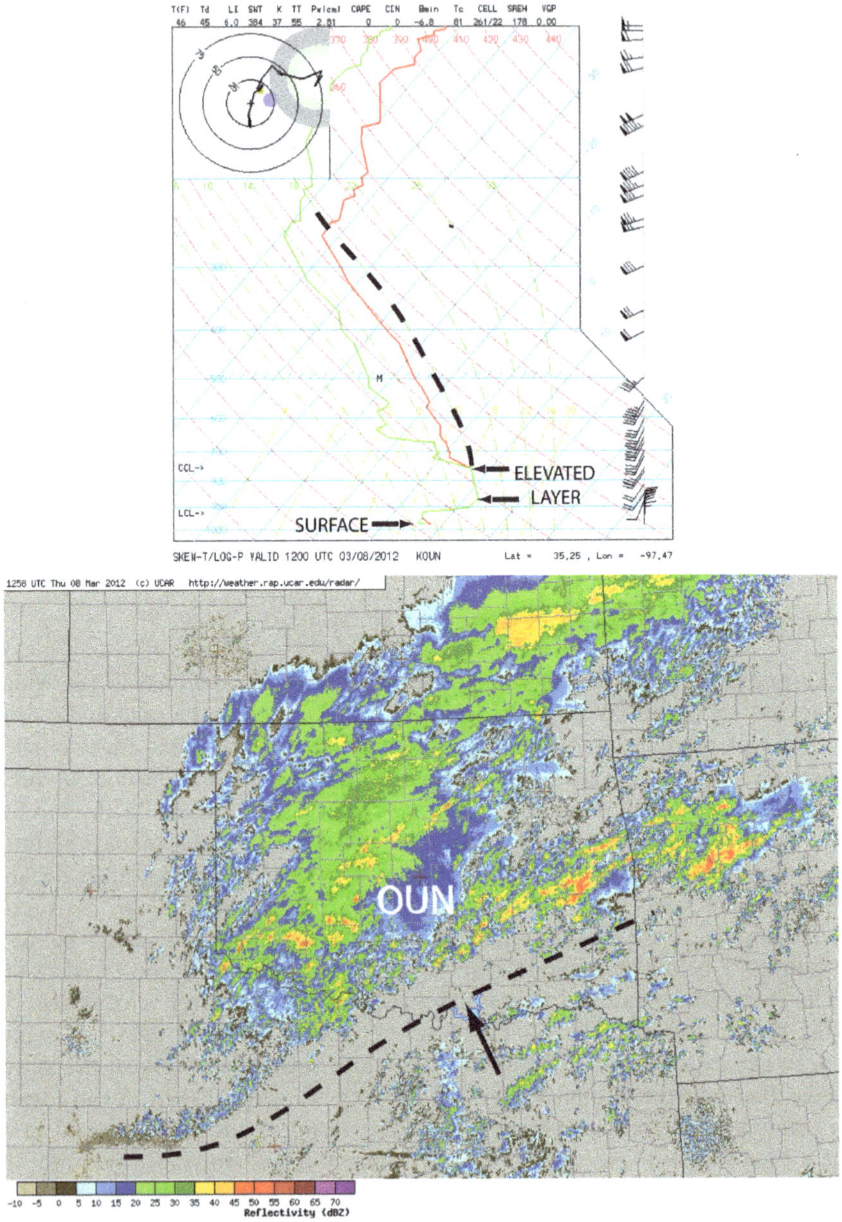

Figure 3.2a. Elevated convection. (Top) Sounding at Norman, OK (OUN) at 12:00 UTC on March 8, 2012; elevated thunderstorms as depicted by WSR-88D composite radar reflectivity in dBZ (bottom) occurring north of a cold front (dashed line at the bottom) as air flowing from the southeast (arrow in the bottom panel) rises over the frontal zone (stable layer between 920 and 870 hPa. In the top panel the elevated LCL is indicated as a saturated layer between 880 and 750 hPa. Surface parcels lifted will not attain an LCL. Above the elevated LCL, air parcels follow the moist adiabat (dashed line), where CAPE is realized.

Figure 3.2b. (Top) Altocumulus castellanus over Norman, OK on September 28, 1977 and (bottom) July 26, 1978. The bases of the clouds are flat and above the boundary layer. The clouds are also aligned in streets, possibly in response to lifting by bores (photographs by the author).

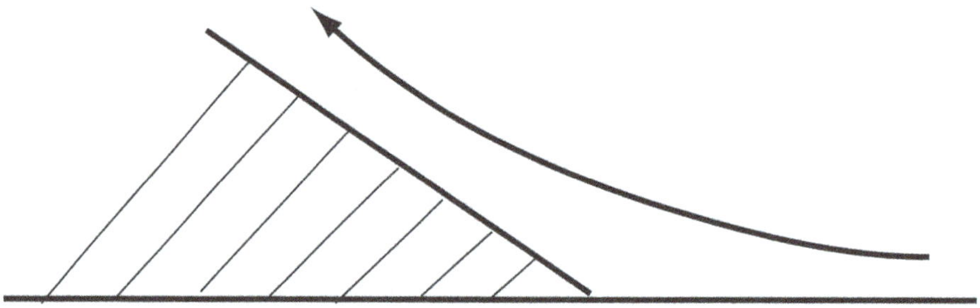

Figure 3.3. Schematic illustrating forced lift over a surface front, an outflow boundary, or orography. Vertical cross section showing air flow (streamline) and cold side of front or outflow boundary, or area between the surface of the rise in ground level and the surrounding ground level (hatched area).

Figure 3.4. Illustration of entrainment of environmental air into a cloud (curled streamlines into curling, rolling, cumulus elements) on July 2, 2011 in Boulder, CO. Air enters the cloud base from below and from the sides. As a result of entrainment, the mass flux of air detraining from the cloud at the top exceeds the mass flux into the cloud base below (photograph by the author).

Figure 3.5. Cumulonimbus anvil on September 9, 2009, in Boulder, CO. Three separate bursts of convection are evident (each burst indicated by an arrow) (photograph by the author).

reach the upper troposphere is referred to as "convective initiation", the qualifier "deep" being understood and that the result is typically a cumulonimbus.

To reach the LFC, convective inhibition (CIN) must be overcome. CIN is defined as

$$\mathrm{CIN} = g\int_{z_0}^{\mathrm{LFC}} [T_c(z') - T_0(z')]/T_0(z')\,dz' \tag{3.4}$$

where z_0 is the height or representative height of the air parcel that is forcibly lifted to its LFC. Because the air is originally stable, the temperature difference between the air parcel (T_c) and its surroundings (T_0) is negative, so that CIN is negative.

One can use (3.4) and (2.70), less the dynamic vertical perturbation pressure gradient terms, and buoyancy defined by (2.19) to estimate what initial upward vertical velocity (in the rising branch of mesoscale circulations) is necessary so that there is enough kinetic energy available to reach the LFC (in the absence of friction, entrainment, and vertical dynamic pressure gradient forces)

$$Dw/Dt = g[T_c(z') - T_0(z')]/T_0(z') \tag{3.5}$$

Then, if w is steady state, at the center of the updraft

$$w_0 = \left\{-2g\int_{z_0}^{\mathrm{LFC}} [T_c(z') - T_0(z')]/T_0(z')\,dz'\right\}^{1/2} = (-2\,\mathrm{CIN})^{1/2} \tag{3.6}$$

if $w(z = \text{LFC}) = 0$. For $w(z = \text{LFC}) > 0$, $w_0 > (-2\,\text{CIN})^{1/2}$. For example, a CIN of just $-50\,\text{j}\,\text{kg}^{-1}$ requires an initial upward velocity burst of at least $10\,\text{m}\,\text{s}^{-1}$ to reach the LFC with a nonzero upward velocity. Very often a low-level moist boundary layer is topped by an inversion, which acts as a cap to boundary layer–based thermals and is responsible for CIN (Figure 3.1). In nature, the mesoscale lift is much weaker ($\sim 1\,\text{m}\,\text{s}^{-1}$ or even $10\,\text{cm}\,\text{s}^{-1}$).

If the environmental air that is entrained is unsaturated, some of the cloud particles may evaporate and cool, so that buoyancy is even further reduced. When the air outside the cloud is nearly saturated, the reduction in buoyancy is negligible as a result of evaporation. In the tropics, the relative humidity of air in many instances is high, so that the potential for evaporation and cooling is low; in mid-latitudes, when the relative humidity in many instances is low, the potential for evaporation is high. Thus, not only is the environmental lapse rate important, but the humidity of the environment and its variation with height are too.

In an environment in which the water vapor mixing ratio decreases with height (this is typically the case, since the source of water vapor is water at the Earth's surface; however, the water vapor mixing ratio is often layered, with a number of relative maxima), lift on the mesocale or synoptic scale, dynamic or orographic, tends to increase humidity aloft in the "environment": convective clouds that form in an environment of ascent have a better chance of growing into deep convective clouds than those that form in an environment of descent, in which humidity in the environment is decreased.

In addition, the environmental lapse rate is changed when there is vertical motion (cf. the static stability tendency equation in Bluestein, 1992) owing to the combined effects of vertical advection of temperature and of adiabatic cooling/heating due to the expansion/contraction of air. At low levels, rising motion is associated with horizontal convergence, which increases the lapse rate as potential temperature isotherms (isentropes) are spread apart in the vertical (Figure 3.6).

It has generally been observed that convectively enhanced environments are those that sustain rising motion on the synoptic scale and convectively suppressed environments are those that sustain sinking motion. Since synoptic-scale vertical motion is two–three orders of magnitude ($\text{cm}\,\text{s}^{-1}$ vs. $\text{m}\,\text{s}^{-1}$) less than vertical motion in a convective cloud and that needed to overcome CIN, synoptic-scale motion is not directly implicated in initiating convection; instead, it makes the environment more or less susceptible to convection by moistening the environment and thereby reducing the deleterious effects of entrainment and reducing static stability, all of which act to reduce CIN rather than overcome it.

3.1.3 Observed life cycle and vertical velocity

After deep convection has been initiated, the cloud grows until the "equilibrium level" (EL) is reached (Figure 3.1), at which the rising air parcel's positive buoyancy has been reduced to zero. Kinetic energy of the rising air parcel is nonzero, so the cloud keeps growing upward until its kinetic energy has been used up. It

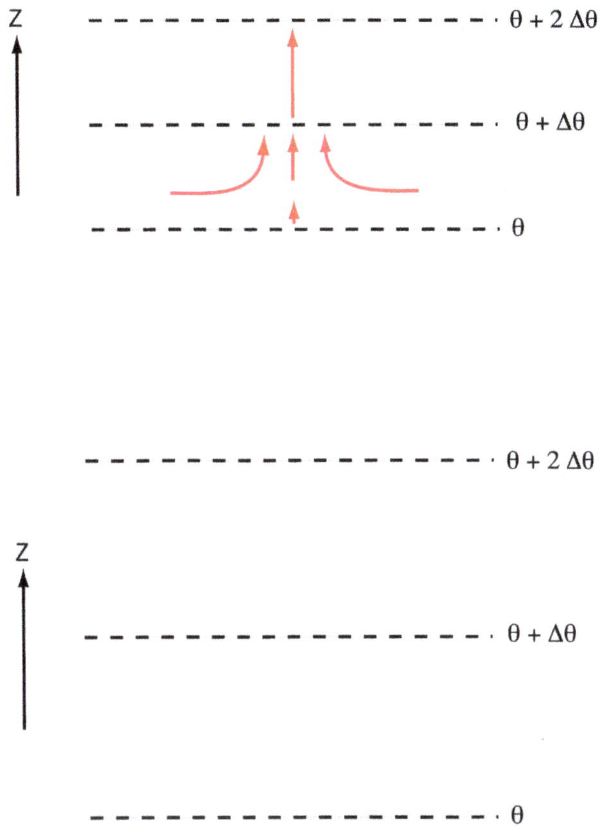

Figure 3.6. Illustration of how low-level convergence reduces static stability in a stable atmosphere. Vertical cross sections of initial (top) and subsequent (bottom) potential temperature (dashed lines) when there is convergence and rising motion (streamlines shown in red in the top panel). The vertical spacing between the isentropes is greater in the bottom panel than in the top panel.

thus overshoots the equilibrium level as a "penetrating top", "overshooting top", or "anvil dome" (Figure 3.7). A penetrating top can entrain very dry stratospheric air and sublimate, thus reducing its remaining buoyancy even more quickly. The cirrus cloud material visible is sometimes called "splashing cirrus" because it looks like water splashing against a shoreline. Since the air is pushing upward against a stable layer, it is likely that gravity waves are triggered (Figure 3.8) and may be responsible for some of the appearance of anvils when seen from above (Figure 3.9). Penetrating tops, though they are not solid bodies, can act like solid bodies and air is forced to flow over them and as such may generate gravity waves that are like the waves forced in the lee of an isolated mountain. Penetrating tops can behave like solid bodies because air outside the storm that moves relative to the updraft (catches it up) decelerates as it hits the updraft (Figure 3.10).

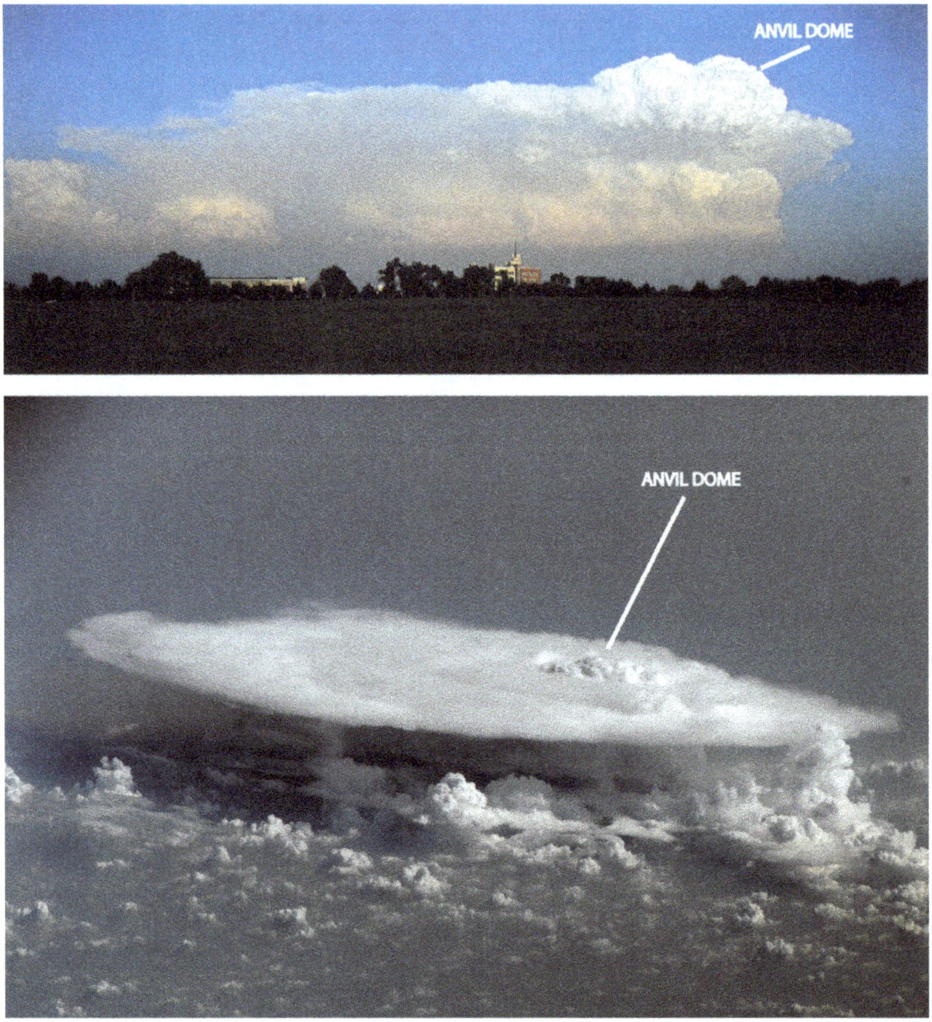

Figure 3.7. Anvil dome (penetrating top) at the top of a convective storm. (top) Tornadic supercell southeast of the National Weather Center, Norman, OK, May 21, 2011 (photograph by the author); (bottom) convective storm viewed from above by International Space Station on February 5, 2008, over Mali in western Africa (from NASA).

There have been many studies using satellite imagery at various wavelengths to detect penetrating tops and to determine when they grow and when they decay. The main motivation for these studies is to use satellite imagery alone to infer the likelihood of severe weather at the surface, hidden from visual view of the satellite: it is assumed that the stronger the updraft, the deeper the penetration of the cloud above the tropopause and the more likely there will be severe weather below. Rings or horseshoe-shaped ("cold U/V or enhanced V") regions of relatively cold

Figure 3.8. Schematic of airflow over a dome at the top of a convective storm in eastern Oklahoma on May 1, 1980. Streamline shown is indicated for flow near the top of the storm when the storm is propagating more slowly than the wind speed. Air is lifted over the dome and then undergoes stable vertical oscillations downstream from the dome (photograph by the author).

Figure 3.9. Waves in the anvil of a convective storm (caused by gravity waves) on July 9, 2009, over the upper Midwest of the U. S., as seen by the NOAA 15 satellite (from Martin Setvak).

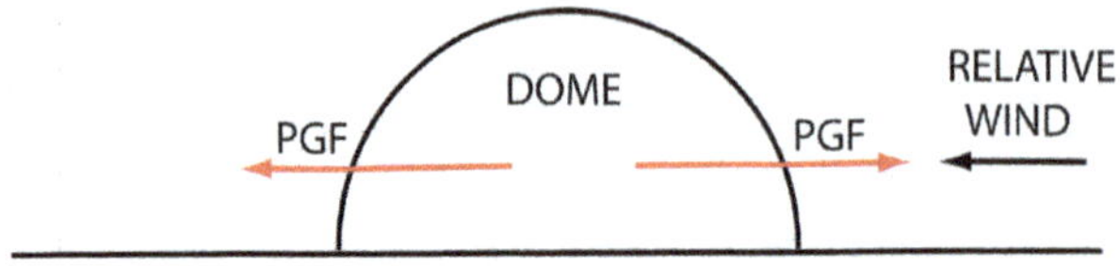

Figure 3.10. Idealized illustration of dynamic horizontal perturbation pressure gradient forces (red arrows) along the sides of an anvil dome. Air decelerates as it approaches from the right (flows over the top as in Figure 3.8) and then accelerates again as it leaves the dome on the left. Air with less momentum from below is brought to the storm top in the dome; air with higher momentum approaches the dome from the right, but must be slowed down. It is the adverse pressure gradient (i.e., in the direction opposite to the flow) that slows the air down and thus the fluid dome acts like a solid body.

infrared (IR) brightness temperatures surrounding a relatively warm region are often seen in infrared satellite imagery (Figure 3.11). Alas, cloud-top brightness temperature does not necessarily indicate the cloud-top temperature because the top edge of the cloud, even if the cloud is opaque, may not be in thermal equilibrium with its environment. Spaceborne radar and lidar observations have been used to correlate IR signatures with penetrating tops. It has been found that pene-

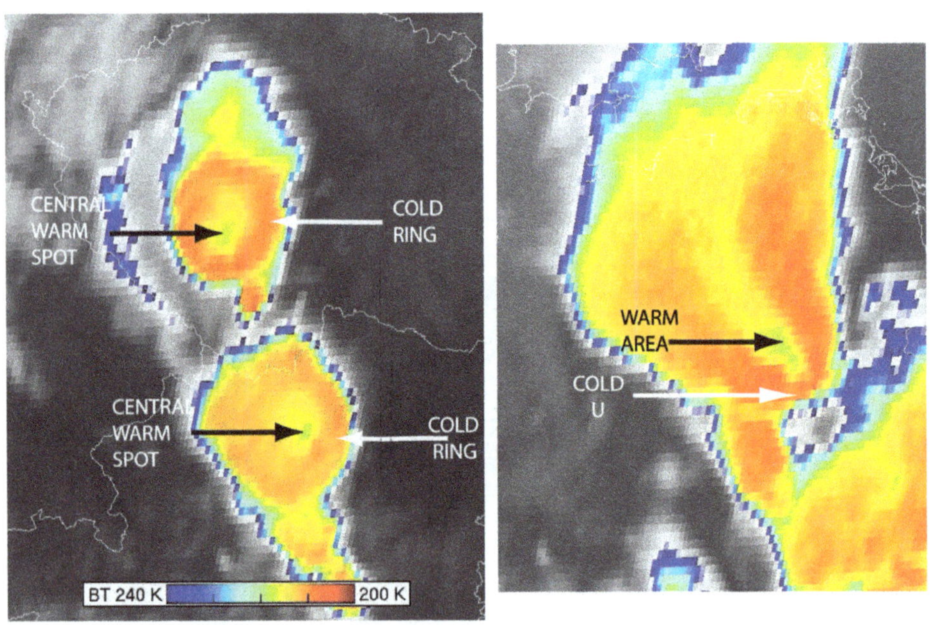

Figure 3.11. Color-enhanced infrared signatures at anvil top. (Left) Cold rings in two convective storms over central Europe as detected by METEOSAT-8. (Right) Cold UV signature in a convective storm over Germany on May 26, 2007, as depicted by METEOSAT-9 (from Martin Setvak).

trating tops can be detected from the blackbody temperature difference between the water vapor and IR bands, but the regions where the blackbody temperature difference is greatest may only locate the "coldest" regions—not necessarily the penetrating tops. The "warm spots" downstream of penetrating tops are thought to be created by gravity waves (wave breaking) or other effects. The difference in brightness temperature between the cold areas and warm areas is several K or greater. There is some model evidence that the ring signature occurs when vertical shear is relatively weak and the U/V signature occurs when vertical shear is relatively strong.

The vertical velocity at the center of a column of rising air at a given altitude may be estimated thermodynamically, using the inviscid, steady-state version of the vertical equation of motion (2.70) without any p'_d or p'_b (according to "parcel theory" in which an air parcel's effect on its environment is neglected), as

$$w(z) = [2(\mathrm{CAPE}(z))]^{1/2} \tag{3.7}$$

where CAPE, the "convective available potential energy", is the vertically integrated energy acquired by the rising air as a result of the upward buoyancy force acting on it:

$$\mathrm{CAPE}(z) = \int_{\mathrm{LFC}}^{z} B\,dz' = g\int_{\mathrm{LFC}}^{z} [T_c(z') - T_0(z')]/T_0(z')\,dz' \tag{3.8}$$

where $T_0(z')$ is the vertical profile of temperature in the environment of the cloud, and $T_c(z')$ is the vertical profile of temperature inside the cloud. The accuracy of CAPE is increased if moisture and cloud liquid water content is accounted for by using virtual temperature or cloud virtual temperature (cf. (2.21)–(2.23)) for temperature.

Estimation of CAPE using (3.8) depends on how the LFC is computed. When surface-based air parcels are used, CAPE is called "surface-based CAPE" or SBCAPE. When the mixed-layer or some other boundary-layer mean is used, CAPE is called the "mixed-layer" or "mean-layer" CAPE or MLCAPE. When the highest combination of temperature and dew point are used, CAPE is called the "most unstable" CAPE or MUCAPE.

In many instances the LFC is also at cloud base, but not necessarily so. When the LFC is higher than the LCL, the outside of the cloud tends to assume a smooth, laminar appearance (Figure 3.12a), suggestive of laminar lift of a stable air mass, like air being lifted in an orographic wave cloud (Figure 3.12b) (cf. the discussion on why mammatus look smooth in 3.2.1.3). Strictly speaking, (3.7) is valid when there is no mixing, the atmosphere is resting (there is no vertical or horizontal wind shear in the environment), and the air parcel is infinitely narrow. However, the wider the air parcel, the less the importance of mixing. So, when the air parcel is infinitely narrow, we can neglect the effects of vertical gradients of p_b, but mixing must be accounted for, and vice versa. One would expect that there is an intermediate range of aspect ratio for which the deleterious effects of both mixing and upward-directed perturbation pressure gradient force are each minimized with respect to each other (cf. (2.75) and Section 2.5.3).

(a)

(b)

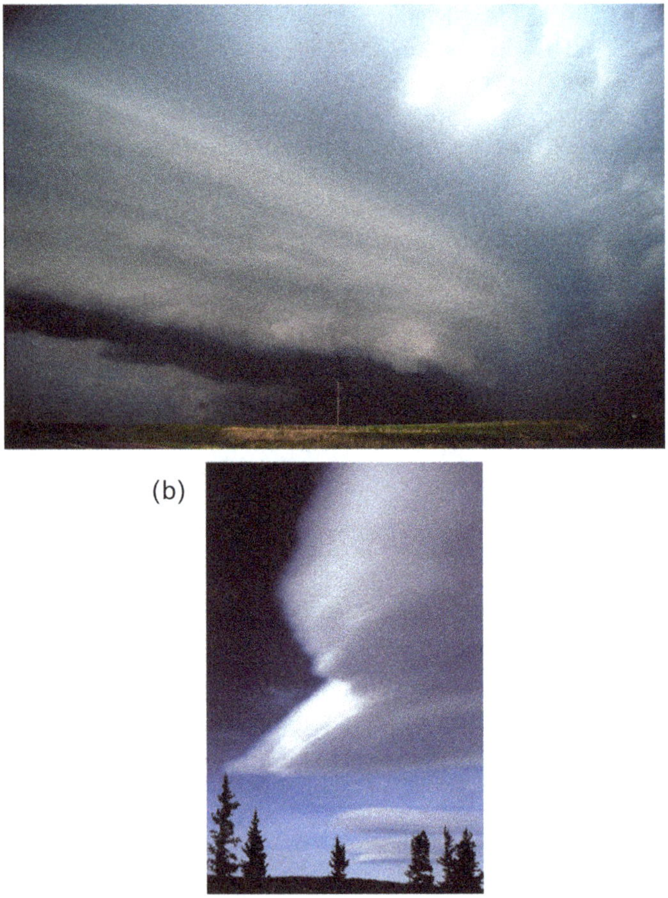

Figure 3.12. (a) Cloud base in a supercell that has striations like an orographic wave cloud, on May 29, 2008, north central Kansas. (b) Orographic wave clouds in the lee of the Rockies just west of Ward, CO on January 2, 2008 (photographs by the author).

Total CAPE is CAPE(EL) and $w(\text{EL}) = [2\,\text{CAPE(EL)}]^{1/2}$; CAPE(EL) is usually just referred to as "CAPE". High CAPE is favored by relatively cool air aloft and relatively warm air at low altitude. Stable layers above the LFC act to reduce CAPE from it would have otherwise been. Stable layers in mid-levels of the troposphere are typically associated with sinking motion on the synoptic or mesoscale. Middle and upper-tropospheric fronts are characterized by stable layers, and as such act to reduce CAPE. Storm-chasers colloquially describe mid-level stable layers as "CAPE robbers" (Figure 3.13).

When there is a low-level, capping inversion (or stable layer), deep convection will not be initiated until the "cap is broken" (i.e., the LFC is attained through cooling at cap level via mesoscale lift and/or warming below via surface heating). Since the cap suppresses convective development everywhere except where it is

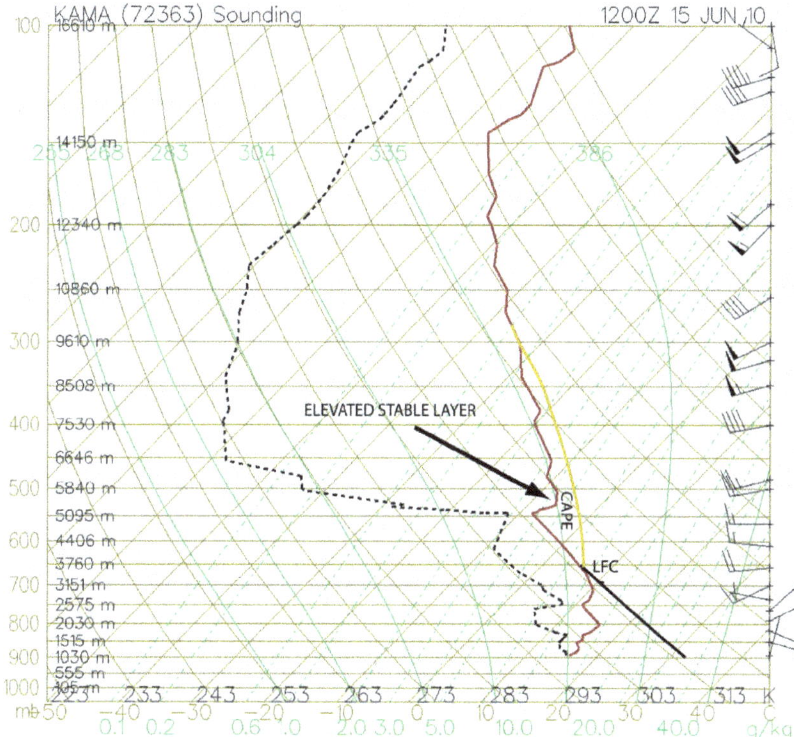

Figure 3.13. Example of a "CAPE robber". Sounding at Amarillo, TX on June 15, 2010 at 12:00 UTC (from Plymouth State College archives and graphics). While the convective temperature is about 32°C and a substantial amount of heating is necessary to realize this temperature, an elevated stable layer just below 500 hPa would "rob" the sounding of quite a bit of potential CAPE, even if it were to be realized, as seen by the area under the yellow curve (moist adiabatic at the LFC) and the red curve (temperature).

broken, convection may be initiated only locally and CAPE used up only locally. It generally takes a relatively long time for CAPE to be built up via horizontal advection on the synoptic scale (or mesoscale) of temperature and moisture, and via temperature changes due to vertical motions on the synoptic scale (or mesoscale).

On the other hand, where there is no or a weak capping inversion, convection may be initiated over a broad region and CAPE used up over a broad region. In the Southern Plains of the U. S. during the spring, synoptic and mesoscale vertical motion and advection build up CAPE gradually, but the CAPE is depleted quickly as convection is initiated. Soundings with high values of CAPE, but capping inversions that suppress convective initiation are called "loaded gun" soundings (Figure 3.1). When the rate at which CAPE is built up is approximately the rate at which it is depleted, there is said to be a "statistical equilibrium" condition. Statistical equilibrium occurs frequently in the tropics, where CAPE is not large and cold pools are not as intense as at mid-latitudes, owing to comparatively high relative humidity in midlevels of the troposphere and a warm ocean surface that heats the low levels.

CAPE can be thought of as the integrated potential buoyancy in a convective cloud; the buoyancy is only "potential" because it may or may not be realized in entirety. Equation (3.8) yields CAPE for the environment: for calculating vertical velocity in a convective cloud, one needs also to include both the dynamic and buoyancy-related vertical perturbation pressure terms, precipitation loading, and turbulent mixing. In a resting atmosphere, the latter two act to reduce buoyancy, so CAPE in the environment, as calculated from a proximity sounding, may be thought of as an upper limit. Thus, the vertical velocity calculated from (3.7) may be thought of as an *upper bound* in most cases and used to estimate only in a very qualitative way the intensity of convection that might occur. Estimates of vertical velocity using (3.7) are in accordance with parcel theory, in which the assumptions noted above are used.

In highly sheared environments, it is possible that the vertical perturbation pressure gradient acts upward and is substantial, so that CAPE may not be a good indicator at all of the intensity of convection. This is the case, for example, in the environment of landfalling tropical cyclones, when low-level vertical shear is strong, but CAPE is relatively low. A discussion of this situation is delayed until Chapter 4, when supercell dynamics are discussed.

Estimates of the vertical velocity in convective storms using (3.7) are typically $O(10\,\mathrm{m\,s^{-1}})$; in the strongest storms, vertical velocity may exceed $50\,\mathrm{m\,s^{-1}}$. In numerical models such high vertical velocities are found in convective updrafts. Verifying such high vertical velocities in nature is more difficult. A limited number of measurements have been made using storm-penetrating aircraft (Figure 3.14). Other estimates have been made from radiosondes released directly into storm updrafts (Figure 3.15); it is assumed that updraft velocity is given by the balloon ascent rate (determined either from the rate of decrease in pressure or the rate of increase of GPS height) less the neutral (considering the buoyancy of the balloon alone less its drag) ascent rate of the balloon. Icing can slow the balloon down and it is not known whether or not the balloon passed through the center of the updraft.

Measurements have been made of updrafts using Doppler radar data. In most instances, however, the beams from Doppler radars are directed in nearly perpendicular directions to the vertical, so that vertical velocity must be inferred kinematically from the divergence of the horizontal wind (cf. Appendix); errors in divergence accumulate when vertical velocity is estimated from upward integrations. Downward integrations are difficult unless one knows for sure the upper boundary condition and at what level it is valid: Updrafts may penetrate the stratosphere, so that one would have to know the equilibrium level to be able to set a zero upper boundary condition (which would be above the EL). Another serious problem with multiple Doppler radar–based, kinematically computed vertical velocities is that it is unlikely that multiple radars sample the identical volume at the same time, owing to the scanning patterns of the radars and the spreading

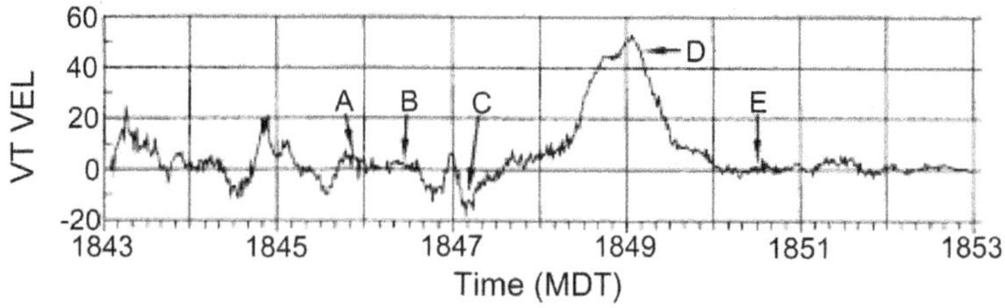

Figure 3.14. Vertical velocity measurements made in a hailstorm in southeastern Montana by the T-28 instrumented aircraft, on August 2, 1981, during CCOPE (Cooperative Convective Precipitation Experiment). Updraft "D" was in excess of $50\,m\,s^{-1}$ (from Musil *et al.*, 1991).

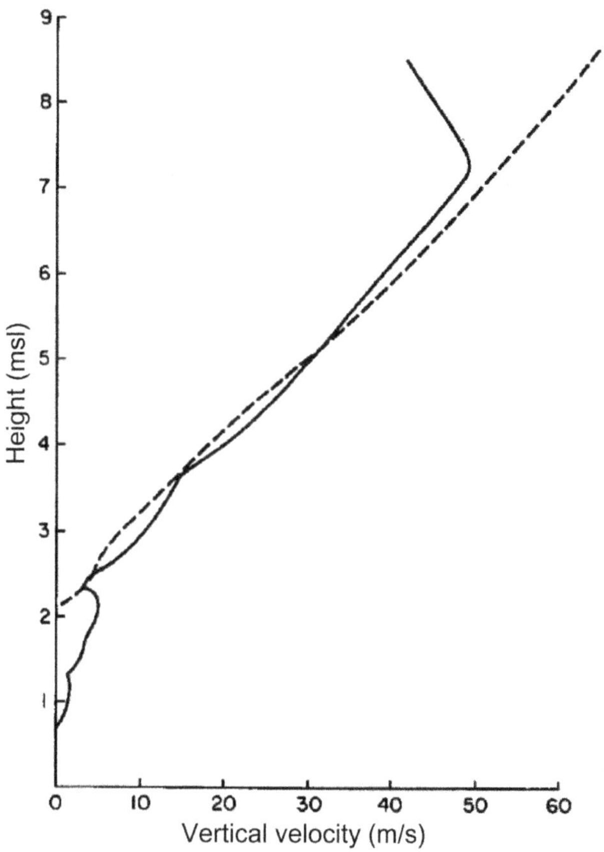

Figure 3.15. Measurements of vertical velocity, based on the ascent rate of a radiosonde (solid line), inside a tornadic supercell in the Texas Panhandle on May 7, 1986 and a comparison with estimated vertical velocity based on parcel theory (dashed line) (from Bluestein *et al.*, 1988). It is probably a coincidence that parcel theory agrees so well with the measurements because vertical perturbation pressure gradients and precipitation loading were not taken into account.

of the radar beams with increasing range, which is different for identical radars when the storm sampled is at different ranges to each radar; thus, error is introduced, which makes accurate estimation of vertical velocity even more difficult. Downward-looking Doppler radars mounted on an aircraft can be used to measure vertical velocities by flying over the tops of convective storms. Gerry Heymsfield *et al.* at NASA have used downward-looking Doppler radars mounted in a high-altitude (20 km) aircraft. The advantage of downward-looking radars is that vertical air motion can be estimated directly (not kinematically), while two-dimensional motions in the plane of the aircraft track can be resolved using pseudo dual-Doppler techniques.

Vertically pointing Doppler radars may be used to measure vertical velocities if they can be placed underneath a convective storm updraft. In any case, the terminal fall speed of hydrometeors backscattering the radar beam must be known so that it can be removed. To do so, simplified models relating the radar reflectivity factor to terminal fall size are used. A better method would be to use polarimetric data to classify hydrometeors based on fuzzy logic and then use a terminal fall velocity relationship between hydrometeor type and characteristics and fall velocity. In the latter case, however, the use of polarimetric radar data is optimum at low elevation angles and becomes more difficult, if not impossible, at higher elevation angles because typically used parameters such as differential reflectivity Z_{DR} and specific differential phase K_{DP} are defined from horizontally and vertically polarized signals; at high elevation angles, the polarimetric parameters do not measure Z_{DR} or K_{DP} well, owing to the geometry. In summary, obtaining accurate estimates of vertical velocity in convective storms using multiple Doppler radar analyses is challenging because identical radar-sampling volumes at the exact same time are usually not available and there is uncertainty in the terminal fall speed of the scatterers in the volume. It is possible that electronically scanning (rapid scan) radars with very narrow beams could improve our ability to estimate vertical velocity because they could sample at high elevation angle and decrease errors due to non-simultaneity.

Before the updraft reaches the equilibrium level, precipitation may form. At this stage, radars detect precipitation suspended *aloft*. If there is substantial liquid water loading, owing to suspended, supercooled drops whose terminal velocity just equals the updraft speed, then the equilibrium level is lower, near or below the tropopause. The stronger the updraft, the less likely precipitation will have had enough time to form and the suspended supercooled drops are higher up, along with the EL. The appearance of radar echoes aloft is a good indicator of the beginning of a convective storm. For this reason, it is necessary for operational radars to scan up to a number of relatively high elevation angles.

When the terminal fall speed exceeds the updraft speed, then precipitation begins to fall out into the updraft if there is no vertical shear and the updraft is destroyed, and a downdraft is produced as buoyancy becomes negative. The physics of downdrafts are different from the physics of updrafts, in large part because microphysics becomes a more important consideration. Also, buoyant updrafts are saturated so that the lapse rate is moist adiabatic. If precipitation falls through unsaturated air, then evaporative cooling (or melting cooling, if frozen precipitation melts, or sublimation cooling, if frozen precipitation sublimates, or a combination of all of the aforementioned) further reduces buoyancy (increases negative buoyancy) and the downdraft is further strengthened. If the condensed water substance is in the form of ice crystals, however, the terminal fall velocity is smaller and, as upward vertical velocity decreases with increasing height, the ice particles remain aloft and diverge horizontally to form an *anvil*.

When there is vertical shear, the anvil may be advected downstream far ahead of the storm (Figure 3.16) and forewarn it. When there is only weak or no vertical

(a)

(b)

Figure 3.16. Long anvils. (a) Cumulonimbus with a long anvil downstream, as viewed from an aircraft over southwestern Nebraska on July 19, 2009. (b) Developing cumulonimbus in Oklahoma on April 30, 2003 (photographs by the author).

Figure 3.17. Cumulonimbus (developing supercell) with a symmetrical, mushroom-like anvil, on May 26, 1997 in eastern Oklahoma (photograph by the author).

wind shear, or if there is strong vertical shear but the storm is viewed from the rear, the anvil takes on a mushroom-like appearance (Figure 3.17). When anvil material in the form of ice crystals is left behind after the updraft that was responsible for it has dissipated, the cloud mass is referred to as "orphaned anvil" (Walter Hitschfeld at McGill is credited with this moniker from a paper published in 1960) (Figure 3.18).

The observed life cycle of an ordinary-cell convective storm (Figure 3.19) is less than an hour, the time it takes buoyant air to travel from the boundary layer (the source of the updraft air; if the source is above the boundary layer, then the convective storm is said to be "elevated" and the observed life cycle should be shorter) all the way to the tropopause and then fall to the ground as precipitation (the total time elapsed is the "advective" time scale). The updraft region in a convective storm is called a "cell". The identical terminology is used to describe the precipitation region detected by a meteorological radar (radar echo). Since each region of precipitation was once associated with an updraft, one-to-one correspondence can be made between the updraft cell and the radar-observed precipitation cell, even though in the latter there may not be any updraft remaining.

Horace Byers and Roscoe Braham were the first to describe the life cycle of an ordinary-cell convective storm[1] based on observations during the Thunderstorm

[1] See Ludlam (1963, Fig. 1) for a summary of earlier conceptual models going as far back as the late 19th century.

Figure 3.18. Orphan (or "orphaned") anvil on March 15, 2012, Ft. Lauderdale, FL. (Top) Convective storm is dissipating, with precipitation falling; (bottom) anvil has become disconnected from the cloud below and most of the precipitation has ended (photographs by the author).

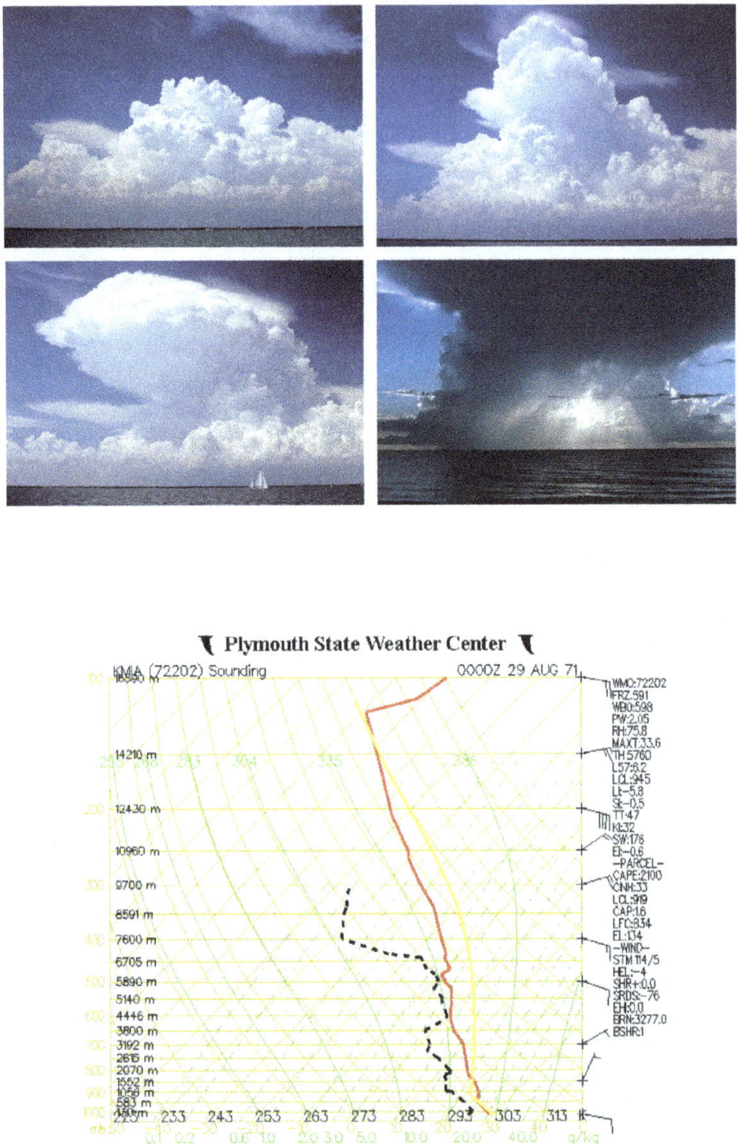

Figure 3.19a. Stages in the life of a Byers–Braham, ordinary-cell convective storm. (Top left) Cumulus congestus stage, on August 28, 1971, over the South Florida peninsula, south of Miami; (top right) cumulus congestus about to hit the tropopause and producing a pileus, on August 28, 1971; (second row, left) mature stage, with an anvil, on August 28, 1971; (second row, right) dissipating stage, off the west coast of South Florida, on August 28, 1973 (photographs by the author). (Bottom) Sounding at Miami, Florida, 00:00 UTC on August 29, 1971, which is a representative environmental sounding for the convective storm on August 28, 1971 (from the Plymouth State College archive and graphics). Vertical shear is weak and CAPE moderate.

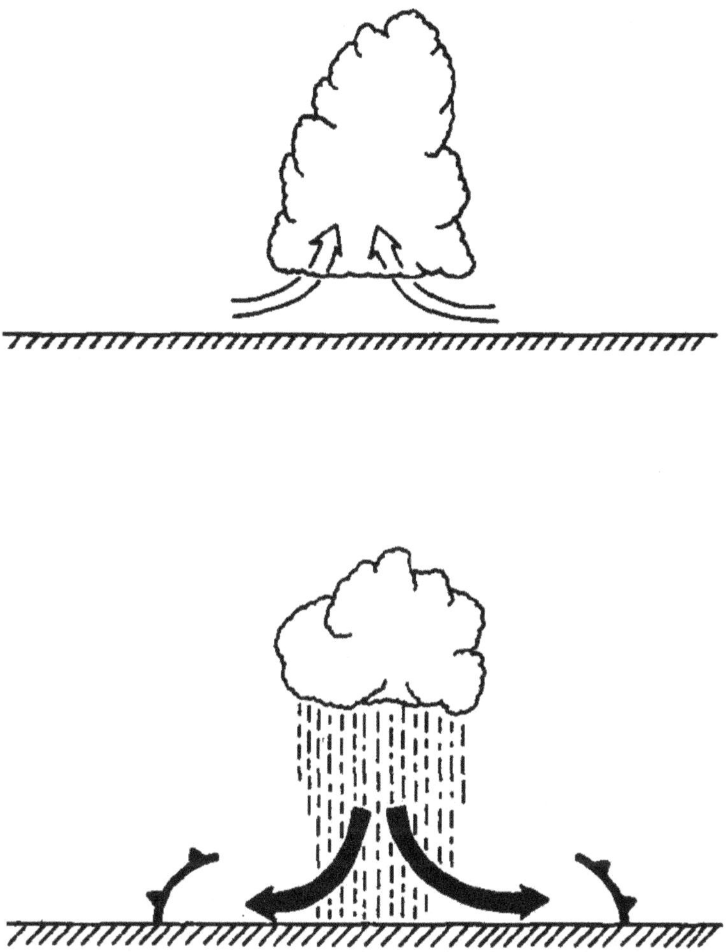

Figure 3.19b. Byers–Braham conceptual model of an ordinary-cell convective storm. (Top) Cumulus stage, characterized by a buoyant updraft; (bottom) dissipating stage, characterized by precipitation, downdraft, and gust fronts (from Rotunno *et al.*, 1988)

Project in the 1940s. A cell's lifetime is therefore at least $\sim 2(12\,\text{km}/10\,\text{m}\,\text{s}^{-1}) =$ 2,400 s (i.e., $\sim$40 min) for a 12 km deep troposphere; additional time may be needed to account for the complete blocking off by the downdraft of potentially unstable air into the updraft and for storm initiation. Cells that last for much longer periods of time than the time it takes air parcels to travel to the tropopause and then downward to the ground will be discussed a bit later. Storms that behave like ordinary cells and consist of only one cell are sometimes referred to as "pulse"-type, single-cell convective storms. They occur in environments of moderate CAPE and weak vertical wind shear.

3.2 GUST FRONTS AND DOWNDRAFTS

3.2.1 Gust fronts in the absence of vertical wind shear

When a downdraft hits the ground, it is forced to spread out laterally, owing to the decrease in height of the downdraft ($\partial w/\partial z < 0$). Gusty winds mark the leading edge of the cooler air, which is called a *gust front*. Observational studies of gust fronts have made use of surface instruments, especially in networks, instrumented towers, and Doppler radars. Instrumented towers are valuable because they provide information not only about the wind, but also thermodynamic variables, as a function of height and time. However, they are useful only when they are taller than the depth of the cold air mass. In addition, the vertical resolution of the wind and thermodynamic vertical profiles determined from tower data is limited by the number of instruments that can be placed on the tower. Doppler radars (both ordinary and wind profilers) can provide high spatial and temporal resolution of wind variables, but not of thermodynamic variables, unless the thermodynamic profiles are retrieved from wind observations or a radio acoustic sounding system (RASS) is used.

Typically, the wind shifts after gust front passage, becomes gusty, and the temperature falls and the pressure rises (Figure 3.20). The amount of cooling behind a gust front depends upon the dryness of the air, the depth over which evaporation (or melting or sublimation) takes place, and the sizes of the water drops and hailstones, if present. Numerical cloud models at the current time have difficulty reproducing temperature deficits behind gust fronts accurately, mainly owing to uncertainties in the amount of and drop-size distributions of the precipitation and cloud droplets. Drop-size distributions vary according to the origin of the air; for example, drop-size distributions in the tropics are different from those in continental regions of mid-latitudes. Our knowledge of drop-size distributions in convective storms using direct measurements is limited, owing to the impossibility of placing an instrument at all locations in a storm all the time.

3.2.1.1 Microbursts

In mid-latitudes over land, when clouds build upward into relatively dry environmental air, the potential for evaporative cooling is high and, hence, very strong downdrafts and gust fronts are possible. When both the winds and the vertical shear are weak, the downdraft may be circularly symmetric and regions of very strong lateral gradients in wind can be produced near the ground, about the center. Horizontal vorticity is created baroclinically along the leading edges of the downdraft of negatively buoyant air (cf. (2.51)) so that a ring of horizontal vorticity is produced (Figure 3.21). A vortex ring may also be induced by the vertical shear of the horizontal wind associated with surface divergent flow (Figure 3.22).

Very strong downdrafts (of 10 to 20 m s^{-1} or more) that reach the ground are called "microbursts" (Figure 3.20 bottom); they can be very hazardous to aircraft landing or taking off: aircraft that enter a microburst experience a brief period when the airflow is opposite that of the aircraft motion, followed by a brief period when the airflow is in the same direction as that of the aircraft motion. Thus, the aircraft experiences a brief period of enhanced lift, followed by a period of diminished lift. Too much overcompensation for the period of enhanced lift can result in stalling and crashes as the aircraft pulls away from the center of the microburst. Microbursts may be strong enough to cause "straight-line" wind damage at the surface, especially if the downdraft is intense and narrow or if there is a strong enough component of the ambient wind near the ground (Figure 3.23). Damage from microbursts is typically associated with diffluent patterns in ground debris.

The foundation of our knowledge about microbursts is the early investigations of aircraft accidents by Ted Fujita. The crash of Eastern Airlines Flight 66 on June 24, 1975 at John F. Kennedy Airport in New York City, in which 112 people were killed and 12 injured, was the seminal disaster that spurred research into microbursts. Fujita originally termed the event a *downburst*, subdividing the downburst into microbursts when they are 4 km or less across and *macrobursts* when they are wider than 4 km. At issue and a source of controversy was whether they caused airplanes to crash owing to strong downdrafts or otherwise, since vertical velocity weakens to zero at the ground. Three major field experiments were conducted to study microbursts: NIMROD (Northern Illinois Meteorological Research on Downbursts) held in the spring and summer of 1978, during which an event (at Yorkville, Illinois) was captured by Doppler radar, JAWS (Joint Airport Weather Studies project) held in the Denver area during the summer of 1982, and MIST (Microburst and Severe Thunderstorm Project) held near Huntsville, Alabama during the summer of 1986. In the latter, dual-polarization radar was used for the first time to probe microbursts and an event at Monrovia, Alabama was studied intensively.

Microbursts have been classified as being "dry" or "wet". Dry microbursts occur over relatively arid terrain when the cloud base is relatively high (Figure 3.24), around 3 km AGL (i.e., when the lapse rate is nearly dry adiabatic and the boundary layer relatively dry). Thus, the potential for evaporative cooling is great as water drops and droplets fall through unsaturated air for a relatively long time and may evaporate completely producing "virga". Even light precipitation can produce strong downdrafts. On the other hand, wet microbursts occur when the atmosphere is relatively moist and the cloud base is relatively low, so that the potential for evaporative cooling at low levels is relatively low. In this case, negative buoyancy is created mainly from water loading in intense precipitation (Figures 3.22 and 3.25) or evaporative cooling aloft. Cooling from the melting of ice particles (graupel) on the way down in a region of precipitation may enhance negative buoyancy in both dry and wet microbursts.

Mark Hjelmfelt found in JAWS that microbursts were associated with descending reflectivity cores seen by radar, as would be expected when new cells form and precipitation falls out. An area of convegence at mid-levels in a convective storm, near the region of minimum equivalent potential temperature, is observed in microburst-producing storms. Results from early microphysical

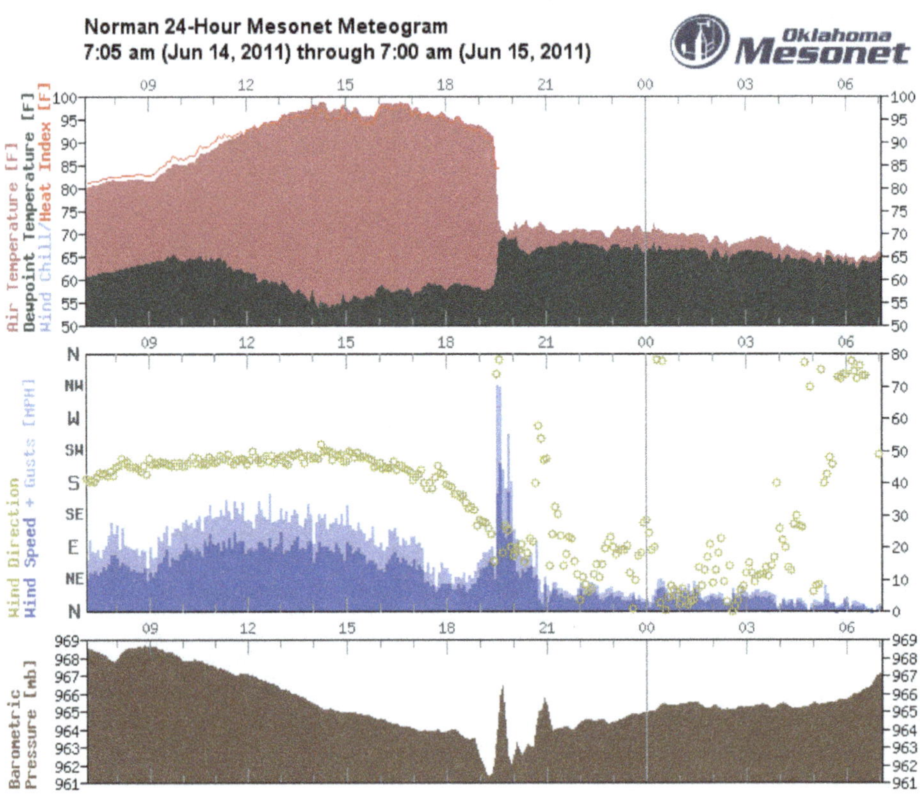
Norman 24-Hour Mesonet Meteogram
7:05 am (Jun 14, 2011) through 7:00 am (Jun 15, 2011)
Oklahoma Mesonet
Air Temperature [F]
Dewpoint Temperature [F]
Wind Chill/Heat Index [F]
Wind Direction
Wind Speed + Gusts [MPH]
Barometric Pressure [mb]
09 12 15 18 21 00 03 06
N NW W SW S SE E NE N

◀ **Figure 3.20.** Gust front passage. (Top) Visualized from the air (on board a NOAA P-3 aircraft) off the east coast of Florida on August 25, 1993; the precipitation curtain is seen to the left, spreading out at the sea surface. The white dashed line marks the leading edge of the gust front, visualized as a change in the color of the sea surface. (Bottom) Temperature, wind, and pressure as a function of time during passage of a strong gust front in central Oklahoma on June 15, 2011 (Oklahoma Mesonet data courtesy of Chris Fiebrich). The gust front passage is marked by a drop in temperature from the mid-90s F to less than 70°F over a time period of less than 10 min, a wind shift from south-southwesterly to northerly and east-northeasterly, a pressure fall followed by a general rise, with smaller scale pressure fluctuations and a gust of about 70 mph. The strong gust represents a microburst (photograph by the author).

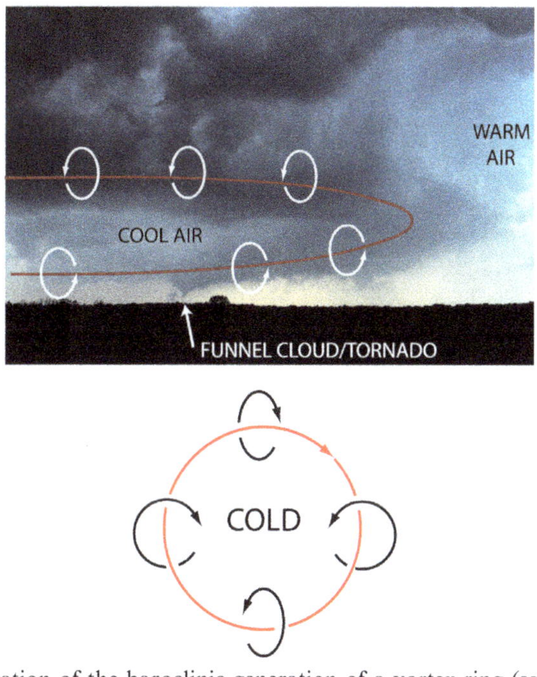

Figure 3.21. Illustration of the baroclinic generation of a vortex ring (sense of rotation indicated by white arrows) about a region of evaporatively cooled air (edge of cool air indicated by solid red line) embedded within an ambient region of warm air, (top) in a tornadic supercell in eastern Oklahoma on May 26, 1997 (photograph by the author). (Bottom) Plan view of a vortex ring induced solenoidally by a circular cold pool. Sense of vortex ring given by red line; circulations induced in the vertical plane denoted by black curved arrows.

modeling by Ramesh Srivastava (while he was at NCAR), in which ice processes were not included, indicated that the strength of downdrafts depends on the lapse rate, the rainwater mixing ratio at the top of the downdraft, decreasing raindrop size, and, counter-intuitively, the humidity of the environment. When the environment is more humid, the virtual temperature is higher, so buoyancy in the downdraft is more negative.

Figure 3.22. Sense of horizontal vorticity (curved arrows) near the ground underneath a precipitation-laden downdraft in a convective storm over Oklahoma City, Oklahoma on July 26, 1978. The vorticity depicted is part of a vortex ring like the one shown at the bottom of Figure 3.21 (photograph by the author).

Ramesh Srivastava later (at the University of Chicago) suggested that in wet microbursts, when ice is included in modeling studies, the breaking up of larger raindrops into smaller raindrops, which evaporate more quickly than larger raindrops, may also be an important mechanism for microburst formation, in addition to melting of graupel. He also found that as the environmental lapse rate decreases, more precipitation and more ice particles in particular are necessary to form intense downdrafts. It is noted that melting of ice particles can occur completely in a fall through just a few kilometers in a layer above freezing, while raindrops of the same size do not evaporate completely during a fall through the same layer: Although the absorbed latent heat of freezing during melting is much less than the absorbed latent heat of condensation during evaporation, more cooling is effected by the melting, so the melting of graupel in downdrafts is a significant contributor to intense downdrafts.

When the lapse rate is very high (near dry adiabatic), both dry and wet microbursts can occur; when the lapse rate is relatively low, only wet microbursts are possible, and they must have significant amounts of ice-phase precipitation. While one can forecast dry microbursts when convective storms are possible (i.e., there is CAPE and the LCL/LFC are attainable with expected daytime heating or mesoscale ascent) and the LCL/LFC is relatively high, differentiating between the

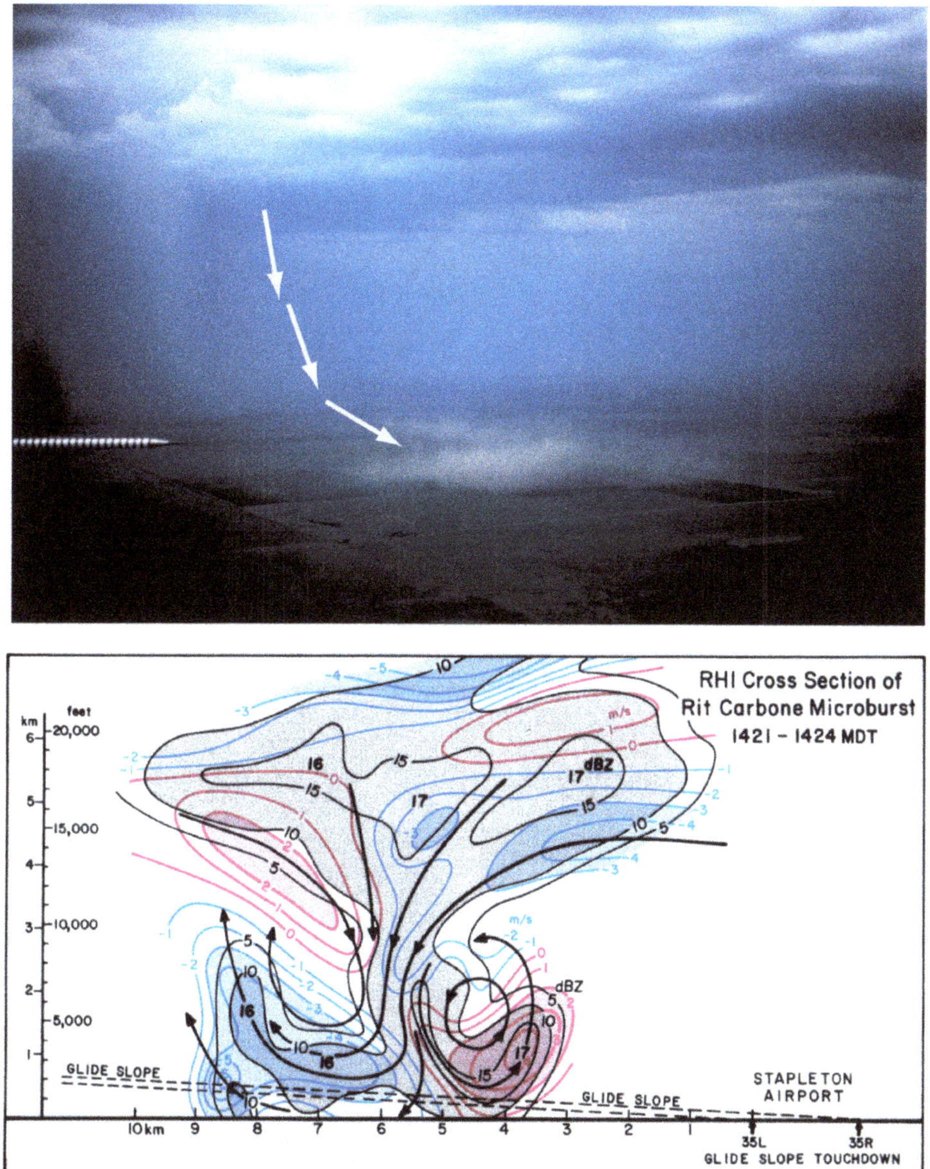

Figure 3.23. (Top) Dry microburst over southwestern Kansas on May 31, 1994, as viewed from the NOAA P-3 aircraft. Precipitation curtain is seen to the left, while a bowed segment of blowing dust is visible at and just above the ground. The qualitative air motion is indicated by the white vectors (photograph by the author). (Bottom) Ted Fujita's single-Doppler analysis of a dry microburst that caused two commercial airlines to abort while landing at Denver's Stapleton Airport. The black contours are the radar reflectivity factor and the red contours are receding Doppler velocities and blue approaching. Streamlines are the heavy black lines (from Wilson and Wakimoto, 2001).

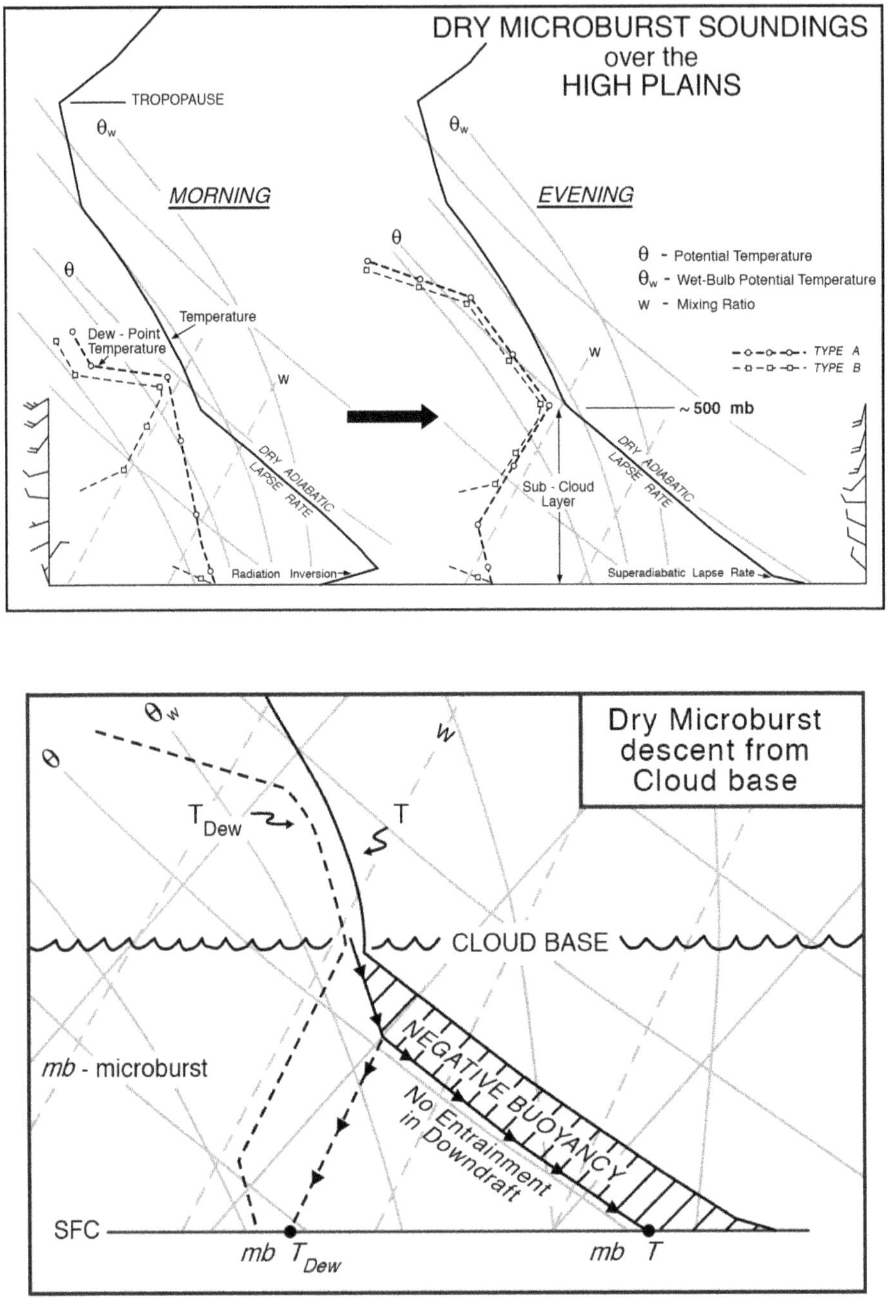

Figure 3.24. (Top) Model of the characteristics of the morning and evening soundings favorable for dry-microburst activity over the High Plains. (Bottom) Model of the thermodynamic descent of a dry microburst from cloud base. Entrainment of air from the environment into the downdraft is neglected (from Wakimoto, 1985).

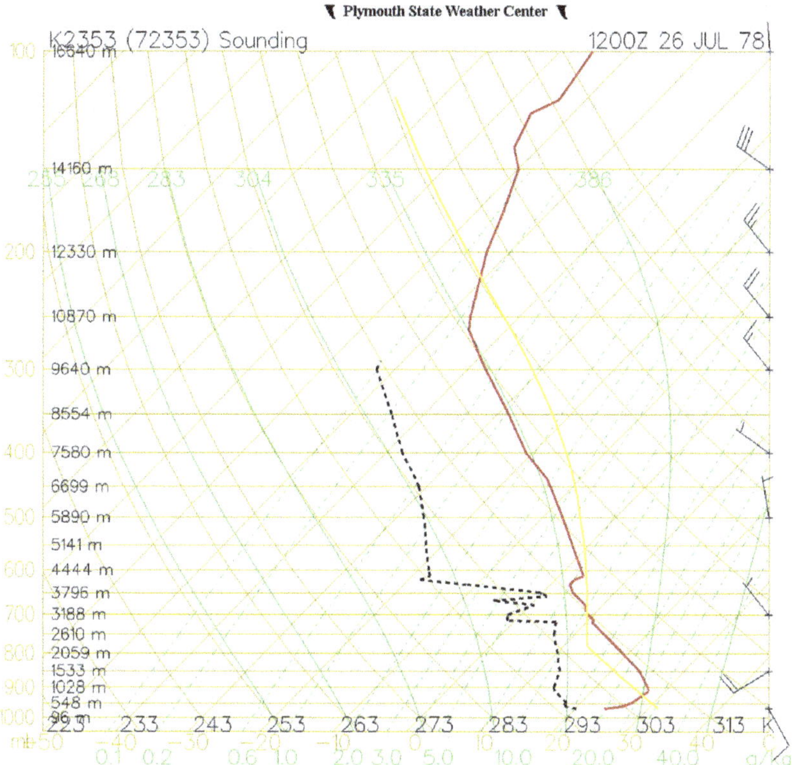

Figure 3.25. Wet-microburst sounding at Oklahoma City, OK at 12:00 UTC on July 26, 1978 (see the storm in Figure 3.22). The notable characteristics of this sounding are the relatively dry air above 600 hPa and the relatively moist air below 700 hPa. There is thus the potential for a relative wet convective storm with heavy water loading and dry air being entrained aloft leading to evaporative cooling aloft.

wet-microburst environment and the non wet-microburst environment is more difficult. Wet microbursts do not always occur when convective storms are possible and the LCL/LFC is relatively low. Nolan Atkins and his mentor Roger Wakimoto at UCLA found that when the difference between the maximum (usually at or near the surface) and the minimum equivalent potential temperature in the afternoon is in excess of 20 K, wet microbursts are likely. When the difference is less than or equal to 13 K, wet microbursts are unlikely. Such an empirical finding supports the obvious notion that mixing of dry, cool air aloft in a storm is conducive to a high evaporation rate, cooling, and production of a negatively buoyant downdraft, and the cool air aloft supports the formation of ice particles, which can subsequently melt and cool as they fall into a layer above freezing.

In the U. S., the Federal Aviation Administration (FAA), with the aid of work at the MIT Lincoln Laboratory, developed a network of C-band (5 cm

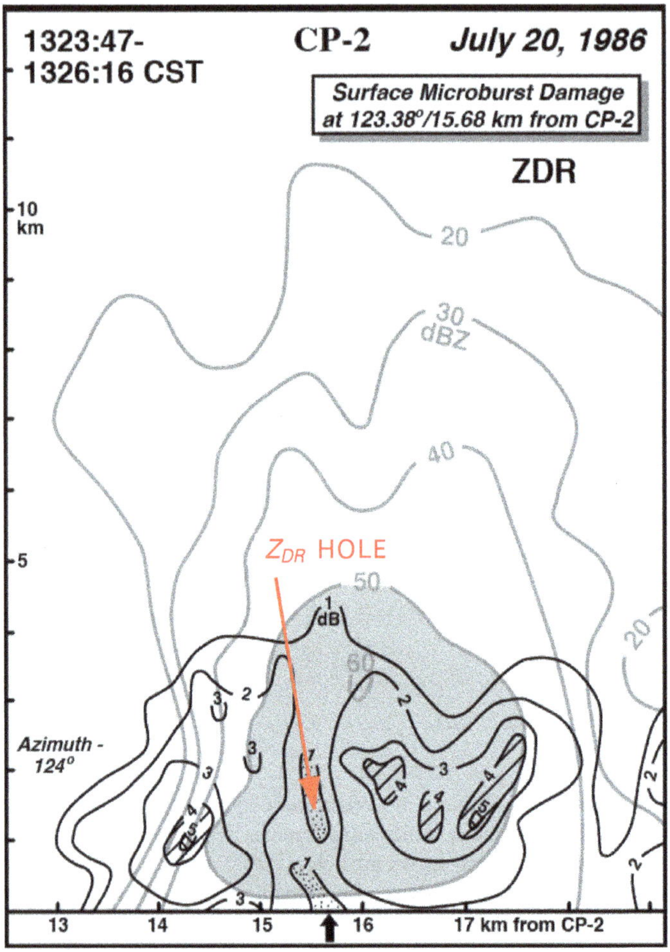

Figure 3.26. Z_{DR} hole in a microburst in Alabama as depicted by data from the NCAR CP-2 Doppler radar (at S-band) during MIST (Microburst and Severe Thunderstorm Project). The radar reflectivity factor is given by thin lines and labeled in dBZ; differential reflectivity Z_{DR} is given by thick lines and labeled in dB. The location of damage is indicated by the arrow at the surface (from Wakimoto and Bringi, 1988).

wavelength) Doppler radars (the Terminal Doppler Weather Radar, TDWR), and installed them at major airports to aid in warnings of microbursts. With the advent of polarimetric radars, the melting of graupel and resultant water-coated ice could be detected as a "hole" in differential reflectivity (Z_{DR}) (Figure 3.26). Raindrops tend to have relatively high differential reflectivity because they flatten as they fall. When low Z_{DR} is found just above a region of high Z_{DR} near the melting layer, it is likely that melting of precipitation is occurring. A narrow region of high reflectivity aloft and low differential reflectivity is indicative of a column of ice particles falling which, when descending, turns into a region of high reflectivity and high differential reflectivity, and is thus suggestive of a developing microburst.

The vertical dynamic pressure gradient force can also play a role in microburst dynamics. At the ground, where there are strong gradients in the wind the squared terms in the diagnostic equation for perturbation pressure (the divergence equation for the Boussinesq system (2.62)) are substantial. From (2.62), we see that this term is associated with a positive perturbation pressure. Above the ground, where the flow is not as divergent or non-divergent and vertical velocity does not change as much with height, the perturbation pressure is much smaller in magnitude. Thus, there is an upward perturbation pressure gradient force that counteracts the downward negative buoyancy force in the microburst near the ground.

Microbursts/downbursts sometimes rotate when there is vertical shear in the environment. A discussion of this topic is delayed until the next chapter, when supercell dynamics is discussed.

3.2.1.2 Heat bursts

When a dry microburst produces sinking air that contains enough momentum to break through a shallow surface inversion, a gust of wind at the surface will be warmer than the surrounding air and in some instances much warmer, by as much as 4–10°C or slightly more (Figure 3.27). Such events, during which relative humidity drops, temperature rises, wind speed increases and becomes gusty, and pressure falls (Figure 3.27), are called "heat bursts"; changes in temperature, dew point, wind, and pressure do not necessarily occur simultaneously. Once thought to be very rare because the chance of one striking a station from the operational synoptic-scale network is very small, they have been documented in the last few decades much more frequently in more closely meshed surface meso-networks.

Since surface inversions are common at night when surface winds are not too strong, heat bursts occur most often at night, especially during the convective season (i.e., spring through summer and early autumn). It can be very startling to experience the temperature at night suddenly rising to daytime levels or even higher. Heat bursts can occur during the day when a surface stable layer is present, but not one due to radiation (e.g., a microburst may hit a shallow cold pool or the top of a shallow front and instigate a heat burst). It is also possible that when a microburst hits a stable layer, gravity waves are triggered. The most dangerous effect of heat bursts is their strong surface winds—not the warm temperatures. Heat bursts tend to be driven by convective systems (but not in every case) and not always as a result of microbursts emanating from them. In Oklahoma, where there has been a long-operating surface mesonetwork, heat bursts have been found underneath or adjacent to weak radar echoes, mostly less than 30 dBZ in reflectivity factor.

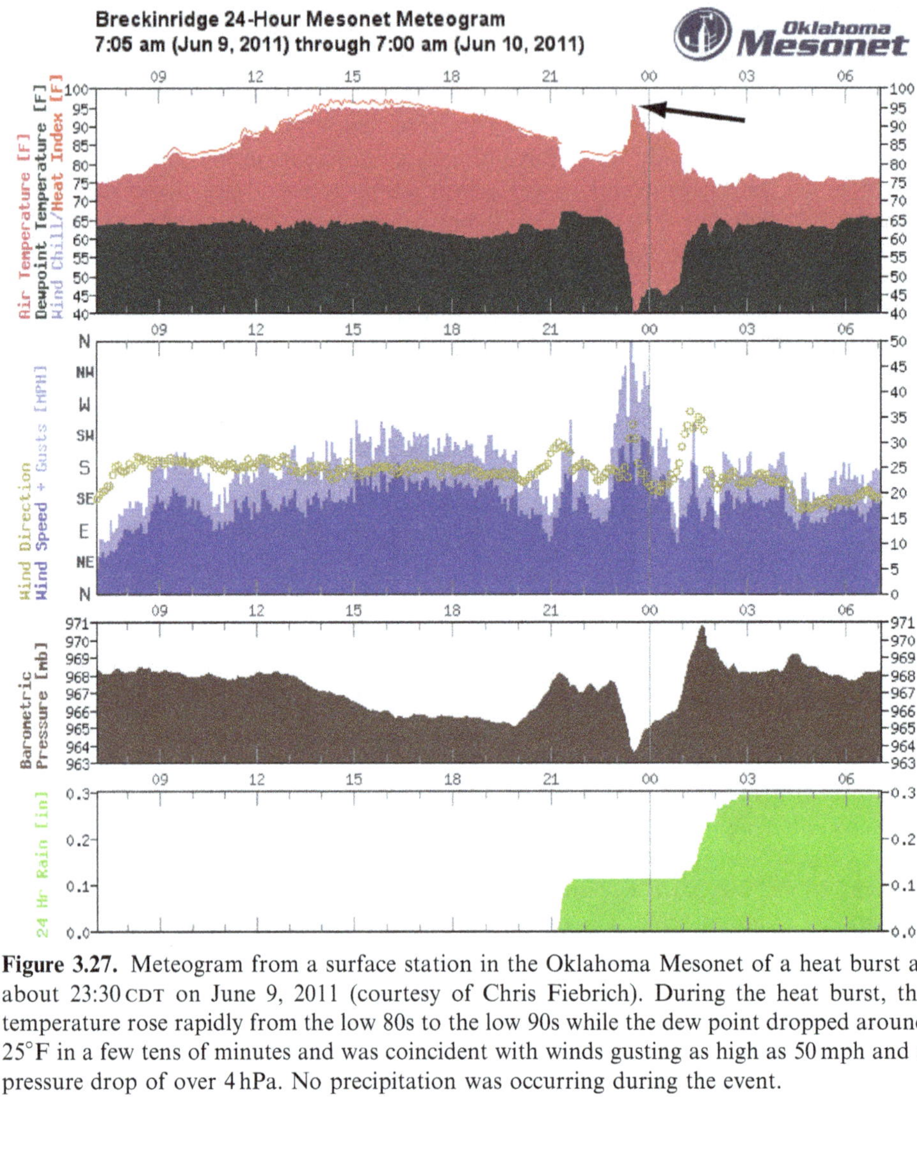

Figure 3.27. Meteogram from a surface station in the Oklahoma Mesonet of a heat burst at about 23:30 CDT on June 9, 2011 (courtesy of Chris Fiebrich). During the heat burst, the temperature rose rapidly from the low 80s to the low 90s while the dew point dropped around 25°F in a few tens of minutes and was coincident with winds gusting as high as 50 mph and a pressure drop of over 4 hPa. No precipitation was occurring during the event.

Heat bursts are therefore triggered when negatively buoyant air, which has evaporated/sublimated all its water substance and become unsaturated, overshoots its equilibrium level on the way down and penetrates a low-level stable layer, where potentially warm air is forced to the surface (Figure 3.28). There is also evidence that the downdrafts that initiate heat bursts can also occur not only when they are forced on the convective scale by negative buoyancy in unsaturated air, but also when they are forced by mesoscale descent at the rear of convective systems. The nature of mesoscale vertical circulations in convective systems will be discussed in Chapter 5.

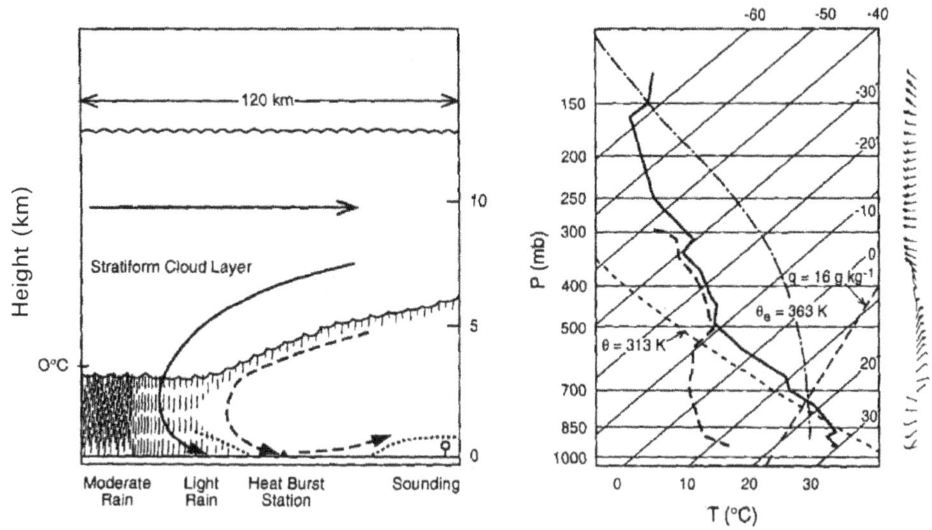

Figure 3.28. Conceptual model of a heat burst as a deformation of a shallow, cool, stable layer at the surface by a descending current of warm, dry air from aloft. Solid arrows (left) represent winds; dashed arrow represents heat burst wind. Dotted lines represent the upper boundary of the stable layer. Sounding (right) at Russell, KS, near a heat burst event. It is seen in the sounding that if saturated air near 500 hPa is cooled by the wet-bulb process until it becomes unsaturated, then below 500 hPa if it is forced downward by mesoscale circulation, it would continue to descend along the 313 K isentrope driven by negative buoyancy and, in the absence of mixing, become positively buoyant at low levels, but still strong enough to penetrate down to the surface (from Bernstein and Johnson, 1994).

3.2.1.3 Cumulonimbus mammatus

Sometimes downdrafts are initiated very high above the ground, from the base of the anvil of a convective storm or above the base of the anvil. The downdrafts may make the underside of the anvil look like an upside-down cumuliform cloud, but smoother and with longer protuberances (Figure 3.29). These features are called in general "mamma" or "mammatus". The latter is short hand for the adjectival modifier applied to the parent cumulonimbus, which in Latin comes after the noun. It is tempting, when looking at them, to believe that they are a consequence of upside-down Rayleigh–Bénard convection. Mammatus may appear to be smoother than upright convection because negatively buoyant thermals continue to move downward beneath their equilibrium level into stable air. It has also been hypothesized that mammatus appear smooth when precipitation falls at the same rate as the downdraft (i.e., when the downdraft and hydrometeor size are both constant) and therefore the precipitation particles remain together. A third hypothesis is that when there is spectrum of ice crystal sizes the smaller crystals sublimate first, mostly along the leading edge of the mamma, leaving the larger crystals behind to appear opaque and smooth: The tops of upright convective elements, on the other hand, do not appear smooth

Figure 3.29a. Cumulonimbus mammatus. (Top) May 4, 1989, Texas Panhandle; (bottom) July 29, 1992, Indian Peaks, CO (photographs by the author).

because the hydrometeors there are small; it may be that the asymmetry due to gravity acting downward on hydrometeors is what makes mammatus appear smoother than the tops of upright convective elements.

It was once thought that mammatus were associated with severe weather and/or tornadoes. A popular book that has been around since the author was very young still perpetuates this myth. Mammatus are sometimes observed not only in cumulonimbus clouds, but also at the base of other stratiform clouds, in jet contrails, and in volcanic ash clouds, which may be composed of both ice crystals and ash particles.

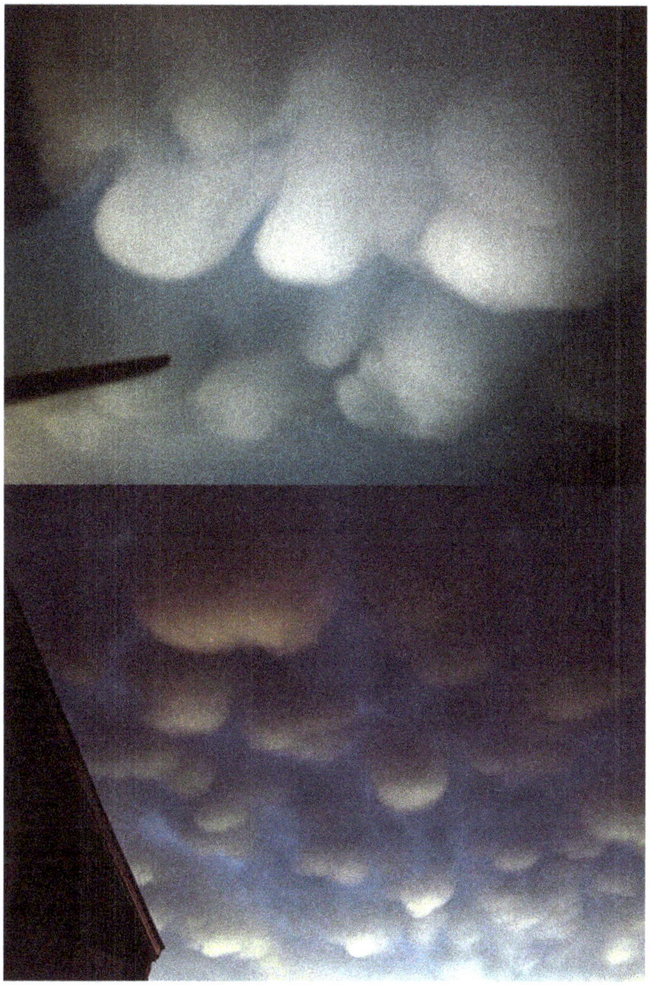

Figure 3.29b. (Top) A close-up view on May 26, 1991, western Oklahoma, from the NOAA P-3 aircraft; (bottom) July 10, 2011, Boulder, CO (photographs by the author).

What we know about mammatus comes mainly from serendipitous visual observations, aircraft penetrations and airborne in situ measurements, and Doppler radar and lidar observations, both airborne and ground based (Figures 3.30 and 3.31). Mammatus in cumulonimbus anvils are usually ~100 m–1 km wide and ~500 m deep. Not all anvils contain mammatus and when mammatus are observed they sometimes occupy only a portion of the anvil. The typical lifetime of one individual mammatus is ~10 min. There are a number of measurements of ice crystals in mammatus, but there are also some observations of liquid water and a mixture of both liquid water and ice. Air descends in the core of mammatus and rises with less speed around the mammatus (Figure 3.31). The appearance of

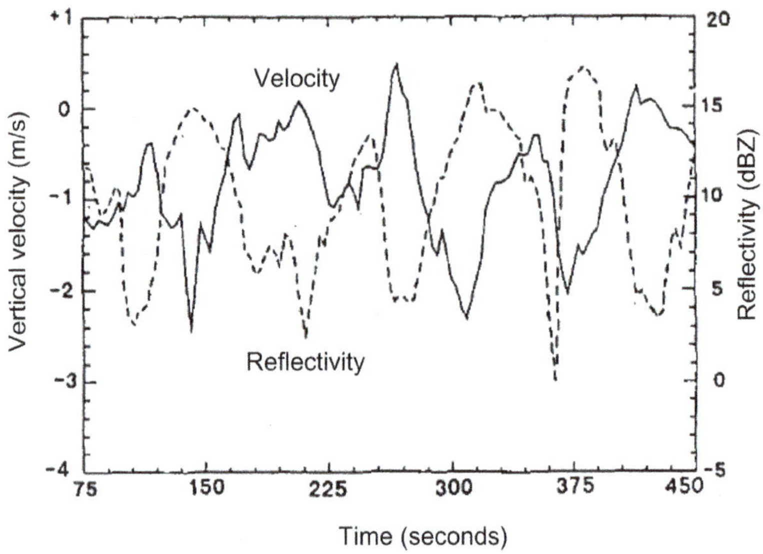

Figure 3.30. Ground-based, Ka-band, vertically pointing Doppler radar observations of cumulonimbus mammatus in north central Manitoba, Canada during the early morning of August 2, 1994. The vertical velocity and radar reflectivity factor are negatively correlated: highest magnitudes of radar reflectivity factor correspond to the strongest descending motion, while the lowest magnitudes of radar reflectivity factor correspond to the weakest descending motion. Since the cloud was translating by, one can convert time to space and infer the spacing between mammatus elements as on the order of a kilometer (from Martner, 1995).

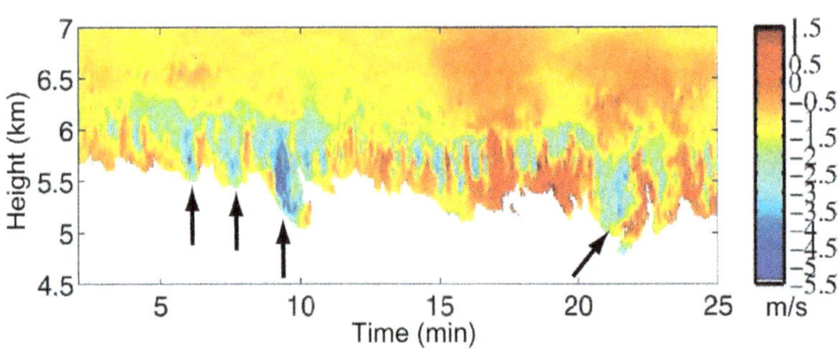

Figure 3.31. Ground-based, W-band, vertically pointing Doppler radar observations of cumulonimbus mammatus in South Florida on July 22, 2002. The mammatus were pendant from a 6 km thick cirrus anvil. Doppler velocity is shown, with upward velocities given as positive, as a function of time; since the cloud was translating by, one can convert time to space and note the similarility in the cloud edge to the pouch-like visual appearance of mammatus, which are spaced about every 200–600 m. The Doppler velocities represent the air motions added to the fall speed of the ice particles. Arrows highlight some individual mammatus elements in which the center of the mammatus is descending (from Kollias *et al.*, 2005).

Figure 3.32. Mammatus under the anvil of a cumulonimbus cloud which are attached to striations or are organized in lines. (Top) Boulder, CO, June 27, 2010. The striations appear to be two-dimensional, linearly symmetric features, as if an individual mammatus pouch were extended into a line. (Bottom) May 8, 2005, Norman, OK (photographs by the author).

mammatus is sometimes preceded by linear striations under the anvil (Figure 3.32), which suggests that gravity waves near the tropopause in the anvil may sometimes trigger mammatus. These gravity waves may be triggered when updrafts near the EL hit the stable tropopause. Aircraft penetrations have found both turbulence and relatively smooth airflow, though turbulence has been found aloft inside the anvil.

There are a number of theories for the generation of mammatus and some combinations of them may be responsible for their appearance. (a) If the air in an anvil subsides, then the air within the anvil warms at the moist-adiabatic rate, while the air being pushed down under the anvil warms at the faster dry-adiabatic rate. The result of these two differing rates of warming is that the lapse rate at the base of the anvil steepens, perhaps to the point of initiating convection. (b) When ice crystals sublimate or raindrops evaporate as they fall from the anvil into unsaturated air below, the air cools and may become negatively buoyant, leading to upside-down convection. (c) When ice crystals fall out of an anvil and pass below the melting layer, latent heat is released and the lapse rate is steepened, perhaps to the point of initiating upside-down convection. (d) When there are local variations in the mass of hydrometeors, the more massive ones fall more quickly and air motion around the leading edge of the falling hydrometeors descends more rapidly, creating segments of upward-flowing air to compensate for the air being moved out of the way (imagine Figure 2.10 upside down). (e) When ice crystals in the anvil flow over unsaturated air below, there is mixing at the interface between the anvil base and the clear air below. Kerry Emanuel in the early 1980s showed that when all the anvil material evaporates (or sublimates), air at its base and underneath may become negatively buoyant. This type of instability is termed "cloud-base detrainment instability" (CDI), which is analogous to "cloud-top detrainment instability", which can occur at the top of clouds when the air above the clouds is cool and dry. If the relative humidity of the air underneath the anvil is too high, evaporative cooling (sublimation cooling) is too slow to counteract adiabatic warming due to subsidence and, if the relative humidity of the air underneath the anvil is too dry, evaporative cooling (sublimation cooling) is too fast and the lapse rate becomes stable. A necessary condition for CDI is that the liquid water or ice static energy[2] ($C_pT + gz + Lr_{l/i}$) of the dry subcloud air is greater than that of the cloud air above (in the anvil). The initial source of mixing might be Kelvin–Helmholtz instability, which occurs when the Richardson number (Ri) is small. The Ri is small when vertical shear is large, as it should be under the base of an expanding anvil and when the lapse rate is low. (f) Radiative cooling at cloud top (due to longwave radiation loss to space) may result in a steep lapse rate within the top of the anvil, leading to convective eddies, some of which reach downward through the anvil, reaching its base. Radiative heating at cloud base (due to upwelling longwave radiation from the ground that cannot pass upward through the anvil) may lead to a steep lapse rate within the bottom portion of the anvil. (g) Double-diffusive convection (due to "biconstituent diffusion"), which has been used to explain some convection in salt water and unsuccessfully to forcing along the dryline, might be used to explain mamma formation: the cloud particles and the air are the two constituents, the former contributing to negative buoyancy and the latter to either positive or negative buoyancy. The diffusion rate of heat is greater for cloud particles than for air. Kathy Kanak and Jerry Straka at OU (University of Oklahoma) found in recent idealized, high-resolution, numerical simulation experiments that CDI is a necessary but not sufficient condition for mammatus formation, and that

[2] For liquid water, r_l is used; for ice, r_i is used.

sublimation is also a necessary condition for mamma formation: the importance of sublimation and dry air underneath the anvil were emphasized. Neither liquid water nor radiative effects were necessary for the process of forming mammatus clouds.

Mammatus are little more than "embroidery" under the anvil and do not appear to play an important role in the dynamics or behavior of severe convective storms. They are included in this text only because they are sometimes prominent visual features and may be of concern to aviation interests.

3.2.1.4 Cold pools and density currents

The depth of a pool of cold air near the ground is important dynamically, because it determines the motion of the leading edge of the cold pool. Generally, the depth of the cold pool behind a gust front ranges from several hundred meters to several kilometers. The deeper and colder a cold pool is, the greater the hydrostatic pressure excess behind the cold pool. At the leading edge of the cold pool, a hydrostatic pressure gradient force is directed from the cold side to the warm side. The leading edge of the cold pool is then forced toward the warm air (Figure 3.33). Much of the cold pool moves as a material surface like a "density current", also known as a "gravity current". The density current is forced by the pressure gradient force acting along it from the denser, colder side to the less dense, warmer side, but is retarded by the drag of the flow along the ground and a dynamic pressure gradient force just along the leading edge.

The reader is referred to the great Boston molasses flood of 1919, when a wall of molasses (surrogate for cold, dense air) from a storage tank swept through a part of the city: This event was probably the ultimate density current, albeit a non-meteorological one. Perhaps an even more dramatic, but fictional, account is that of a "living" density current, the alien life form in the 1958 science fiction/horror movie *The Blob*. More recently, engineers have studied oil spills that behave like density currents. Other examples include gust fronts in thunderstorms in arid regions when almost all of the precipitation has evaporated and the winds have kicked up a wall of blowing dust called a "haboob" (Figure 3.34).

In the simplest model of a density current, the air behind the gust front is assumed to be at rest (there is no density current-relative flow behind the density

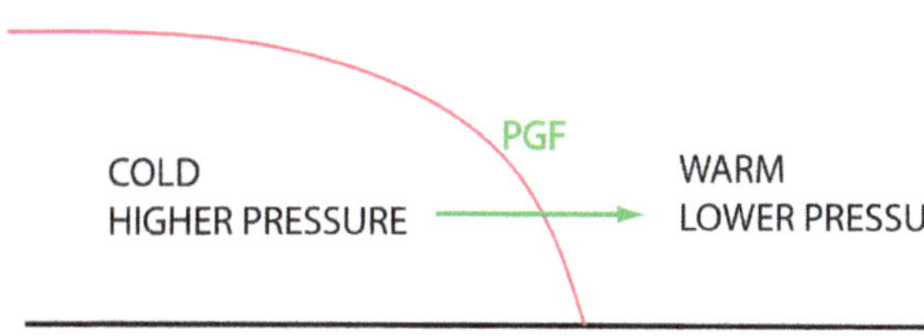

Figure 3.33. Idealized vertical cross section across a cold pool of air near the surface. A horizontal hydrostatic pressure gradient force (PGF) is directed from the cold side to the ambient, warm side at low levels.

Figure 3.34. Haboob in Arizona, August 3, 1978. Note how the leading edge (left side) is wedge shaped (photograph by the author).

current), the cold air mass within the cold pool does not mix with the ambient air outside the cold pool, and the cold air mass is held at a constant temperature. In reality, there is vigorous mixing along the interface of the cold, dense air with the ambient air as a result of the strong vertical shear there and there is some surface drag. Some of the ambient air does mix in with the cold pool and dilute it somewhat, just as environmental air can be entrained into a cumulus cloud and dilute its buoyancy, and Kelvin–Helmholtz waves can occur as a result of the vertical shear and stratification, as $\mathrm{Ri} = (g/\theta\, \partial\theta/\partial z)/(\partial u/\partial z)^2 < \frac{1}{4}$, where u is the wind speed (Figure 3.35). The region of turbulent mixing as the waves break can make for a very bumpy aircraft ride when flying near the top of a gust front.

There are a number of ways of estimating the speed of the density current. The one presented here is based on the seminal work of T. B. Benjamin published in 1968; Mitch Moncrieff and his collaborators in the U. K. in the 1970s and early 1980s extended his work to find analytic nonlinear solutions to two-dimensional, steady-state, squall lines, the simplest possible model of a squall line. Employing this approach, it is assumed that the gust front is two dimensional and steady state. In nature, gust fronts do have some three-dimensional structure and do exhibit a life cycle, but we will neglect any three-dimensional aspects, which are not usually of primary importance, and consider what happens when a gust front is mature and quasi-steady. The Boussinesq equations of motion and continuity are used to find constraints on the "system" (both the gust front and its environment). The "atmosphere" of the fluid is not continuously stratified, but is composed simply of two homogeneous layers: the cold, dense air of density ρ_1 and the ambient, environmental air of density ρ_0, neither of which mix with the other.

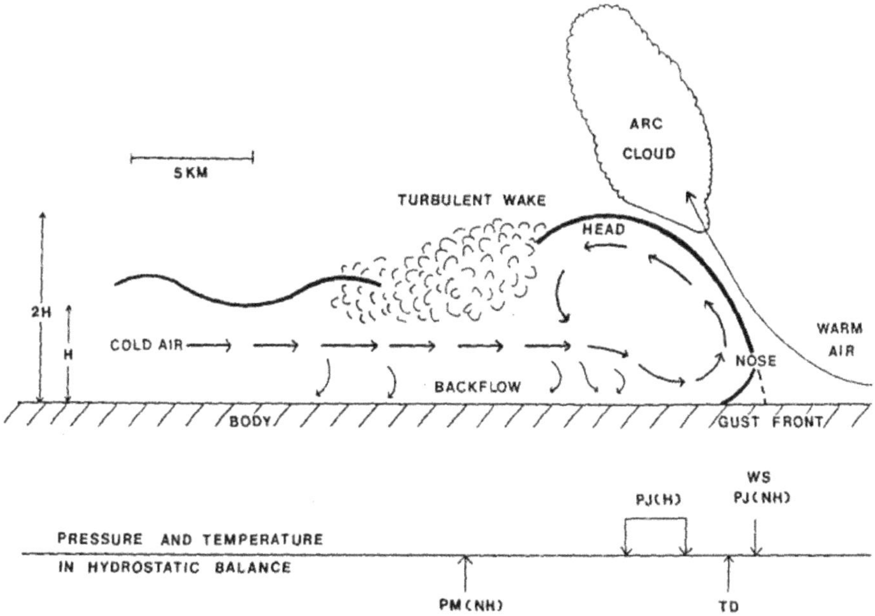

Figure 3.35. Idealized representation of features in an atmospheric density current associated with a gust front in a convective storm, as seen in a vertical cross section and the corresponding changes in meteorological parameters at the surface, where WS is the wind shift and PJ(NH) is the nonhydrostatic pressure jump at the leading edge, PJ(H) is the hydrostatic pressure jump at the leading edge, TD is the temperature drop behind the leading edge, and PM is the pressure minimum (NH, non-hydrostatic) under the turbulent wake (from Droegemeier and Wilhelmson, 1987).

Consider a density current h length units deep[3] embedded in an environment H length units deep ($H \gg h$).

In two dimensions, the Boussinesq equation of continuity (2.35) in the x–z-plane is

$$\partial u/\partial x + \partial w/\partial z = 0 \tag{3.9}$$

We integrate (3.9) over the domain shown in Figure 3.37, so that

$$\iint \partial u/\partial x\, dx\, dz + \iint \partial w/\partial z\, dx\, dz = 0 \tag{3.10}$$

[3] In nature, the top of the cold pool is not necessarily level; in fact, the leading edge of a gust front often has a deeper "head" or nose (Figure 3.35) and the depth of the cold air increases with distance behind the leading edge and so the gust front assumes a wedge shape; Figure 3.36 illustrates the way in which, since the hydrostatic pressure excess behind the density current decreases with height, the cold side/warm side–directed pressure gradient force decreases with height, so that a layer of vertical shear caused by baroclinically generated horizontal vorticity (in terms of $\partial B/\partial x$) at the leading edge is advected up and over the leading edge, leading to a rise and suppression of the height of the cold air mass.

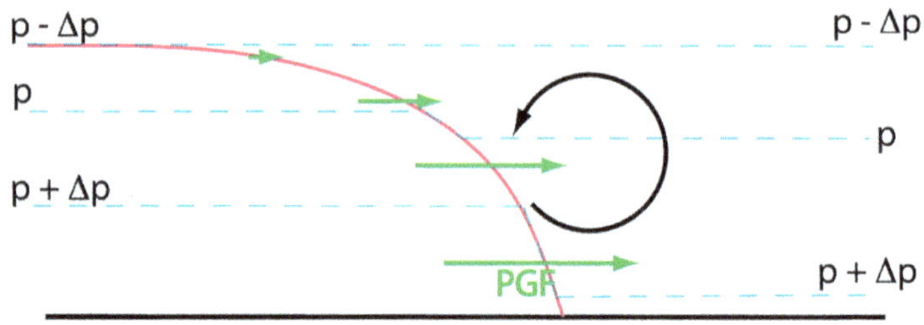

Figure 3.36. Illustration of how the cold side/warm side–directed, hydrostatic pressure gradient force (green vectors) decreases with height in a density current. Vertical cross section showing edge of cold pool (solid red line), hydrostatic pressure (dashed thin blue line), and sense of rotation induced at the leading edge (curved arrow). The cold pool lies to the left of the solid red line, as in Figure 3.33.

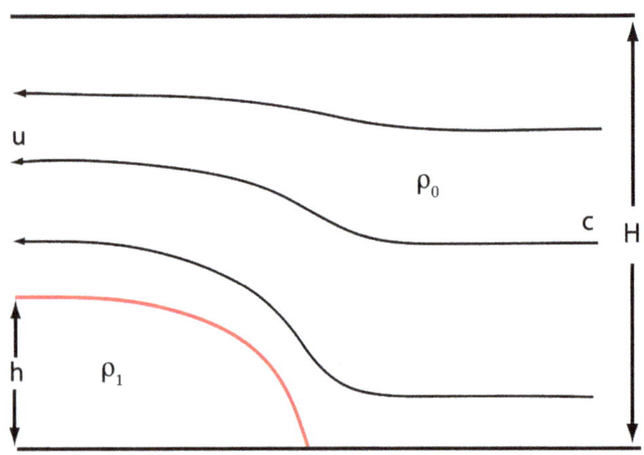

Figure 3.37. Vertical cross section of idealized density current, in which a cold (dense) pool having a density ρ_1 and depth h (edge of which is shown as a solid red line) propagates into an environment of less density ρ_0 and depth H. The density current (cold pool) relative flow from right to left is depicted by the streamlines; at the right edge, the density current relative flow is c to the left; at the left edge above h the density current relative flow is u to the left.

After integrating the first term with respect to x and the second term with respect to z, this equation becomes

$$\int_0^H u\Big|_{-\infty}^{\infty} dz + \int_{-\infty}^{\infty} w\Big|_0^H dx = 0 \tag{3.11}$$

But we set $w(H) = 0$ and $w(0) = 0$ (there is no flow into the domain from the surface or out from the top), so

$$\int_0^H u_\infty \, dz - \int_0^H u_{-\infty} \, dz = 0 \tag{3.12}$$

In the reference frame of the density current (which in Figure 3.37 is moving from left to right) it follows from (3.12) that

$$cH = u(H - h) \tag{3.13}$$

where c is the density current-relative air speed ahead of it (and the speed of the density current); and u is the density current-relative air speed above it. In other words, the air must speed up as it ascends over the density current and is forced into a narrower channel in the vertical. Or the flux of air entering the domain from the right is the same as the flux of air exiting the domain on the left, above the dense, cold air. In nature, the air behind the gust front is not necessarily "resting" and there could be (in most instances there in fact is) vertical shear, so u is a function of z.

For simplicity, it is assumed that the atmosphere is hydrostatic far ahead of the density current and surface friction is ignored (turbulent friction is ignored also above the ground), even though we know that there is surface drag. In terms of specific volume, where $\alpha_0 = 1/\rho_0$, the vertical equation of motion in the resting (i.e., $\bar{u} = \bar{v} = \bar{w} = 0$) density current, for $z < h$, is

$$\alpha_0 \, \partial p'/\partial z = g(\rho_0 - \rho_1)/\rho_0 \tag{3.14}$$

Above the density current, for $z \geq h$,

$$\alpha_0 \, \partial p'/\partial z = 0 \tag{3.15}$$

So, inside the density current we find p' by integrating (3.14) from height z to height h that

$$\alpha_0 p' = \alpha_0 p_t + g[(\rho_1 - \rho_0)/\rho_0](h - z) \tag{3.16}$$

where $p_t = p'(h)$. Above the density current, we find from (3.15) that

$$\alpha_0 p' = \alpha_0 p_t \tag{3.17}$$

At $z = 0$ (where the kinematic lower boundary condition is $w = 0$) in the density current the steady-state, two-dimensional ($\partial/\partial y = 0$), inviscid, horizontal equation of motion is

$$u \, \partial u/\partial x + \alpha_0 \, \partial p'/\partial x = 0 = \partial/\partial x(\tfrac{1}{2}u^2 + \alpha_0 p') \tag{3.18}$$

Equation (3.18) is a Bernoulli-like equation, in that the quantity $\frac{1}{2}u^2 + \alpha_0 p'$ is constant, anywhere along the x-coordinate: In the reference frame of the density current, air approaches along the ground (along the x-axis) from the right and must come to a grinding halt as it encounters the wall of cold air, the "nose" of the advancing cold air (x decreases, so p' must increase); therefore the air parcel encounters an adverse pressure gradient force and it decelerates (Figure 3.38). The reader is cautioned that the relationship between p' and $\mathbf{v}$ is *not* causal: one does not cause the other; each must *accompany* the other, as a consistency argument. The pressure field is disturbed not just at the leading edge, but also *ahead* of the leading edge of the density current in an "action at a distance"-like manner (the air ahead of the leading edge of the advancing density "knows" about the existence of the density current before encountering it as a result of sound waves

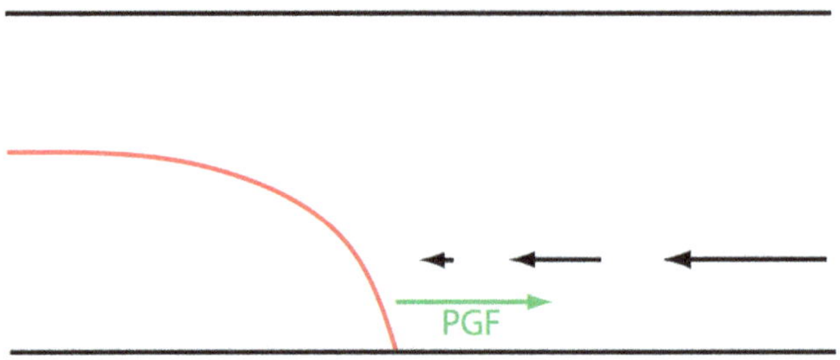

Figure 3.38. Illustration showing how gust front relative flow (black vectors) is decelerated as it encounters an adverse, dynamic (non-hydrostatic) pressure gradient force (green vector labeled PGF). The gust front relative wind comes to a grinding halt at the edge of the cold pool (solid red line) because we have assumed that there is no mixing between the cold pool air and the ambient air and that we are in the reference frame of the density current.

traveling at infinite speed—a real fast-talking gust front!). The wedge of an immiscible cold air mass advancing against a resting air mass is like an airfoil advancing against a resting air mass: air slows down and flows up and over the air foil (Figure 3.39) and lift is created as an upward-directed perturbation pressure gradient.

Another way to diagnose the perturbation pressure is to use the inviscid form of the divergence equation (2.62) applied to this simple model

$$1/\rho\,\nabla^2 p' = -(\partial u/\partial x)^2 - (\partial w/\partial z)^2 - 2(\partial w/\partial x\,\partial u/\partial z) + \partial B/\partial z \qquad (3.19)$$

Consider first what happens just ahead of the density current, where $B = 0$. Since $\partial u/\partial x$ and, by continuity, $\partial w/\partial z$, are both nonzero just ahead of the leading edge of the density current, it follows that the first two terms on the right-hand side of (3.19) are negative and that the last term is zero. Also, $\partial w/\partial x < 0$, but $\partial u/\partial z = 0$. So, $p' > 0$. Far ahead of the density current $\partial u/\partial x$ and $\partial w/\partial z$ are both zero, so $p' = 0$. Just above the density current, $\partial B/\partial z > 0$, so there is a negative contribution to p'; this is also the case when the leading edge of the density current is wedge shaped, so that $\partial B/\partial z > 0$ everywhere at its interface.

The perturbation pressure high ahead of the leading edge of the density current, which is dynamically induced, falls off with distance ahead of the density current. If it did not exist, then there would be a discontinuity in pressure across the leading edge of the density current, which would violate the dynamic boundary condition, and the horizontal pressure gradient at the leading edge, which is hydrostatic, would be infinite. The non-hydrostatic high pressure allows the pressure to be continuous across the leading edge.

We now integrate (3.18) from the far left, well behind the leading edge of the gust front (i.e., the leading edge of the dense air mass), where $u(-\infty) = 0$, to the far right, well ahead of the gust front, where $u(\infty) = -c$ and $p' = 0$: We find that

$$\tfrac{1}{2}c^2 = \alpha_0 p_t + gh[(\rho_1 - \rho_0)/\rho_0] \qquad (3.20)$$

(a)

(b)

Figure 3.39. (a) Illustration of how air flowing up and over a cold pool behaves like air flowing up and over an airfoil. (Top) Vertical cross section showing airflow (streamline) relative to an airfoil shaped like a cold pool wedge; a dynamic high-pressure area slows the air down as it encounters the airfoil (in the reference frame of the moving airfoil) and is lifted upward. (Bottom) Shelf (or arcus) cloud produced as stable air is lifted to saturation at the leading edge of a gust front (advancing cold pool), early in the morning of May 27, 1977, in Norman, OK (photograph by the author). (b) Underside of air flowing up and over a cold pool: the "Whale's Mouth" in north central Oklahoma on May 15, 2009. The underside is ragged and looks turbulent, as there is mixing at the interface between the cool air below and the warmer air aloft (photograph by the author).

The above is not a satisfactory equation for finding c because we have an unknown constant p_t. To eliminate p_t, we must find an independent equation in which p_t appears. We consider the horizontal equation of motion again, but this time we consider it everywhere—not just at $z = 0$. The inviscid, steady-state, two-dimensional equation of motion for this case is

$$u\, \partial u/\partial x + w\, \partial u/\partial z + \alpha_0\, \partial p'/\partial x = 0 \tag{3.21}$$

which is similar to (3.18), except that the vertical advection term is retained. Combining (3.21) with the Boussinesq two-dimensional equation of continuity

$$\partial u/\partial x + \partial w/\partial z = 0 \tag{3.22}$$

we find that

$$\partial u^2/\partial x + \partial/\partial z(uw) + \alpha_0\, \partial p'/\partial x = 0 \tag{3.23}$$

We now integrate (3.23) over the x–z-plane from $x = -\infty$ to $x = \infty$ and from $z = 0$ to $z = H$ (i.e., over the entire domain), making use of (3.16) and (3.17), and eliminate u using the mass conservation relation (3.13), to find that

$$c^2 Hh/(H - h) = -\alpha_0\, p_t H - \tfrac{1}{2} gh^2(\rho_1 - \rho_0)/\rho_0 \tag{3.24}$$

Eliminating p_t from (3.24) and (3.20), we find that

$$c^2 = gh[(\rho_1 - \rho_0)/\rho_0]\{2[(H - h)/(H + h)][(H - \tfrac{1}{2}h)/H]\} \tag{3.25}$$

Alas, it is not obvious whether or not (3.25) will work for all values of h/H, so we must find yet another constraint to nail down c. To do so, we now make use of the Boussinesq vorticity (the y-component) equation (2.51) in flux form, subject to the conditions of being steady state and two dimensional ($\partial/\partial y = 0$), and to the continuity equation (3.22):

$$\partial(u\eta)/\partial x + \partial(w\eta)/\partial z = -\partial B/\partial x \tag{3.26}$$

where $\eta = \partial u/\partial z - \partial w/\partial x$, the y-component of the vorticity vector. Integrating (3.26) over the entire domain and using the upper and lower kinematic boundary conditions $w = 0$ at $z = 0$ and $z = H$, and assuming that $w = 0$ at $x = \pm\infty$ (altogether assuming that $w = 0$ along all the edges of the domain), it is found that

$$\tfrac{1}{2} u^2_{H,-\infty} = gh[(\rho_1 - \rho_0)/\rho_0] \tag{3.27}$$

We need to find an expression now for c in terms of $u_{H,-\infty}$ and eliminate the latter. To do so, we use of the conservation of mass relation (3.13), noting that $u \equiv u_{H,-\infty}$ and find that

$$c^2 = gh[(\rho_1 - \rho_0)/\rho_0]2[(H - h)/H]^2 \tag{3.28}$$

We now have two independent equations for c^2. Setting (3.28) and (3.25) equal to each other, we obtain the following relationship between H and h:

$$(H - h)^2/H^2 = [(H - h)/(H + h)][(H - \tfrac{1}{2}h)/H] \tag{3.29}$$

This equation has solutions $h = H/2$, $h = H$, and $H = \infty$. The solution $h = H$ is non-physical, because there no longer is a density current: the cold, dense air mass takes up the entire depth of the atmosphere and there is no place for advancing air to flow left of the edge of the cold pool or above the domain at the top, so that $c = 0$. This case has been referred to as the "lock exchange" problem. Klemp *et al.* in 1994 showed that it requires time-dependent solutions (the cold, dense air does not remain locked up, but is released, like water held up by a dam which is suddenly released, and its structure evolves with time). In any event, it is also a case that is not relevant meteorologically because cold pools in convective storms do not extend up to the troposphere. The solution for the very restrictive case when $h = H/2$ is

$$c^2 = \tfrac{1}{2}gh[(\rho_1 - \rho_0)/\rho_0] \tag{3.30}$$

Klemp *et al.* in 1994 showed that this steady-state solution is also non-physical because energy is not conserved unless turbulent friction is included. The more relevant solution is for the case when $H \rightarrow \infty$ (i.e., or when $h \ll H$ and the cold pool is relatively shallow) so that

$$c^2 = 2gh[(\rho_1 - \rho_0)/\rho_0] \tag{3.31}$$

This formula appears frequently in the literature and is used very often. The speed c of the *shallow* cold pool, in the absence of surface drag is therefore

$$c = \{2gh[(\rho_1 - \rho_0)/\rho_0]\}^{1/2} \tag{3.32}$$

If surface drag is included and its effect is assumed to slow down the movement of the cold pool, then

$$c = K\{gh[(\rho_1 - \rho_0)/\rho_0]\}^{1/2} \tag{3.33}$$

where K is an empirical constant ~1–1.5. The formula (3.32) is similar to that for the phase speed of a shallow-water gravity wave, even though in the case of the latter net mass is not transported along with the wave, while in the case of the former mass is transported by the gust front. If a density current enters a region where there is a strong stable layer, then gravity waves (a bore) may be generated, but one can still distinguish between the density current and gravity waves above it. Formulas (3.32) and (3.33) may be expressed more conveniently for use with meteorological data by using the expression for buoyancy in terms of potential temperature (2.148) (instead of in terms of density) as follows:

$$c = \{2gh[(\theta_0 - \theta_1)/\theta_0]\}^{1/2} \tag{3.34}$$

and

$$c = K\{gh[(\theta_0 - \theta_1)/\theta_0]\}^{1/2} \tag{3.35}$$

where θ_0 is the average potential temperature of the environmental air; and θ_1 is the cooler, average potential temperature of the cold pool.

In the reference frame of the moving density current, ambient air slows down as it approaches the leading edge and rises up and over the leading edge (Figure 3.38). When the ambient air is unsaturated and stable for the displacements it undergoes as it rises over the cold pool, a laminar shelf or arcus cloud may form

(Figure 3.39). Evidence of turbulent mixing underneath the arcus is often seen and colloquially known by storm-chasers as the "whale's mouth", since it has the texture of the inside of the mouth of the whale (actually simply the "Terrible Dogfish") in Disney's animated version of the children's tale *Pinocchio*.

Because air moves up and over the cold, low-level pool of air associated with the density current, the air acts as if it is being lifted over a mountain range. Of course, in this case it is not really moving over a solid body and there is some mixing at the interface between the ambient air and the colder air below. Under the right conditions, lifting of the air triggers gravity waves, which behave like waves in the lee of a mountain range. Such waves, depending on static stability, the strength of the wind, and vertical shear, may be vertically propagating or trapped. The reader is referred elsewhere for detailed analyses of mountain waves. Above a density current, gravity waves may modulate the production of clouds and also provide sources of horizontal vorticity parallel to the leading edge of the density current.

Studies of density current behavior for conditions that depart from the highly idealized ones we have assumed have been done using both analytical and numerical models. A basic problem is to determine the *transient* behavior of the cold pool under idealized conditions (two dimensions, inviscid, etc.). That is, we start out with a cold pool and let it go such that it eventually may reach a quasi-steady state, as happens in convective storms: A cold pool forms and then begins to propagate and may reach a steady state. In nature, of course, the cold pool itself takes time to be built up as well. Microphysics is obviously very important and so is the nature of the airflow. What happens before the cold pool reaches a steady state, if it ever does? This problem has been referred to as the dam-breaking problem, as noted previously, owing to its analogy to a wall of water constrained by a dam, which is suddenly opened and involves an analysis of propagating gravity waves generated by the density current, which is beyond the scope of our introductory discussion.

In our analysis, we have assumed that the density current is stagnant and that there is no environmental flow ahead of it. The motion, relative to the movement of the updraft(s) in a storm, of a density current generated in a convective storm is important for the subsequent evolution of the storm: If the density current outruns the updraft, then the updraft will no longer have potentially unstable low-level environmental air available to feed it. If the density current moves along with the updraft, at the same speed that the updraft is moving, then the updraft may be long lived. Strong opposing flow in the environment slows down the motion of the density current and may even stop it. Of equal or greater importance for the initiation of convection is environmental vertical shear.

3.2.2 Gust fronts in the presence of vertical shear: RKW theory

Typically, in the absence of vertical wind shear, a cold pool will spread out (like a pancake) at the surface and new convective cells will not be initiated along the periphery of the cold pool in response to the lifting of ambient air, unless the

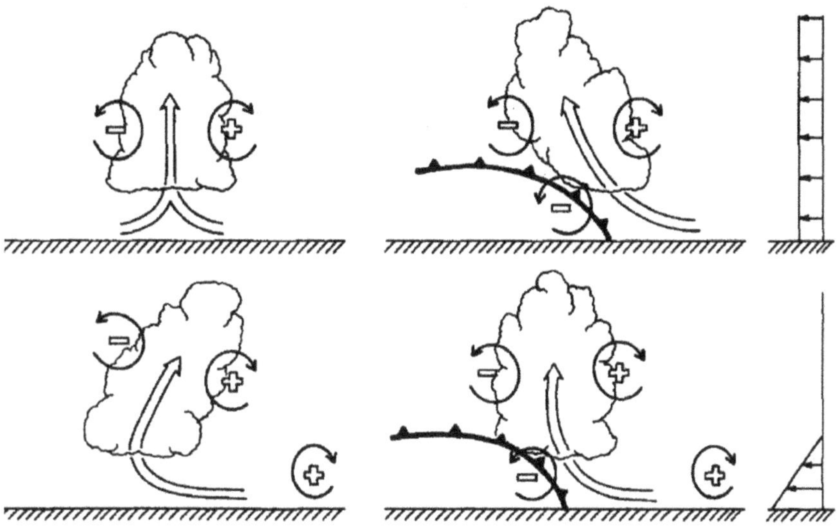

Figure 3.40. Idealized illustration of how a buoyant updraft may be influenced by vertical wind shear or a cold pool or both. (Top left) Positive and negative vorticity are induced by horizontal gradients in positive buoyancy near the right and left edges of the cloud. Positive and negative signs refer to vorticity vectors pointing into and out of the page, respectively. With no environmental vertical shear and no cold pool, the axis of the updraft forced by the thermally created, symmetric horizontal vorticity distribution is vertically oriented. (Top right) With a cold pool, the distribution of horizontal vorticity is biased by the negative horizontal vorticity induced at the leading edge of the underlying cold pool and causes the updraft to lean in the downshear (with respect to shear produced by horizontal buoyancy gradients, not the environmental shear, which in this case is zero; this is indicated by the vertical distribution of winds in the environment at the upper right) direction. (Bottom left) With vertical shear in the environment at low levels as indicated at the lower right, but no cold pool, the distribution of horizontal vorticity is biased toward positive vorticity (into the page) and the updraft leans in the downshear (with respect to the environmental shear) direction. (Bottom right) With both a cold pool and vertical shear, the two effects may negate each other, allowing the formation of an erect updraft (from Rotunno *et al.*, 1988). (The flow fields depicted in the left-hand panels do not evolve into those at the right after precipitation develops; the left- and right-hand panels are independent of each other and used for illustrative purposes only.)

magnitude of the CIN is very small. However, if the vertical wind shear vector over the depth of the cold pool is oriented so that the horizontal vorticity vector associated it with has a substantial component in the direction opposite to that of the baroclinically generated horizontal vorticity at its leading edge (cf. the buoyancy term in (2.51)), then the lift along the leading edge of the cold pool will be more vertically oriented (Figure 3.40) so there is a greater likelihood that air will be lifted enough to reach its LCL and trigger a new convective cell.

This behavior can be understood in terms of the horizontal vorticity equation (2.51) expressed in flux form and for a steady state as (3.26). Remember, also, we are neglecting, for simplicity, Earth's rotation and friction. We integrate (3.26) over a specific portion of the domain in the vertical plane (Figure 3.41): from

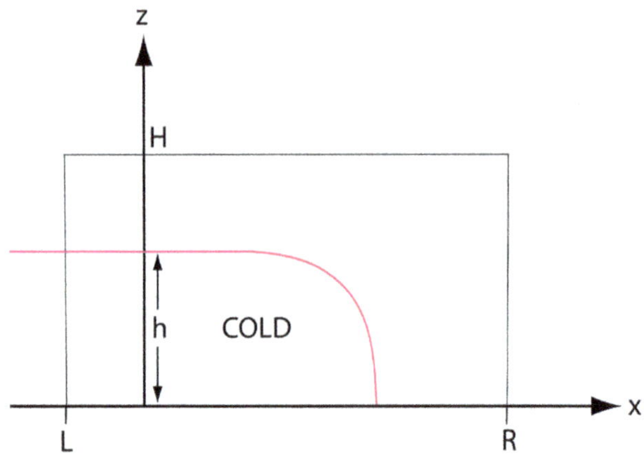

Figure 3.41. Vertical cross section across the leading edge of the cold pool, showing the domain used over which the steady-state, frictionless, horizontal vorticity equation in flux form is integrated.

$x = L$ (behind, to the left of the leading edge of the cold pool) to $x = R$ (ahead, to the right of the leading edge of the cold pool), and from $z = 0$ (the surface) up to $z = H$ (the top of the atmosphere, well above the top of the cold pool at $z = h$). We find that

$$\int_0^H \int_L^R \partial/\partial x(\eta u)\, dx\, dz + \int_L^R \int_0^H \partial/\partial z(\eta w)\, dz\, dx = -\int_0^H \int_L^R \partial B/\partial x\, dx\, dz \quad (3.36)$$

The physical meaning of (3.36) is that, when there is no net generation of horizontal vorticity η locally, the net baroclinic generation of η (RHS) must be balanced by the net flux divergence of η (LHS): Horizontal vorticity gets "out of Dodge" as quickly as it is generated. (We will use (3.36) with slightly different limits at the top of the domain shortly.) Note that (3.36) is similar to (3.26) integrated over a domain in the x–z-plane as before, except that the limits of integration in the x-direction are from L to R—not $-\infty$ to $+\infty$.

Now, suppose that there is (constant and unidirectional) vertical shear in the environment (Figure 3.42) such that

$$\Delta u \equiv u_{R,h} - u_{R,0} \quad (3.37)$$

where the winds in the reference frame of the density current are directed from right to left. In other words, there is low-level shear over the depth of the cold pool of $\Delta u/h$, where $\Delta u > 0$. We let the shear be restricted to the depth of the cold pool only, so that

$$u_{R,H} = 0 \quad (3.38)$$

In addition, *suppose that airflow is straight up at the leading edge of the density current* (Figure 3.42), which is consistent with the condition that

$$u_{L,H} = 0 \quad (3.39)$$

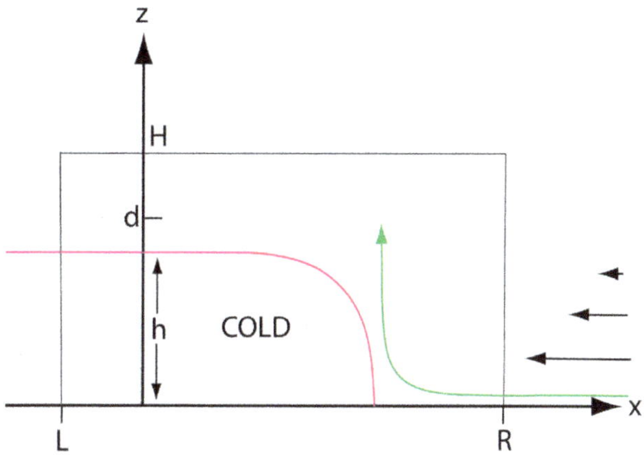

Figure 3.42. Demonstration of "optimum" orientation of flow normal to a cold outflow (gust front, density current) from a precipitating convective cloud when environmental shear is as indicated by the vertical profile of gust front relative winds below $z = h$ at the lower right. Vertical cross section across the leading edge of the cold pool. The red line encloses the cold outflow; the green streamline indicates the vertically upright flow along the leading edge of the outflow. Other symbols are explained in the text.

Under what circumstance(s) is this possible? To find out, we integrate the steady-state, inviscid, flux form of the horizonal vorticity equation as we did for (3.36), but this time only over that part of the domain that extends up to a height that is higher than the top of the cold pool, but not higher than the top of the previous domain; we integrate only up to $z = d < H$, but $z > h$. It follows, then, from (3.37), (3.38), and (3.39) (Figure 3.42) that

$$u_{R,d} - u_{R,0} = \Delta u \equiv u_{R,h} - u_{R,0} \tag{3.40}$$

and that

$$u_{L,d} = 0 \tag{3.41}$$

We first show that the second term on the LHS of (3.36) vanishes. Integrating with respect to height first, we find that

$$\int_L^R (w\eta)\Big|_0^d dx = \int_L^R (w\eta)_d \, dx \tag{3.42}$$

because $w(z = 0) = 0$. Now,

$$\eta_d = (\partial u/\partial z)_d - (\partial w/\partial x)_d = -(\partial w/\partial x)_d \tag{3.43}$$

because it is seen in Figure 3.42 that

$$(\partial u/\partial z)_d = 0 \tag{3.44}$$

If there is an erect updraft at the leading edge of the density current and $w = 0$ at $z = 0$ at $x = R$ and $x = L$, then we may assume that the updraft is (locally) symmetric about the leading edge of the cold pool so that

$$(w\eta)_d = -w_d(\partial w/\partial x)_d \tag{3.45}$$

is *anti*-symmetric about the updraft. Then the integral of (3.45) from L to R vanishes. The first term on the LHS of (3.36) integrated also up to only $z = d$ is

$$\int_0^d \int_L^R \partial/\partial x(\eta u)\, dx\, dz = \tfrac{1}{2}u_{R,d}^2 - \tfrac{1}{2}u_{R,0}^2 - (\tfrac{1}{2}u_{L,d}^2 - \tfrac{1}{2}u_{L,0}^2) \tag{3.46}$$

where it is recognized that $\partial w/\partial x\, u|_L^R = 0$ except right near the leading edge of the density current, but certainly not at $x = L$ and $x = R$. Furthermore, if the density current is stagnant, $u_{L,0} = 0$; also, recall that $u_{L,d} = 0$ (3.41). From Figure 3.42 we see that

$$u_{R,d} = 0 \tag{3.47}$$

Then from (3.40), and recognizing that $B_R = 0$, we find that (3.36) integrated to $z = d$ becomes

$$-\tfrac{1}{2}u_{R,0}^2 = -\tfrac{1}{2}(-\Delta u)^2 = -\tfrac{1}{2}(\Delta u)^2 = B_L h = -g[(\rho_1 - \rho_0)/\rho_0]h \tag{3.48}$$

so that

$$(\Delta u)^2 = 2gh[(\rho_1 - \rho_0)/\rho_0] \tag{3.49}$$

But from (3.31) and (3.49) it follows that

$$c = \Delta u \tag{3.50}$$

In other words, to get a vertically erect updraft at the leading edge of a steady-state, frictionless, stagnant density current, the speed of the density current must be identical to the difference in horizontal wind across the shear layer that encompasses the depth of the cold pool. Such a situation is said to be "optimal" for triggering new convection at the leading edge. Physically, this condition means that when the rate of generation of horizontal vorticity baroclinically ($-\partial B/\partial x$) at the leading edge of the density current is counterbalanced by the advection of (import from) horizontal vorticity from vertical shear in (from) the environment [$-u\,\partial/\partial x(\partial u/\partial z)$], there is a maximum in upward motion along the leading edge of the density current and the probability of triggering a discrete new cell is increased (Figure 3.40). This behavior is described in a theory known as the "RKW theory", after the NCAR scientists Rich Rotunno, Joe Klemp, and Morris Weisman, who proposed it in the late 1980s.

We now *include* low-level vertical shear ($\Delta u/h_0$) whose orientation is parallel to density current, in an analytical model of a steady-state, two-dimensional, inviscid density current and seek an expression for the speed of the density current in terms of the dimensions and strength of the cold pool. We are particularly interested in how it differs from (3.32). The model setup (Figure 3.43) is similar to the one for no shear, but there are now *three* airstreams—not just two—and we do not require the airflow to be erect at the leading edge of the density current. There is still the stagnant/resting cold pool, but the flow regime ahead of and above the density current is broken down into two sub-airstreams. First, the low-level shear

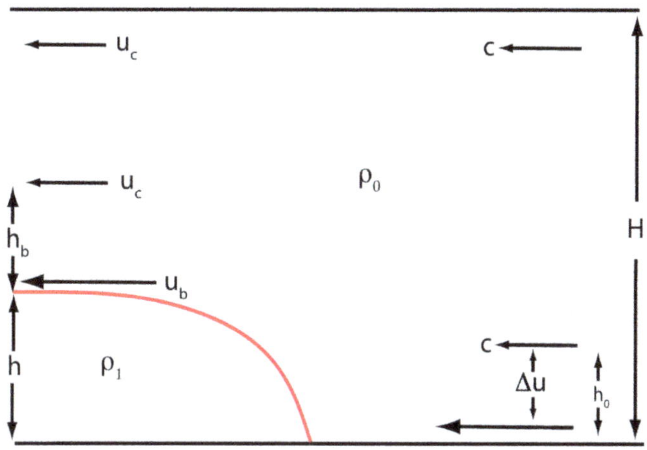

Figure 3.43. As in Figure 3.37, but at the right edge of the domain the flow is uniform and from right to left at speed c, above $z = h_0$, while the wind speed increases linearly by Δu with decreasing height until at the surface ($z = 0$) c is $c + \Delta u$ from the right; at the left edge of the domain, the wind $u = u_c$ (from the right) above $z = h + h_b$, while the wind speed increases linearly with decreasing height to u_b at the top of the cold pool, at $z = h$. If there is a shear layer on the right-hand side, there must also be one on the left-hand side because, following the streamlines which do not pass through the cold pool, horizontal vorticity is conserved (based on an illustration in Emanuel, 1994).

layer moves up and over the density current (in the reference frame of the density current). Second, there is an airstream having no vertical shear that lies above the shear layer. In the case of low-level vertical shear, the depth of the shear layer way ahead of the cold pool is h_0, which is not necessarily the same as h (the depth of the cold pool) and the depth of the shear layer above the density current is h_b; the flow relative to the density current is u_b in the shear layer and u_c above, in the no-shear layer ($u = 0$ in the cold pool, as before).

The basic equation is (3.26), the flux form of the vorticity equation. Integrated over the entire domain of the density current, we find that:

$$\tfrac{1}{2}(c + \Delta u)^2 - \tfrac{1}{2}c^2 + \tfrac{1}{2}u_c^2 = gh[(\rho_1 - \rho_0)/\rho_0] \tag{3.51}$$

where $w = 0$ at $z = 0$ and H, $\partial w/\partial x = 0$; and $B = 0$ at $x = \pm\infty$, $u_{\infty,H} = -c$, $u_{\infty,0} = -(c + \Delta u)$, and $u_{-\infty,H} = -u_c$, and $u_{-\infty,0} = 0$. Conservation of mass for the no-shear layer is found again by integrating (3.9) over the x–z-plane:

$$u_c[H - (h + h_b)] = c(H - h_0) \tag{3.52}$$

We could have used as an additional constraint the macroscale conservation of mass in the shear layer, but it is not needed for our analysis of the special case of a "shallow" density current that we are going to consider. Eliminating u_c from (3.51) and (3.52), we find that

$$c^2[(H - h_0)/(H - h - h_b)]^2 + 2c\,\Delta u + (\Delta u)^2 = 2gh[(\rho_1 - \rho_0)/\rho_0] \tag{3.53}$$

This quadratic equation can be solved for c, but one would need to make use of additional constraints on H, h, h_0, and h_b and to find what combinations of these parameters allow for steady-state solutions. It is most informative for pedagogical purposes to consider the simplified, but common case of a shallow density current (i.e., when $H \rightarrow \infty$, or when $h \ll H$).

For a shallow density current, the coefficient in the leading (quadratic) term for c approaches unity, so the solution to (3.53) is

$$c = -\Delta u + \{2gh[(\rho_1 - \rho_0)/\rho_0]\}^{1/2} \tag{3.54}$$

The negative root is rejected as a solution because it is non-physical in that the density current incorrectly would move from the warm side to the cold side.

This expression for speed (3.54) is similar to that without shear (3.32), but now includes a "$-\Delta u$" term. When

$$\Delta u = \{2gh[(\rho_1 - \rho_0)/\rho_0]\}^{1/2} \tag{3.55}$$

$c = 0$, so that *low-level shear effectively negates the motion of the cold pool* into the ambient warmer air mass. We get the same result (i.e., $c = 0$) with *no* vertical shear, however, if we impose an opposing current from the right that is equal to c.

Imagine an air parcel (in the reference frame of the leading edge of the cold pool) approaching the leading edge of the cold pool from the right in the presence of low-level shear in the left-to-right direction (Figure 3.43). The shear necessary to slow it down or even stop it is in the same sense as that required by RKW theory to make parcel displacements at the leading edge of the cold pool more erect; however, the steady flow pattern assumed by RKW theory, with the erect updraft (Figure 3.42), is not identical to the steady flow pattern assumed in the derivation of (3.55) (Figure 3.43), so that we should be careful not to confuse these two findings.

The effects of varying the depth of a low-level layer of constant shear (with respect to the depth of the cold pool) and of varying the magnitude of the shear have been studied numerically by Ming Xue and collaborators with free slip upper and lower boundary conditions in a confined vertical channel. In our previous analyses, we considered only steady-state, inviscid solutions. In nature we find *transient* Kelvin–Helmholtz waves generated above the leading edge of the density current, which then propagate rearward (from right to left). These eddies are generated baroclinically at the leading edge, and are swept rearward by the front-to-rear flow. In nature, cold pools encounter environments with deep vertical shear, not just vertical shear over the depth of the cold pool, and vertical shear which changes direction with height, not just vertical shear that is normal to the cold pool. In addition, periodic bursts of precipitation can lead to cooling due to time-varying amounts of evaporation or melting, which lead to changes in the cold pool strength and depth. Also, it is not clear how deep the model should be. Does the tropopause confine the behavior of the model as a rigid lid, or should we include air exchange in penetrating tops? Finally, if the cold pool persists for many hours, the Coriolis force can play a role. Our idealized models must then be used with extreme caution when extrapolating to the more complicated conditions in real convective storms.

3.2.3 Gravity waves forced by a density current

Since, relative to the leading edge of the gust front, air is forced up and over it, gravity waves may be triggered there when the necessary conditions for gravity wave propagation are satisfied. The reader is referred elsewhere for details on gravity wave propagation. When the Scorer parameter ℓ, defined for an incompressible atmosphere as

$$\ell^2 = N^2/u^2 - 1/u\, d^2u/dz^2 \tag{3.56}$$

decreases rapidly with height, where u is the wind speed in a unidirectional wind profile, it is possible that gravity waves may propagate vertically and then be reflected back downward, leading to resonance and a series of "trapped gravity waves", which do not tilt with height like pure internal gravity waves. In the atmosphere, the Scorer parameter decreases rapidly with height when N decreases with height (there is low lapse rate at low altitude and high lapse rate at high altitude), u increases with height, and when d^2u/dz^2 is large and negative at low altitude and zero at higher altitude. The latter condition occurs when there is a low-level jet. The ascending branches of the gravity waves may lift air to saturation and produce clouds or even trigger elevated convection above the gust front.

3.3 MULTICELL CONVECTIVE STORMS

Keith Browning, in an influential hailstorm monograph published in 1977, proposed that there is *a continuum of storm types based on the lifetime* of convective cells. His proposal was pre-dated by work done by him and his collaborators in England, by A. Chisholm and colleagues, who studied Alberta hailstorms, by John Marwitz, who studied hailstorms in Alberta, and by A. S. Dennis and colleagues, who studied hailstorms in South Dakota. Sometimes discrete new "secondary" cells form periodically along the leading edge of a gust front on the right, rear flank of the storm and reach a quasi-equilibrium state, as cells mature and decay on the storm's left, front flank (Figure 3.44). At other times, under other conditions (when low-level vertical shear is weak, according to RKW theory), convective cells simply decay as the gust front spreads out and no secondary cells are triggered. In the following chapter, the case in which convective cells persist for a long time period is discussed in detail. In the case of "single" Byers–Braham cells, precipitation is the "brake" acting on a cell's life. It was postulated that the brake is released somewhat when there is enough vertical shear in the environment so that precipitation does not fall back into the updraft. It was thought, however, based on the Thunderstorm Project, that too much shear inhibits storm formation because it rips an updraft apart so that precipitation does not have a chance to form.

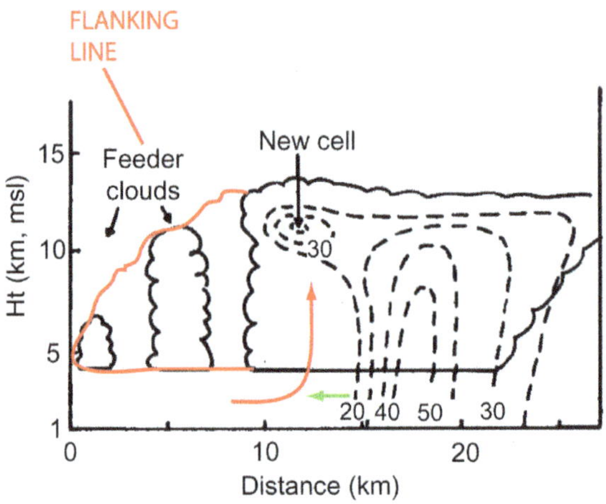

Figure 3.44. (Top) Idealized depiction of the vertical cross section of clouds and radar echoes in a multicell convective storm, as seen in the plane of the mean vertical shear vector or, equivalently, from the right side of the storm with respect to its motion (in the Northern Hemisphere). What were originally called "feeder clouds" are now recognized as a continuous "flanking line"; mature cells are on the right and a new cell is seen aloft in the center. Lines of constant radar reflectivity factor in dBZ are given by the dashed lines; low-level inflow into the updraft is indicated by the red streamline and the cold surface outflow is indicated by the green vector (adapted from Dennis *et al.*, 1970). (Bottom) Multicell convective storm in eastern Colorado on July 13, 1996. New cumulus growth is seen being forced along the edge of a gust front denoted by the dashed line (photograph by the author).

A succession of ordinary cells, each of which is independent of each other and initiated along the gust front from an ordinary convective cell, is referred to as a "multicell" convective storm (Figure 3.45). In some numerical simulations (see papers by Rob Fovell and coauthors), new cells grow approximately once every 15 min. The periodic nature of new-cell growth is linked to the temporary suppression of new-cell growth by sinking motion on the flanks of the existing updraft; these sinking regions are associated with the horizontal circulations induced by the gradient of buoyancy at the edges of the buoyant updraft (Figure 3.46). Also,

Figure 3.45. Photographs of multicell convective storms. (Top) August 22, 1988 in the Oklahoma Panhandle; (bottom) August 27, 1993 over the extreme southwestern part of the Florida Peninsula, over the Everglades, viewed from a NOAA helicopter. In these cases, vertical shear is very weak (photographs by the author).

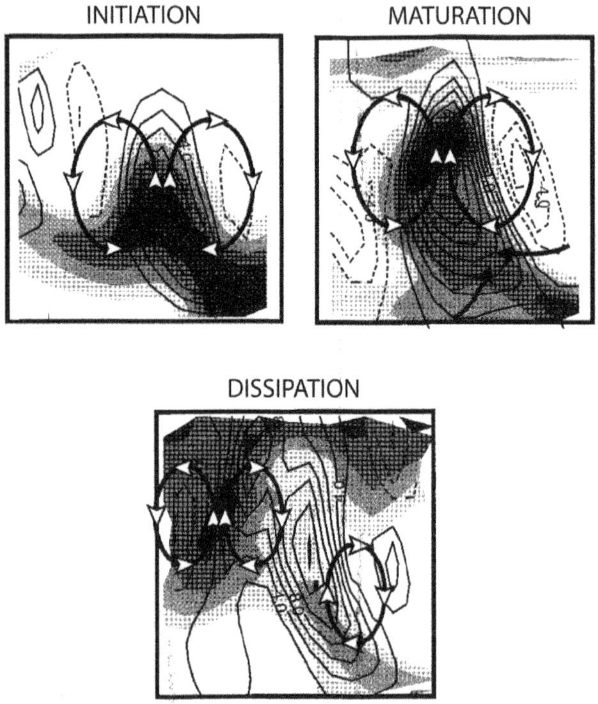

Figure 3.46. Conceptual model of the three stages in the life cycle of an ordinary convective cell within a mature, multicellular, squall line mesoscale convective system. Equivalent potential temperature (gray shaded) and vertical velocity (contoured) in a reference frame fixed to the cell's principal updraft. Solid (dashed) contours indicate rising (sinking) motion. Buoyancy-induced vertical circulations are depicted by elliptical streamlines. During the initiation stage, the buoyancy-induced vertical circulation ingests potentially warm air from below. During the maturation stage, the cell's buoyancy-induced vertical circulation acts to weaken forced lifting as subsidence occurs on the flanks of the cell, cutting off the cell from any older, earlier cell; stable, potentially cold air mixes into the cell's inflow region from below, reducing buoyancy. It is the rearward movement of the cell with respect to the leading edge of the cold pool that results in the cell's inflow being cut off from the potentially warmest air. During the dissipation stage, the cell's buoyancy-induced vertical circulation on the front-facing flank (with respect to the front-to-rear motion of subcloud air, which is dominant during the mature stage when the cold pool's circulation is dominant) weakens as potentially cooler air is ingested. The potentially warmest air in the cell detrains from the updraft and some moves in the direction the cold pool is moving, reducing buoyancy there and thus suppressing upward motion on the cell's rear-facing side (adapted from Fovell and Tan, 1998).

the ingestion of cooler air by the vertical circulations suppresses convective growth for a while. When the new updraft has propagated back toward the rear side of the convective storm, if it ever does, then a new updraft may be triggered as the suppression of new-cell growth ceases. Moreover, in some simulations the moist boundary layer is deepened upstream from the region where new convective growth is suppressed and new-cell growth is accelerated.

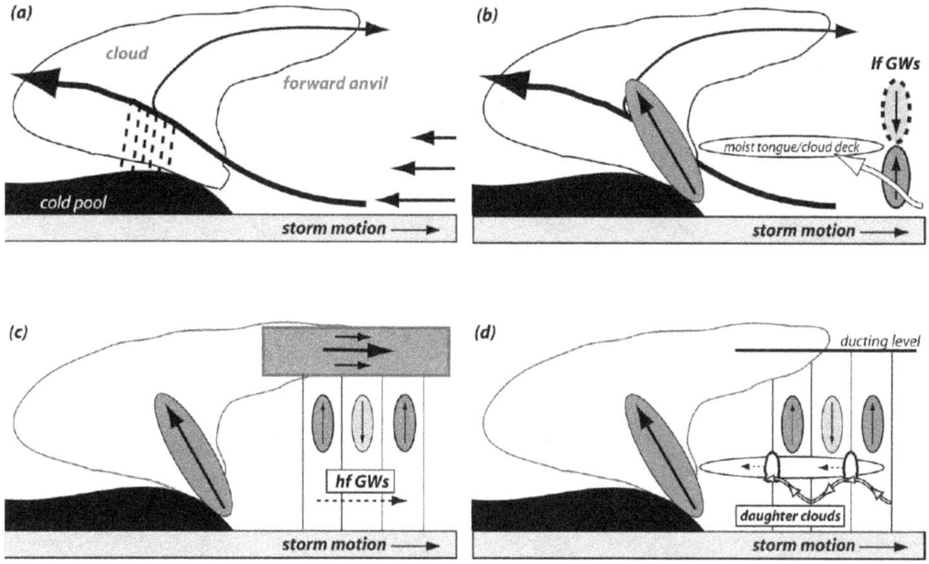

Figure 3.47. Conceptual model of stages in the discrete propagation of a multicellular squall line in the forward direction (to the right) as high-frequency gravity waves (hf GWs) are forced by the squall line and are trapped (below the "ducting level") beneath the forward anvil (owing to reduced static stability and jet-like flow in the anvil, which results in the Scorer parameter decreasing with height); the gravity waves then trigger new cell growth ("daughter clouds") ahead of the storm in regions where there is enhanced lift (from Fovell *et al.*, 2006).

An alternative to Fovell's explanation of multicell periodicity is that gravity waves are triggered as air is lifted over the stable cold pool and new-cell growth is initiated in the ascending regions of the wave, as proposed by Bob Houze and his colleagues. Fovell and his colleagues have found observational and numerical evidence that gravity waves can be ducted ahead of the storm when the boundary layer is capped by a stable layer, expecially at night, under the anvil outflow, and that low-frequency gravity waves may trigger new cells far ahead of the storm. In this case, the storm may appear to jump ahead of itself (Figure 3.47).

When the period of new-cell development is very short, the convective cells are no longer independent of each other. Such an evolution is termed "weak evolution". Multicell storms propagate in part by cell translation and in part by "discrete propagation" (Figure 3.48) along an advancing gust front. When the period of new-cell development is very short, the cells propagate *continuously* (Figure 3.49). When new cells form, either by discrete or continuous propagation, the entire storm as a whole propagates to the right of the mean wind (Figure 3.50) when new-cell growth is along the right flank of the storm.

According to RKW theory, the optimum condition for new-cell growth is that Δu, the low-level shear multiplied by the depth of the shear layer, equals c, the speed of the cold pool. With time, more and more precipitation falls into unsaturated air and cools evaporatively and frozen precipitation falls below the

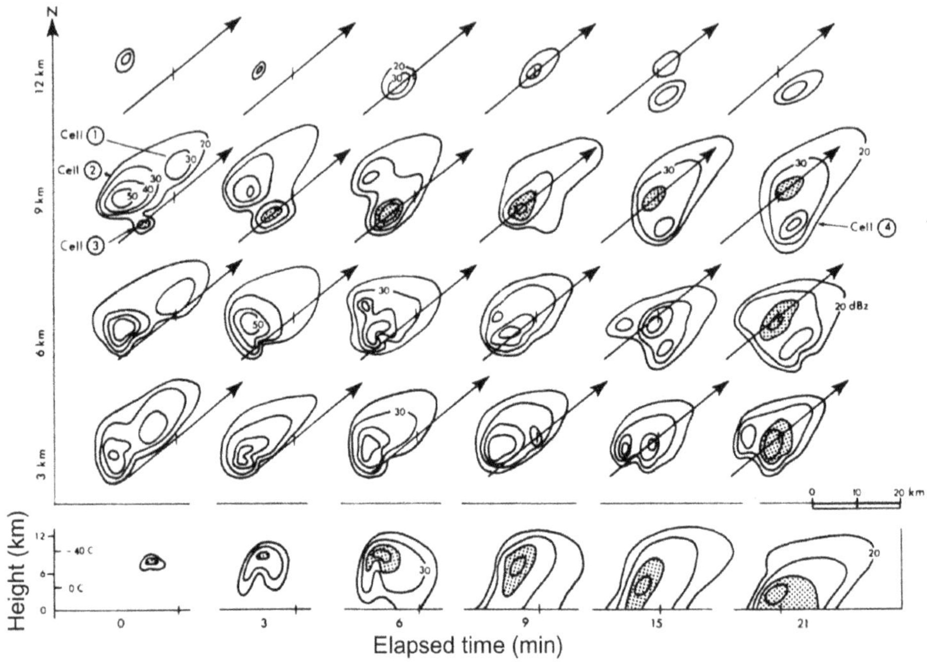

Figure 3.48. Illustration of discrete propagation in a multicell convective storm. (Top) PPI (plan view of) radar reflectivity factor contours (in dBZ) at various altitudes as a function of time. The arrow shown in each panel indicates the direction of cell motion. (Bottom) Vertical cross section of radar reflectivity contours for "Cell 3", which is shaded. The storm motion is to the northeast and discrete propagation occurs on the right flank of the storm (from Chisholm and Renick, 1972).

melting layer and cools even more, so that the cold pool deepens or becomes cooler or both. If Δu remains constant, eventually c increases (cf. (3.32)), so that $c > \Delta u$ and the RKW condition for optimum growth no longer is met. The vertical circulation at the leading edge of the cold pool then leans rearward with height (Figure 3.40, top right), so that new cells propagate rearward with respect to the leading edge of the gust front (cold pool) and discrete, multicell growth is possible; *the phase shift between growing vertical circulations and the vertical circulation along the leading edge of the cold pool is essential to the process* (Figure 3.51). If the shear is increased so that $\Delta u \rightarrow c$, new-cell growth becomes more continuous and "weakly evolving".

3.4 GENERAL MONOGRAPHS AND BOOKS

The reader is referred to p. 24 for a list of relevant general monographs and books.

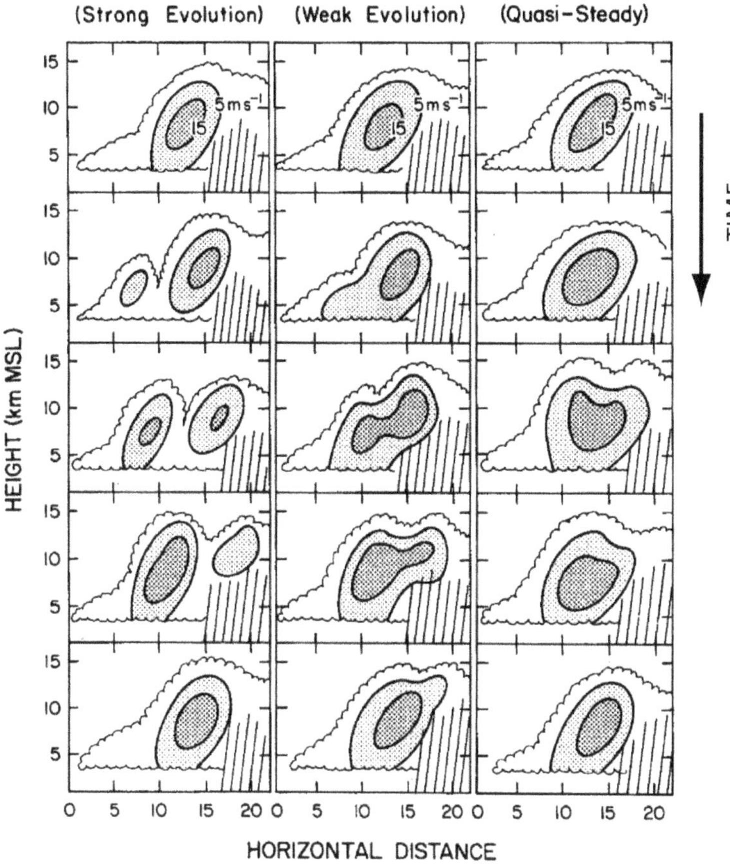

Figure 3.49. Idealized illustration of "strong" evolution (left column, showing vertical cross section of updraft speed contoured and shaded, cloud outlined by scalloped line, and precipitation by hatching; time between successive frames is ~3–5 min.), "weak" evolution (center column), and "quasi-steady" evolution in a convective storm (adapted from Foote and Frank, 1983). Strong evolution is the mode of multicell ordinary-cell convection.

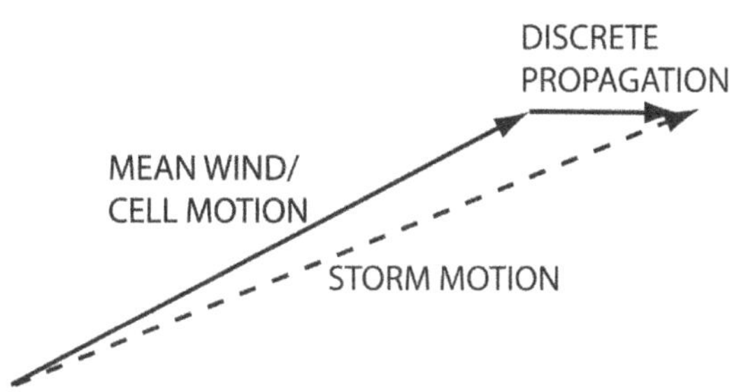

Figure 3.50. Example of a case in which the cell motion is the same as that of the mean wind (in a vertical layer), and discrete propagation occurs along a gust front oriented in a meridional direction, so that the storm motion vector lies to the right of the mean wind.

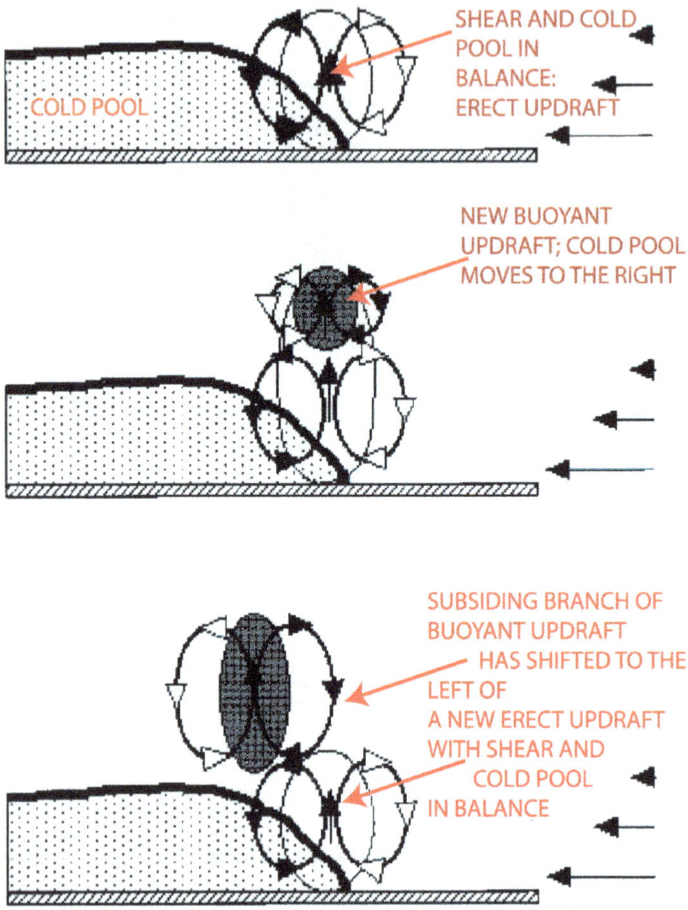

Figure 3.51. Illustration of how the movement of a cold pool relative to the movement of a growing cell results in a phase shift of the buoyancy-induced vertical circulation of the growing cell with respect to the buoyancy-induced vertical circulation of a new cell, such that it is suppressed from above by a subsiding branch of the older cell. Idealized streamlines, etc., shown in the reference frame of the cold pool, which moves to the right (adapted from Fovell and Tan, 1998).

3.5 REFERENCES AND BIBLIOGRAPHY

Adler, R. F. and R. A. Mack (1986) Thunderstorm cloud top dynamics as inferred from satellite observations and a cloud top parcel model. *J. Atmos. Sci.*, **43**, 1945–1960.

Atkins, N. T. and R. M. Wakimoto (1991) Wet microburst activity over the Southeastern United States: Implications for forecasting. *Wea. Forecasting*, **6**, 470–482.

Bedka, K., J. Brunner, R. Dworak, W. Feltz, J. Otkin, and T. Greenwald (2010) Objective satellite-based detection of overshooting tops using infrared window channel brightness temperature gradients. *J. Clim. Applied Meteor. Climatol.*, **49**, 181–202.

Benjamin, T. B. (1968) Gravity currents and related phenomena. *J. Fluid Mech.*, **31**, part 2, 209–248.

Bernstein, B. C. and R. H. Johnson (1994) A dual-Doppler radar study of an OK PRE-STORM heat burst event. *Mon. Wea. Rev.*, **122**, 259–273.

Bluestein, H. B. (1992) *Synoptic-Dynamic Meteorology in Midlatitudes, Vol. 1: Principles of Kinematics and Dynamics*. Oxford University Press, New York, 431 pp.

Bluestein, H. B., E. W. McCaul, Jr., G. P. Byrd, and G. R. Woodall (1988) Mobile sounding observations of a tornadic storm near the dryline: The Canadian, Texas storm of 7 May 1986. *Mon. Wea. Rev.*, **116**, 1790–1804.

Browning, K. A. (1977) The structure and mechanism of hailstorms. *Hail: A Review of Hail Science and Hail Suppression*, AMS Monograph 16, American Meteorological Society, Boston, pp. 1–43.

Browning, K. A., J. C. Fankhauser, J.-P. Chalon, P. J. Eccles, R. G. Strauch, F. H. Merrem, D. J. Musil, E. L. May, and W. R. Sand (1976) Structure of an evolving hailstorm, Part V: Synthesis and implications for hail growth and hail suppression. *Mon. Wea. Rev.*, **104**, 603–610.

Byers, H. R. and R. R. Braham, Jr. (1949) *The Thunderstorm*, U.S. Government Printing Office, Washington, D.C., 287 pp.

Charba, J. (1974) Application of gravity current model to analysis of squall-line gust front. *Mon. Wea. Rev.*, **102**, 140–156.

Chisholm, A. J. and J. H. Renick (1972) *The Kinematics of Multicell and Supercell Alberta Hailstorms*, Alberta Hail Studies Rep. 72-2, Research Council of Alberta Hail Studies, Edmonton, Alberta, pp. 24–31.

Crook, N. A. (2001) Understanding Hector: The dynamics of island thunderstorms. *Mon. Wea. Rev.*, **129**, 1550–1563.

Dennis, A. S. *et al.* (1970) Characteristics of hailstorms of western South Dakota. *J. Appl. Meteor.*, **9**, 127–135.

Doswell, C. A., III (1985) *The Operational Meteorology of Convective Weather, Vol. II: Storm Scale Analysis*, NOAA Tech. Memo. ERL ESG-15, National Oceanic and Atmospheric Administration, Silver Springs, MD, 240 pp.

Doswell, C. A., III (2008) Comments on the mysteries of mammatus clouds: Observations and formation mechanisms. *J. Atmos. Sci.*. **65**, 1093–1094.

Doswell, C. A., III and E. N. Rasmussen (1994) The effect of neglecting the virtual temperature correction on CAPE calculation. *Wea. Forecasting*, **9**, 625–629.

Droegemeier, K. K. and R. B. Wilhelmson (1985) Three-dimensional numerical modeling of convection produced by interacting thunderstorm outflows, Part I: Control simulation and low-level moisture variations. *J. Atmos. Sci.*, **42**, 2381–2403.

Droegemeier, K. K. and R. B. Wilhelmson (1985) Three-dimensional numerical modeling of convection produced by interacting thunderstorm outflows, Part II: Variations in vertical wind shear. *J. Atmos. Sci.*, **42**, 2404–2414.

Droegemeier, K. K. and R. B. Wilhelmson (1987) Numerical simulation of thunderstorm outflow dynamics, Part I: Outflow sensitivity experiments and turbulence dynamics. *J. Atmos. Sci.*, **44**, 1180–1210.

Emanuel, K. A. (1981) A similarity theory for unsaturated downdrafts within clouds. *J. Atmos. Sci.*, **38**, 1541–1557.

Emanuel, K. A. (1994) *Atmospheric Convection*, Oxford University Press, New York, Chapters 1–6, 9–11.

Fankhauser, J. C. and C. G. Mohr (1977) Some correlations between various sounding parameters and hailstorm characteristics in northeast Colorado, preprint. *10th*

Conference on Severe Local Storms, Omaha, NE, American Meteorological Society, Boston, pp. 218–225.
Foote, G. B. and H. W. Frank (1983) Case study of a hailstorm in Colorado, Part III: Airflow from triple-Doppler measurements. *J. Atmos. Sci.*, **40**, 686–707.
Fovell, R. G. and P. S. Dailey (1995) The temporal behavior of numerically simulated multicell-type storms, Part I: Modes of behavior. *J. Atmos. Sci.*, **52**, 2073–2095.
Fovell, R. G. and Y. Ogura (1989) Effect of vertical wind shear on numerically simulated multicell storm structure. *J. Atmos. Sci.*, **46**, 3144–3176.
Fovell, R. G. and P.-H. Tan (1998) The temporal behavior of numerically simulated multicell-type storms, Part II: The convective cell life cycle and cell regeneration. *Mon. Wea. Rev.*, **126**, 551–577.
Fovell, R. G., G. L. Mullendore, and S.-H. Kim (2006) Discrete propagation in numerically simulated nocturnal squall lines. *Mon. Wea. Rev.*, **134**, 3735–3752.
Fujita, T. T. (1981) Tornadoes and downbursts in the context of generalized planetary scales. *J. Atmos. Sci.*, **38**, 1511–1534.
Fujita, T. T. (1982) Principle of stereoscopic height computations and their applications to stratospheric cirrus over severe thunderstorms. *J. Meteor. Soc. Japan*, **60**, 355–368.
Garrett, T. J., C. T. Schmidt, S. Kihlgren, and C. Cornet (2010) Mammatus clouds as a response to cloud base radiative heating. *J. Atmos. Sci.*, **67**, 3891–3903.
Heymsfield, G. M., G. Szejwach, S. Schotz, and R. H. Blackmer, Jr. (1983) Upper level structure of Oklahoma tornadic storms on 2 May 1979, Part 2: Proposed explanation of V pattern and internal warm region in infrared observations. *J. Atmos. Sci.*, **40**, 1756–1767.
Heymsfield, G. M., S. W. Bidwell, I. J. Caylor, S. Ameen, S. Nicholson, W. Boncyk, L. Miller, D. Vandemark, P. E. Raette, and L. R. Dod (1996) The EDOP radar system on the high-altitude NASA ER-2 aircraft. *J. Atmos. Oceanic Technol.*, **13**, 795–809.
Hitschfeld, W. (1960) The motion and erosion of convective storms in severe vertical wind shear. *J. Meteor.*, **17**, 270–282.
Hjelmfelt, M. R. (1987) The microburst of 22 June 1982 in JAWS. *J. Atmos. Sci.*, **44**, 1646–1665.
Hjelmfelt, M. R. (1988) Structure and life cycle of microburst outflows observed in Colorado. *J. Appl. Meteor.*, **27**, 900–927.
Jin, Y., S. E. Koch, Y.-L. Lin, F. M. Ralph, and C. Chen (1996) Numerical simulations of an observed gravity current and gravity waves in an environment characterized by complex stratification and shear. *J. Atmos. Sci.*, **53**, 3570–3588.
Johnson, B. C. (1976) The heat burst of 29 May 1976. *Mon. Wea. Rev.*, **111**, 1776–1792.
Johnson, R. H., S. Chen, and J. J. Toth (1989) Circulations associated with a mature-to-decaying midlatitude mesoscale convective system, Part I: Surface features—Heat bursts and mesolow development. *Mon. Wea. Rev.*, **117**, 942–959.
Kanak, K. M. and J. M. Straka (2002) An unusual reticular cloud formation. *Mon. Wea. Rev.*, **130**, 416–421.
Kanak, K. M., J. M. Straka, and D. M. Schultz (2008) Numerical simulation of mammatus. *J. Atmos. Sci.*, **65**, 1606–1621.
Klemp, J. B., R. Rotunno, and W. C. Skamarock (1994) On the dynamics of gravity currents in a channel. *J. Fluid Mech.*, **269**, 169–198.
Kollias, P., I. Jo, and B. A. Albrecht (2005) High-resolution observations of mammatus in tropical anvils. *Mon. Wea. Rev.*, **133**, 2105–2112.
Lee, W.-C., R. E. Carbone, and R. M. Wakimoto (1992) The evolution and structure of a bow-echo-microburst event, Part I: The microburst. *Mon. Wea. Rev.*, **120**, 2188–2210.

Lee, W.-C., R. M. Wakimoto, and R. E. Carbone (1992) The evolution and structure of a bow-echo-microburst event, Part II: The bow echo. *Mon. Wea. Rev.*, **120**, 2211–2225.

Lemon, L. R. (1998) The radar three-body scatter spike: An operational large-hail signature. *Wea. Forecasting*, **13**, 327–340.

Li, L., G. M. Heymsfield, P. E. Racette, L. Tian, and E. Zenker (2004) A 94-GHz cloud radar system on a NASA high-altitude ER-2 aircraft. *J. Atmos. Oceanic Technol.*, **21**, 1378–1388.

Ludlam, F. H. (1963) Severe local storms: A review. *Severe Local Storms*, Meteorological Monograph 27, no. 5, American Meteorological Society, Boston, pp. 1–30.

Ludlam, F. (1966) Cumulus and cumulonimbus convection. *Tellus*, **18**, 687–698.

Martner, B. E. (1995) Doppler radar observations of mammatus. *Mon. Wea. Rev.*, **123**, 3115–3121.

Marwitz, J. D. (1972) The structure and motion of severe hailstorms, Part I: Supercell storms. *J. Appl. Meteor.*, **11**, 166–179.

Marwitz, J. D. (1972) The structure and motion of severe hailstorms, Part II: Multicell storms. *J. Appl. Meteor.*, **11**, 180–188.

Marwitz, J. D. (1972) The structure and motion of severe hailstorms, Part I: Severely sheared storms. *J. Appl. Meteor.*, **11**, 189–201.

McCann, D. W. (1983) The enhanced-V: A satellite observable severe storm signature. *Mon. Wea. Rev.*, **111**, 887–894.

McPherson, R. A., J. D. Lane, K. C. Crawford, and W. G. McPherson (2010) A climatological analysis of heatbursts in Oklahoma (1994–2009). *Int. J. Climatology*, Roy. Meteor. Soc., 14 pp (*http://onlinelibrary.wiley.com/doi/10.1002/joc.2087/pdf*).

Moncrieff, M. W. (1981) A theory of organized steady convection and its transport properties. *Quart. J. Roy. Meteor. Soc.*, **107**, 29–50.

Musil, D. J., S. A. Christopher, R. A. Deola, and P. L. Smith (1991). Some interior observations of southeastern Montana hailstorms. *J. Appl. Meteor.*, **30**, 1596–1612.

Newton, C. W. (1950) Structure and mechanism of the prefrontal squall line. *J. Meteor.*, **7**, 210–222.

Parsons, D. B. and R. A. Kropfli (1990) Dynamics and fine structure of a microburst. *J. Atmos. Sci.*, **47**, 1674–1692.

Parsons, D. B. and M. L. Weisman (1993) A numerical study of a rotating downburst. *J. Atmos. Sci.*, **50**, 2369–2385.

Peterson, R. E. (1984) A triple-Doppler radar analysis of a discretely propagating multicell convective storm. *J. Atmos. Sci.*, **41**, 2973–2990.

Roberts, R. D. and J. W. Wilson (1989) A proposed microburst nowcasting procedure using single-Doppler radar. *J. Appl. Meteor.*, **28**, 285–303.

Rotunno, R., J. B. Klemp, and M. L. Weisman (1988) A theory for strong, long-lived squall lines. *J. Atmos. Sci.*, **45**, 463–485.

Schultz, D. M., K. M. Kanak, J. M. Straka, R. J. Trapp, B. A. Gordon, D. S. Zrnic, G. H. Bryan, A. J. Durant, T. J. Garrett, P. M. Klein, and D. K. Lilly (2006) The mysteries of mammatus clouds: Observations and formation mechanisms. *J. Atmos. Sci.*, **63**, 2409–2435.

Simpson, J. E. (1997) *Gravity Currents in the Environment and the Laboratory*, Second Edition, Cambridge University Press, Cambridge, U. K., 244 pp.

Srivastava, R. C. (1985) A simple model of evaporatively driven downdraft: Application to microburst downdraft. *J. Atmos. Sci.*, **42**, 1004–1023.

Srivastava, R. C. (1987) A model of intense downdrafts driven by the melting and evaporation of precipitation. *J. Atmos. Sci.*, **44**, 1752–1773.

Stommel, H. (1947) Entrainment of air into a cumulus cloud. *J. Meteor.*, **4**, 91–94.

Thorpe, A. J., M. J. Miller, and M. W. Moncrieff (1982) Two-dimensional convection in non-constant shear: A model of mid-latitude squall lines. *Quart. J. Roy. Meteor. Soc.*, **108**, 739–762.

Wakimoto, R. M. (1985) Forecasting dry microburst activity over the High Plains. *Mon. Wea. Rev.*, **113**, 1131–1143.

Wakimoto, R. M and V. N. Bringi (1988) Dual-polarization observations of microbursts associated with intense convection: The 20 July storm during the MIST Project. *Mon. Wea. Rev.*, **116**, 1521–1539.

Wakimoto, R. M., C. J. Kessinger, and D. E. Kingsmill (1994) Kinematic, thermodynamic, and visual structure of low-reflectivity microbursts. *Mon. Wea. Rev.*, **122**, 72–92.

Wang, L. and K. Sassen (2006) Cirrus mammatus properties derived from an extended remote sensing dataset. *J. Atmos. Sci.*, **63**, 712–725.

Weisman, M. L. and J. B. Klemp (1986) Characteristics of isolated convective storms. In: P. S. Ray (Ed.), *Mesoscale Meteorology and Forecasting*, American Meteorological Society, Boston, pp. 331–358.

Weisman, M. L. and R. Rotunno (2004) A theory for strong-lived squall lines revisited. *J. Atmos. Sci.*, **61**, 361–382.

Wilson, J. W. and R. M. Wakimoto (2001) The discovery of the downburst: T. T. Fujita's contribution. *Bull. Amer. Meteor. Soc.*, **82**, 49–62.

Winstead, N. S., J. Verlinde, S. T. Arthur, F. Jaskiewicz, M. Jensen, N. Miles, and D. Nicosia (2001) High-resolution airborne radar observations of mammatus. *Mon. Wea. Rev.*, **129**, 159–166.

Xu, Q., M. Xue, and K. K. Droegemeier (1996) Numerical simulations of density currents in sheared environments within a vertically confined channel. *J. Atmos. Sci.*, **53**, 770–786.

Yang, M.-J. and R. A. Houze (1995) Multicell squall-line structure as a manifestation of vertically propagating trapped gravity waves. *Mon. Wea. Rev.*, **123**, 641–661.

4

Supercells

"Lear: Blow, winds, and crack your cheeks! rage! blow!
You cataracts and hurricanoes, spout
Till you have drench'd our steeples, drown'd the cocks!
You sulphurous and thought-executing fires,
Vaunt-couriers to oak-cleaving thunderbolts,
Singe my white head! And thou all-shaking thunder
Smite flat the thick rotundity o' the world!"

William Shakespeare—Act III, Scene II, *King Lear*

In the previous chapter we considered the behavior of ordinary cells and multicell convective storms, which consist of a succession of ordinary cells for which positive buoyancy is of primary importance. This positive buoyancy is realized by air parcels accelerating upward in a conditionally unstable atmosphere, which is counteracted by a downward-directed perturbation pressure gradient force that depends on the aspect ratio of the spatial dimensions of the air parcel, and by precipitation and cloud–particle loading, and by the entrainment of cooler or drier environmental air.

In addition, density currents near the ground driven by evaporative cooling and melting of precipitation or cloud material interact with the low-level environmental shear to initiate new convective cells adjacent to earlier cells or to cut off any new convective growth. Multicell storms may be *either* two dimensional or three dimensional and the individual cells that make up the multicell complex *do not typically achieve any steady state* (multicell storms that become organized on the mesoscale, however, can achieve a steady state and will be discussed in Chapter 5).

When the magnitude of the (mainly deep-layer) vertical shear in the environment exceeds some threshold, *dynamic* vertical perturbation pressure gradients ((2.63) and (2.70)) begin to play a significant role and affect the location and intensity of vertical accelerations. In this chapter we will consider what this threshold is and how the behavior of convective storms is modified by dynamic vertical perturbation pressure gradients. For a restricted range of environmental conditions (vertical shear and potential buoyancy, the latter represented by CAPE) a convective cell (storm) that is relatively *long lived, three dimensional, has a rotating updraft (vertical vorticity is correlated with vertical velocity), and is quasi-steady* can form. In these storms, "supercells", some of their behavior and dynamics are identical to that of ordinary-cell/multicell storms, but the unique aspects of the supercell are a result mainly of the effects of strong (deep) vertical shear. While both ordinary cells and supercells can produce severe weather, including tornadoes, supercells are more prolific in terms of inflicting damage and often the most damaging. In short, supercells are relatively long lived and have a rotating updraft, while ordinary-cell/multicell storms are not long lived and do not have a rotating updraft.

4.1 SUPERCELLS AND THE BULK RICHARDSON NUMBER

We now consider under what restricted range of environmental conditions (vertical shear and potential buoyancy—CAPE) an isolated convective cell (storm) can develop into a supercell. To do so, we first consider the frictionless, vertical equation of motion in terms of buoyancy and the dynamic and buoyancy vertical perturbation pressure terms (2.70)

$$Dw/Dt = -1/\bar{\rho}\,\partial p'_d/\partial z + [(-1/\bar{\rho})\,\partial p'_b/\partial z + B] \tag{4.1}$$

Suppose that the vertical dynamic perturbation pressure gradient term $(-1/\bar{\rho}\,\partial p'_d/\partial z)$ is approximately the same magnitude as the inertial term on the LHS of (4.1), and also the same order of magnitude as the "effective" buoyancy term $([(-1/\bar{\rho})\,\partial p'_b/\partial z + B])$. If the aspect ratio of the buoyant air parcel in the storm is on the order of unity, then it follows from (2.75) that

$$|Dw/Dt| \sim W/(H/W) \sim B/2 \tag{4.2}$$

where W is the scale of vertical velocity; H is the vertical scale (the depth of the tropopause, which is the typical vertical scale for deep convection); and the advective time scale is H/W. (For a spherical bubble, our analysis showed that the RHS of (4.2) is $\frac{2}{3}B$, cf. (2.133); if water loading is considered, then the factor multiplying B on the RHS of (4.2) is even less, so a factor of $\frac{1}{2}$ is reasonable.) Then

$$B \sim 2W^2/H \tag{4.3}$$

From (3.7) and the above, it is seen that when the LFC is not too high,

$$W^2 \sim \tfrac{1}{2}BH \sim 2\,\mathrm{CAPE} \tag{4.4}$$

So, the buoyancy term is

$$B \sim 4\,\mathrm{CAPE}/H \tag{4.5}$$

Suppose now that the vertical perturbation pressure gradient term is as large as the effective buoyancy term. From the horizontal equation of motion (2.13) we recall from (2.15) that

$$p' \sim \bar{\rho} U^2 \tag{4.6}$$

where U is the scale of the horizontal wind component, so that

$$|-1/\bar{\rho}\,\partial p'/\partial z| \sim \partial/\partial z\, U^2 \sim 2U^2/H \tag{4.7}$$

The ratio of the approximate magnitude of the acceleration due to the vertical perturbation pressure gradient force to the acceleration due to buoyancy, which is a measure of the relative importance of dynamic effects, is given by

$$|-1/\bar{\rho}\,\partial p'/\partial z|/|B| \sim 2(U^2/H)/(4\,\mathrm{CAPE}/H) \sim \tfrac{1}{2}U^2/\mathrm{CAPE} \tag{4.8}$$

This ratio is like a Froude number because it represents the ratio of the magnitude of the inertial term to that of the term involving gravity (namely, CAPE). (Since CAPE $\sim W^2/2$, (4.8) represents the ratio of the kinetic energy of the horizontal component of the wind to the kinetic energy of the vertical component of the wind that is induced by buoyancy.) So, when $\frac{1}{2}U^2$ is comparable with the CAPE, the effects of the dynamic perturbation pressure gradient are comparable with the effects of effective buoyancy.

The empirical dimensionless parameter called the *bulk Richardson number* (R), was first described in a seminal paper by Morris Weisman and Joe Klemp based on numerical simulation experiments in homogeneous environments of varying vertical shear and CAPE, and based on an analytical formulation of two-dimensional steady convection in an environment of vertical shear by Mitch Moncrieff and J. S. A. Green in 1972:

$$R = \mathrm{CAPE}/(\tfrac{1}{2}U^2) \tag{4.9}$$

In this case, U is the approximate magnitude of the storm-relative inflow velocity (generally, it can include both u and v components). In practice, U is computed from the difference between the pressure-weighted mean wind vector in the moist boundary layer and the pressure-weighted mean wind vector in the lowest 6 km. The depth of the boundary layer is usually taken to be 500 m, so $U \sim$ the speed of the vector difference between mean wind in the lowest 6 km and the mean wind in the lowest 500 m. If the mean wind in the lowest 6 km represents storm motion, an assumption that is supported by many observations, then U represents the storm-relative wind speed of the inflow layer in the storm.

Since fundamental storm dynamics should not depend on our reference frame, we should be able to analyze storm dynamics in the reference frame of the storm and get the same "answers" to questions about dynamics as we would in the ground-relative frame. From (4.8) and (4.9) we see that if U in (4.8) is interpreted as the same as the storm-relative wind speed (just transform the coordinate system to one moving along with the storm, which is one moving along at a constant

Figure 4.1. Cumulus congestus being sheared off as it develops during several spurts of growth in central Oklahoma on (top) April 27, 1983 and (bottom) October 12, 2011 (photograph by the author).

velocity), then for *low* values of R the dynamic vertical perturbation pressure gradient force is at least as important as the buoyancy effects. We will soon see what the dynamic consequences of this condition are in detail.

The bulk Richardson number can be interpreted physically in terms of energy and integrated vertical shear. The scale of the horizontal component of the storm-relative wind (U) is a measure of the vertical shear of the environmental wind (multiplied by the difference between the altitude of the level under consideration and the mean altitude of the subcloud boundary layer). The denominator of (4.9) is therefore a measure of the square of the vertically integrated vertical shear, or the kinetic energy associated with the *horizontal* component of the wind in an air parcel that goes up in a storm's updraft. The numerator is a measure of the kinetic energy associated with the *vertical* component of the wind of an air parcel (converted from potential energy associated with buoyancy) that has risen to the top of a storm. When R is low, the kinetic energy associated with the horizontal wind component of the air going upward in the updraft is greater than the kinetic energy associated with the vertical wind component in the updraft due to buoyancy. If R is too low, however, the amount of kinetic energy going into the updraft is so small that the updraft gets sheared over so much that the updraft aloft gets separated from the updraft below (Figure 4.1) and a storm will not persist. This latter finding will be discussed later on in more detail in terms of horizontal vorticity dynamics.

Severe storm forecasters use the bulk Richardson number criterion to determine whether or not supercell convection is possible. CAPE is computed from the LFC up to the equilibrium level (EL). It is noted that R is a qualitative—not quantitative—parameter: it does not take into account moisture or water loading in the calculation of CAPE or CIN, and may not include a good measure of U if the depth of the boundary layer is substantially different from

500 m or if the environmental wind profile is such that U is not representative of storm-relative inflow or if gust front propagation also makes U not representative of storm-relative inflow. It is generally accepted that when

$$10 < R < 50 \tag{4.10}$$

supercell convection is possible. However, (4.10) expresses a necessary, but not a sufficient condition and ordinary cells/multicells may also be present. In practice, severe storm forecasters are usually concerned more with the upper bound and not with the lower bound.

It is seen in idealized numerical simulation experiments that the upward acceleration due to the vertical perturbation pressure gradient force is sometimes much greater than that due to buoyancy, especially at low levels when buoyancy is not high and when vertical shear is very strong, as it sometimes is, for example, in the environment of landfalling hurricanes (Figure 4.2). Storms in landfalling hurricanes can have dynamically driven updrafts that are as strong as updrafts in supercells in the Great Plains of the U. S.

It is possible that R can be small enough to satisfy the supercell criterion (4.10) when CAPE is very low and vertical shear is relatively high, but still not very strong in an absolute sense. When CAPE is low, buoyancy is also low, so that the vertical accelerations are weak in an absolute sense and it is not possible to produce a supercell because updrafts are simply not very strong, even though they are due, in significant part, to dynamic vertical pressure gradients.

On the other hand, when shear is extremely strong, even though CAPE is relatively high, but R is small, it is difficult for a convective storm to develop because *initially* shear makes the updraft lean over so much that the top of the storm may become detached from the updraft. The use of R for forecasting must be applied with caution and it is probably more useful to consider CAPE, shear, and R all together.

We end this section with a *gedunken* ("thought") experiment to show how the dynamical vertical pressure gradient force can act to *augment* the updraft intensity beyond that predicted from CAPE alone when R is small (i.e., when the effects of vertical shear are significant). Consider what happens when the vertical dynamic perturbation pressure gradient force $(-1/\bar{\rho}\,\partial p'_d/\partial z)$ augments the net effect of upward-directed buoyancy and downward-directed buoyancy perturbation pressure gradient force $([(-1/\bar{\rho})\,\partial p'_b/\partial z + B])$ (4.1). To do so, we derive a Bernoulli-like equation for steady-state, frictionless flow.

The steady-state, frictionless form of the three-dimensional equation of motion is as follows:

$$D\mathbf{v}/Dt = (\mathbf{v}\cdot\nabla)\mathbf{v} = -\alpha_0\,\nabla p' + B\mathbf{k} \tag{4.11}$$

where α_0 is the mean specific density. The advective part of the inertial term $((\mathbf{v}\cdot\nabla)\mathbf{v})$ may be expressed using a vector identity as $\nabla(\frac{1}{2}\mathbf{v}\cdot\mathbf{v}) + (\nabla\times\mathbf{v})\times\mathbf{v}$, where the first term is called the "Bernoulli" term and the second term is called the "Lamb" term. We define a differential position vector

$$d\mathbf{r} = dx\,\mathbf{i} + dy\,\mathbf{j} + dz\,\mathbf{k} \tag{4.12}$$

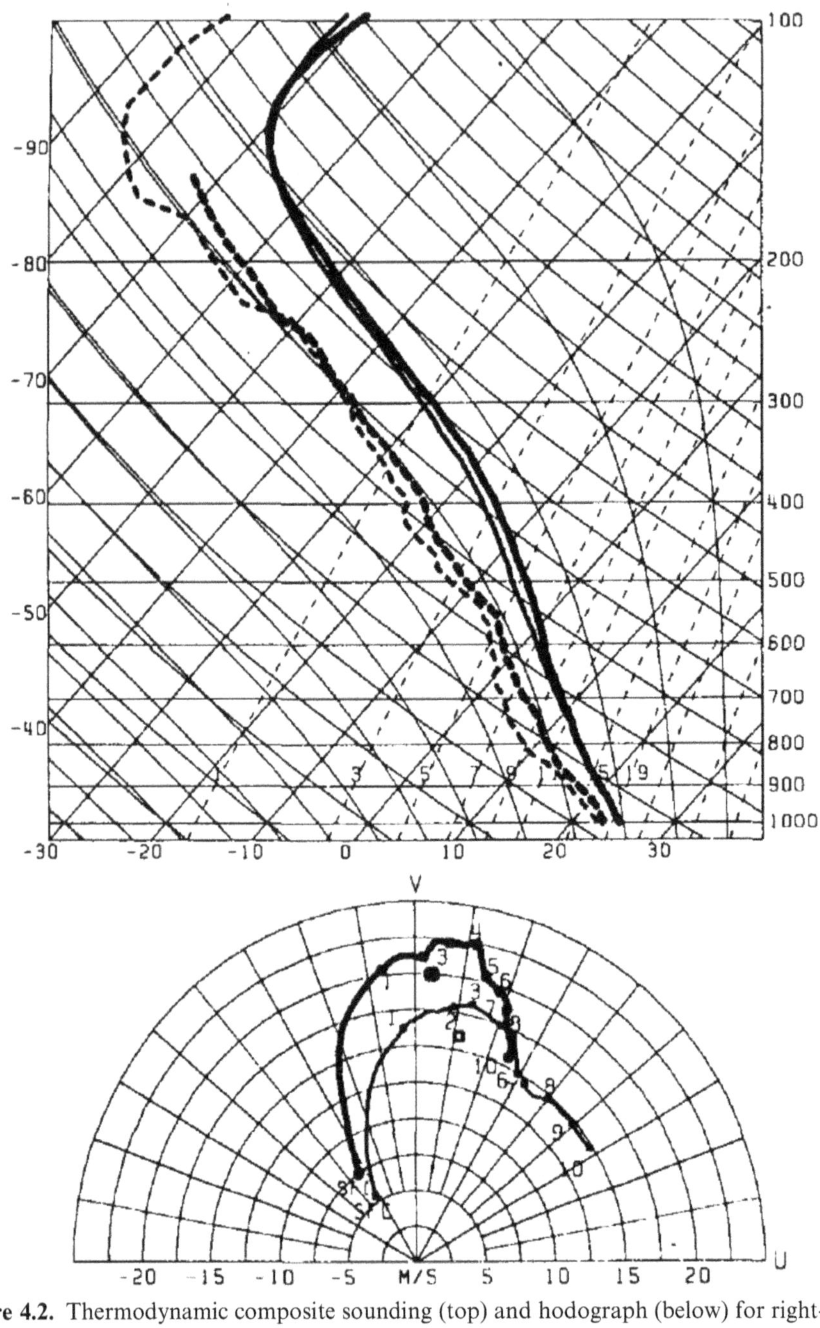

Figure 4.2. Thermodynamic composite sounding (top) and hodograph (below) for right-front quadrant in hurricanes. Skew T–$\log p$ soundings within 300 km of hurricane center plotted by heavy lines, soundings beyond 300 km of hurricane center plotted as light lines. In soundings, solid lines are temperature (°C) and dashed lines are dew point (°C); pressure plotted at right in hPa. In hodographs, height MSL plotted in km (from McCaul, 1991).

Taking the dot product of the equation in (4.11) with the differential position vector defined by (4.12), we find that

$$D\mathbf{v}/Dt \cdot d\mathbf{r} = \nabla(\tfrac{1}{2}\mathbf{v}\cdot\mathbf{v})\cdot d\mathbf{r} + [(\nabla \times \mathbf{v}) \times \mathbf{v}]\cdot d\mathbf{r} = -\alpha_0 \nabla p' \cdot d\mathbf{r} + B\,dz \qquad (4.13)$$

Integrating (4.13) along a streamline ($\mathbf{v}$), it follows that the Lamb term $[(\nabla \times \mathbf{v}) \times \mathbf{v}]$ is zero, because $[(\nabla \times \mathbf{v}) \times \mathbf{v}]\cdot\mathbf{v} = 0$, and so

$$\int \partial/\partial x(\tfrac{1}{2}|\mathbf{v}|^2)\,dx + \int \partial/\partial y(\tfrac{1}{2}|\mathbf{v}|^2)\,dy + \int \partial/\partial z(\tfrac{1}{2}|\mathbf{v}|^2)\,dz$$
$$= -\int \alpha_0(\partial p'/\partial x\,dx + \partial p'/\partial y\,dy + \partial p'/\partial z\,dz) + \int B\,dz \qquad (4.14)$$

Since the LHS of (4.14) may be expressed as $\int d(\tfrac{1}{2}|\mathbf{v}|^2)$, it follows that

$$\int d(\tfrac{1}{2}|\mathbf{v}|^2) + \alpha_0 \int dp' - \int B\,dz = 0 \qquad (4.15)$$

Integrating (4.15) from an initial location (denoted by the subscript "i") to a final position (denoted by the subscript "f"), we find that

$$\tfrac{1}{2}|\mathbf{v}_f|^2 - \tfrac{1}{2}|\mathbf{v}_i|^2 + \alpha_0 p'_f - \alpha_0 p'_i - \int B\,dz = 0 \qquad (4.16)$$

Now, suppose an air parcel moving at constant altitude, at mid-levels, in a storm-relative reference frame catches up with a convective storm and air flows around the updraft, but forks to the left side (Figure 4.3). Outside the updraft of the storm $B = 0$ and far upstream of the storm $p'_i = 0$. Following an argument Kerry Emanuel reproduced in his text, suppose that the air speeds up as it flows around the edge of updraft such that

$$|\mathbf{v}_f|^2 \approx 2|\mathbf{v}_i|^2 \qquad (4.17)$$

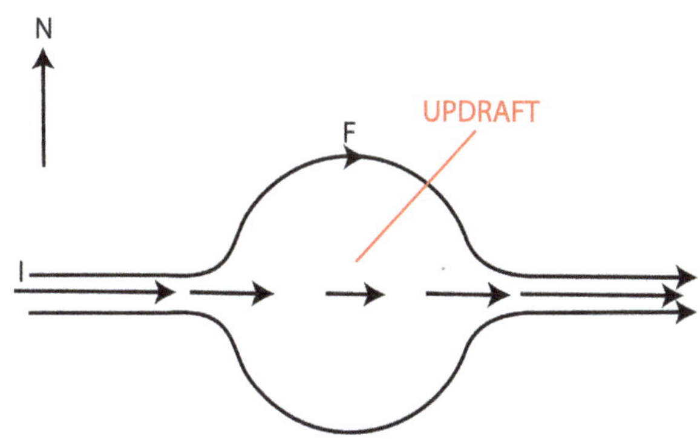

Figure 4.3. Idealized illustration of air at mid-levels catching up with and flowing around an updraft inside which lower values of westerly momentum have been advected upward. Air begins at I and is decelerated as it approaches the updraft; it then speeds up as it flows around the left side of the updraft and ends up at F.

We can justify this approximation qualitatively if the horizontal momentum of air flowing into the storm's updraft is approximately conserved as it rises up into the storm without mixing with environmental air (and neglecting water vapor and water substance loading). Consider a simple case when the vertical shear is unidirectional; in the example shown in Figure 4.3 the shear is westerly. So in mid-levels air approaches the storm from the west, but slows down as it encounters easterly momentum brought up from below; to conserve momentum, it must be diverted around the updraft and, furthermore, speeds up as it passes through the channel to the left of the updraft. Then, from (4.16) and (4.17) it follows that

$$\alpha_0 p'_f = -\tfrac{1}{2}|\mathbf{v}_i|^2 < 0 \tag{4.18}$$

so that there is a negative perturbation pressure at mid-levels on the left side of the updraft. Now, consider another air parcel, but this one begins from the boundary layer, far ahead of the storm, so once again $p'_i = 0$. Imagine that this air parcel, however, ascends in the updraft and ends up on the north side of the updraft, just where the previous air parcel had ended its journey. Using (4.16) and (4.18) we see that

$$\tfrac{1}{2}w_m^2 = \tfrac{1}{2}|\mathbf{v}_0|^2 + \tfrac{1}{2}|\mathbf{v}_m|^2 + \int_{\text{LFC}}^{\text{mid-level}} B\,dz \tag{4.19}$$

where w_m is the updraft in the storm at mid-levels; $|\mathbf{v}_0|$ is the storm-relative wind speed in the boundary layer upstream from the storm; $|\mathbf{v}_m|$ is the storm-relative wind speed at mid-levels on the left side of the updraft, and CAPE at mid-levels is

$$\text{CAPE}_m = \int_{\text{LFC}}^{\text{mid-level}} B\,dz \tag{4.20}$$

It follows that

$$w_m = (2\,\text{CAPE}_m + |\mathbf{v}_0|^2 + |\mathbf{v}_m|^2)^{1/2} \tag{4.21}$$

So, when there is no vertical shear, both the storm-relative winds in the boundary layer and at mid-levels are zero, and (4.21) is equivalent to that predicted by parcel theory from buoyancy alone (3.7). However, when there is vertical shear, there is storm-relative wind in the boundary layer or at mid-levels or both. The stronger the vertical shear, the stronger the updraft (4.21).

Finally, consider the special case in which the storm moves along with the pressure-weighted mean wind from the surface to 6 km and that the storm-relative boundary-layer wind is from the east at $U\,\text{m s}^{-1}$. Then the vertical shear in the lowest 6 km is U/6 km. From (4.21), we see that in the reference frame of the storm $|\mathbf{v}_0|^2 \equiv U^2$ and $|\mathbf{v}_m|^2 = 0$. Therefore, the dynamic contribution to vertical velocity (w_m) is equivalent to that from buoyancy $[(2\,\text{CAPE})^{1/2}]$ when

$$2\,\text{CAPE} = U^2 \tag{4.22}$$

(i.e., when CAPE $= U^2/2$, or when $R = 1$). For $R < 1$, $U^2/2 >$ CAPE, and the contribution from dynamic perturbation pressure gradients dominates the buoyancy contributions. We will shortly analyze the details of what contributes to the dynamic vertical perturbation pressure gradient force. We now take a step backward and look at a brief history of supercell research.

4.2 OBSERVED SUPERCELL BEHAVIOR AND EARLY THEORIES

In the late 1950s and early 1960s a few storms[1] were observed on conventional (i.e., non-Doppler, reflectivity only) radar that persisted for a much longer period of time than most other storms, whose individual cells lasted only on the order of the advective time scale. Furthermore, ordinary cells move along approximately with the pressure-weighted (i.e., mass-weighted) mean wind in the layer in which they are embedded, while most long-lived cells instead propagate to the *right* of the mean wind. In 1962, Keith Browning named these convective storms "supercells"[2] mainly owing to their longevity. It is fascinating to view time lapse radar reflectivity imagery in which a field of convective cells propagate in one direction and evolve quickly, while one, two, or several iconoclastic cells move more slowly and off to the right of the others, and maintain their appearance longer. (It is noted, however, that in multicell storms in which *new-cell development occurs along the right flank* of the storm, the motion of the storm *also* deviates to the right of the mean wind.)

Browning in 1977 proposed that the two basic building blocks of convective storms are ordinary cells and supercells. While the definitions of these two convective building blocks depends on cell behavior (longevity, motion with respect to the environmental wind profile, internal flow characteristics), it will be shown that one could equally well define these two building blocks in terms of their underlying, governing dynamics. Although the paradigm of two different types of convective building blocks is useful for pedagogical purposes, in nature there is more of a continuous spectrum of storm behavior; supercell behavior does not suddenly begin and ordinary-cell behavior does not suddenly end when a threshold of environmental parameters is crossed.

Without Doppler radar it was difficult to determine precisely how the wind field in supercells differed from that in ordinary cells. However, Ted Fujita and collaborators inferred from analyses of time series of radar reflectivity and analyses of wind data collected from aircraft located outside storms that the main updraft in a supercell rotated and it was suggested that this characteristic was responsible for their "deviant" motion and, at least in part indirectly, for their longevity. Some supercells produced tornadoes and it was therefore thought that there must be a connection between storm-scale rotation and the much smaller scale tornado. Early analyses of supercell dynamics drew upon an analogy between the interaction between spinning solid bodies and the airflow around them, like baseballs curving through the air when spin is imparted to them (Figure 4.4). Supercells, however, are not solid bodies embedded in the airflow: They are part of the airflow itself and air circulates up, through them, and then out from them. Furthermore, these early theories did not consider thermodynamics or precipitation microphysics. However, it was recognized that the

[1] For example, the Wokingham storm of July 9, 1959 in the U. K., and the Geary storm of May 4, 1961 in Oklahoma.

[2] The term "supercell" first appeared in the refereed literature in 1964, but had been used in 1963 at an American Meteorological Conference and in 1962 in the journal *Meteorological Magazine*. Browning was initially criticized for using such a "vulgar" term.

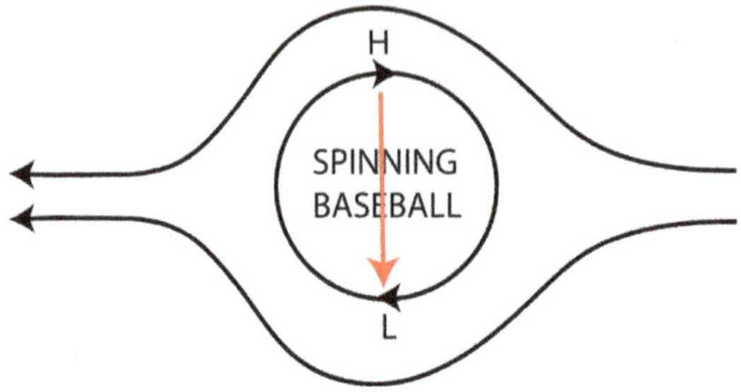

Figure 4.4. Illustration of the "Magnus effect" as a clockwise-spinning baseball experiences a force that deflects it to the right. In the figure, the baseball is shown in its reference frame, as it is thrown from left to right (relative streamlines are shown). Air approaches the baseball and is deflected about either side. Air is diverted to the right but is slowed down by an adverse pressure gradient as a result of spinning, so there is relative high pressure on the left side, while air is diverted to the left but is accelerated by a pressure gradient acting in the direction of the flow as a result of spinning: the net result is a force acting to the right of the baseball's path (red arrow).

vertical shear of the wind in the environment of the storm plays an important role in their longevity.

During a hail project in Alberta conducted in the late 1960s and another in northeast Colorado in 1970 conducted by the University of Wyoming and by NCAR, John Marwitz found that supercell storms form in an environment of much stronger vertical shear than that of ordinary cells and in an environment of stronger shear than that of multicell storms, whose existence requires at least low-level shear. Early theories correctly pointed out that in supercells, owing to vertical shear, precipitation falls out away from the main updraft, allowing the updraft not to weaken and dissipate as precipitation falls back into it. Since then, there have been many observational studies that have demonstrated the relationship between a high-shear environment and supercells.

In 1963, Yoshi Ogura, writing in a monograph while at MIT, and Keith Browning and C. R. Landry, writing in a radar conference proceeding, were the first to suggest that the tilting of low-level horizontal vorticity might be responsible for creating cyclonic vortices (such as tornadoes) in convective storms. Stan Barnes at NSSL in 1968 in an NSSL Technical Memorandum on a study of a number of storms and in 1970 in the analysis of a specific storm was the first to present detailed observational evidence that the source of storm-scale rotation in them was in fact due to the tilting of horizontal vorticity in the environment (Figure 4.5). This horizontal vorticity is associated with the vertical shear of the environmental wind. Col. Robert Miller, in his classic severe storm forecasting manual revised in the early 1970s, implicitly recognized observationally the importance of vertical shear in that it was noted that when a low-level jet intersects an upper-level jet at a substantial angle this synoptic situation was a necessary condition for severe weather outbreaks.

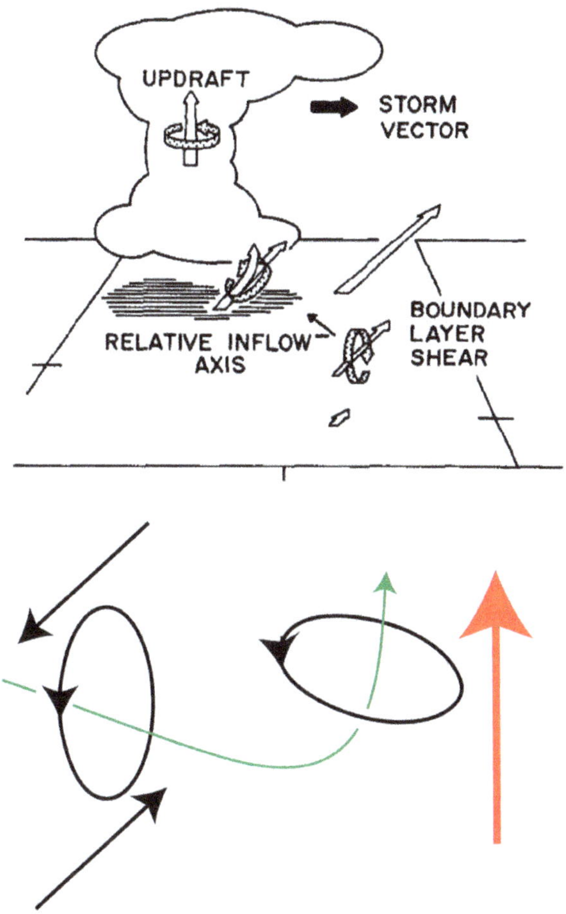

Figure 4.5. (Top) One of the first illustrations of how boundary-layer horizontal vorticity associated with vertical shear could be tilted onto the vertical as fluid parcels in the boundary layer are tilted upward when they enter an updraft in a convective storm (from Barnes, 1970). (Bottom) Idealized depiction of horizontal vorticity (green streamline points in the direction of the 3D vorticity vector) associated with vertical shear of the horizontal wind being tilted onto the vertical by an updraft (red vector).

According to the *thermal wind relation*, which is a consequence of the observed approximate (*geostrophic*) balance between the large-scale (synoptic-scale) pressure gradient force and the Coriolis force, and of hydrostatic balance, the magnitude of the vertical shear of the geostrophic wind—which is approximately the same as the vertical shear of the total wind (i.e., of the geostrophic + the ageostrophic wind)—is proportional to the horizontal gradient of temperature normal to the vertical shear vector

$$\partial \mathbf{v}_g / \partial z \approx g/fT \, \mathbf{k} \times \nabla_h T \tag{4.23}$$

where $\mathbf{v}_g$ is the geostrophic wind; f is the Coriolis parameter ($2\Omega \sin \phi$, where Ω is the rotation rate of the Earth about its axis and ϕ is the latitude); and the subscript h by the gradient operator denotes (horizontal) derivatives computed on a surface of constant height. Thus, supercells should be found preferentially when there are strong horizontal temperature gradients (e.g., near surface fronts and baroclinic waves in the upper troposphere) and when there is the potential for localized, strong, buoyant updrafts. On the synoptic scale, the horizontal temperature gradient is $\sim$10 K/1,000 km, so that in mid-latitudes

$$|\partial \mathbf{v}_g/\partial z| \sim [10\,\mathrm{m\,s^{-2}}/10^{-4}\,\mathrm{s^{-1}}(300\,\mathrm{K})]10\,\mathrm{K}/10^6\,\mathrm{m} \sim 10^{-3}\,\mathrm{s^{-1}}$$

In the vicinity of fronts, where the horizontal temperature gradient is an order of magnitude greater

$$|\partial \mathbf{v}_g/\partial z| \sim 10^{-2}\,\mathrm{s^{-1}}$$

Moreover, vertical shear in the boundary layer is also a possible source of horizontal vorticity that can be tilted onto the vertical in a convective storm.

4.3 OBSERVED SUPERCELL STRUCTURE: CLOUD FEATURES, PRECIPITATION DISTRIBUTION, POLARIMETRIC RADAR–OBSERVED PARAMETERS, AND WIND AND TEMPERATURE FIELDS

A large leap in our understanding of supercells occurred during the 1970s thanks to storm-chasers, who first systematically documented the visual cloud structure of supercells (Figure 4.6), the advent of the use of Doppler radar which led to detailed depictions of the wind field in supercells in the late 1970s and early 1980s, especially in dual-Doppler studies done by Gerry Heymsfield at OU (University of Oklahoma), and Ed Brandes and Peter Ray at NSSL (Figure 4.7), and the nearly simultaneous advent of three-dimensional, non-hydrostatic "cloud" models, which could be used to conduct controlled experiments. Not being able to control nature, we must settle for whatever the atmosphere reveals to us when studying it observationally. In the mid and late 1970s, Robert (Bob) Schlesinger at the University of Wisconsin at Madison, Joe Klemp at NCAR, and Bob Wilhelmson did pioneering experiments, in which the three-dimensional aspects of supercells (rotation, deviant motion, etc.) were simulated for the first time (Figure 4.8). More recently, polarimetric radars have afforded a more complete look at the spatial distribution of hydrometeors, etc. in supercells (and in multicells).

4.3.1 The main updraft in supercells

Supercells are prolific producers of large hail; it was hypothesized that large hail is associated with a very strong updraft because large hailstones have a high terminal fall velocity and because water substance must be recycled into and out of

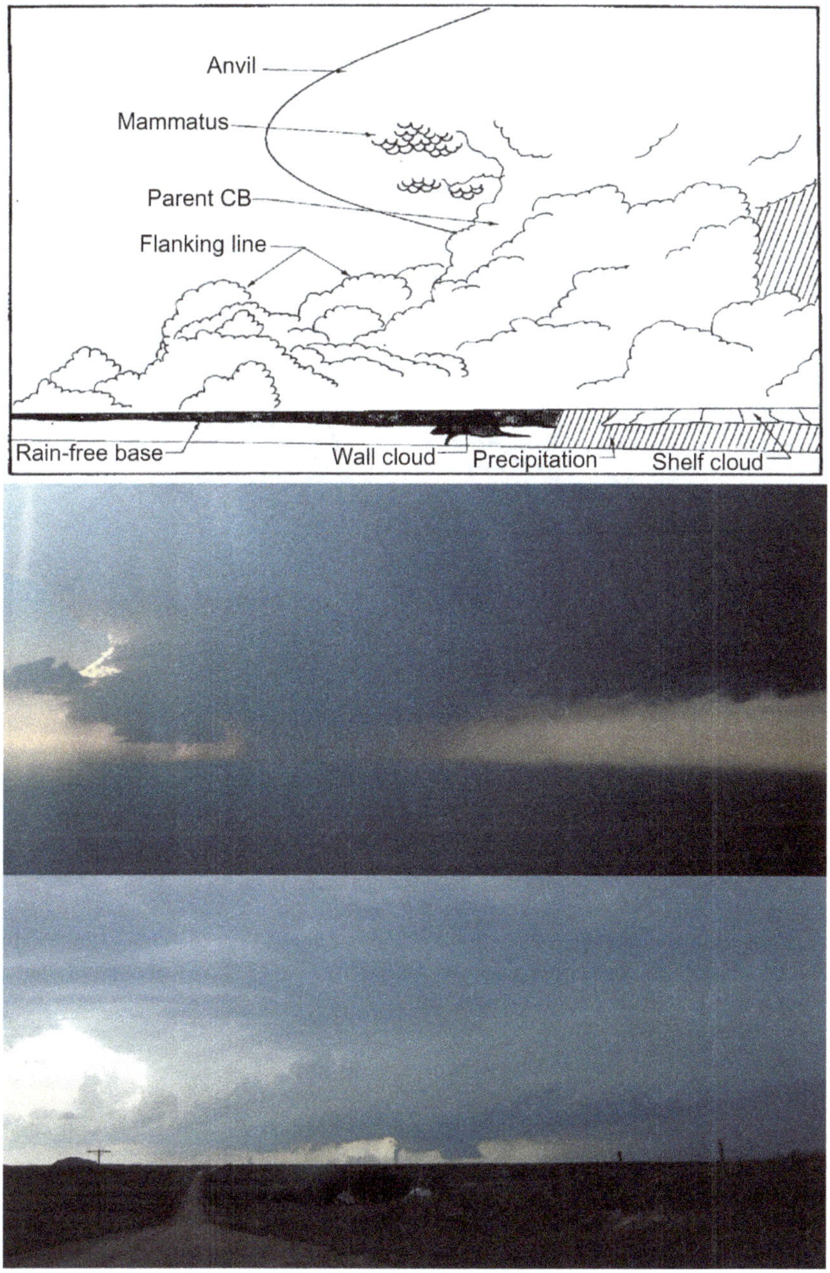

Figure 4.6. Supercell, as viewed from ahead and approximately to the right of storm motion. (Top) Idealized sketch inspired by observations by Chuck Doswell (from Moller, 1978); (middle) tornadic supercell over southwest Kansas on May 26, 1991, as viewed from a NOAA P-3 aircraft; (bottom) tornadic supercell over southeastern Wyoming on June 5, 2009, during VORTEX2 (photographs by the author).

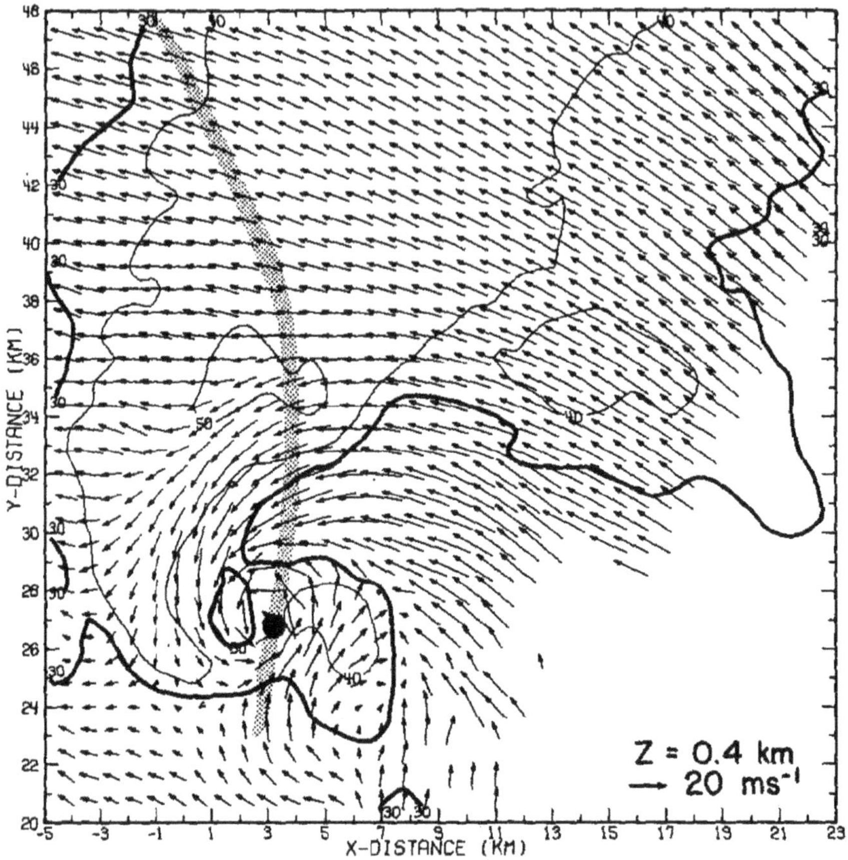

Figure 4.7. Storm-relative winds at 400 m AGL, synthesized from an early dual-Doppler radar analysis using fixed site radars operated by NSSL on May 20, 1977 in central Oklahoma. The tornado damage path is indicated by a stippled line; the tornado at the time of the analysis is indicated by a black dot. Solid lines are radar reflectivity factor (dBZ) from one of the radars; thick solid line is the 30 dBZ contour (from Brandes, 1981).

the updraft. The existence of a strong updraft was inferred indirectly from the *weak-echo region (WER)* and *bounded weak-echo region (BWER)* seen in radar reflectivity constant elevation angle data or constant altitude data[3] (i.e., in a quasi-horizontal plane) (Figures 4.9 and 4.10). From the WER and BWER it was inferred that large-enough (i.e., radar-detectable) particles did not form until relatively high up in the cloud: It takes time for cloud droplets to grow into precipitation-size particles and, in a very strong updraft, it does not take long for air coming from low altitudes to reach very high altitudes. Surrounding the core of the updraft, where the updraft strength is weaker, radar-detectable particles form

[3] Looked at as a vertical cross-section it was named a "vault".

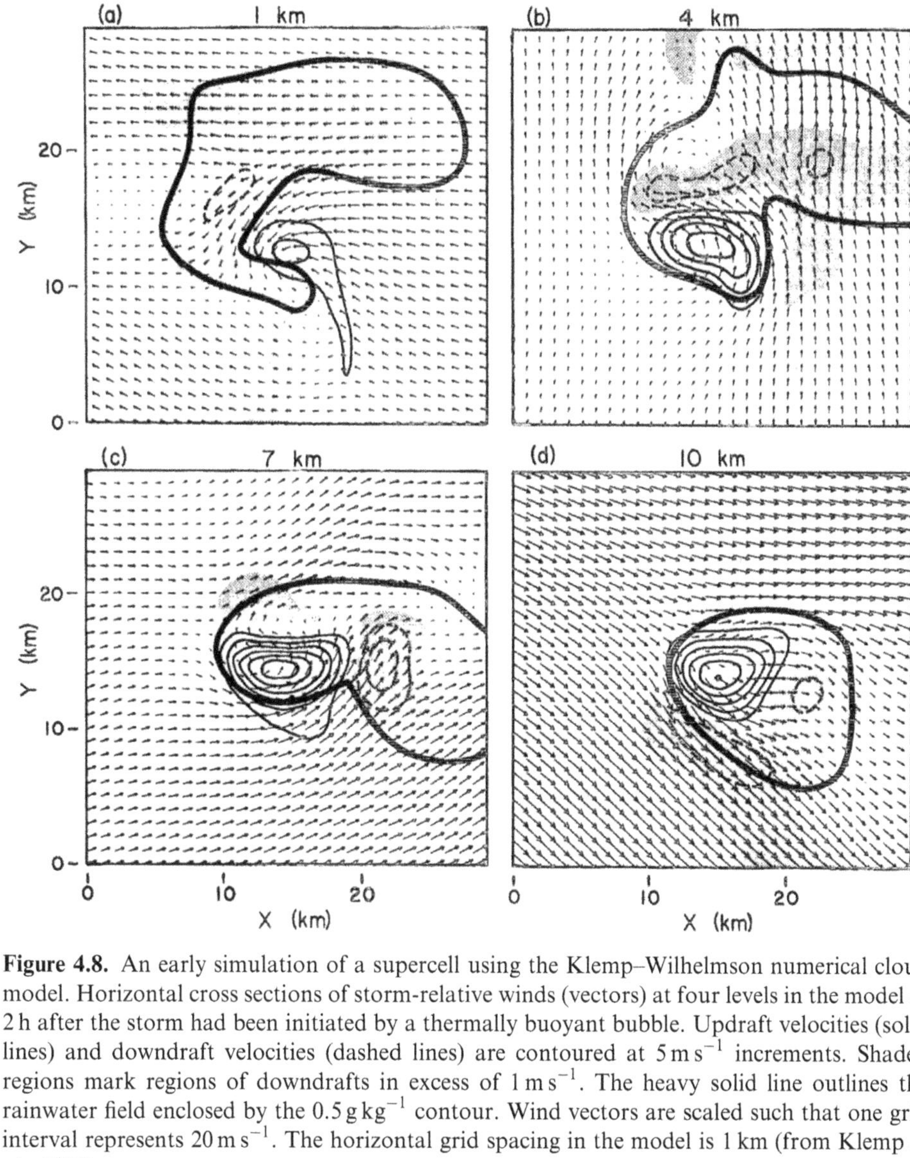

Figure 4.8. An early simulation of a supercell using the Klemp–Wilhelmson numerical cloud model. Horizontal cross sections of storm-relative winds (vectors) at four levels in the model at 2 h after the storm had been initiated by a thermally buoyant bubble. Updraft velocities (solid lines) and downdraft velocities (dashed lines) are contoured at $5\,\mathrm{m\,s^{-1}}$ increments. Shaded regions mark regions of downdrafts in excess of $1\,\mathrm{m\,s^{-1}}$. The heavy solid line outlines the rainwater field enclosed by the $0.5\,\mathrm{g\,kg^{-1}}$ contour. Wind vectors are scaled such that one grid interval represents $20\,\mathrm{m\,s^{-1}}$. The horizontal grid spacing in the model is 1 km (from Klemp *et al.*, 1981).

at lower altitude. The BWER sometimes assumes an arc shape (Figure 4.11), indicative of a curved region of updraft (U shaped, horseshoe shaped, or crescent shaped). The high intensity ($\sim 50\,\mathrm{m\,s^{-1}}$) of updrafts in supercells has been confirmed (Figures 3.14 and 3.15) from measurements made by a storm-penetrating, armored aircraft, from the ascent rate of balloons launched into updrafts, and corroborated with measurements in numerically simulated storms by cloud models.

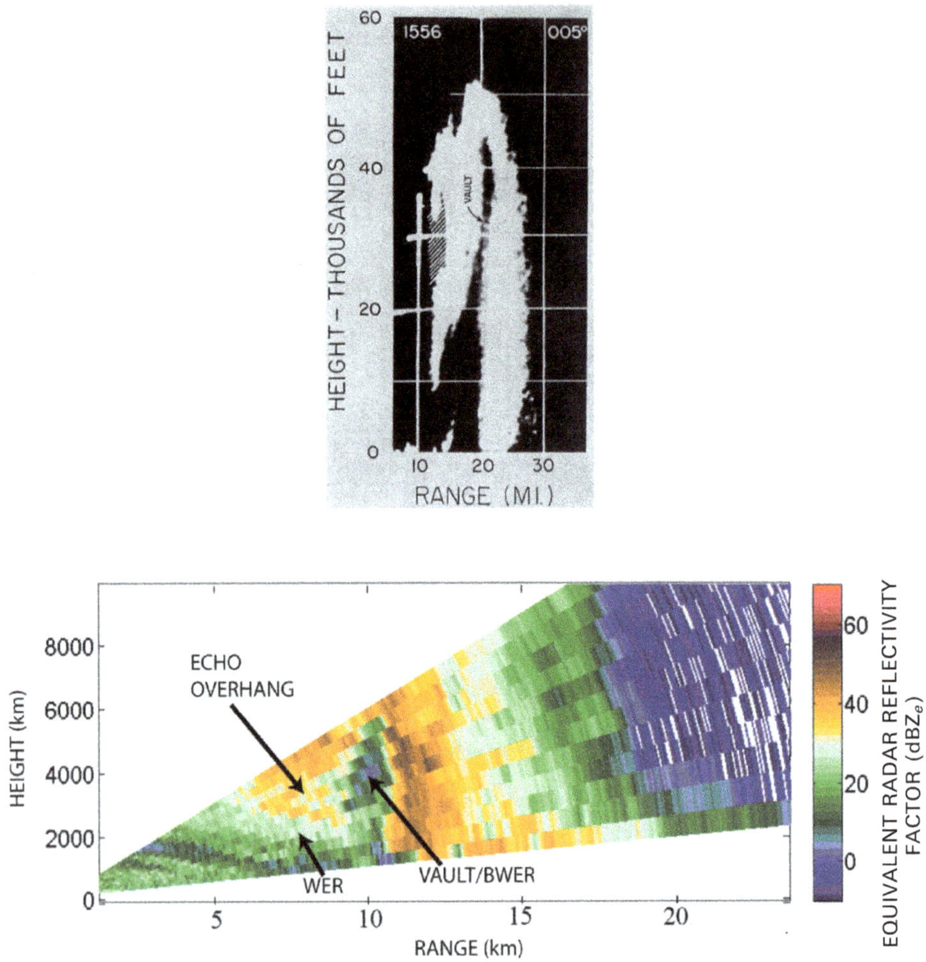

Figure 4.9. (Top) Vault observed in a supercell by a radar in central Oklahoma on May 26, 1963: Vertical cross section (RHI) of radar echo; the vertical scale is exaggerated (from Browning, 1965b). (Bottom) Vault/BWER, WER, and "echo overhang" observed in a supercell in north central Oklahoma on May 24, 2008 by a mobile Doppler radar from the University of Massachusetts from a synthesis of sector scans (courtesy of Jeff Snyder).

When the updraft is strong enough to bring up relatively large raindrops, water-coated hailstones, or melted ice particles to high elevations, recycled from falling raindrops (or water-coated hailstones or melted ice particles) outside the center of the updraft, where it is well below freezing, the raindrops, which are flattened, backscatter much more horizontal radiation than vertical to radars that are probing them. The "differential reflectivity" is defined as

$$Z_{\mathrm{DR}} = 10 \log Z_H / Z_V \tag{4.24}$$

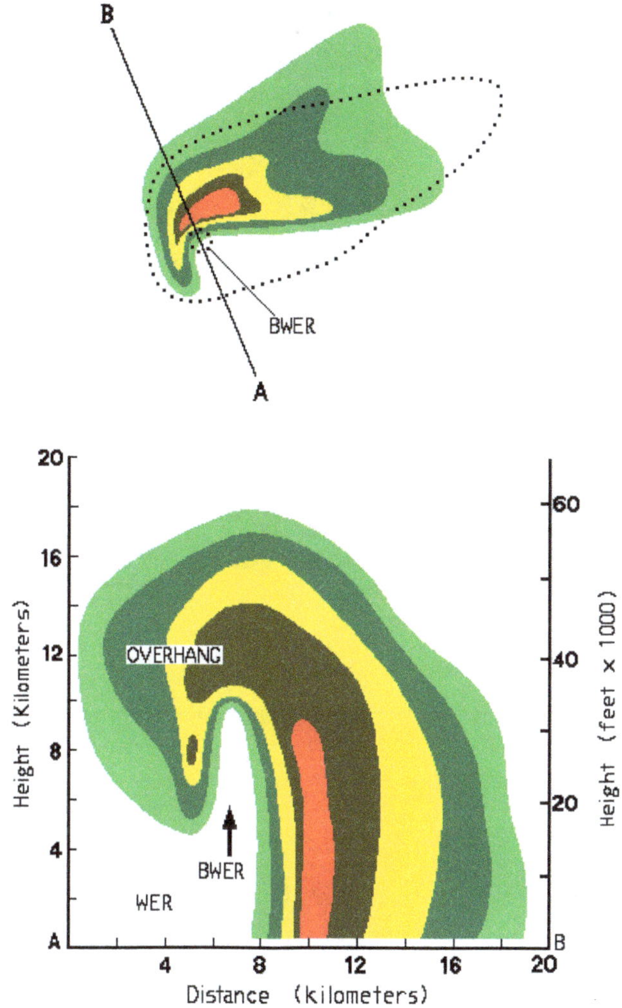

Figure 4.10. Idealized model of the WER, BWER, and echo overhang in a supercell. (Top) Horizontal cross section of color-coded radar reflectivity factor, with warm colors the most reflective and cool colors the least at low levels; mid-level echo is outlined by dotted line. (bottom) Vertical cross section across the line A–B at the top (from the National Weather Service *Storm Spotter Glossary*).

where Z_H and Z_V are the radar reflectivity factors of horizontally and vertically polarized radar beams (it is assumed here that the horizontally polarized outgoing beam is received by the horizontally polarized antenna and that the vertically polarized outgoing beam is received by the vertically polarized antenna). There is a column of relatively high values of Z_{DR} (called a "Z_{DR} column") at locations displaced from the center of the updraft, which is thought to be associated with

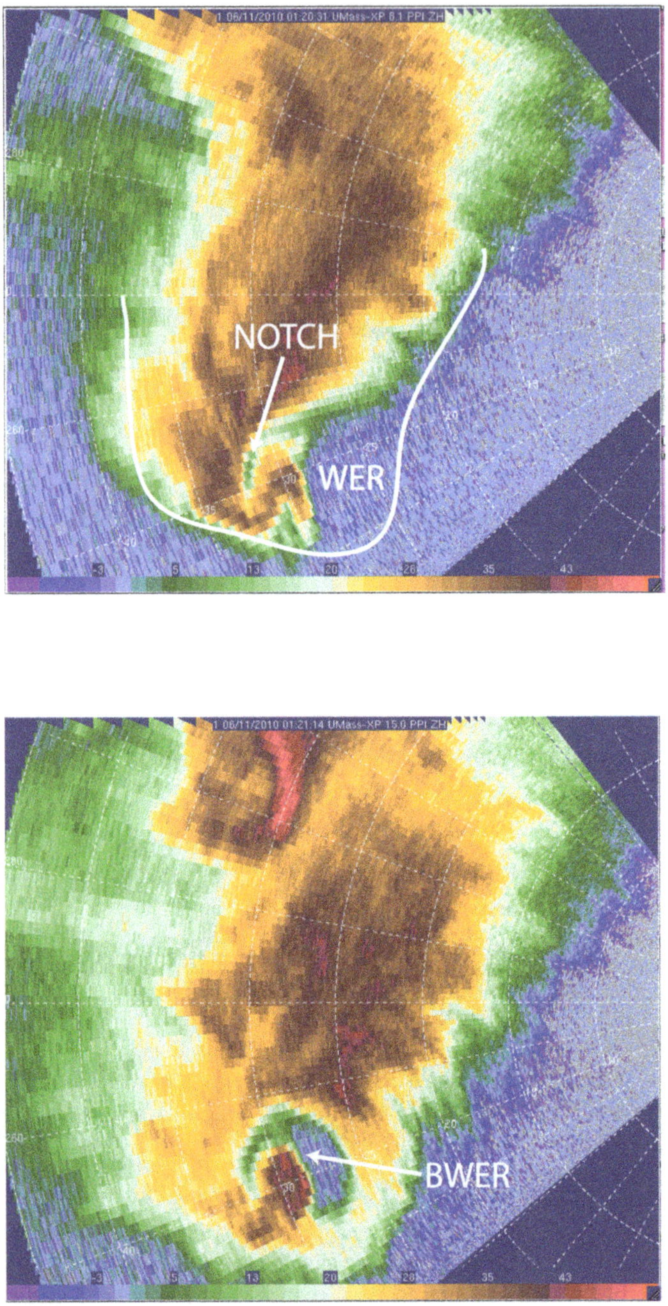

Figure 4.11a. Example of a crescent-shaped BWER in a supercell in eastern Colorado on June 10, 2010 (radar reflectivity color-coded in dBZ_e). (Top) 8° elevation angle scan; solid white line is approximate location of $20\,dBZ_e$ contour at 15° elevation angle. (Bottom) 15° elevation angle scan (from the U. Mass. X Pol radar).

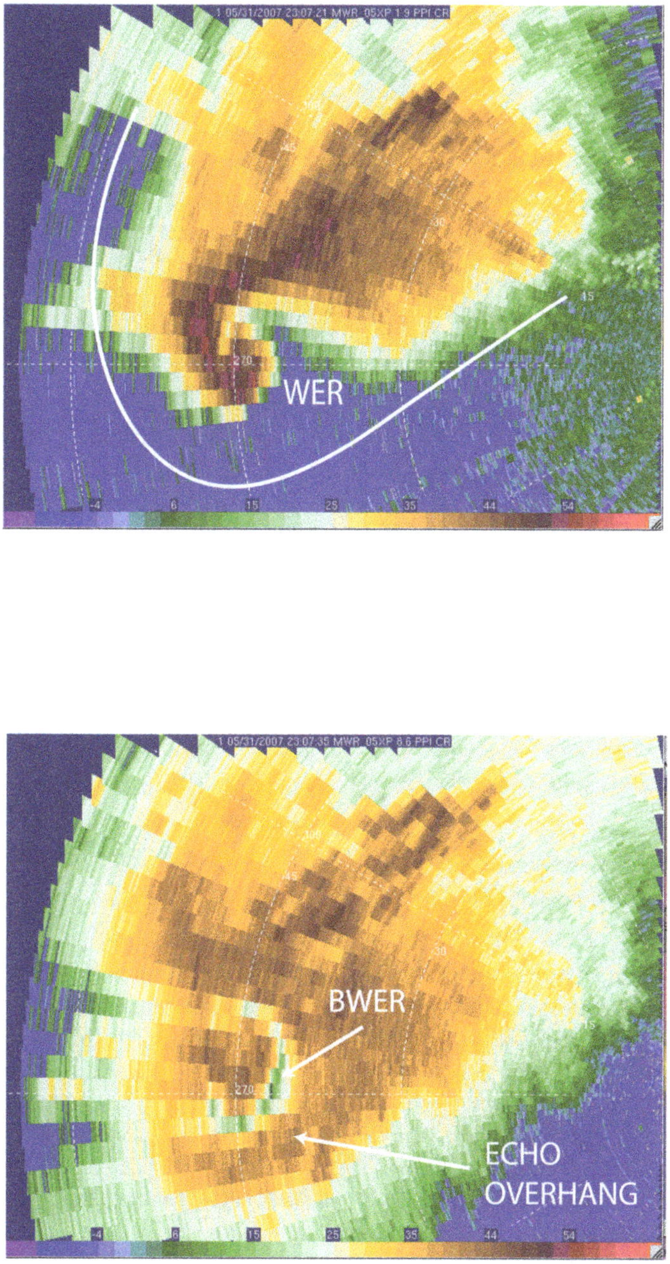

Figure 4.11b. Example of a crescent-shaped BWER in a supercell in the Oklahoma Panhandle on May 31, 2007. (Top) 1.9° elevation angle scan. Solid white line is the approximate location of the 20 dBZ_e contour at 8.6° elevation angle. (Bottom) 8.6° elevation angle scan (from the MWR-05XP radar). The BWER, which is seen aloft, is crescent shaped because it merges with the narrow notch, which is seen at low elevation angles.

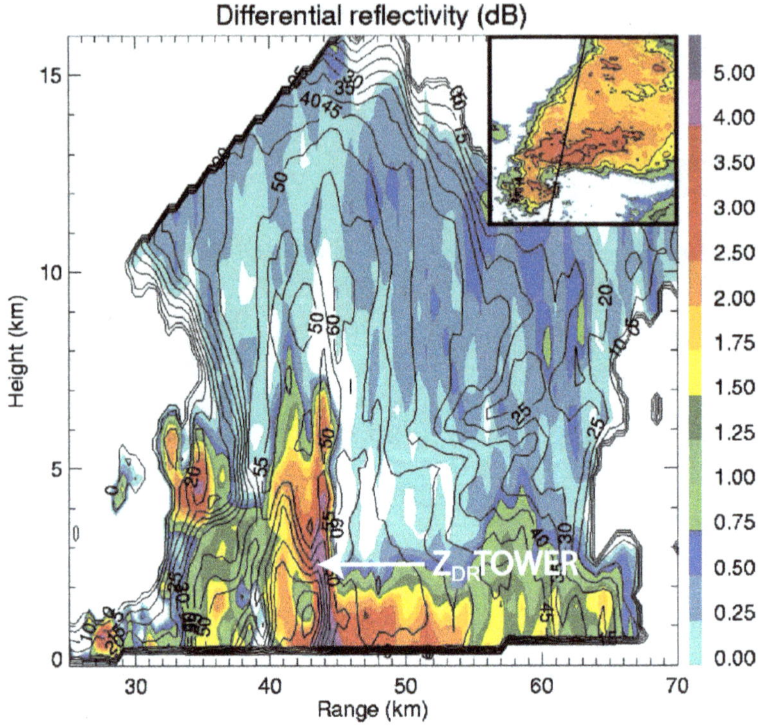

Figure 4.12. Example of a Z_{DR} column in a vertical cross section through a supercell (see solid line in the inset in the upper right-hand corner of the radar reflectivity factor at low elevation) on May 9, 2003 in central Oklahoma, as detected by the S-band KOUN radar. Differential reflectivity, Z_{DR} (dB) is color-coded; radar reflectivity factor (dBZ) is contoured by thin solid lines (from Kumjian and Ryzhkov, 2008)

updraft strength. Therefore, the change of Z_{DR} in a column (Figure 4.12) is an indicator of changes in updraft intensity, all other factors remaining unchanged.

When electromagnetic radiation propagates through rain, it slows down (compared with how quickly it propagates through air). The "propagation" phase delay (compared with the initial phase) upon being backscattered to the radar when it is slowed down depends, in part, on the total amount of liquid water in its path that is traversed. When the shape of raindrops is not spherical (i.e., as is the case for large raindrops, which are oblate), then radiation in one plane propagates more slowly than it does in the orthogonal plane, so there is a difference in phase delay between horizontally and vertically polarized beams of radiation: this phase difference is called ϕ_{DP} or "differential phase". For most radar systems receiving horizontally polarized radiation, the differential phase is positive for raindrops, which are flattened (oblate spheroids), because horizontally polarized radiation is slowed more than vertically polarized radiation, since more water is encountered. The "total differential phase" is actually different, owing to the added "backscatter" differential phase, which is a result of resonance effects, which are most prominent in shorter wavelength radar systems. The longer the path length through non-spherical scatterers such as large raindrops, the higher the ϕ_{DP}. To determine if there are non-spherical scatterers in a volume the quantity

$$K_{DP} = \tfrac{1}{2}\partial\phi_{DP}/\partial r \tag{4.25}$$

the "specific differential phase", which is proportional to the radial gradient of the differential phase, is useful (the factor of $\frac{1}{2}$ appears in (4.25) because phase shifts occur in both outgoing and backscattered radiation). So, when radiation enters a region of many raindrops, ϕ_{DP} suddenly increases, so K_{DP} increases and stays relatively constant, while ϕ_{DP} continues to increase. K_{DP} generally increases with raindrop diameter for Rayleigh scatterers (i.e., those whose diameter is much less than that of the wavelength of the radar used). When the wavelength of the radar is decreased so that the diameter of the raindrops is a larger fraction of the wavelength, "resonance" effects make the dependence of K_{DP} on raindrop diameter more complicated, especially for shorter wave radars (5 cm and 3 cm wavelengths, for example), and in fact non-monotonic with respect to small changes in raindrop diameter. K_{DP} is a useful quantity, however, especially when there is a lot of attenuation or beam blockage, which affects the measurement of Z_{DR}, but does not affect phase measurements.

Columns of relatively high K_{DP} in supercells are found to the left (with respect to storm motion) of the updraft (Figure 4.13). There is evidence that in "K_{DP} columns" there is a mixture of rain and wet hail or wet graupel and is indicative of drops of water shed from hailstones.

4.3.2 Downdrafts: forward-flank downdraft and the rear-flank downdraft

Idealized conceptual models and radar imagery depict the relationship between the main updraft, which assumes a U shape, and the two main downdrafts (forward-flank downdraft or FFD and rear-flank downdraft or RFD), storm-relative wind flow, and radar reflectivity structure (Figures 4.14–4.15). While isolating the two different locations for the downdrafts is useful in terms of defining storm structure, there may not be two separate downdrafts and the two may actually be contiguous. The FFD, however, forms before the RFD does.

It is likely that water substance conversion processes such as evaporative cooling and/or sublimation and/or melting cooling and/or precipitation loading play a major role(s) in driving the FFD and the RFD. When the boundary layer is relatively dry, evaporative cooling/sublimational cooling is enhanced, but is limited when the boundary layer is relatively moist. Evidence for precipitation loading in some instances is found from what has been named the "descending reflectivity core" (DRC), a blob-like protuberance of precipitation above the weak-echo region and underneath the echo overhang (Figure 4.16) by Erik Rasmussen and colleagues and has been hypothesized to play a role in tornado formation. Evidence for the RFD is also found in photographs, movies, and videos of the cloud base associated with a low-level mesocyclone, in which the

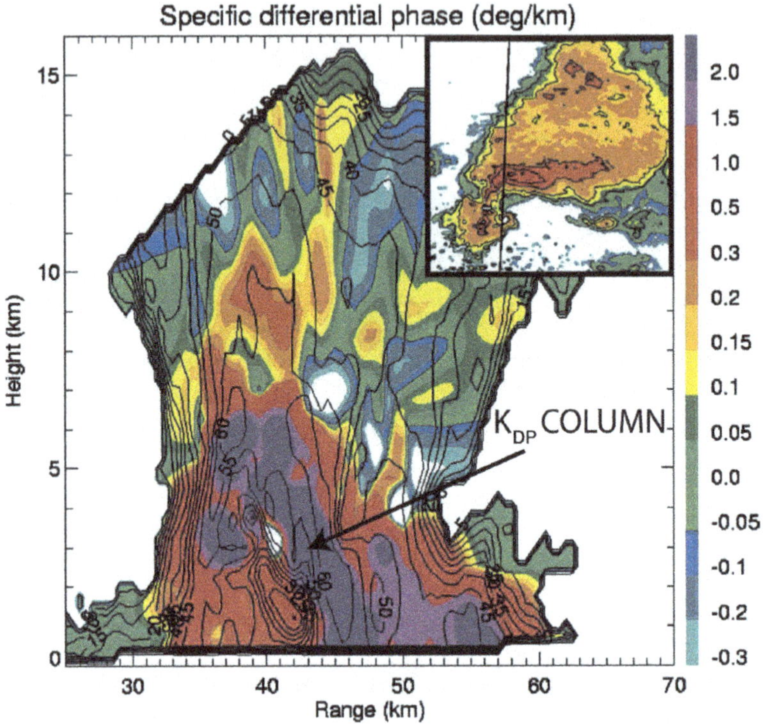

Figure 4.13. As in Figure 4.12, but for a K_{DP} column. Specific differential phase, K_{DP} (deg km^{-1}) is color-coded (from Kumjian and Ryzhkov, 2008).

disappearance of cloud material and what Al Moller, Les Lemon, and Chuck Doswell, among others, in the mid-1970s named a "clear slot"[4] are seen. Eventually, air from the rear-flank downdraft reaches the ground and curves around the mesocyclone, effectively cutting off the supply of ambient, moist, relatively warm, potentially buoyant air into the main updraft of the supercell. Such a process has been referred to as an "occlusion" (not to be confused with the occlusion downdraft to be described later), analogous to the occlusion process in synoptic-scale, extratropical cyclones. The curved, bulging nature of the leading edge of the RFD produces a curved band of convergence and rising motion along it, which is shaped like a horseshoe, or letter "U" (Figure 4.15) and sometimes a bounded weak-echo region that is also U shaped (Figure 4.11, bottom panels). The separation of the mesocyclone at low levels by the RFD into a region of rising motion from one of sinking motion has been termed its "divided structure" (Figure 4.15).

Part of the main updraft is located within the deepest convective cloud, above a lowered cloud base named the "wall cloud" (Figure 4.17) by Ted Fujita, because the side of the cloud feature is vertical like a wall. The wall cloud forms when

[4] Like a "dry slot" in synoptic-scale cyclones.

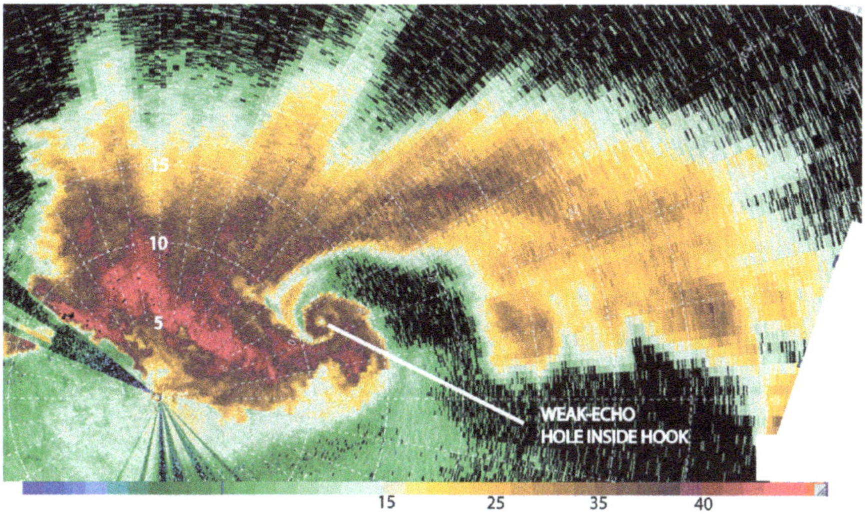

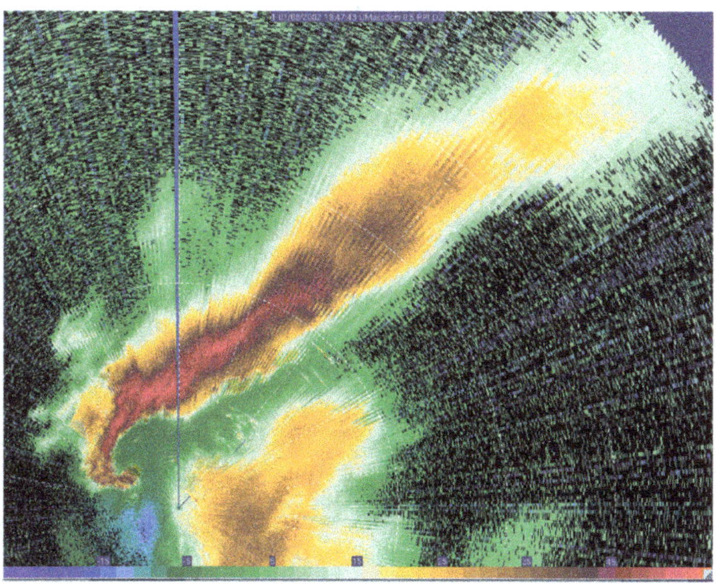

Figure 4.14a. Hook echoes in supercells. Radar reflectivity factor only, in dBZ_e, (top) for a supercell in the Texas Panhandle on May 29, 2001 (as detected by the U. Mass. X-Pol mobile radar); (bottom) for a tornadic supercell in Kansas on May 8, 2003 (range rings given every 5 km). Attenuation has reduced the thickness of the radar echo (bottom), making it look artificially elongated.

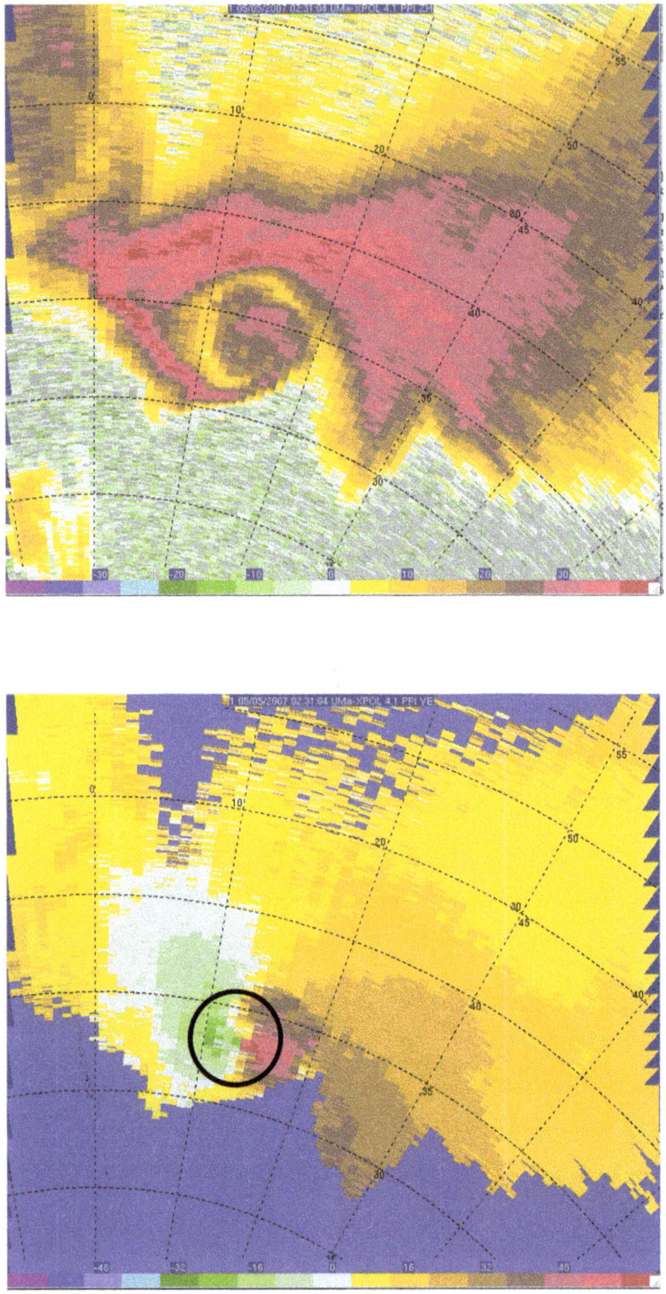

Figure 4.14b. Hook echoes in supercells. Radar reflectivity factor (top) and Doppler velocity in m s^{-1} (bottom) for a tornadic supercell near Greensburg, KS during the early evening of June 4, 2007 (as detected by the U. Mass. X-Pol mobile radar). The cyclonic vortex signature associated with the parent vortex of the tornado is marked by a circle.

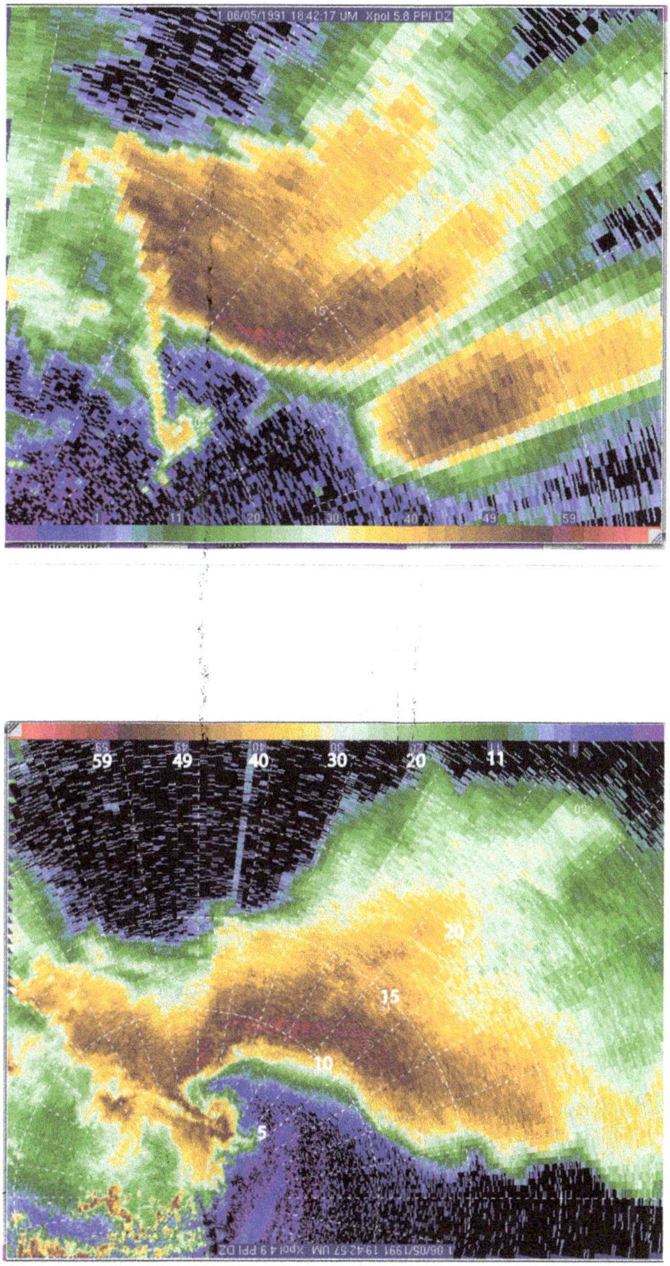

Figure 4.14c. Hook echoes in supercells. Radar reflectivity factor for supercells in northwest Oklahoma on June 5, 2001 (as detected by the U. Mass. X-Pol mobile radar; range rings are shown every 5 km). There are weak-echo holes inside the hook echoes in the supercells of May 29, 2001, June 5, 2001, and May 4, 2007. The hook echo (top) is very thin, while the hook echo in Figure 4.14a (top) is relatively thick.

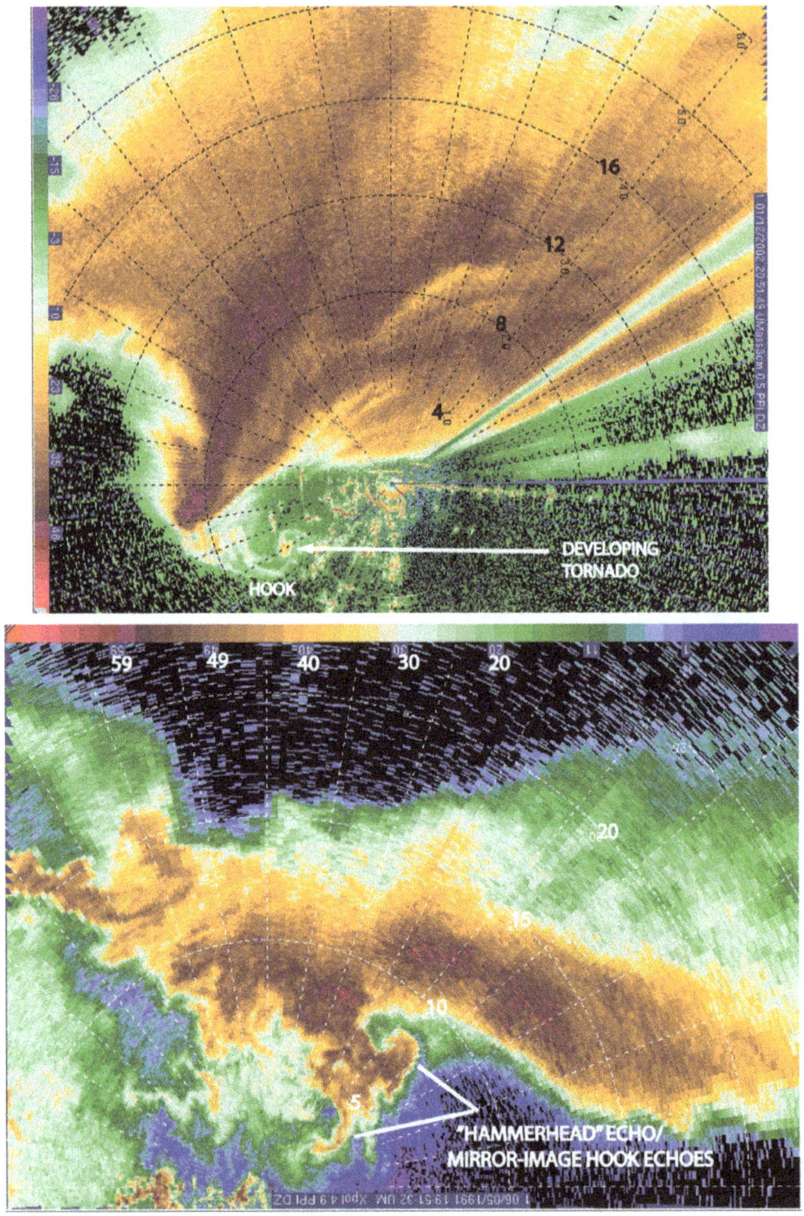

Figure 4.14d. Hook echoes in supercells. (Top) Very thin and weak hook echo on May 12, 2004 near Attica, KS, as detected by the U. Mass. X-Pol mobile radar; small debris ball (arrow) at the center of a developing tornado and spiral band curving the other way, as the tornado is located at an inflection point of the rear-flank gust front. (Bottom) "Hammerhead" echo (as it is colloquially known) in a supercell in northwest Oklahoma on June 5, 2001 (as detected by the U. Mass. X-Pol mobile radar); the "hammerhead" echo consists of a pair of mirror image hook echoes.

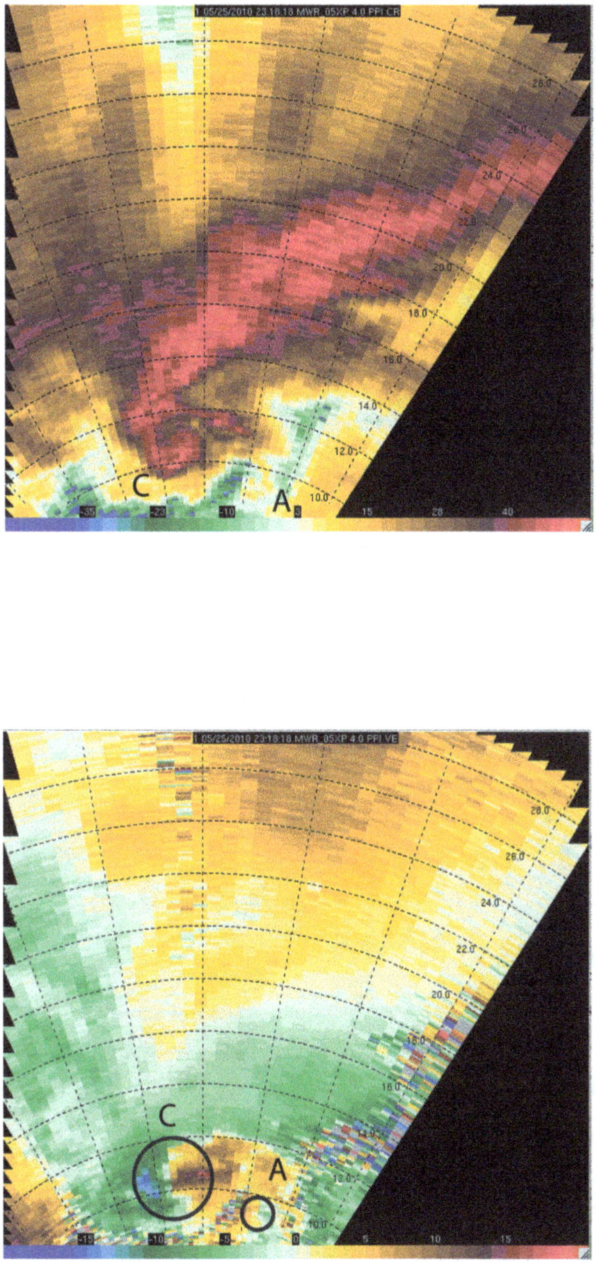

Figure 4.14e. Hook echoes in supercells. Cyclonically rotating (circle labeled "C") and anticyclonically rotating (circle labeled "A") mirror image hook echoes in a tornadic supercell on May 25, 2010 in far western Kansas (as detected by the MWR-05XP, mobile, X-band, phased array Doppler radar; range rings shown every 5 km). (Top) Radar reflectivity factor in dBZ_e and (bottom) Doppler velocity in $m\,s^{-1}$ (range rings shown every 2 km).

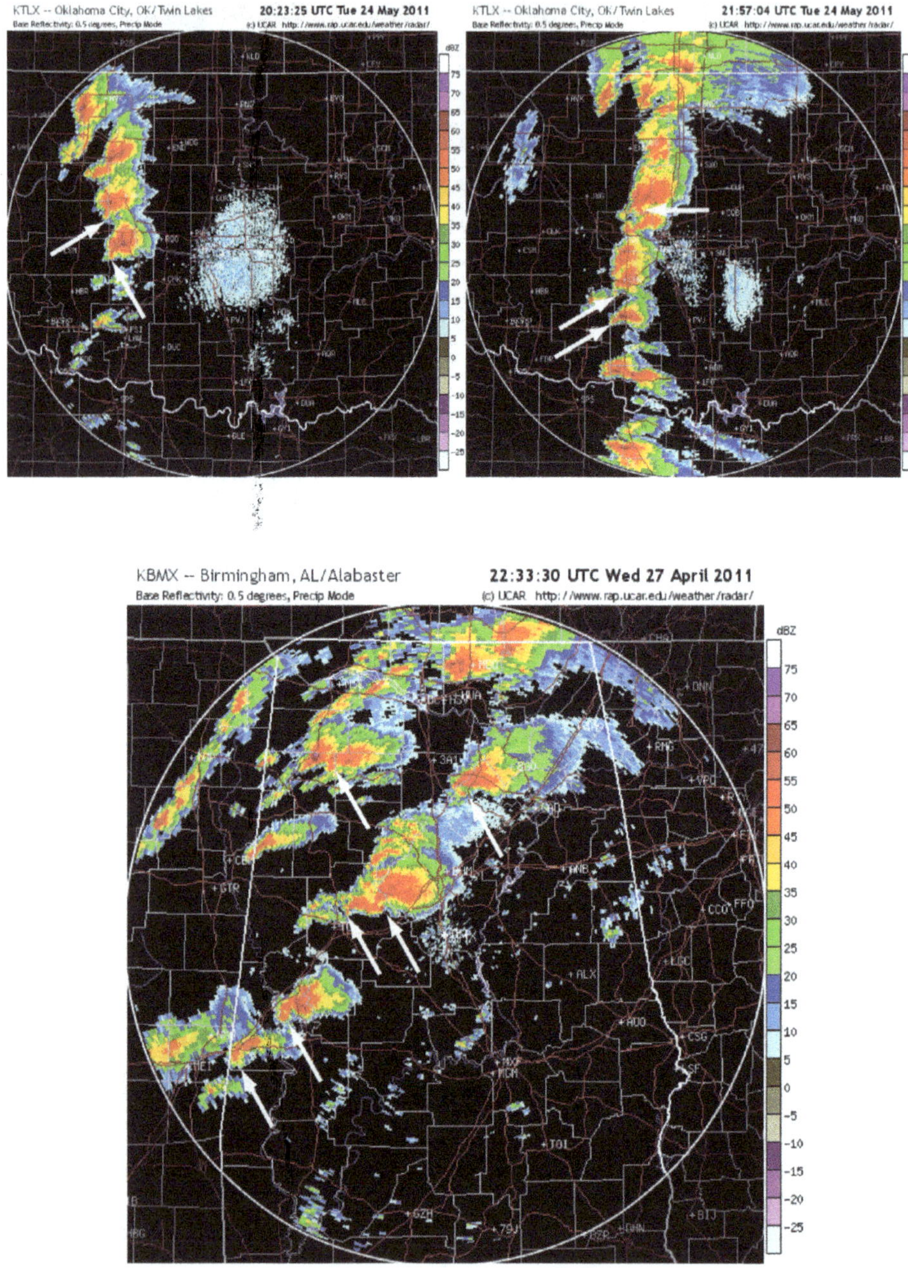

Figure 4.14f. Hook echoes in supercells. Multiple supercells during tornado outbreaks (as viewed on WSR-88D radar) (Top) Radar reflectivity as depicted from the radar at Oklahoma City, OK on May 24, 2011, showing a string of supercells along the dryline (hook echoes are highlighted with arrows). (Bottom) Radar reflectivity as depicted by the radar at Birmingham, AL on April 27, 2011 (hook echoes are highlighted with arrows).

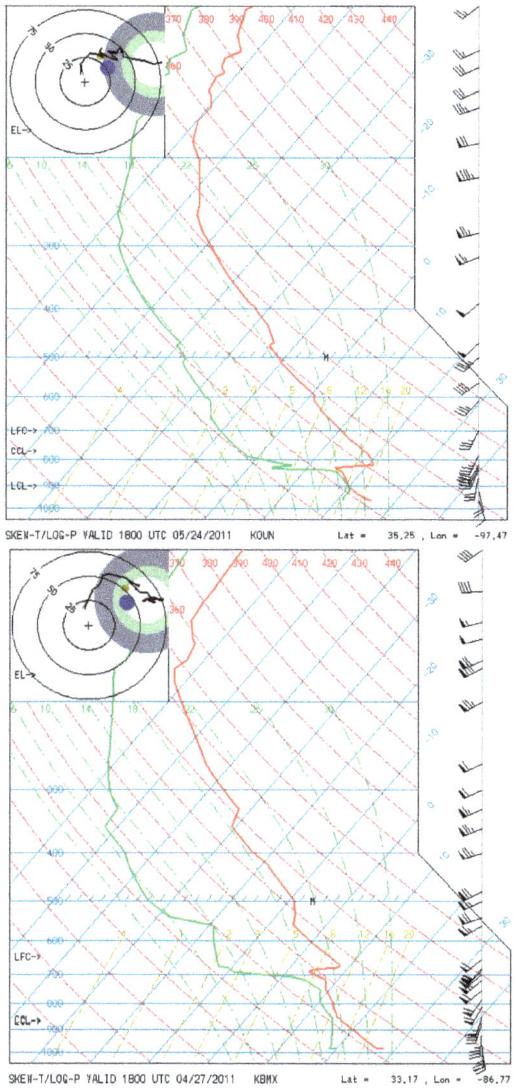

Figure 4.14g. Soundings on days when there were tornado outbreaks (see Figure 4.14f). Sounding at (top) Norman, OK on May 24, 2011 at 18:00 UTC and (bottom) at Birmingham, AL on April 27, 2011 at 18:00 UTC. Soundings may vary substantially in space and time near tornadic supercells. The most representative soundings may be those released on days when widespread, long-lived, tornadic supercells occur, such as those shown here. In both, about three hours before tornadic supercells were mature, there was a moist boundary layer, topped by a capping inversion. In the case of the Birmingham sounding, the cap is elevated, indicative of some lifting having occurred. In both some CIN must have been overcome for low-level air to reach its LFC. In each, the hodograph (insets at the upper left) is "long" and there is a bend in the vertical shear vector near the altitude of the interface between the moist boundary layer and the region above (soundings from NCAR/EOL).

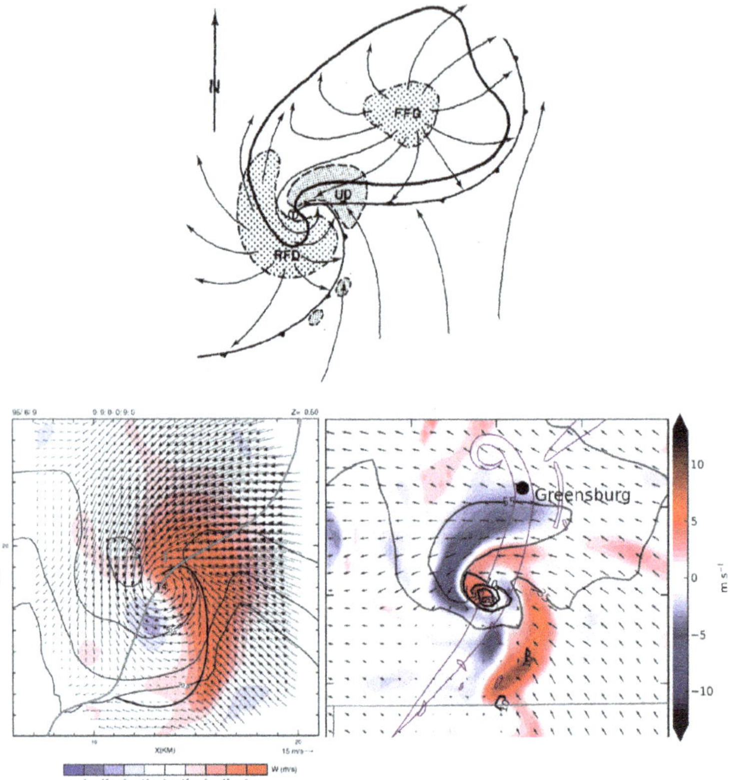

Figure 4.15. (Top) Conceptual model of the major vertical air currents in a supercell (from Lemon and Doswell, 1979), showing the rear-flank gust front (cold front symbol) ahead of the rear-flank downdraft (RFD), the main updraft (UD), the forward-flank downdraft (FFD), and a warm/stationary/cold front–like boundary (with the appropriate symbols) separating air underneath the FFD from ambient air (streamlines indicate the flow field near the ground). Tornado location is indicated by the "T". Updrafts are indicated by finely stippled areas; downdrafts are indicated by coarsely stippled areas. Radar echo associated with the storm outlined by solid line. (Bottom, left) Pseudo dual-Doppler analysis of storm-relative winds in a tornadic supercell in the Texas Panhandle on June 8, 1995 from ELDORA data, at 550 m AGL. Vertical velocity is color-coded. "T4" indicates the tornado location at the time of the analysis. The solid black line indicates the leading edge of the rear-flank gust front. The solid gray line indicates the tornado track. The solid contours are radar reflectivity in dBZ_e. Note the RFD (purple) and main updraft (red). There is no general FFD or windshift along the leading edge of an FFD as in the Lemon and Doswell conceptual model (courtesy of David Dowell). (Bottom, right) Ensemble mean vertical velocity color-coded and storm-relative ensemble mean wind (plotted every 2 km) from mobile Doppler radar data from the U. Mass. X-Pol radar and data from the WSR-88D radar at Dodge City, KS, on May 4, 2007, in south central Kansas, assimilated into a numerical cloud model. Tornado paths are indicated by thin purple lines and radar reflectivity factor (dBZ) is contoured at 35 and 55 dBZ in gray. There again is no windshift along the edge of an FFD. In this case, however, the RFD is connected to the FFD, while both the RFD and the main updraft are broken up into two sections (courtesy of Robin Tanamachi).

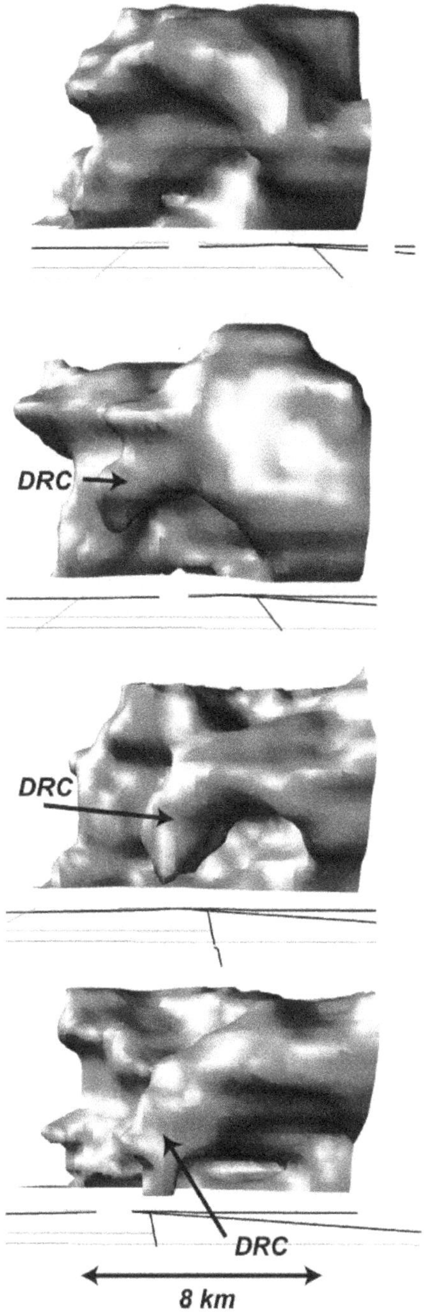

Figure 4.16. Example of a "descending reflectivity core" in a supercell. Three-dimensional view of the 40 dBZ isosursface in a supercell prior to tornado formation in the Texas Panhandle on June 2, 1995. Time increases by 5 min with each panel below. Radar data from the WSR-88D radar at Lubbock, TX (from Rasmussen *et al.*, 2006).

Figure 4.17. Wall clouds. (Top) May 26, 1978 in the Texas Panhandle with a developing tail cloud feeding into the bottom of the wall cloud from the right; (bottom) May 26, 1991 in southwest Kansas (as seen from the NOAA P-3 aircraft). A scud cloud is attaching itself to the lowered cloud base above. (Wall clouds are also seen in Figures 4.6, 4.23, 4.25, and 6.2b, upper right) (photographs by the author).

cooler, but more humid air from the adjacent FFD enters the updraft and lowers the condensation level. There is some evidence that additional lowering of the pressure is necessary to explain the observed lowering of the cloud base and that it is a result of dynamic effects (i.e., the rotation associated with a mesocyclone). The dynamical effects of vorticity on the perturbation pressure field will be discussed subsequently. So, the appearance of a wall cloud is indicative of both an updraft and rotation, though it is not clear which is the predominant characteristic. Visually, one often sees wall clouds that are not rotating or are rotating only very slowly.

Tornadoes, when they occur, are typically observed in the vicinity of the wall cloud (Figure 4.6). The RFD forms and impinges on the gust front that is located adjacent to the "hook echo", on the upshear side of the main body of the storm, creating a sharp wind shift or zone of confluence that bears resemblance to a synoptic-scale cold front and its spatial relationship to a synoptic-scale extratropical cyclone. Sometimes the hook echo is not apparent owing to a lack of scatterers, but there is a narrow appendage of enhanced radar echo. The hook echo may also be accompanied by a thin, bowed line of radar echo connected to it (Figure 4.14d, top), which marks the leading edge of the gust front associated with the RFD; this feature is called the "rear-flank gust front" (RFGF). As noted previously, the RFD may be caused, in part, by the same processes (negative buoyancy) that produce downdrafts in ordinary cells; in addition, however, it may be forced in part by dynamic effects, which will be discussed subsequently. The contribution from the negatively buoyant component may be particularly complex, since it involves not only the trajectory of unsaturated air underneath the precipitation that is falling out, but also the precise nature of the precipitation particles, which determines the rate and total amount of evaporational cooling and melting cooling.

Polarimetric radars provide an estimate of the three-dimensional distribution of the types of hydrometeors present in a storm. Only direct aircraft penetrations can confirm the existence of the types of hydrometeors actually present, while surface observations from instrumented vehicles or pods can quantify the degree of cooling. For many years an armored aircraft, the T-28, flew into severe convective storms. At the time of this writing in 2012, there are plans for a new armored aircraft, the A10, to do the same. When the RFD is driven mostly by dynamical effects there may not be much if any temperature gradient across the rear-flank gust front or, even in the presence of evaporative cooling and melting, the effects of cooling may be counteracted by enough subsidence-induced warming that there is little temperature gradient across the RFGF. The RFD, like ordinary downdrafts, exhibits temporal variations and there can be a series of more than one RFD surge, as has been noted in dual-Doppler analyses (Figure 4.18).

There is anecdotal evidence (from storm-chasers) that when the difference between the surface temperature and dew point temperature exceeds 25°F the evaporation of raindrops is so intense that the RFD produced via evaporative cooling is strong enough that the speed of the RFGF far exceeds the speed of the parent storm and new convective growth is suppressed as warm, moist, ambient air is cut off from the updraft. One may use RKW theory also to infer that new-cell growth will not occur when the temperature difference between that of the cold pool and that of the ambient air exceeds some threshold for a given amount of low-level vertical wind shear normal to the leading edge of the cold pool. However, detailed theoretical studies of gust front behavior in a supercell (i.e., when there is rotation and significant three dimensionality), at the time of this writing, are lacking.

The FFD, unlike the RFD, is not usually associated with a well-defined wind shift (Figure 4.15, bottom panels; Figure 4.18). It is therefore not directly analogous to the warm front in a synoptic-scale, extratropical cyclone (of course, neither do all extratropical cyclones have warm fronts). However, as will be dis-

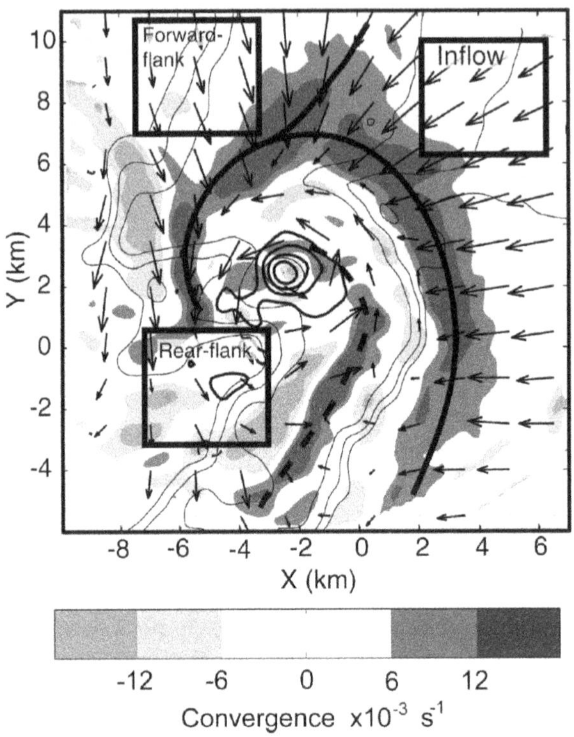

Figure 4.18. Multiple RFD surges in a tornadic supercell, as documented by two DOW mobile Doppler radars in northwest Texas on April 30, 2000. Storm-relative wind field depicted by vectors. Convergence is coded by gray shading. Solid lines represent primary forward-flank and rear-flank gust fronts, though only a gradual wind shift is evident in the forward flank. The secondary rear-flank gust front is represented by a dashed line. Equivalent radar reflectivity factor is contoured every 5 dBZ_e, beginning at 20 dBZ_e (from Marquis *et al.*, 2008).

cussed later, the width and magnitude of the surface temperature gradient at the right edge of the FFD and the associated wind field may play a role in low-level mesocyclogenesis and in tornadogenesis. Much of the FFD originates as precipitation falls from the downshear anvil region. The origin of the FFD is therefore a natural consequence of any convective storm in the presence of vertical shear. However, the RFD's appearance on the upshear side of the storm is suggestive of the importance of storm rotation, because air must be driven backwards against the environmental flow, and the importance of vertical shear, which promotes storm rotation and can be responsible for evaporation as dry environmental air catches up with a more slowly moving storm and is entrained into a region of precipitation on the rear side of the storm (Figure 4.19).

The differential reflectivity Z_{DR} is often high near the edge of the forward-flank portion of the radar reflectivity core, forming a "Z_{DR}" arc (Figure 4.20). Matt Kumjian at OU has shown that this feature may be created when the

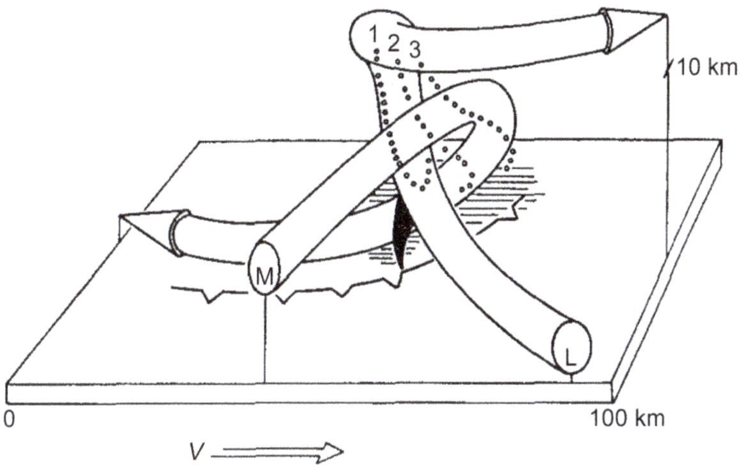

Figure 4.19. Three-dimensional conceptual model of the storm-relative airflow in a cyclonically rotating, right-moving supercell, showing how the mid-level airstream (M) catches up with the storm and descends behind the rear-flank gust front ("L" refers to the low airstream) (from Browning, 1964). Browning's early attempts to show that some storms do not fit the Byers–Braham model were controversial. An early version of this conceptual model was referred with skepticism as the "pretzel diagram".

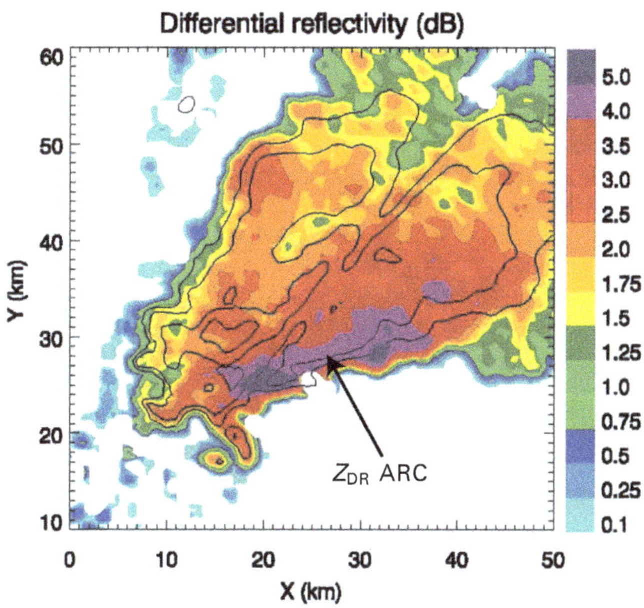

Figure 4.20. Example of a differential reflectivity, Z_{DR}, arc along the edge of the right-front flank of a supercell's FFD, on May 8, 2003, in central Oklahoma, as depicted by the S-band, KOUN Doppler radar, at 1.5° elevation angle. Contours of radar reflectivity factor are overlaid at 30 (outer contour), 40, 45, and 50 dBZ (from Kumjian and Ryzhkov, 2008).

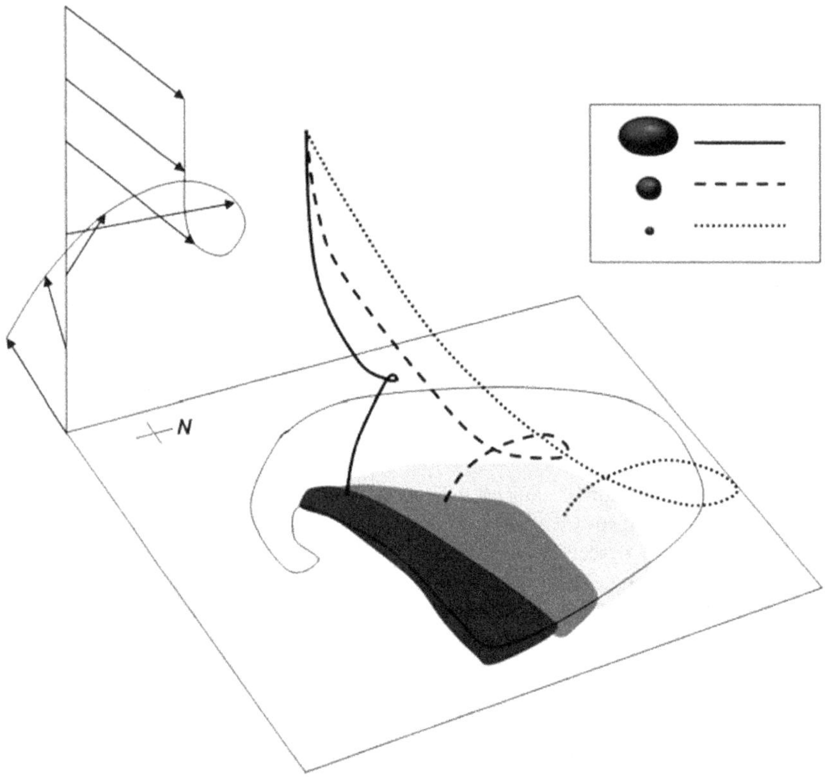

Figure 4.21. Idealized illustration of how a supercell that formed in an environment of a clockwise-turning hodograph with height can lead to the enhancement of differential reflectivity Z_{DR} along the edge of the FFD on the right-front flank of the storm. The causes of this polarimetric signature are differential fall speeds in an environment of strongly curved vertical shear. The largest water drops (thick solid line) fall out (from the common location noted) and land near the edge of the radar echo (shaded dark gray); the medium-size drops (dashed line) fall out and land farther inside the radar echo (shaded medium gray); the small drops (dotted line) fall out and land farthest inside the radar echo (shaded light gray). Since small drops are "seen" as a region of relatively small differential reflectivity, while large drops are "seen" as a region of relatively high differential reflectivity, there is a differential reflectivity gradient on the right-forward flank of supercells at low levels, owing to the vertical shear profile (if hail falls out, the Z_{DR} arc is disrupted) (from Kumjian *et al.*, 2009).

smallest raindrops are carried farther into the core before falling out than the larger, more widely spaced raindrops, a size-sorting process that is associated with both vertical directional and speed shear and vertical velocity (Figure 4.21). A summary of the prominent polarimetric signatures seen in supercells is shown in Figure 4.22. (The tornado-debris signature is discussed in Chapter 6, while the Z_{DR} and K_{DP} rings are not discussed here.)

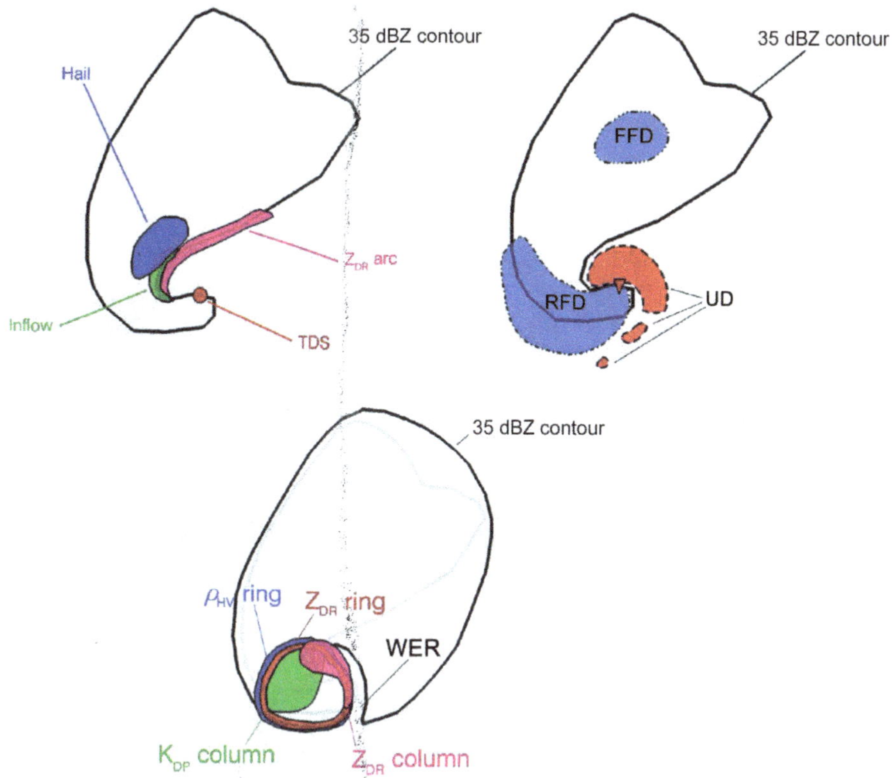

Figure 4.22. Idealized illustration of some polarimetric signature in supercells and their locations within the storm. (Top left) Hail (to be discussed), Z_{DR} arc, and tornado debris signatures (TDS) at low levels. (Top right) Lemon and Doswell conceptual model of main updrafts and downdrafts in a supercell (see Figure 4.15, top panel); (bottom) Z_{DR} and K_{DP} columns and rings at mid-levels; gray outline shows the location of the low-level 35 dBZ contour (from Kumjian and Ryzhkov, 2008).

The *flanking line*, a band of convective clouds adjacent to the tallest cloud towers which are associated with the main updraft, is often present along the rear-flank gust front. Since the cloud surface on the downshear side (i.e., in the direction of a vector that represents the difference between the wind vector at an altitude above and the wind vector at an altitude below) of the storm often appears smooth (Figure 4.23), it is inferred that the air is stable with respect to lifting by a finite upward displacement (the cloud surface would otherwise appear bubbly, more like the surface of a cauliflower; see also Figures 3.12 and 3.39a). The U-shaped updraft is sometimes associated with a deep convergence zone (DCZ) that extends upward to 3 km or so (Figure 4.24). Towers in the flanking line are forced by updrafts in the DCZ. The DCZ will be looked at in slightly more depth in Section 4.6.

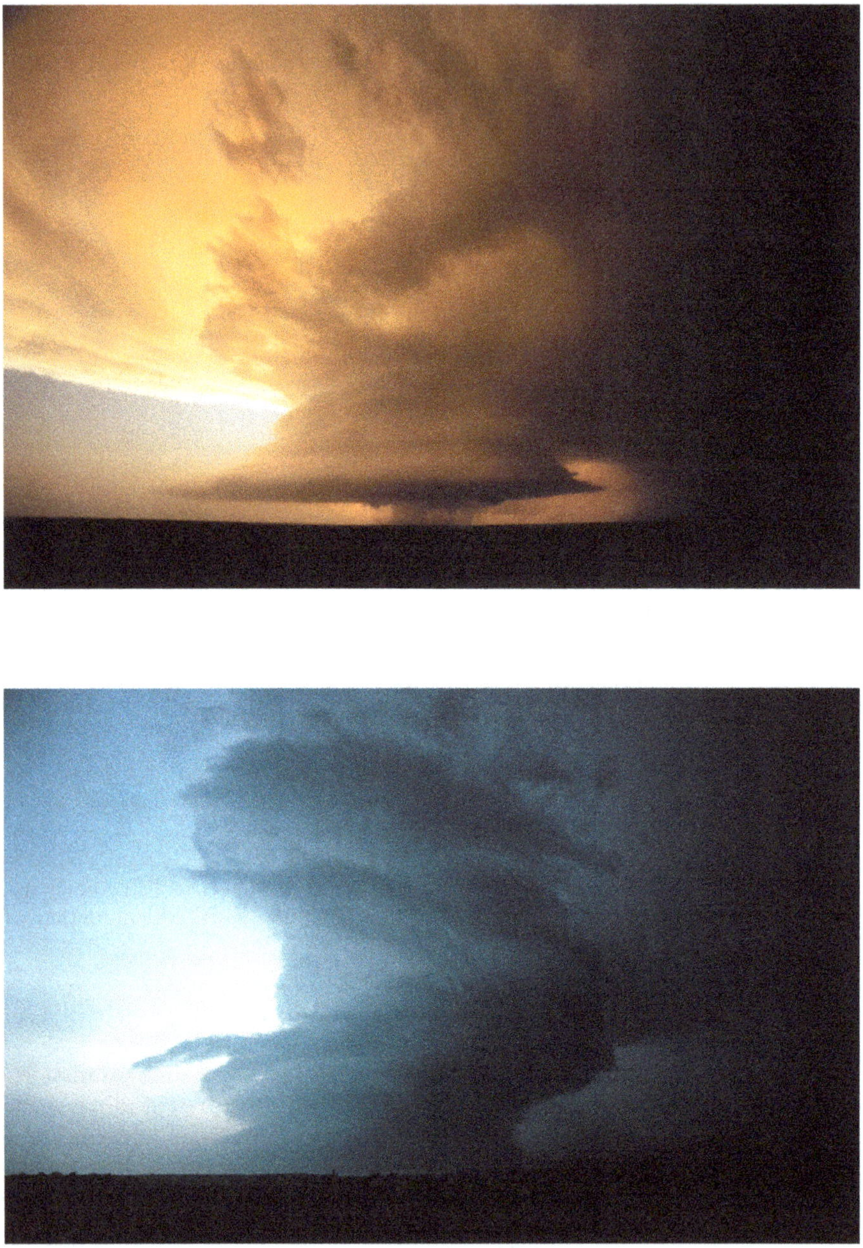

Figure 4.23a. The smooth, striated, laminar appearance of the downshear side of the updraft tower of supercells, as viewed from ahead and to the right of the storm movement. It is thought that this laminar appearance is due to the forced lifting of stable air. (Top) In northwest Kansas on June 10, 2008; (bottom) in central Oklahoma on May 3, 1999. In these photographs, mid-level clouds are seen streaming in around the right-forward flank of the storm (photographs by the author).

Figure 4.23b. As for Figure 4.23a. (Top) In central Oklahoma on May 26, 2004; (bottom) in southwest Kansas on June 9, 2009 (photographs by the author).

Figure 4.23c. As for Figure 4.23a. In southwest Kansas on June 9, 2009 (photograph by the author).

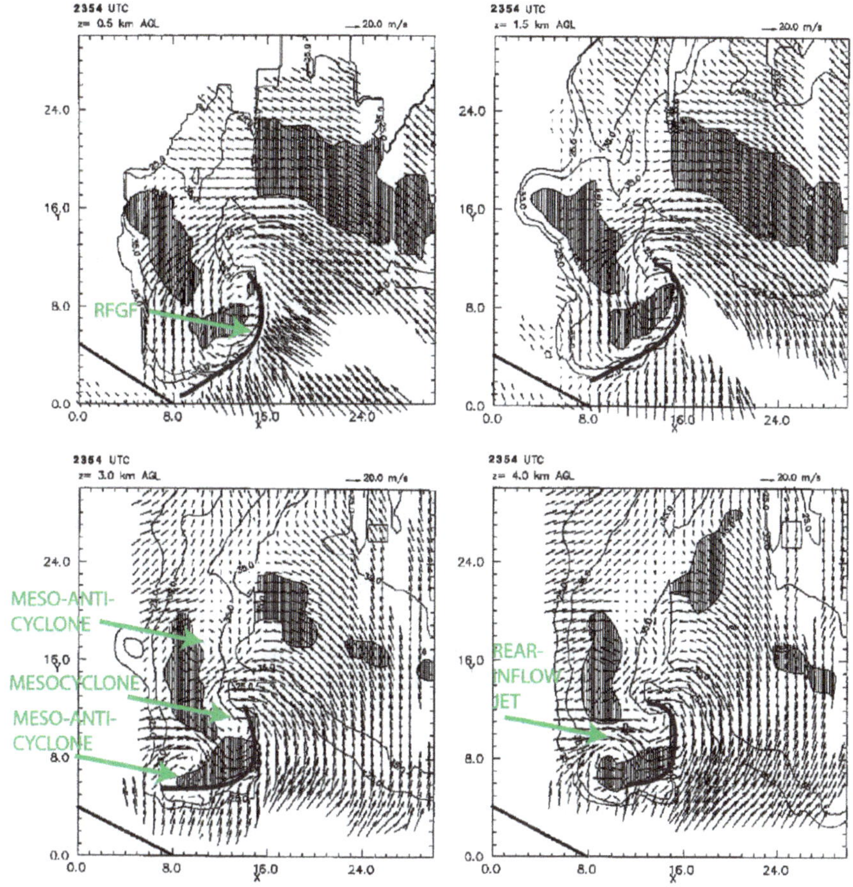

Figure 4.24. Example of a DCZ (deep convergence zone), denoted by a thick black line, in a supercell in the Texas Panhandle on May 22, 1995. Storm-relative wind field from a pseudo dual-Doppler analysis of data from ELDORA at four different levels, from 500 m AGL (upper left) to 4 km AGL (lower right). Equivalent radar reflectivity factor in dBZ_e. Shaded areas mark radar reflectivity in excess of $45\,dBZ_e$. At mid-levels, the DCZ is flanked by a mesocyclone–meso-anticyclone couplet, with a rear-inflow jet in between the vortices. There is also a meso-anticyclone at mid-levels located to the northwest of the mesocyclone. At low levels, the rear-flank gust front (RFGF) is evident. The DCZ extends from the RFGF at low levels upward to mid-levels (from Bluestein and Gaddy, 2001).

4.3.3 Precipitation type and distribution

The most intense precipitation is found in the downshear direction from the main updraft (to the right of the wall cloud, when viewed from a location to the right of the direction in which the storm is moving). The most intense precipitation and highest radar reflectivity are frequently co-located with an optically translucent region, while less intense precipitation is located in an optically opaque region nearby (Figures 4.6, 4.25–4.28). This observation is interpreted as meaning that the region of heaviest precipitation is composed of widely scattered, large raindrops and hailstones, and the region of less intense precipitation is composed of more densely packed smaller raindrops and hailstones.

Storm-chasers have noticed that the region behind the RFD is sometimes optically translucent and contains little if any precipitation and the region where there is typically the most intense precipitation is also optically translucent and contains almost no rain, but some hail (Figure 4.25). The only rain observed falls out from the anvil, relatively far from the storm's main updraft. Such storms are called "low-precipitation (LP)" supercells, the name originating with the author of this text; Don Burgess and Bob Davies-Jones at NSSL first called these storms "dryline storms" because they tended to be found near the dryline: LP supercells do not require a dryline for their existence and not all supercells near the dryline are of the LP type.

On the other hand, the region behind the RFD is sometimes optically opaque and contains an abundance of precipitation and the region where there is typically the most intense precipitation is also optically opaque and contains rain and hail (Figure 4.26). Such storms are called "high-precipitation (HP)" supercells, the name being given originally by Al Moller and collaborators. LP and HP supercells are the extreme ends of a spectrum of a variation of supercell types in which precipitation coverage near the rear-flank gust front and wall cloud (i.e., near the updraft) is variable. The idealized visual model depicts the *classic* supercell (i.e., a supercell in which precipitation efficiency is greater than that of an LP supercell, but less than that of an HP supercell; Figure 4.27).

Interesting questions concerning differences in the thermodynamics of LP and HP storms arise in the context of tornado formation and are addressed in a later section. When there is little if any rain, the potential for the production of an evaporatively cooled pool of air near the ground is very low; when there is a lot of rain that falls out into relatively dry air, the potential for the production of an evaporatively cooled pool of air is very high. Thus, the temperature gradient near the surface across the rear-flank gust front and near the edge of the forward-flank downdraft region should be greatest in HP supercells and lowest in LP supercells, when air is unsaturated near the ground.

The reason(s) precipitation efficiency varies so widely in supercells is (are) not known very well because details of precipitation processes are not very well understood. When vertical shear is relatively weak or nonexistent at high levels, ice particles from the anvil can fall back into the storm's updraft and seed growing convective towers, so that the precipitation process is enhanced; on the other hand, when shear is relatively strong at high levels, ice particles from the anvil are blown far downstream and do not seed the same storm in which the ice particles were formed. Thus, the character of high-level vertical shear is thought to be important in determining whether or not a supercell belongs to the LP, classic, or HP archetype. It is possible, however, for ice particles produced in an anvil in an environment of strong upper shear to fall out into the updraft of a nearby supercell and seed it, thereby enhancing precipitation in the adjacent storm. Such

Figure 4.25. Low-precipitation (LP) supercells. (Top) In northeast Colorado on July 10, 1996; (middle) in western Kansas on June 10, 2008; (bottom) in northeast Colorado on July 21, 2000, with a tornado (photographs by the author).

Figure 4.26. High-precipitation (HP) supercells. (Top) In the Texas Panhandle on May 23, 2002; (middle) in the Oklahoma Panhandle on May 31, 2007; (bottom) in southeastern Colorado on June 11, 2009 (photographs by the author).

Figure 4.27. Classic supercell. (Top) May 31, 1990 in the northern Texas Panhandle. The area to the right of the tornado is translucent, but has widely spaced hail falling, which on radar appears as an area of high reflectivity. (Bottom, left) Clear slot in eastern Colorado on June 8, 1994, as seen from an NOAA P-3 aircraft; (bottom, right) clear slot over Redesdale, Victoria, Australia, on October 12, 2010, which is the mirror image of a clear slot in the Northern Hemisphere (cf. with the image to the left in Colorado) (photographs by the author, except the last one, which is courtesy of John Allen).

a hypothesis may explain why LP and HP supercells are sometimes observed in close proximity (when the environments are similar) or why LP storms sometimes become transformed into HP supercells. A summary of the types of supercells is given in Figure 4.28.

4.4 THE PRODUCTION OF MID-LEVEL ROTATION

When a buoyant updraft rises in an environment of vertical shear, which represents horizontal vorticity (Figure 4.29), some of the latter is converted into

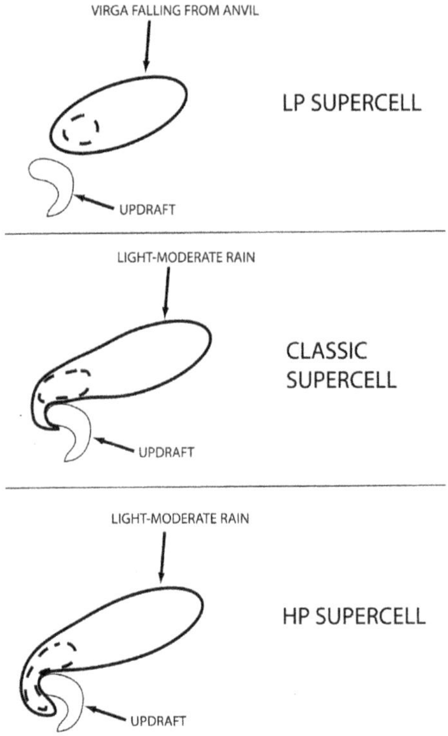

Figure 4.28. Idealized representation of a horizontal cross section at low levels of features in (top) an LP supercell, (middle) a classic supercell, and (bottom) an HP supercell. Dashed line encloses hail and heavy rain. The region under the updraft base is translucent, as is much of the storm, in an LP supercell. The region under the updraft base shows a limited precipitation shaft and is partially translucent in a classic supercell. The region under the updraft base is opaque in an HP supercell and precipitation has wrapped almost or all the way around the tip of the updraft, where a tornado might be found, but mostly hidden from view, except perhaps in the notch just ahead of the nose of the updraft, which is not a safe place to be (after Rasmussen and Straka, 1998).

cyclonic (*vertical*) vorticity and some is converted into anticyclonic (*vertical*) vorticity as a result of tilting (cf. (2.50)) along the edges of the updraft, in a direction with respect to the updraft that is normal to the shear vector. We first consider, for simplicity, an atmosphere in which the shear profile (i.e., vertical variation of shear) is unidirectional (shear does not change direction with height) and is constant (shear magnitude does not vary with height). For westerly wind shear, which according to the thermal wind relationship (4.23) is associated with a north to south–directed temperature gradient (cold to the north, warm to the south), a cyclonic vortex will form on the equatorward side of the updraft and an anticyclonic vortex will form on the poleward side.

Analysis of the production of mid-level rotation may also be easily visualized using "vortex line" analysis. Consider the frictionless form of the three-

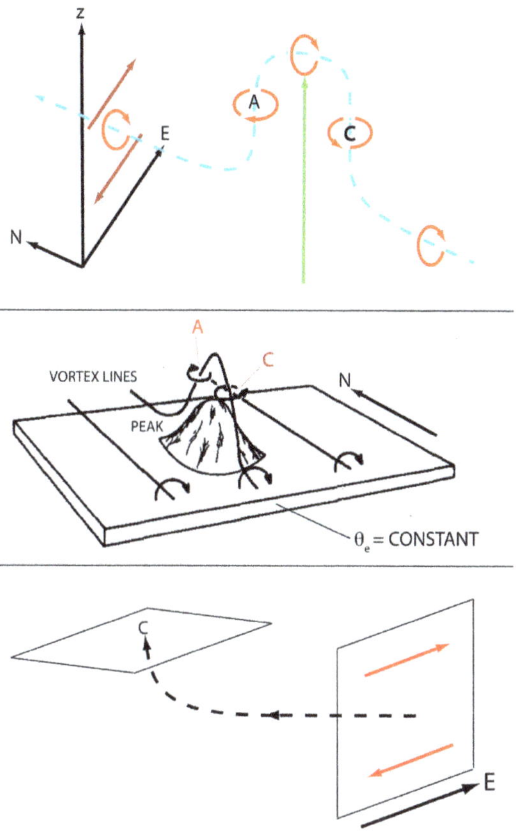

Figure 4.29. (Top) Idealized illustration of how an updraft (green vector) in an environment of westerly vertical shear (e.g., easterly winds below, westerly winds aloft, represented by red vectors) tilts a vortex line pointing towards the north (dashed streamline) so that horizontal vorticity is converted into cyclonic (C) vorticity south of anticyclonic (A) vorticity north of the updraft. (Middle) Idealized illustration of how an updraft that deforms a θ_e surface upward so that there is a bulge/peak, also deforms a vortex line upward because the vortex line must always lie on a surface of constant θ_e (adapted from Davies-Jones, 1984). (Bottom) Idealized illustration of how circulation (associated with vertical shear) in the vertical plane is advected and tilted upward to produce cyclonic circulation (C) in the horizontal plane at mid-levels. The dashed streamline indicates the motion of the vertical plane so that it becomes the horizontal plane at the left.

dimensional vorticity equation (2.49) for a barotropic ($\nabla B = 0$) and Boussinesq ($\mathbf{\nabla} \times (1/\bar{\rho}\, \nabla p') = 0$ because $\bar{\rho}$ is not a function of x or y) atmosphere

$$\begin{aligned} D/Dt(\mathbf{\nabla} \times \mathbf{v}) &= \partial/\partial t(\mathbf{\nabla} \times \mathbf{v}) + u\,\partial/\partial x(\mathbf{\nabla} \times \mathbf{v}) + v\,\partial/\partial y(\mathbf{\nabla} \times \mathbf{v}) + w\,\partial/\partial z(\mathbf{\nabla} \times \mathbf{v}) \\ &= [(\mathbf{\nabla} \times \mathbf{v}) \cdot \mathbf{\nabla}]\mathbf{v} \end{aligned} \tag{4.26}$$

The term on the RHS of (4.26) represents tilting and stretching. So, in a frictionless, barotropic atmosphere, vorticity is changed only through tilting and stretching. A vortex line is oriented along the three-dimensional vorticity vector. It therefore moves along with the wind (advective terms on the LHS) and is tilted (tilting terms on the RHS). Stretching does not alter the orientation of vortex lines. According to Bob Davies-Jones, at NSSL, "*Barotropic* vortex lines are 'frozen' into the fluid and behave like elastic strings that the flow moves, stretches, and reorients." If baroclinic effects (i.e., gradients in buoyancy) are considered, then vortex lines may change as a result and the analysis is more complicated; thus, we typically ignore baroclinic effects for simplicity and consider qualitatively the consequences of baroclinic effects: vortex lines cannot be broken in a barotropic, frictionless atmosphere.

So, an initially horizontally oriented vortex line that points to the pole is distorted by the updraft so that it is deformed into an upside-U shape; the vortex line has a component that points upward on the equatorward side and downward on the poleward side. Thus, the vertical component of vorticity on the equatorward (poleward) side has a component in the direction of (in the direction opposite to that of) the rotation vector of the Earth (Figure 4.29, top panel).

Another way to analyze the production of vertical vorticity in a vertically sheared environment by an updraft is to make use of the conservation of Ertel's potential vorticity (2.137 and 2.138). Inside a cloud, where the air is saturated, we replace θ with θ_e, so that

$$Z \sim \nabla \times \mathbf{v} \cdot \nabla \theta_e \tag{4.27}$$

In other words, the component of three-dimensional vorticity in the direction of the equivalent potential temperature gradient must remain constant.

Suppose that initially the three-dimensional vorticity vector is associated with vertical shear alone due to a westerly thermal wind and that there is conditional instability (i.e., that θ_e decreases with height, but much more rapidly than it does with y). It is seen in Figure 4.29 (middle panel) that $Z = 0$ in this case, so that vortex lines must always lie on surfaces of constant θ_e. (We will not be concerned here if the potential vorticity vanishes so that the atmosphere is neutral with respect to symmetric instability or if $\partial\theta_e/\partial z > 0$ so that the atmosphere is convectively stable.)

In the absence of diabatic heating and friction, when a localized updraft forms, surfaces of constant θ_e bulge upward (because θ_e is conserved for adiabatic processes) and so must the vortex lines, which in this case have a component toward the north. We note that using Ertel's potential vorticity in our analysis, we do not have to constrain ourselves to barotropic cases: Vortex lines always remain on isentropic surfaces regardless of whether or not there is baroclinicity. However, just as vortex lines cannot be broken in a barotropic atmosphere, surfaces of constant potential vorticity cannot be fractured in a frictionless, adiabatic atmosphere.

When air approaches and enters the updraft at low levels from the equatorward side, it begins with its vorticity vector pointing towards the pole. It is therefore seen that the three-dimensional vorticity vector must change from being directed from the equator to the pole only to having a component directed vertically: thus, in order that the potential vorticity vector remain zero, the vorticity vector must become directed upward (and poleward) on the equatorward side and downward (and poleward) on the poleward side (Figure 4.29, middle panel).

Finally, one can use circulation analysis to understand mid-level vortices in supercells. For example, consider a circuit in a horizontal plane encompassing a mesocyclone at mid-levels. How did this configuration come about? There must have been a gradient in vertical motion and the plane had been advected so that a circuit in the vertical plane had been tilted upward onto the horizontal (Figure 4.29, bottom panel). In the case of a mid-level mesocyclone, circulation analysis, potential vorticity analysis, and vortex line analysis are all useful.

In nature, cyclonic–anticyclonic couplets are observed in Doppler radar observations of the mid-levels of supercells. When a (single) Doppler radar scans a supercell at mid-levels, the signature of a cyclonic–anticyclonic couplet is usually evident (Figure 4.30). These cyclonic–anticyclonic couplets are usually most pronounced at mid-levels because (1) updrafts in supercells are strongest at upper levels in the troposphere, so that horizontal vertical velocity gradients are also strongest there and (2) vertical shear is usually strongest in the lower half of the troposphere. The net result is that the tilting of horizontal vorticity is strongest at mid-levels. When these vortices are intense and long lived they are called "mesocyclones" and "meso-anticyclones"; mesocyclones are usually the focus of attention rather than meso-anticyclones because they are associated more frequently with severe weather.

4.5 INTERACTION OF VERTICAL SHEAR WITH UPDRAFTS/ DOWNDRAFTS FORCED BY BUOYANCY: LINEAR AND NONLINEAR PRESSURE EFFECTS

In the previous section the vorticity equation (and vortex line analysis and potential vorticity analysis) was used to explain the formation of a counter-rotating vortex pair when a strong updraft interacts with horizontal environmental vorticity. The divergence equation (2.62) is now used to examine the effects of the interaction between a buoyant updraft and environmental vertical shear *on the pressure field*. It turns out these counter-rotating vortices play an important role in updraft propagation. An analysis of (2.62) can be used to explain how and why supercell updrafts propagate because in regions of upward-directed dynamic perturbation pressure gradients, air is accelerated upward and updrafts may be triggered when the LFC is reached, while in regions of downward-directed perturbation pressure gradients air is accelerated downward and updrafts are suppressed or downdrafts are forced.

The propagation velocity of updrafts triggered by upward-directed dynamic perturbation pressure gradients can be determined using Petterssen's formula for the motion of extrema in scalar fields (see pp. 47–54 of Bluestein, 1992), which depends on the horizontal gradient of the field of the vertical dynamic

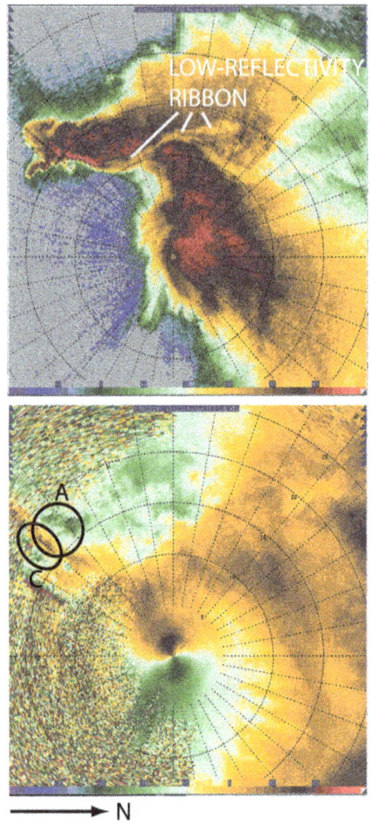

Figure 4.30. A cyclonic (C, ellipse)–anticyclonic (A, circle) Doppler velocity shear couplet at mid-levels in a supercell in southwestern Oklahoma on May 23, 2011, from data from the RaXPol mobile Doppler radar. The radial at which the cyclonic–anticyclonic couplet is seen is oriented approximately along the mean vertical shear vector at mid-levels. (Top) Equivalent radar reflectivity factor color-coded in dBZ_e. (Bottom) Doppler velocity color-coded in $m\,s^{-1}$. A "low-reflectivity ribbon" is also seen dividing the storm in half.

perturbation pressure gradient

$$c_x = -\partial/\partial x[\partial/\partial t(-\partial p'_d/\partial z)]/\partial^2/\partial x^2(-\partial p'_d/\partial z) \tag{4.28}$$

$$c_y = -\partial/\partial y[\partial/\partial t(-\partial p'_d/\partial z)]/\partial^2/\partial y^2(-\partial p'_d/\partial z) \tag{4.29}$$

where c_x and c_y are the x and y-components of forced upward motion centers due to upward-directed dynamic perturbation pressure gradient forces. Updrafts propagate against the horizontal gradient in the tendencies of vertical dynamic perturbation pressure gradients (Figure 4.31). (The effects of buoyancy and gust fronts are not included in this analysis here so we can isolate the effects of dynamic perturbation pressure gradients.) Since updrafts are kinematically associated with low-level convergence, and convergence acting on existing vertical

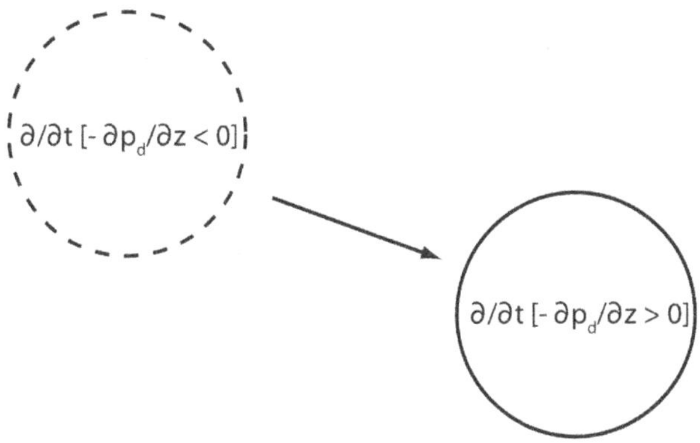

Figure 4.31. Illustration of how updrafts propagate (propagation vector shown) from where the downward-directed (solid circle) dynamic pressure gradient is increasing the most (dashed circle) to where the upward-directed dynamic pressure gradient is increasing the most.

vorticity increases vorticity, updraft propagation is also important in vorticity amplification at low levels in a storm.

The divergence equation (2.62) may be expressed as:

$$\alpha_0 \nabla^2 p' = -[(\partial u/\partial x)^2 + (\partial v/\partial y)^2 + (\partial w/\partial z)^2] - 2[\partial u/\partial y\, \partial v/\partial x + \partial w/\partial x\, \partial u/\partial z + \partial w/\partial y\, \partial v/\partial z] + \partial B/\partial z \quad (4.30)$$

where $\alpha_0 \equiv 1/\bar{\rho}$, which is treated as a constant for the base state atmosphere (cf. (2.4)). The terms "$[(\partial u/\partial x)^2 + (\partial v/\partial y)^2 + (\partial w/\partial z)^2]$" are called "fluid extension terms" and the terms "$[\partial u/\partial y\, \partial v/\partial x + \partial w/\partial x\, \partial u/\partial z + \partial w/\partial y\, \partial v/\partial z]$" are called "shear terms". The shear terms may be expressed in terms of "$\frac{1}{2}[|D|^2 - |\omega|^2]$", as may be verified by brute force and obstinacy using simple algebra, where $|D|$ is the magnitude of resultant three-dimensional deformation and $|\omega|$ is the magnitude of three-dimensional vorticity (ω):

$$|D|^2 = D_{xy}^2 + D_{zx}^2 + D_{zy}^2 \quad (4.31)$$

where

$$D_{xy} = \partial v/\partial x + \partial u/\partial y, \text{ deformation in the } x\text{–}y\text{-plane} \quad (4.32)$$

$$D_{zx} = \partial u/\partial z + \partial w/\partial x, \text{ deformation in the } z\text{–}x\text{-plane} \quad (4.33)$$

$$D_{zy} = \partial w/\partial y + \partial v/\partial z, \text{ deformation in the } z\text{–}y\text{-plane} \quad (4.34)$$

and

$$|\omega|^2 = \zeta^2 + \xi^2 + \eta^2 \quad (4.35)$$

where

$$\zeta = \partial v/\partial x - \partial u/\partial y, \text{ vorticity about the } z\text{-axis} \quad (4.36)$$

$$\xi = \partial w/\partial y - \partial v/\partial z, \text{ vorticity about the } x\text{-axis} \quad (4.37)$$

$$\eta = \partial u/\partial z - \partial w/\partial x, \text{ vorticity about the } y\text{-axis} \quad (4.38)$$

So, the divergence equation may be written as

$$\alpha_0 \nabla^2 p' = -[(\partial u/\partial x)^2 + (\partial v/\partial y)^2 + (\partial w/\partial z)^2] - \tfrac{1}{2}[|D|^2 - |\omega|^2] + \partial B/\partial z \quad (4.39)$$

To isolate the effects of the dynamic part of the pressure perturbation field, we consider only that part of the divergence equation associated with the dynamic perturbation pressure (cf. 2.70)

$$\alpha_0 \nabla^2 p'_d = -[(\partial u/\partial x)^2 + (\partial v/\partial y)^2 + (\partial w/\partial z)^2] - \tfrac{1}{2}[|D|^2 - |\omega|^2] \quad (4.40)$$

and do not consider

$$\alpha_0 \nabla^2 p'_b = \partial B/\partial z \quad (4.41)$$

where $p' = p'_d + p'_b$ (cf. (2.63)).

According to (4.40), the shape of the three-dimensional pressure field is determined by terms involving the vertical and horizontal shears of each component of the wind, including deformation and vorticity. With proper boundary conditions, the pressure field can be determined. Since the operator on the LHS of (4.40) is a (three-dimensional) Laplacian, the sign of each forcing function on the RHS of (4.40) is of the opposite sign of its contribution to dynamic perturbation pressure.

To isolate the effects of the updraft/downdraft on its environment, each variable is expressed in terms of the environmental (mean) value and the perturbation (primed) storm value. Thus,

$$u = U(z) + u'(x, y, z, t) \quad (4.42)$$

$$v = V(z) + v'(x, y, z, t) \quad (4.43)$$

$$w = w'(x, y, z, t) \quad (4.44)$$

The reader is cautioned that the perturbation terms in (4.42)–(4.44) are not necessarily small compared with the respective mean terms (in (4.42)–(4.44)), as was the case in our earlier linear stability analyses; in fact, they are typically on the same order of magnitude as the mean terms. In (4.42)–(4.44) it is seen that the environmental horizontal wind field **V** is chosen, for simplicity, to be horizontally homogeneous, varying only as a function of height, and the vertical environmental wind field (vertical velocity in the environment) is zero (i.e., the vertical motion field is "resting"). This environmental wind field may be represented by a hodograph (Figure 4.32), a plot of wind vs. height that is represented by the locus of points marked by the tip of the wind vectors at each height, where each wind vector is plotted at a common origin.

The storm-related wind field, however, varies as a function of three-dimensional space and time. In nature, there are inhomogeneities in the environmental wind field, but they are neglected and usually are considered to be second-order effects, which is not necessarily always the case (e.g., consider what happens when a convective storm approaches, straddles, or crosses a surface front or outflow boundary or other surface boundary). Also, there is often a band of mesoscale ascent where convective storms are triggered, so the average value of

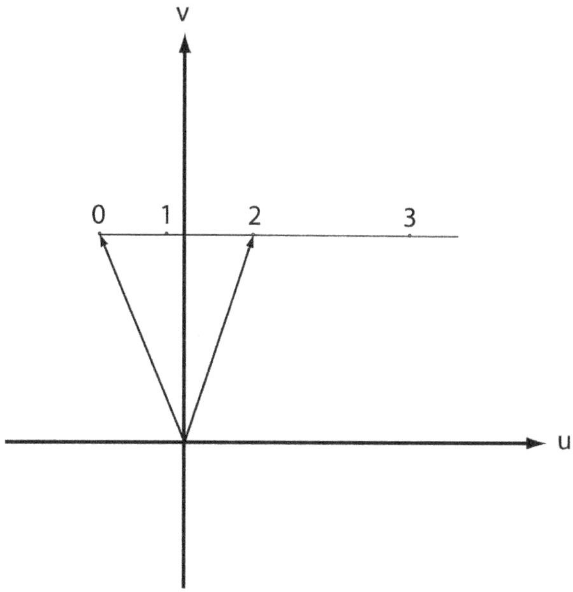

Figure 4.32. Idealized example of a straight (unidirectional) hodograph (thin solid line), which is labeled every kilometer. The wind vectors at 0 and 2 km AGL are shown for purposes of illustration.

vertical velocity is nonzero; the speed of ascending air (e.g., along fronts, outflow boundaries, etc.), however ($\sim$10 cm s^{-1}–1 m s^{-1}), is an order of magnitude or more less than that of buoyant updrafts ($\sim$10 m s^{-1}).

Thus, using (4.42)–(4.44), it is seen that (4.40) may be expressed as

$$\begin{aligned}\alpha_0 \nabla^2 p'_d = &-[(\partial u'/\partial x)^2 + (\partial v'/\partial y)^2 + (\partial w'/\partial z)^2] \\ &- 2[(\partial u'/\partial y\, \partial v'/\partial x + \partial w'/\partial x\, \partial u'/\partial z + \partial w'/\partial y\, \partial v'/\partial z)] \\ &- 2[\partial w'/\partial x\, \partial U/\partial z + \partial w'/\partial y\, \partial V/\partial z] \end{aligned} \quad (4.45)$$

The terms on the RHS are separated into the following linear terms

$$-2[\partial w'/\partial x\, \partial U/\partial z + \partial w'/\partial y\, \partial V/\partial z] = -2\, \partial \mathbf{V}/\partial z \cdot \mathbf{\nabla} w' \quad (4.46)$$

and nonlinear terms

$$-[(\partial u'/\partial x)^2 + (\partial v'/\partial y)^2 + (\partial w'/\partial z)^2] \\ - 2(\partial u'/\partial y\, \partial v'/\partial x + \partial w'/\partial x\, \partial u'/\partial z + \partial w'/\partial y\, \partial v'/\partial z)$$

as first shown by Rich Rotunno and Joe Klemp in the mid-1980s. The nonlinear terms $[(\partial u'/\partial x)^2 + (\partial v'/\partial y)^2 + (\partial w'/\partial z)^2]$ are called "fluid extension terms" (as in (4.30)) and the nonlinear terms $(\partial u'/\partial y\, \partial v'/\partial x + \partial w'/\partial x\, \partial u'/\partial z + \partial w'/\partial y\, \partial v'/\partial z)$ are called "shear terms" (as in (4.30)).

The dynamic perturbation pressure is decomposed as follows into linear and nonlinear parts

$$p'_d = p'_L + p'_{NL} \tag{4.47}$$

We consider the nonlinear terms first because they are often the most important. The nonlinear shear terms in (4.45) can be expressed, like before (4.40), as the following:

$$-[(\partial u'/\partial x)^2 + (\partial v'/\partial y)^2 + (\partial w'/\partial z)^2] - \tfrac{1}{2}[|D'|^2 - [|\omega'|^2] \tag{4.48}$$

where D represents the perturbation (storm-related) resultant three-dimensional deformation; and ω' represents the perturbation (storm-related) three-dimensional vorticity. In particular

$$D'^2 = (\partial w'/\partial y + \partial v'/\partial z)^2 + (\partial u'/\partial z + \partial w'/\partial x)^2 + (\partial v'/\partial x + \partial u'/\partial y)^2 \tag{4.49}$$

and

$$|\omega'|^2 = (\partial w'/\partial y - \partial v'/\partial z)^2 + (\partial u'/\partial z - \partial w'/\partial x)^2 + (\partial v'/\partial x - \partial u'/\partial y)^2 \tag{4.50}$$

The forcing function in (4.48) involving vorticity alone is called "spin". Bob Davies-Jones in 2002 proposed that the nonlinear terms be decomposed as the sum of the fluid extension and shear terms involving deformation, or "splat"; the remaining terms are spin. Davies-Jones argued that this decomposition is more physical because the terms are invariant with respect to rotations of the coordinate axes.

The fluid extension part of the nonlinear term contributes to positive perturbation pressure and the deformation part contributes to positive perturbation pressure, while the spin part contributes to negative perturbation pressure, because the forcing functions associated with the fluid extension and deformation are each negative definite, while that associated with spin is positive definite. The main nonlinear effects are therefore as follows: regions of sharp horizontal gradients in the horizontal wind field, sharp vertical gradients of the vertical component of the wind field, or strong deformation are associated with positive perturbation pressure. Regions of strong vorticity (either cyclonic or anticyclonic in the vertical, or horizontal of any sign) are associated with negative perturbation pressure; cyclones and anticyclones are therefore associated with centers of negative perturbation pressure. To understand this relationship qualitatively without resorting to a divergence equation, just consider the special case of a vortex in cyclostrophic balance (Figure 4.33): An outward-directed centrifugal force must always be balanced by a radially inward-directed pressure gradient force, so that the pressure at the center of a vortex must be relatively low.[5]

We now consider the linear terms. To analyze the dynamics of convective storms when an updraft interacts with vertical shear, it is easiest first to consider an environment characterized by a hodograph. When the hodograph is "straight",

[5] A parcel must always encounter a force normal to its motion if the flow is curved; when the flow, for example, is counterclockwise, there must be a pressure gradient force acting to the left. This situation is consistent with flow about a region of relatively low pressure.

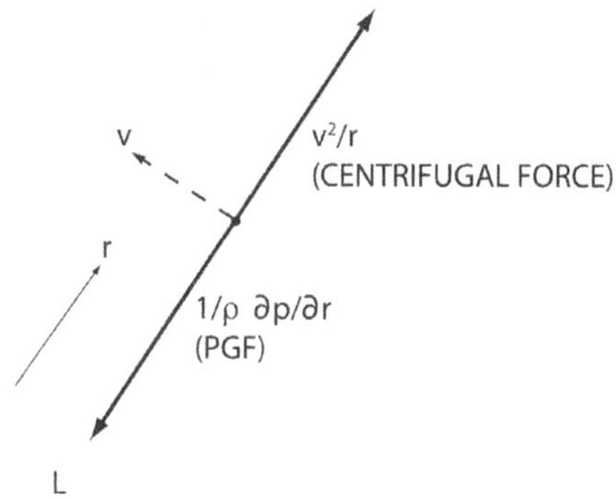

Figure 4.33. Illustration of cyclostrophic balance. The azimuthal wind speed v may be cyclonic as shown or anticyclonic. The pressure gradient must be positive, so that pressure increases radially outward.

the vertical shear vector always points in the same direction; when the hodograph is curved, the vertical shear vector changes direction with height. A straight (unidirectional) hodograph may be associated with a wind profile in which the wind direction changes with height, even though the vertical shear direction does not vary with height (e.g., Figure 4.32). It is assumed in the following analyses that $R < 1$ (see (4.9)), so that the effects of the vertical perturbation pressure gradient force are at least comparable with that of buoyancy and that the buoyancy is substantial.

From (4.46) we see that for a localized updraft, the perturbation pressure is relatively low on the downshear side of the updraft and relatively high on the upshear side (Figure 4.34, top). To understand this physically, consider an atmosphere in which the wind is westerly and increases in speed with height. If horizontal momentum is conserved (we ignore lateral mixing across the updraft), then the weaker westerly momentum from below is advected upwards in an updraft, so that an air parcel approaching the updraft aloft encounters lower values of westerly momentum inside the updraft (Figure 4.34, bottom panel). The air parcel must therefore decelerate as it enters the updraft, which is consistent with an adverse pressure gradient force (i.e., the pressure increases as it enters the updraft, so that the horizontal pressure gradient force is directed in the direction opposite that of the flow). Similarly, an air parcel leaving the updraft encounters higher values of westerly momentum outside the updraft. The air parcel therefore accelerates as it leaves the updraft, which is consistent with a pressure gradient force that is directed along the flow. We now put together aspects of both the linear and nonlinear pressure terms to see what happens to convective storms in various environments.

4.5.1 Convective storm dynamics for straight hodographs

At the onset of convection, a buoyant updraft in an environment of strong vertical shear, most of which is concentrated in the lower half of the troposphere, and that does not change direction (or magnitude) with height produces a couplet of

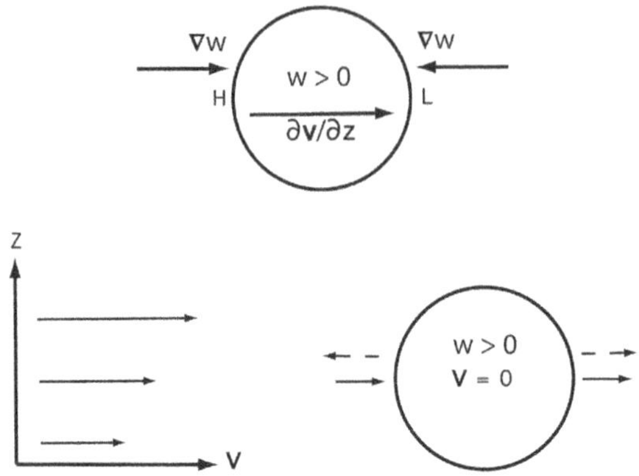

Figure 4.34. Illustration of how the dynamic perturbation pressure is high upshear and low downshear from an updraft in a unidirectionally vertically sheared environment. (Top) Mathematic description: the updraft gradient points in the same direction as the vertical shear vector on the upshear side and in the opposite direction on the downshear side (cf. (4.46)). Thus, the forcing function is negative on the upshear and positive on the downshear side, so that the perturbation pressure is positive on the upshear side and negative on the downshear side. (Bottom) Physical explanation: vertical variation of the wind vector is shown on the left. On the right, it is seen that if the updraft moves along with the mean wind (middle wind vector on the left), then at the top level air approaches the updraft from the west, decelerates to zero (because the momentum characteristic of the wind in the layer below, which is the same as the vertically averaged momentum, has been advected upward), and then accelerates to the east. This wind field is consistent with the pressure gradient forces shown by the dashed vectors.

counter-rotating vortices that are strongest at mid-levels (Figure 4.29, top and middle panels); in the Northern Hemisphere, the cyclonic (anticyclonic) member is found to the right (left) of the updraft, with respect to the orientation of the vertical shear vector. Perturbation low-pressure areas are associated at mid-levels with each of the vortices (nonlinear, spin). Since vortices are strongest at mid-levels and the perturbation pressure deficit is proportional to the square of the perturbation (storm-related) vorticity (4.48), upward-directed perturbation pressure forces are found at lower levels, below the altitude of the strongest vorticity. Thus, new updrafts may be triggered in the off-shear (normal to the shear) direction with respect to the original updraft.

In the absence of any precipitation, the updraft splits into two parts; each new updraft then acts on environmental shear to produce two new updrafts on each flank of the split updrafts (Figure 4.35); the process continues so that the two outer updrafts propagate to the right and left of the shear vector, respectively. The inner updrafts are likely to be situated in a region where precipitation falls, where there is evaporative cooling, and consequently the original updraft decays; in the absence of precipitation, these inner updrafts will propagate towards each other. The net result is that updrafts following the original updraft split and propagate apart. Such behavior is observed in radar imagery and in numerical simulations (Figures 4.36 and 4.37) and in the 1960s, when splitting storms were

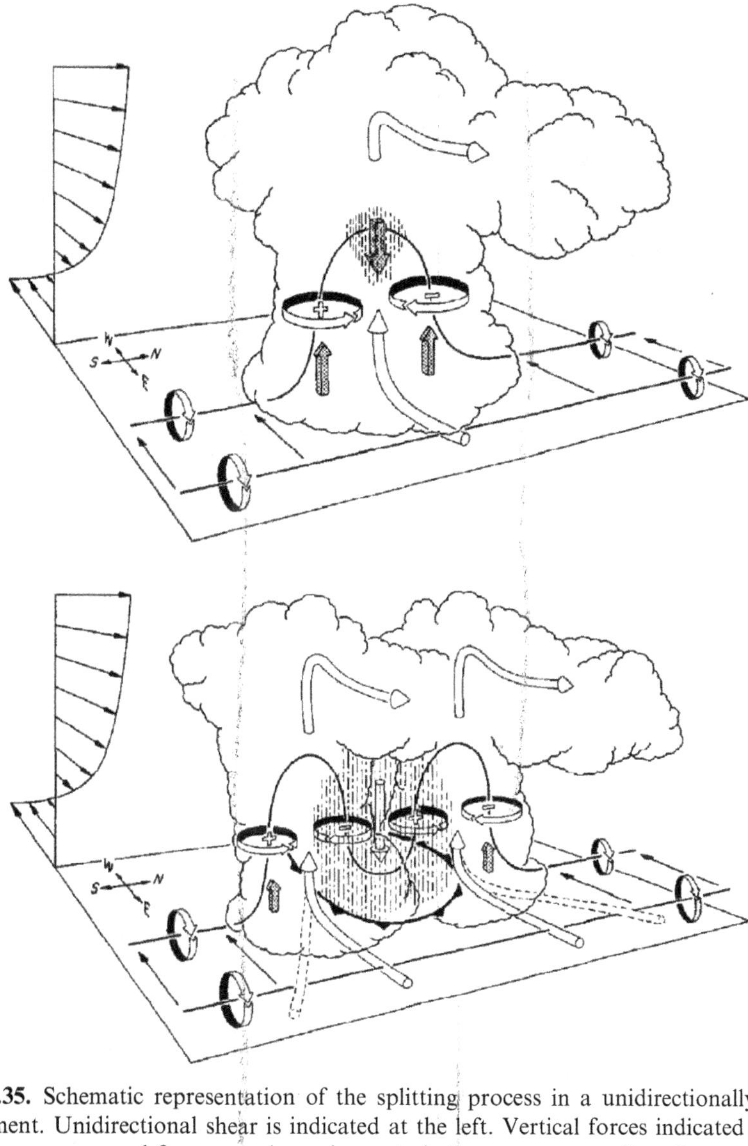

Figure 4.35. Schematic representation of the splitting process in a unidirectionally sheared environment. Unidirectional shear is indicated at the left. Vertical forces indicated by thick, shaded arrows; upward forces are dynamic, perturbation pressure gradient forces associated with mid-level vortices; downward force is due to precipitation loading. Storm-relative airflow is indicated by cylindrical arrows. Cyclonic (+) and anticyclonic (−) vortices are produced at mid-levels. Lines denote vortex lines, which point toward the right, indicating vorticity in the northerly direction, associated with vertical shear. Vertical dashed lines indicate precipitation. Cold front symbol at the ground indicates the leading edge of a cold pool. (Top) An updraft creates counter-rotating vortices, each of which is associated with an upward-directed force that triggers new updrafts on either side of the westerly vertical shear vector. (Bottom) New updrafts in turn produce new, mirror image, cyclonic–anticyclonic couplets, and so on (from Klemp, 1987).

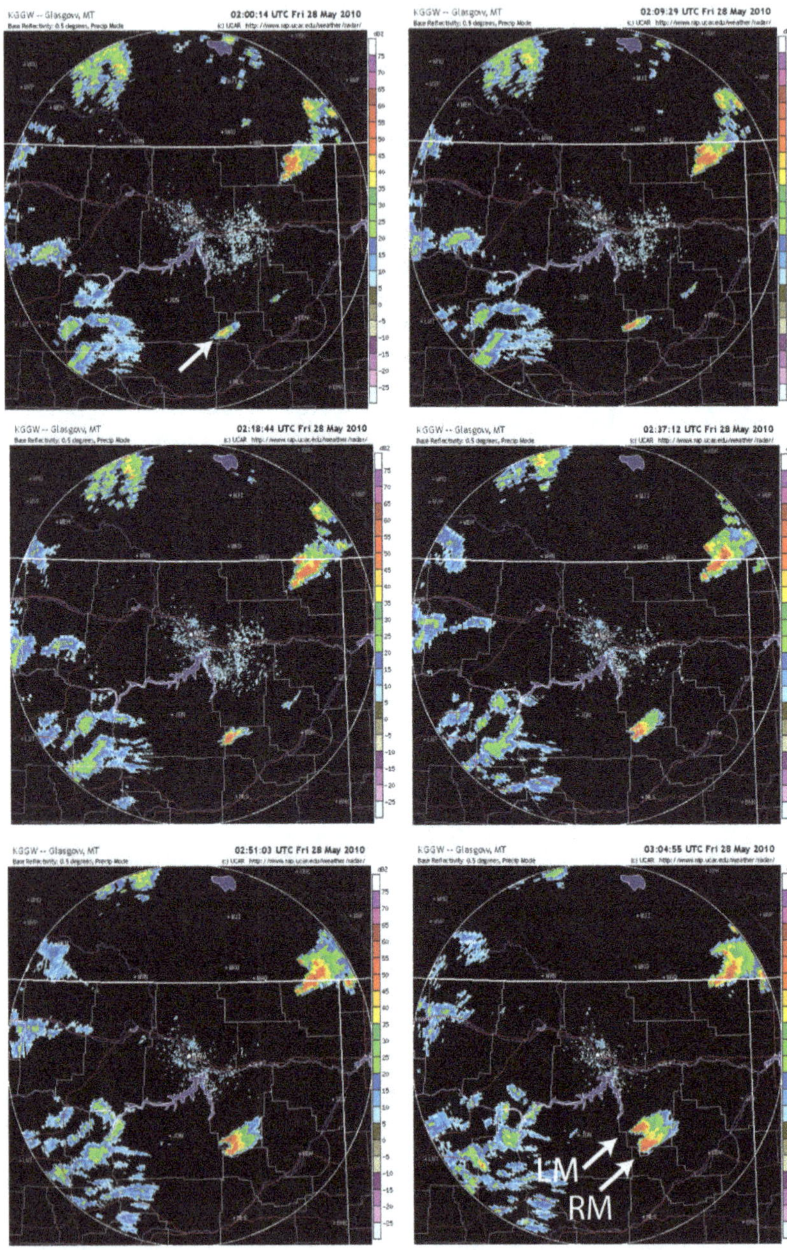

Figure 4.36a. Examples of storm-splitting in eastern Montana and as depicted by a sequence of images of radar reflectivity factor (color-coded in dBZ) at low elevation angle, at approximately 10 min intervals, from the Glasgow, Montana WSR-88D on May 27, 2010. The initial storm (arrow) at 02:00:14 UTC on May 28 splits about 02:37:12 UTC (the time between first cell and splitting is about 1.5 h); the right mover is designated by "RM" and the left mover is designated by "LM".

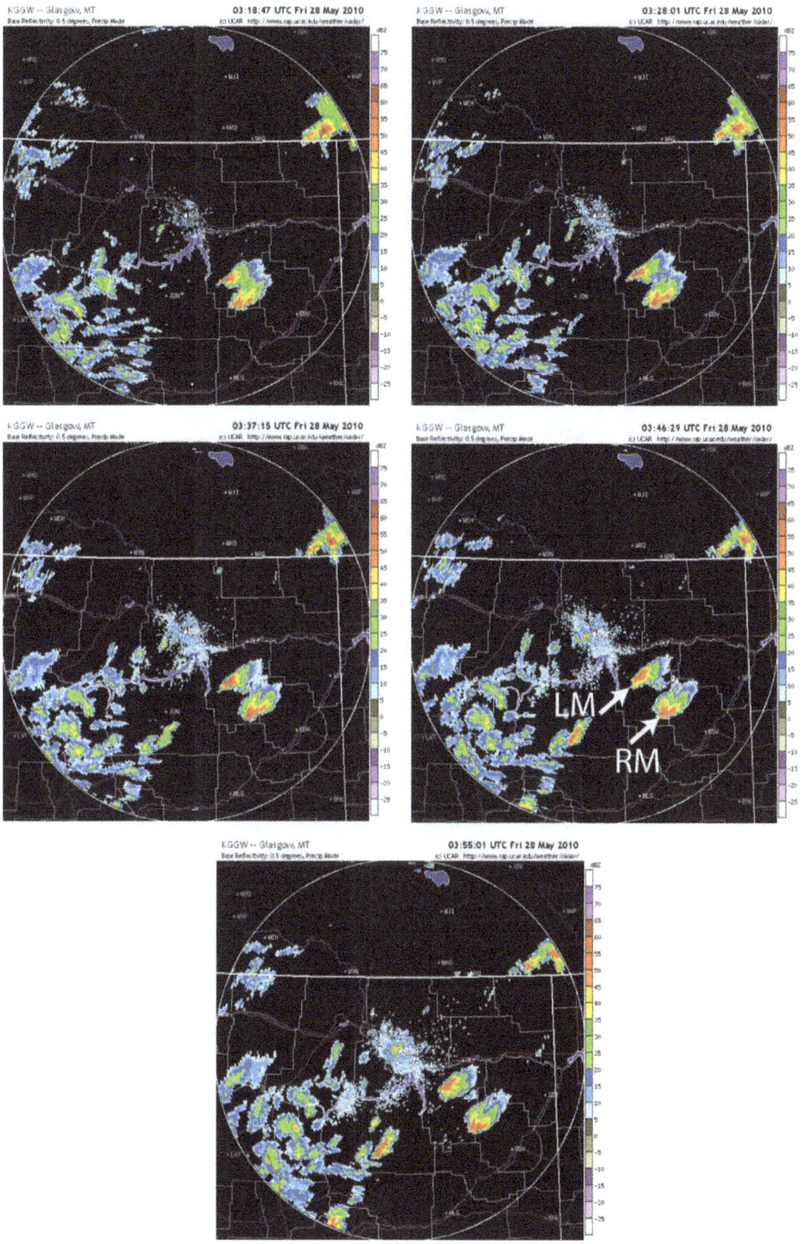

Figure 4.36b. As for Figure 4.36a.

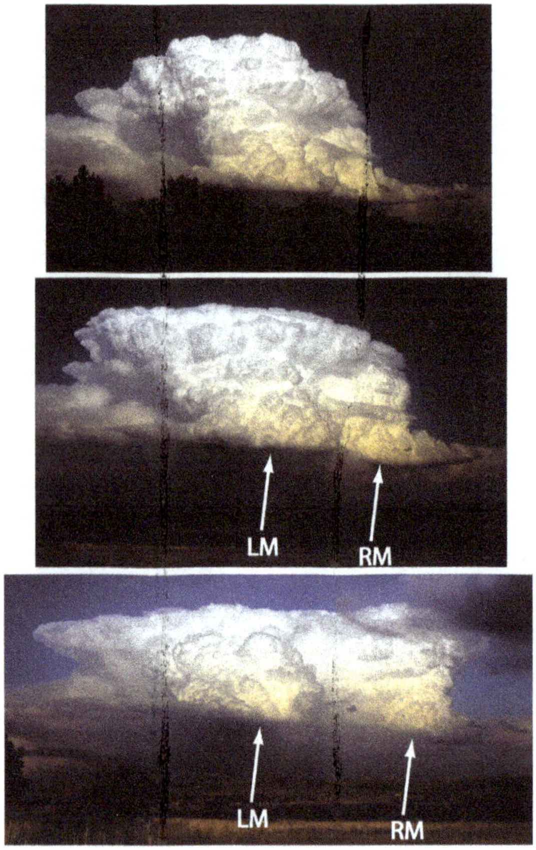

Figure 4.36c. As seen visually, in eastern Colorado, on August 16, 2009, with a view to the east. The storm is about to split in the top image; the middle and bottom images indicate the RM and LM updraft bases. In the bottom image, precipitation can be seen falling in between the updraft bases, as in Figure 4.35 (photographs by the author).

discovered, was considered quite exotic and likened to the behavior of biological mitosis.

In nature, straight (or nearly straight) hodographs are frequently found above the boundary layer, but not in the boundary layer itself, owing to turbulent friction. The vertical variation of vertical shear in the well-known Ekman profile (without baroclinicity) has a marked change in direction with height (Figure 4.38).

Let us now consider how updraft propagation is affected by the linear pressure term. From (4.46) it was shown (Fig. 4.34) that in the upshear (downshear) from an updraft the perturbation pressure is relatively high (low). When the hodograph is straight and most of the shear is below mid-levels, and when the updraft speed increases with height, then there is an upward (downward) directed perturbation pressure gradient force on the downshear (upshear) side. Thus, the

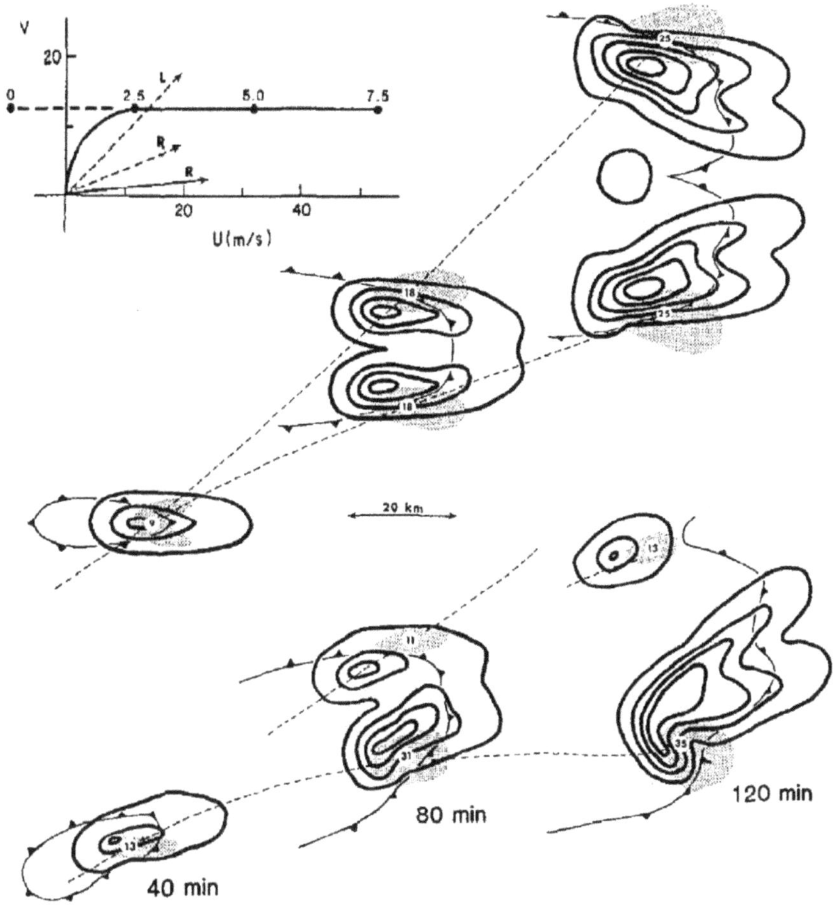

Figure 4.37. Numerical simulation of an isolated, splitting supercell for unidirectional (top sequence) and hybrid curved low, unidirectional aloft (bottom sequence) shear profiles. Low-level rainwater field shown at 40 min intervals (related to radar reflectivity factor) at low levels and contoured at $2\,\mathrm{g\,kg^{-1}}$ intervals; mid-level updrafts indicated by shaded regions; cold pool boundaries denoted by cold front symbols. Hodographs (locations of heads of vectors indicated at km AGL) of environmental winds at upper left: dashed line connecting to straight line indicates unidirectional shear profile; continuously solid line indicates hodograph with clockwise-turning low-level curvature. Dashed "L" and "R" vectors indicate storm motion for unidirectional shear and solid "R" vector indicates storm motion for right mover in hybrid shear simulation. In both simulations, splitting is occurring at 80 min, but the LM persists at and beyond 120 min only in the unidirectional simulation (from Klemp, 1987).

linear effects of shear interacting with an updraft (that increases with height) are to trigger convection on the downshear side and suppress it on the upshear side (Figure 4.39). How the updraft propagates, without consideration of gust front propagation, shading/cooling underneath the anvil downstream (with respect to the flow near the equilibrium level), and other effects, depends on the sum of non-

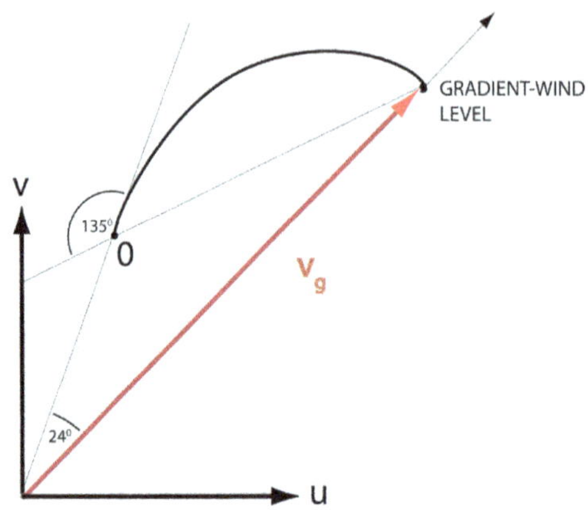

Figure 4.38. Hodograph in the Ekman layer in the absence of baroclinicity (thick solid line) for a boundary-layer geostrophic wind that is southwesterly. The hodograph is shown between the surface (0 km AGL) and the gradient wind level (near, just above, the top of the boundary layer). Note how the hodograph has substantial clockwise turning with height (more than 90°). Thin solid lines are displayed to show geometry for typical values of the drag coefficient, Coriolis parameter (in the Northern Hemisphere), air density, eddy coefficient of viscosity, and a geostrophic wind speed of $\sim 10\,\mathrm{m\,s^{-1}}$ (see Bluestein, 1992).

linear, off-shear propagation and linear, along-shear propagation. The net effect is to promote updraft propagation in the direction of and to the right of the vertical shear vector.

4.5.2 Convective storm dynamics for curved hodographs

When the hodograph is curved, the dynamics of the convective storm are significantly affected by the linear term. Consider the special limiting case, for purposes of illustration, when a hodograph turns 180° in the clockwise direction with increasing height (Figure 4.40) up to the tropopause. If the hodograph is curved as part of a circle, then the horizontal vorticity vector associated with vertical shear (the vertical shear vector, tangent to the hodograph at each level, is perpendicular and to the right of the horizontal vorticity vector) is always oriented in the same direction as the wind at each altitude (Figure 4.40). Such a hodograph depicts "Beltrami" flow, in which the vorticity vector is always oriented in the same direction as the wind. It is seen from this figure that there is an upward (downward) directed perturbation pressure gradient force on the right (left) side ("right" and "left" refer to the concave and convex sides of the hodograph, respectively; Figure 4.41). Thus, the linear term acts to promote updraft propagation to the right of the hodograph and suppresses updraft propagation on the left side. When the hodograph turns 180° in the counterclockwise direction with height (not shown), the linear term acts to promote propagation to the left of the hodograph.

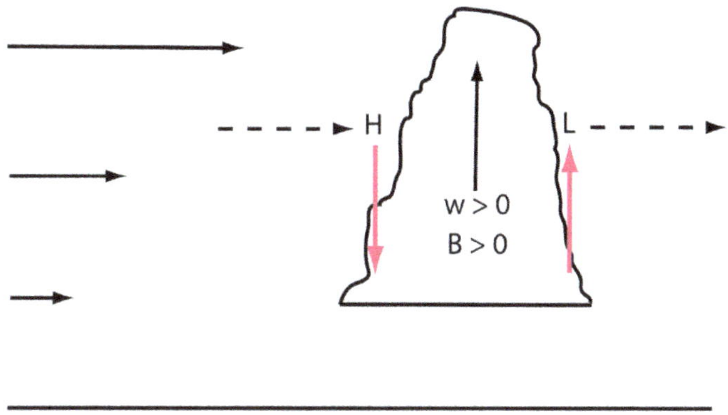

Figure 4.39. Idealized illustration of how a new updraft is encouraged on the downshear (vertical shear vector indicated by dashed vectors) side of an updraft and suppressed on the upshear side of an updraft (cloud outline shown) by linear dynamic vertical perturbation pressure gradient forces (red vectors) in unidirectional shear (wind profile indicated by vectors at the left) in the presence of a buoyant updraft.

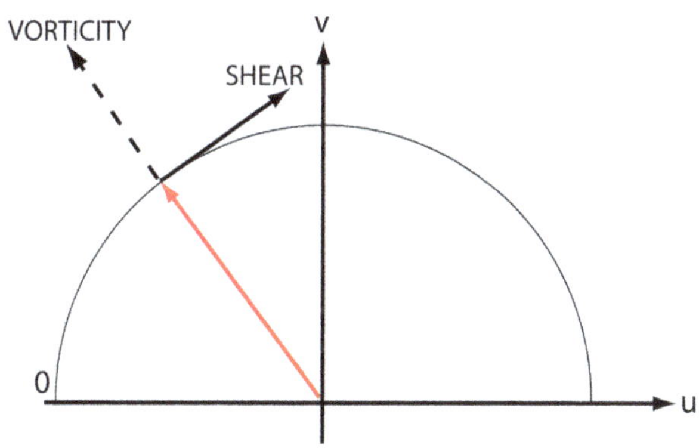

Figure 4.40. Illustration of half of a circle hodograph, for which the wind vector at any height (red vector) points in the same direction as the horizontal vorticity vector (dashed vector) and normal to the vertical shear vector (black vector).

The effect of nonlinear terms when the hodograph turns 180° with increasing height is as follows (Figure 4.42): Counter-rotating vortices are produced through tilting in the normal-to-shear direction. Each vortex is associated with relatively low pressure. Since the shear vector reverses direction aloft, there is relatively low pressure both near the surface and at the top level in the normal-to-shear direction. Since the effects of tilting generally increase with height when the strength of the updraft increases with height, the perturbation pressure is lower aloft than at the surface. Thus, one would expect there to be a *slight* upward-directed perturbation pressure gradient on both sides normal to the low-level and high-level shear vectors. At middle levels, vortices are also produced from tilting in the normal-to-shear direction; they are located along a line normal to the line joining the

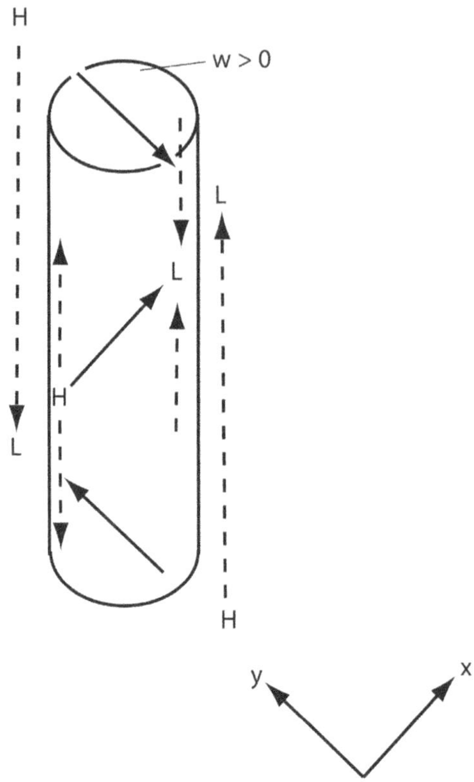

Figure 4.41. Illustration of how the half-circle hodograph in Figure 4.40, for which the hodograph curves in a clockwise manner with height, promotes new updraft growth to the right and suppresses new updraft growth to the left of the mean vertical shear, which is like the vertical shear (vectors) at mid-levels, as a result of vertical perturbation pressure gradient forces (dashed vectors) from the linear dynamic perturbation pressure gradient forces in the presence of an updraft.

low-pressure areas produced at both low levels and upper levels; there is thus an upward-directed perturbation pressure gradient force at low levels and a downward-directed perturbation pressure gradient force aloft normal to the mid-level shear vector. The effect of the nonlinear pressure field should therefore be relatively small and the linear part of the perturbation pressure field more important.

In nature, hodographs frequently have both curved and straight sections. Hodographs are often curved through some depth and straight elsewhere. For example, in many tornado outbreaks there is clockwise curvature in the boundary layer, above which the hodograph is straight (upper-left insets in Figure 4.14g). Enhanced curvature and lengthening of the hodograph at low levels in the Great Plains of the U. S. is often associated with the development of the nocturnal low-level jet. Sometimes hodographs exhibit curvature at low levels and reverse curvature aloft (Figure 4.43). The association of specific types of hodographs with

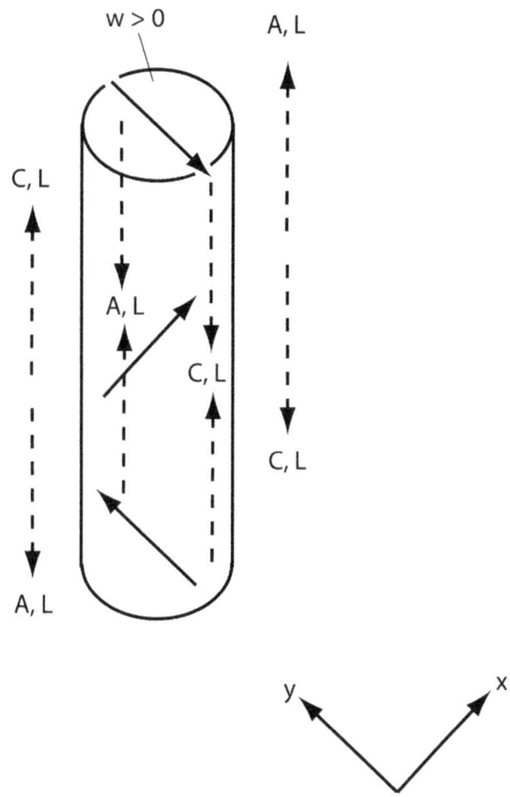

Figure 4.42. As for Figure 4.41, but from nonlinear dynamic perturbation pressure gradient forces.

specific synoptic conditions is not well known, but must be related to the vertical structure of jets (and via the thermal wind relation, to the horizontal temperature gradient) and the temperature advection profile in the vertical. Both cold advection (in which the geostrophic wind (vector) backs (i.e., turns in a counterclockwise direction) with height) and warm advection (in which the geostrophic wind (vector) veers (i.e., turns in a clockwise direction) with height) situations may each be associated with clockwise or counterclockwise-turning hodographs. Relating synoptic situations (as described by the structure of baroclinic waves) to hodographs is an area in which little research has been done, though there have been climatological analyses of hodographs with respect to the quadrant of surface cyclones and anticyclones for each season.

In general, clockwise-turning hodographs favor cyclonically rotating, "right-moving" (RM) supercells and counterclockwise-turning hodographs favor anticyclonically rotating, "left-moving" (LM) supercells. The former are the predominant type of supercell observed, while the latter are observed only on very rare occasions (Figure 6.38).

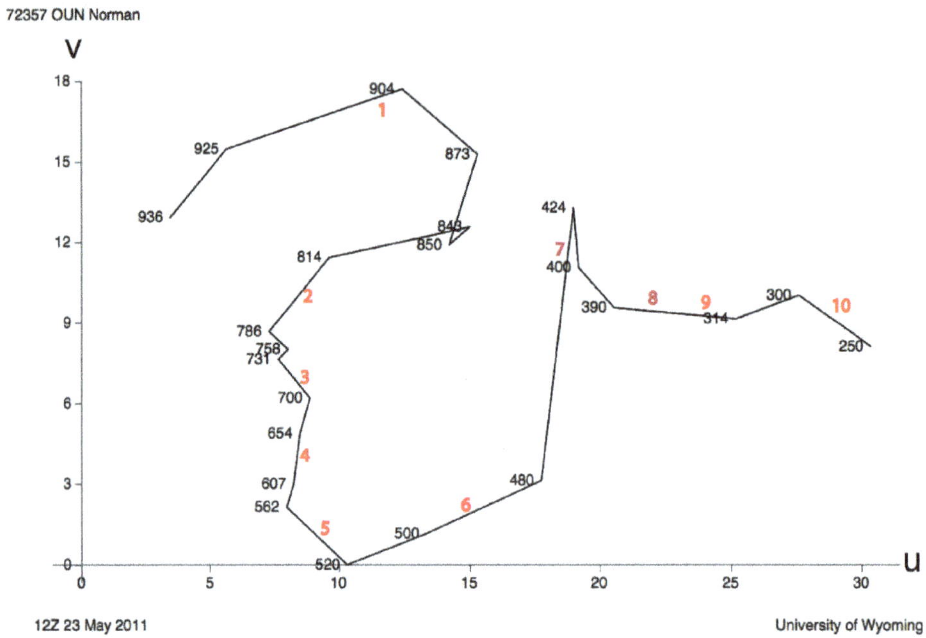

Figure 4.43. Hodograph at Norman, OK at 12:00 UTC on May 23, 2011, showing clockwise curvature from the surface to 2 km (heights shown in red) and counterclockwise curvature from 2 km to 7 km AGL (wind speed in m s^{-1}). Black numbers plotted are pressure (hPa).

4.5.3 Straight vs. curved hodograph dynamics: two paradigms

In the past two decades, the dynamics of supercells have been explained using two paradigms: (a) the "vertical shear perspective" and (b) the "helicity perspective". In the former, the main idea is that the storm propagates because it is rotating; in the latter, the main idea is that the storm generates rotation because it is propagating. According to the vertical shear perspective, the basic physical processes responsible for supercell behavior are the tilting of environmental horizontal vorticity and the subsequent propagation owing to the rotation produced (nonlinear effect) and turning of the hodograph with height (linear effect). This is the perspective that we have taken so far. The problem is that one needs a theory that explains *both* propagation and rotation.

We will now consider the "helicity" perspective to see what its strengths are. The helicity approach was originally sparked by work by Doug Lilly in the early 1980s in which it was hypothesized that rotation in a convective storm reduces turbulent dissipation and increases its stability so that it can be longer lived; the seed idea came from turbulence theory. The equation of motion (4.11), with the restriction of steady state removed, may be expressed as

$$D\mathbf{v}/Dt = \partial\mathbf{v}/\partial t + \nabla(\tfrac{1}{2}\mathbf{v}\cdot\mathbf{v}) + [(\nabla\times\mathbf{v})\times\mathbf{v}] = -\alpha_0\,\nabla p' + B\mathbf{k} \tag{4.51}$$

Now, $\nabla\times\mathbf{v}\equiv\omega$, so when $\omega\times\mathbf{V}=0$ a part of the advective term in the equation of motion that is in part responsible for the downscale cascade of energy through nonlinear scale interactions (the Lamb vector) is absent. Recall also the results from classical Rayleigh–Bénard theory in which it is shown that rotation is a stabilizing effect (cf. (2.266) and (2.267)).

To understand the downscale cascade of energy, consider the Eulerian rate of change of the u-component of the wind as a result of the nonlinear advection term in the x-component of the equation of motion

$$\partial u/\partial t = u\,\partial u/\partial x \tag{4.52}$$

If we represent the variation of u in space as a wave in the x-direction characterized by a wavenumber k, then

$$u \sim e^{ikx} \tag{4.53}$$

It follows from (4.52) that

$$\partial u/\partial t \sim u\,\partial u/\partial x \sim (ik)\,e^{i2kx} \tag{4.54}$$

Since

$$2k = 2(2\pi/L) = 2\pi/(L/2) \tag{4.55}$$

where L is the horizontal wavelength, or scale; then fluctuations on the scale of $L/2$ are introduced. Further action (at later time steps) of the nonlinear advection term produces fluctuations again on even shorter space scales, and "so on to viscosity" as L. F. Richardson once put it.

Helicity (H), which is equivalent to streamwise vorticity, is simply the dot product between velocity and vorticity

$$H = \mathbf{v}\cdot\omega \tag{4.56}$$

When helicity is maximized (i.e., when $\mathbf{v}$ and ω point in the same direction), $\omega \times \mathbf{V} = 0$, so that the term in the equation of motion in part responsible for the downscale cascade of energy disappears.

Helicity has been hypothesized to be important also in understanding the amplification of rotation in supercells. Consider a case in the Northern Hemisphere when there is westerly vertical shear, so that the vorticity vector points to the north. When air enters an updraft at low levels from its western side (i.e., when the storm-relative wind vector is oriented in the direction normal to the vorticity vector), there is purely "crosswise" vorticity (Figure 4.44, top). When air enters the updraft at low levels from the southern side (i.e., when the storm-relative wind vector is oriented in the same direction as that of the vorticity vector), there is purely "streamwise" vorticity (Figure 4.44, bottom panel). The relative amounts of streamwise and crosswise vorticity are determined by storm-relative motion. Storm motion is determined by the mean wind, propagation due to vertical perturbation pressure gradients, propagation due to density current behavior, among other things, and is not easily determined precisely from a hodograph because cloud microphysics, temperature, and moisture stratification also play a role, among other factors. When there is streamwise vorticity, vertical vorticity is stretched by convergence as air parcels are tilted and enter the base of an updraft. Bob Davies-Jones at NSSL, in a classic paper in 1984, demonstrated,

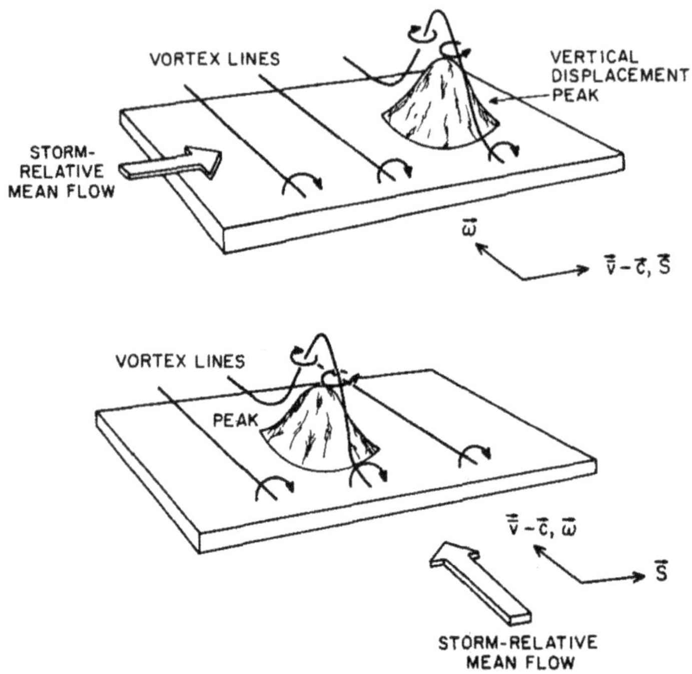

Figure 4.44. Examples of crosswise (top) and streamwise (bottom) vorticity. The vorticity vector is denoted by $\vec{\omega}$ and the vertical shear vector by $\vec{S}$ (see also Figure 4.29, middle) (from Davies-Jones, 1984).

using an idealized, linear model (frictionless, Boussinesq, isentropic, dry, statically unstable, basic state in which there is constant vertical shear), that vertical velocity and vertical vorticity are positively correlated when there is streamwise vorticity. What happens *after* a short period of time, however, depends on nonlinear processes and one needs a numerical simulation to see how the storm will actually behave.

A measure of the correlation between an updraft and vertical vorticity is given by "storm-relative environmental helicity (SREH)", which is similar to (4.56), except that $\mathbf{v}$ is replaced by the storm-relative value of $\mathbf{v} - \mathbf{c}$, where $\mathbf{c}$ is the storm motion vector. In practice, SREH is usually integrated over h, the depth of the "inflow layer" of the storm

$$\text{SREH}(\mathbf{c}) = \int_0^h (\mathbf{v} - \mathbf{c}) \cdot (\nabla \times \mathbf{v})\, dz \tag{4.57}$$

When SREH is relatively high, then the environment is deemed to satisfy a necessary condition for storm rotation intense enough for a mesocyclone to form. Typical values of SREH for $h = 3$ km (a height frequently used by forecasters) associated with supercells are $150\,\text{m}^2\,\text{s}^{-2}$ or higher. It is thought that the higher the SREH, the more probable supercell tornadoes will occur.

An easier way to visualize the magnitude of SREH better than by simply computing the integral in (4.57) is as follows: vorticity in the environment is associated with vertical shear only (i.e., we neglect vertical motions in the environment, which are typically an order of magnitude or more weaker than horizontal motions). Then

$$\nabla \times \mathbf{v} = -\partial v/\partial z\, \mathbf{i} + \partial u/\partial z\, \mathbf{j} \tag{4.58}$$

and therefore

$$\mathrm{SREH}(\mathbf{c}) = \int_0^h -(u - c_x)\, dv/dz\, dz + \int_0^h (v - c_y)\, du/dz\, dz \tag{4.59}$$

where

$$\mathbf{c} = c_x\, \mathbf{i} + c_y\, \mathbf{j} \tag{4.60}$$

Then

$$\mathrm{SREH}(\mathbf{c}) = -\int_0^h u\, dv + \int_0^h v\, du + \int_0^h c_x\, dv - \int_0^h c_y\, du \tag{4.61}$$

The sum of the first two terms on the RHS of (4.61) is ground-relative environmental helicity, SREH(0). The third and fourth terms on the RHS of (4.61) are $c_x\, \Delta v$ and $-c_y\, \Delta u$, respectively.

Now, consider the idealized hodograph in Figure 4.45, with areas A, B, C, D, E, F, G, H, I, and J noted. The first term on the RHS of (4.61) is $-I - J + A$. The second term is $A + 2B + C + D + E + F + G + H$. The third term is $C + I$. The fourth term is $-(F + G + H)$. Since, $D + J = E$, the sum of all four terms is $2(A + B + C + D)$, which *is twice the area swept out by the storm-relative wind vector between the ground and height h*. Thus, even if the hodograph is unidirectional, SREH is high if the updraft movement is far off the hodograph. For a given hodograph, SREH may be increased simply by forcing the tip of the storm motion vector to be farther away from the hodograph (Figure 4.46). When the hodograph is curved in a counterclockwise manner with height or if the storm motion is to the left of the hodograph, negative SREH is created.

Consider the simple example of the application of this geometric formulation illustrated in Figure 4.47. A quarter-circle hodograph from 0 to 3 km, in which the storm-relative wind speed at all levels is $10\,\mathrm{m\,s^{-1}}$ has an SREH of $2(\pi\, 100\,\mathrm{m^2\,s^{-2}})/4 \sim 150\,\mathrm{m^2\,s^{-2}}$, the approximate observation-based threshold for supercells.

From the thermal wind relation (4.23), we may substitute the horizontal temperature gradient for a measure of the vertical shear of the horizontal component of the wind, so that

$$(\mathbf{v} - \mathbf{c}) \cdot \nabla \times \mathbf{v} \sim -(\mathbf{v} - \mathbf{c}) \cdot \nabla_h T \tag{4.62}$$

so that

$$\mathrm{SREH}(0) \equiv H \sim -\mathbf{v} \cdot \nabla_h T \tag{4.63}$$

or, in other words, helicity in the environment is proportional to horizontal temperature advection. Thus, helicity in the environment may be enhanced in the vicinity of warm fronts, stationary fronts, and outflow boundaries, where we tend to find warm advection. (It is assumed that cold fronts are characterized by cold advection.) It is also noted, in accord with the thermal wind relation (4.23), that

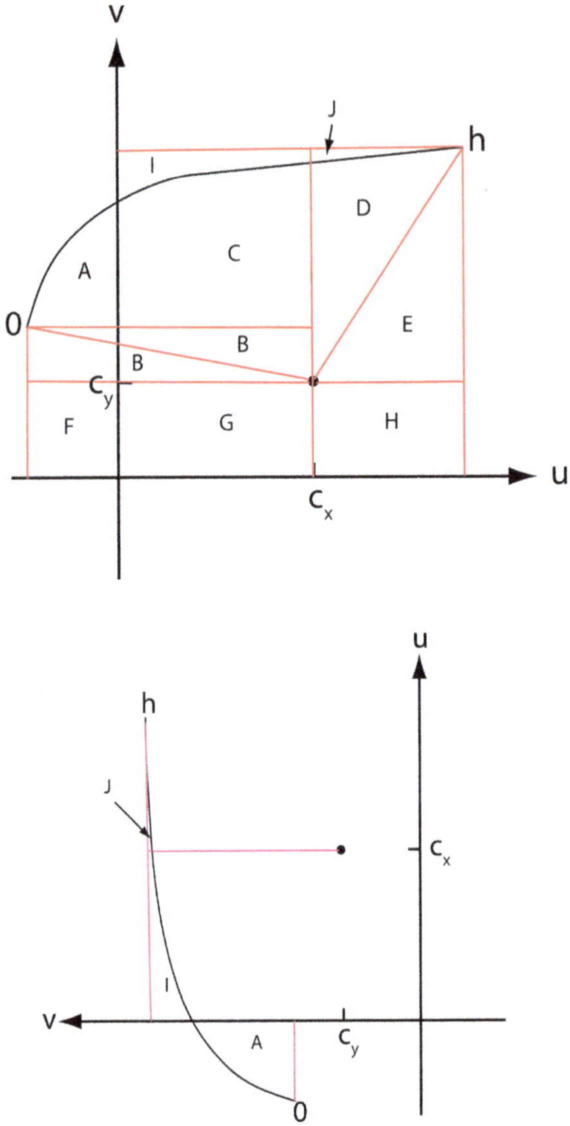

Figure 4.45. Illustration of the relationship between SREH and the area under the hodograph swept out by the storm-relative wind vector from $z = 0$ to $z = h$. (Top) Area elements used in the first, third, and fourth terms on the RHS of (4.61); (bottom) area elements used in the second term on the RHS of (4.61).

fronts and outflow boundaries, which may be associated with relatively high values of vertical shear, also manifest as horizontal vorticity, which is available to be tilted onto the vertical at the edges of zones of rising motion that are found along and near them.

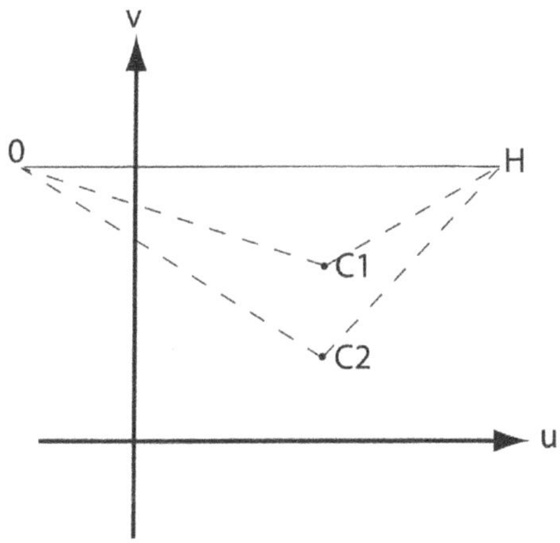

Figure 4.46. Idealized illustration of how moving the storm motion vector away from the hodograph increases SREH. SREH, which is proportional to the area swept out by the storm-relative wind vector between the ground (0) and height H, increases when the storm motion vector defined by the line joining the origin to C1 is moved to C2. In this case, more deviant storm motion to the right of the hodograph is accompanied by increased SREH even for a unidirectional hodograph.

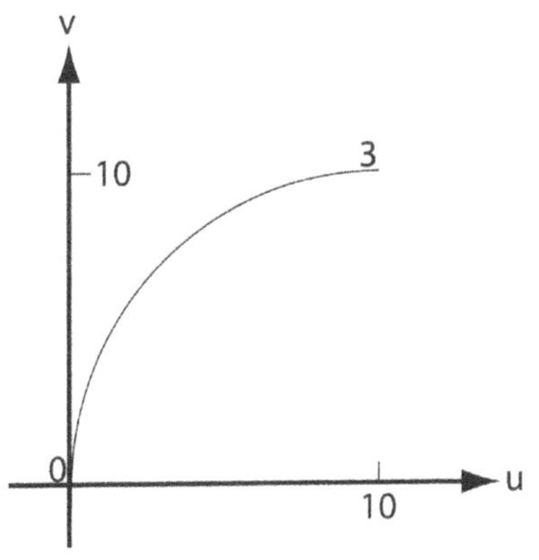

Figure 4.47. Illustration of the computation of SREH between the ground and 3 km AGL for a quarter-circle hodograph (wind speed in $\mathrm{m\,s^{-1}}$) having a radius of $10\,\mathrm{m\,s^{-1}}$ and a storm motion from the west to the east of $10\,\mathrm{m\,s^{-1}}$. The area swept out by the storm-relative wind is $\frac{1}{4}$ the area of a circle having a radius of $10\,\mathrm{m\,s^{-1}}$. Twice this quantity is the SREH.

If a convective storm updraft moves along with the pressure-weighted mean wind and if the hodograph is straight, there is no SREH. If the updraft propagates off the shear vector (e.g., as a result of the nonlinear pressure term, created through the tilting of environmental horizontal vorticity onto the vertical), then SREH develops as the updraft splits; positive and negative SREH develop in the right and left mover, respectively. It is important to note that SREH is not Galilean invariant, since it depends on storm motion. If the vertical shear vector at all altitudes is normal to the storm-relative wind as it is when there is a perfectly circular hodograph, then SREH is the highest it can be for a given shear.

As noted earlier, the problem with the helicity approach to understanding supercell dynamics is that storm motion must be known, but it is not yet predictable from theory. Storm motion is often influenced by the movement of a storm's own gust front, which depends to some extent on cloud microphysics parameters (through evaporative and sublimation cooling) and by factors *external* to the storm such as the motion of outflow boundaries, fronts, the dryline, and orography. An empirical technique for predicting storm motion which blends theory with observations was presented by Matthew Bunkers and colleagues in an oft-referenced paper published in 2000. This method makes use of motion with the mean wind (calculated with respect to height—not to mass—for simplicity, and without an apparent significant loss of accuracy) and normal-to-shear propagation. The contribution from the former is given by the 0–6 km height-weighted mean wind; the contribution from the latter is given by a vector $7.5\,\mathrm{m\,s^{-1}}$ in magnitude, normal and to the right of the 0–0.5 to 5.5–6 km vertical shear vector. It is useful for forecasting, in the absence of other factors such as gust front propagation, etc.

The reader is cautioned that sometimes a supercell may become anchored with respect to a location (e.g., when the motion of an outflow boundary is in the direction opposite to that of propagation due to a storm's internal dynamics). In this case, SREH may change when a storm interacts with the outflow boundary, etc.

We gain some insight into the role of curved hodographs vs. straight hodographs by considering the vertical component of (4.51), the pressure having been separated into the dynamic part (p_d) and the part due to buoyancy (p_b)

$$\partial w/\partial t + \partial/\partial z[(u^2 + v^2 + w^2)/2] - (u\eta - v\xi) = -\alpha_0\, \partial p'_d/\partial z + [B - \alpha_0\, \partial p'_b/\partial z] \tag{4.64}$$

where the components of horizontal vorticity ξ and η are given by (4.37) and (4.38). The second term on the LHS of (4.64) is the Bernoulli term and the third term on the LHS is the Lamb term. Consider forcing in the vertical due to vertical perturbation pressure gradients associated with the Bernoulli and Lamb terms. In 2000, Morris Weisman and Rich Rotunno demonstrated, using idealized numerical simulations, that for supercells grown in environments of both straight and highly curved hodographs, the upward-directed perturbation pressure gradient on the right side of the shear vector, which is responsible for propagation of the updraft in the normal-to-shear direction, is due mainly to the Lamb term. They therefore concluded that *the nonlinear pressure term, which does not depend upon hodograph curvature, is of fundamental importance.*

4.5.4 Sensitivity of simulated supercell structure to environmental thermodynamic and cloud microphysics parameters

While the overall behavior of supercells can be explained qualitatively based on idealized soundings (i.e., on idealized vertical profiles of vertical wind shear and CAPE), significant differences in storm morphology (e.g., in the degree of surface cold outflow) and intensity (updraft speed, peak mid-level, and surface vorticity) are found, especially when CAPE is relatively low. In most supercells in the Great Plains of the U. S., CAPE is relatively high ($>1{,}500\,\mathrm{j\,kg^{-1}}$) and the level at which potential buoyancy is highest is around 6–10 km AGL. When CAPE is relatively low ($<1{,}000\,\mathrm{j\,kg^{-1}}$), but is concentrated at low levels, potential buoyancy at low levels can match shear better (in the sense that $R \sim 1$ for buoyancy and shear at low levels); when shear is relatively weak, but is concentrated at low levels, shear at low levels can match CAPE better at low levels. Thus, supercells can occur that are relatively shallow, as is observed in landfalling tropical cyclones or the outer rainbands in tropical cyclones (Figure 4.48) and in some mid-latitude storms having a cold core aloft and a low tropopause (Figure 4.49). In these storms, potential buoyancy is highest around 3 km AGL. These relatively shallow supercells are sometimes called "mini-supercells" or "low-topped supercells". Like

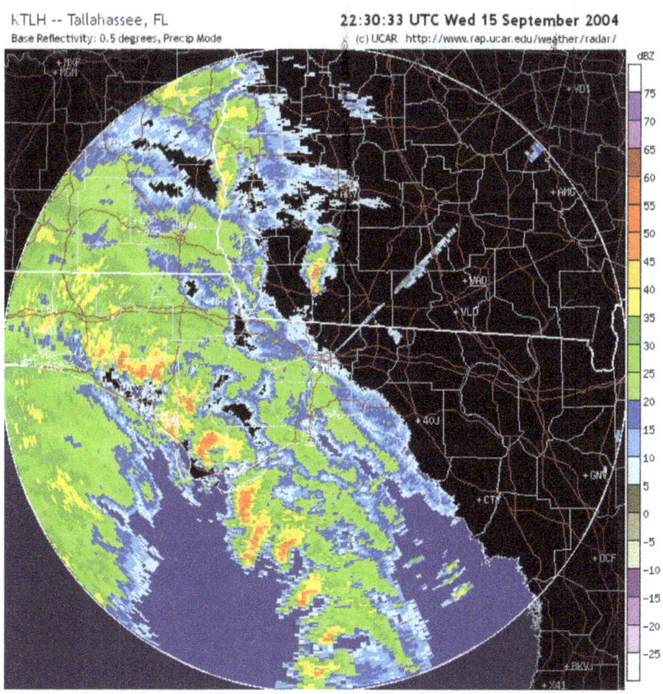

Figure 4.48. Supercells in the Gulf of Mexico off the west coast of Florida, in an outer rainband of Hurricane Ivan on September 15, 2004, as detected by the WSR-88D radar at Tallahassee, FL.

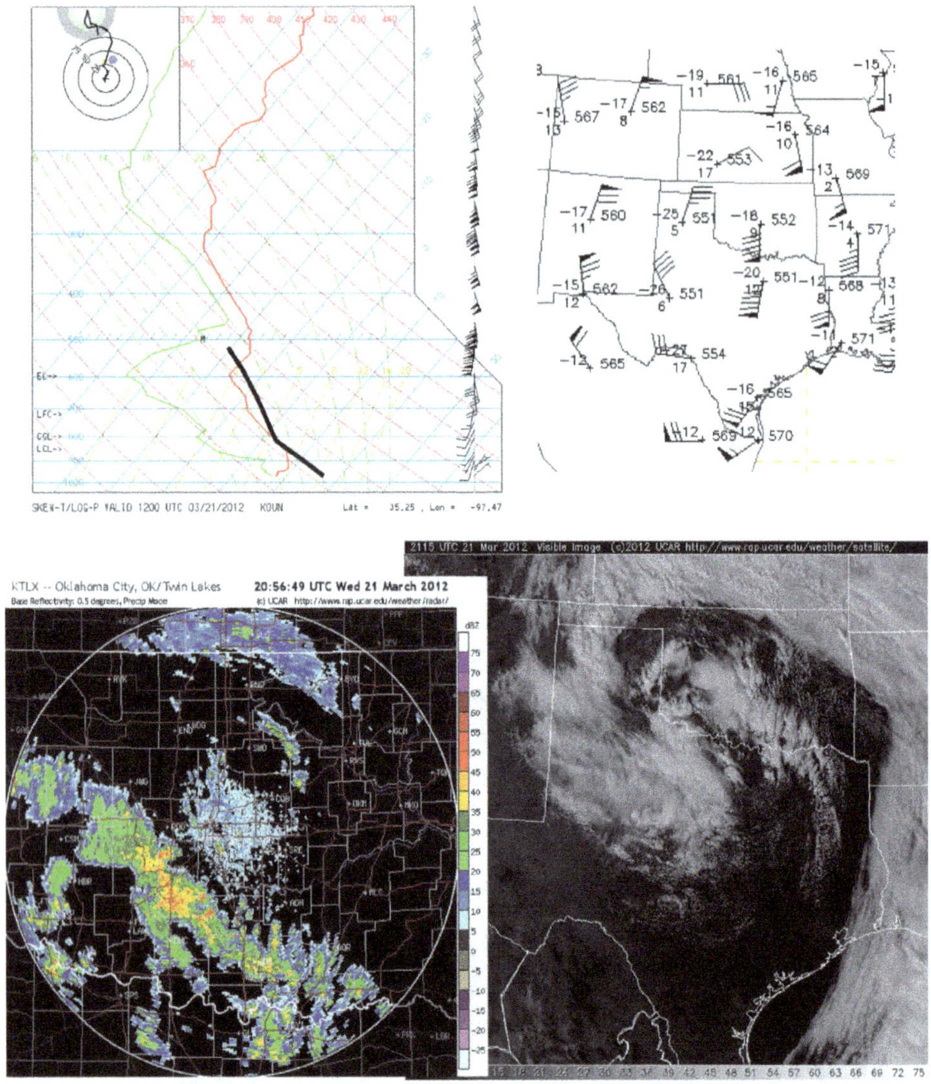

Figure 4.49. An example of convective storms that produced funnel clouds near a cold, upper-level low on March 21, 2012 in Oklahoma. Some funnel clouds may have been associated with convective storms having relatively low tops in an environment of low CAPE, but strong vertical shear in the cloud-bearing layer. These storms may have been mini- or low-topped supercells. (Top left) Sounding at 12:00 UTC at Norman, OK. Solid black lines show the path taken by a surface-based air parcel if lifted to its LCL and LFC: it would rise in the presence of small CAPE until its equilibrium level (only 550 hPa), but would experience strong vertical shear. (Top right) 500 hPa plotted data at 12:00 UTC in the Southern Plains. The center of the cold-core low is near southwestern Oklahoma and northwest Texas. (Bottom left) Radar reflectivity factor from the WSR-88D Doppler radar at Oklahoma City, OK at about 21:00 UTC. (Bottom right) Visible satellite image covering the Southern Plains.

ordinary supercells, they can produce tornadoes (and waterspouts). They have been documented in the Great Plains of the U. S. during the spring and summer, along the Gulf Coast of the U. S. during the hurricane season, in Japan during the typhoon season, and in the cool season in California and Australia, among others.

Numerically simulated supercells have stronger surface outflow when the mid-troposphere is relatively dry; when shear is strong or when dryness is concentrated at higher altitudes, however, this effect is less. The potential for evaporatively cooled downdrafts increases with increasing dryness, especially when vertical shear is not too strong. Strong downdrafts are detrimental to storm longevity: when surface outflow moves at the same speed as the updraft and mid-level mesocyclone, then storms can persist and intensify; when surface outflow moves faster than the updraft and mid-level mesocylone, storms weaken.

The depth of the environmental, moist, boundary layer also significantly affects numerically simulated supercell morphology and behavior. Numerical experiments have been conducted to explain contrasting supercell behavior in the relatively moist environments of the Central Plains of the U. S. from that in the relatively dry environments of the High Plains. For example, in the former, the LCL and LFC are at relatively low altitude, while in the latter the LCL and LFC are relatively high. Therefore, the potential for evaporative cooling near the surface is higher in the latter case, since precipitation has farther to fall through a layer of unsaturated air; so, the dynamics of the cold pool then play a more important role in storm behavior. An interesting finding is that, under some circumstances, storms in a low-CAPE environment may be more intense than storms in a high-CAPE environment, owing to vertical perturbation pressure gradients that act to enhance updraft intensity.

The effects of microphysical parameterization schemes on numerically simulated supercell behavior have been investigated. When the amount of rainfall is high relative to the amount of ice material, evaporation is higher and colder surface cold pools may be generated, thus decreasing the likelihood of storm longevity. When freezing processes occur, warming is increased due to latent heat release, while cooling is increased due to melting.

From studies of the effect of dryness and microphysics, it is concluded that the behavior of supercells is influenced not only by vertical shear and CAPE, but also by the intensity of the surface cold pool and its effect on decoupling surface features from those aloft that are not affected by the cold pool.

Studies that elucidate the behavior of convective storms based on vertical shear and CAPE alone must be regarded with some caution because, for example, the same CAPE at different temperature regimes may be associated with varying amounts of hydrometeors of different types, as temperature and pressure vary. Varying amounts of hydrometeor types can lead to different cold pool strengths. A good place for a student to experiment by simply varying shear and CAPE and exploring the effect on storm type is the COMET (Cooperative Program for Operational Meteorology, Education, and Training) module for the "Convective Storm Matrix" (*http://www.meted.ucar.edu*).

4.6 THE DEEP CONVERGENCE ZONE (DCZ)

Supercells sometimes contain counter-rotating vortices at mid-levels which are not necessarily created through the tilting of *environmental* horizontal vorticity (vertical shear) by the main *updraft* in the storm. They may be created by a downdraft, acting on environmental shear or, more likely, by an updraft along the flanking line band, acting on baroclinically generated horizontal vorticity at the leading edge of the rear-flank gust front (as in some quasi-linear MCSs; see Chapter 5). The latter case is possible only if the air behind the rear-flank gust front is relatively cool. In either case, the anticyclonic member of a couplet is found at mid-levels along the right flank of the flanking line/rear-flank gust front, while a cyclonic member is found in the usual location (the southernmost and middle vortices, respectively, in Figure 4.24). A strong *rear-inflow jet* may be nestled in between the anticyclonic and cyclonic members of the couplet. The structure and dynamics of this rear-inflow jet and vortex couplet may be similar to those found in mesoscale convective systems, to be discussed in more detail in the next chapter.

At the leading edge of the jet, there is a curved band of strong convergence. Typically there is a curved band of convergence at low levels along the flanking line, rear-flank gust front (Figure 4.24). When the band extends up to the mid-troposphere (and perhaps higher), it is referred to as a "deep convergence zone (DCZ)". There may be a vertical wall marking the DCZ, within which there is substantial turbulent mixing.

Another anticyclonic vortex may be found, paired with the cyclonic vortex, but on the left side of the mean vertical shear vector (the northernmost anticyclonic vortex in Figure 4.24). This vortex is formed through tilting by the updraft of environmental shear (as described earlier). The net effect of all the tilting is that there can be a triad of vortices at mid-levels of anticyclonic–cyclonic–anticyclonic vortices.

4.7 THE PRODUCTION OF LOW-LEVEL ROTATION

The production of mesocyclones in supercells at *mid-levels* in the troposphere has been explained as a consequence of the tilting of environmental vertical vorticity along the edges of the main updraft. This explanation cannot explain how mesocyclones form at *low levels* because on a level surface vertical velocity must vanish, as a consequence of the kinematic lower-boundary condition. It has therefore been proposed that vertical vorticity at the surface could be that produced aloft via tilting, and then advected downward by a downdraft. However, at the surface, where vertical velocity vanishes, there must be strong divergence, which acts to decrease vorticity.

Low-level mesocyclones have also been explained as being formed from baroclinically generated horizontal vorticity that has been tilted as air approaches

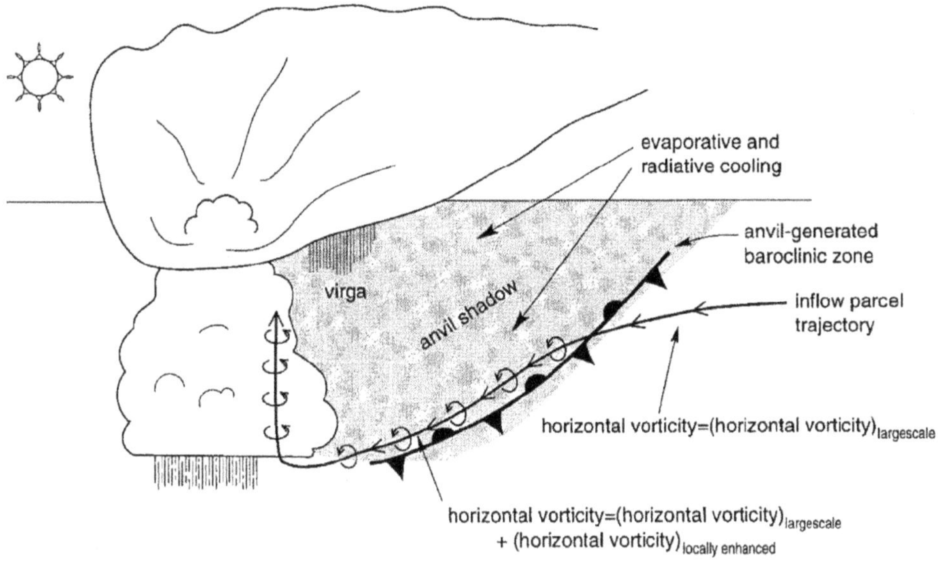

Figure 4.50. Illustration of how low-level horizontal vorticity may be enhanced by an anvil-generated baroclinic zone due to a horizontal gradient in radiation (from Markowski *et al.*, 1998b).

the updraft. However, it has been demonstrated numerically that low-level, *environmental*, horizontal vorticity in the form of low-level shear also affects low-level mesocyclogenesis. In short, the mechanisms for low-level and mid-level mesocyclogenesis are different. However, the two may interact, and this interaction will be discussed in a subsequent section on tornadogenesis.

First, consider a vorticity analysis following air parcel trajectories. Air may enter a low-level mesocyclone from the southeast (for simplicity we consider a supercell in the Northern Hemisphere), which represents a flow of ambient, relatively warm, and moist air. Air entering the low-level mesocyclone from the north or northwest is relatively cold. During the daytime hours, there may also be temperature differences created by anvil shading (Figure 4.50); during the nighttime hours, the baroclinic effect due to anvil shading is reversed. Anvil shading is not a fundamental process, but probably one that can enhance or diminish the horizontal temperature gradient already existing due to differences in the amount of evaporational/melting/sublimation cooling. When a storm and its surrounding area is covered by anvils from other convective storms, there cannot be any anvil-shading contribution to surface baroclinicity.

If there is a horizontal gradient in the coolness of the air underneath the region of precipitation in the FFD area, then horizontal vorticity may be generated as air enters the low-level mesocyclone after having passed through this area. Moreover, air entering the low-level mesocyclone from the east or northeast may have passed over a baroclinic zone separating evaporatively cooled air underneath

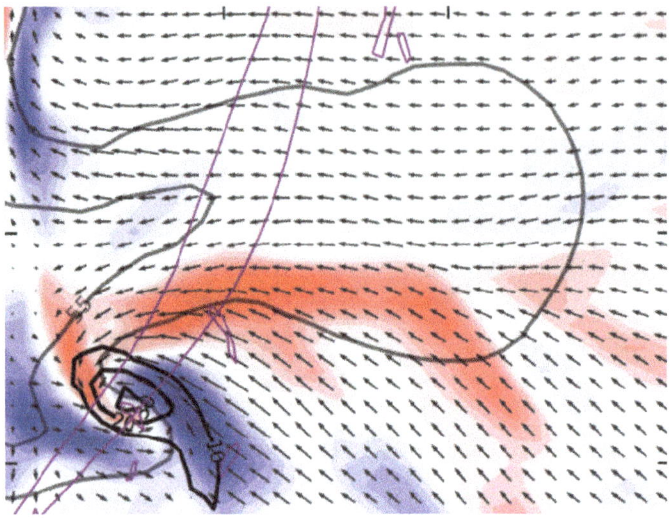

Figure 4.51. Ensemble mean of the rate of baroclinic generation at 750 m AGL of storm-relative streamwise horizontal vorticity ($10^{-4}\ s^{-2}$) (color coded) and storm-relative ensemble mean wind (plotted every 2 km) from mobile Doppler radar data from the U. Mass. X-Pol radar and data from the WSR-88D radar at Dodge City, KS on May 4, 2007 in south central Kansas, assimilated into a numerical cloud model. Tornado paths are indicated by thin purple lines and radar reflectivity factor (dB*Z*) is contoured at 35 and 55 dB*Z* in gray. Vertical vorticity (black contours) shown at intervals of $10^{-2}\ s^{-1}$ beginning at $10^{-2}\ s^{-1}$. Note the region of positive (red) generation of streamwise vorticity along the forward-flank region and the region of negative (purple) generation of streamwise vorticity (or the generation of positive anti-streamwise vorticity) along the RFD region and the leading edge of the rear-flank gust front. The upward tilting of streamwise vorticity or the downward tilting of anti-streamwise vorticity could produce a mesocyclone (courtesy of Robin Tanamachi).

the FFD from ambient air and thus have acquired streamwise horizontal vorticity (Figure 4.51). Air entering the low-level mesocyclone from the southwest or west may also encounter horizontal gradients in temperature and, therefore, horizontal vorticity can also be generated there (Figure 4.51). In addition, *environmental* low-level horizontal vorticity may be advected towards and tilted upward into the low-level mesocyclone from the southeast (Figure 4.52). In summary, a number of vastly different types of trajectories along which changes in vorticity may be quite different are likely.

An alternative and perhaps better way to analyze low-level mesocyclogenesis is to use circulation analysis. Rich Rotunno and Joe Klemp in the early 1980s pioneered this approach for the analysis of low-level mesocyclones in supercells. Consider a material curve surrounding and centered about a low-level mesocyclone (Figure 4.53). This curve lies in the horizontal plane *and cuts across isotherms or, more precisely, across lines of constant equivalent potential temperature* θ_e. If potential vorticity was initially (before the storm's updraft deformed the

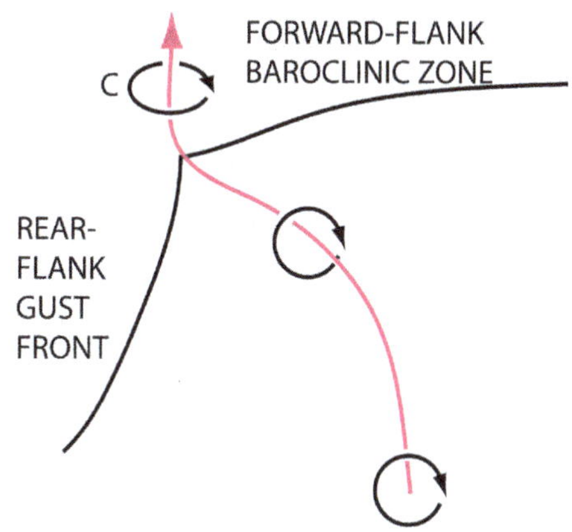

Figure 4.52. Idealized illustration showing how streamwise vorticity associated with low-level vertical shear could be advected toward an updraft and tilted to produce a mesocyclone (C) just above the ground.

horizontal planes of constant θ_e) zero, then vortex lines must always lie on surfaces of constant θ_e, as was shown earlier in this chapter. It follows that on surfaces of constant θ_e there can be *no* circulation, because there can be no component to the vorticity vector normal to the surface of constant θ_e. Therefore, there can be no circulation around any curve lying entirely on a surface of constant θ_e. Circulation analyses from numerically simulated data show how the area of the material surface had been larger and the portion on the "cold" side had been vertically oriented, while the portion on the "warm" side had been horizontally oriented (Figure 4.53). Based on (2.53), at the earlier time applied to the material curve in Figure 4.53, the circulation tendency has zero contributions on the horizontal side because $B = 0$, but on the vertical side $B\mathbf{k}\cdot\mathbf{dl} < 0$ (remember that the convention for line integration is in the counterclockwise direction) to the east, but only for a short distance, while $B\mathbf{k}\cdot\mathbf{dl} > 0$ to the west, so that $DC/Dt > 0$. One concludes, then, that sinking motion on the cold side resulted in the tilting of the surface downward onto the horizontal and that convergence underneath the updraft shrank the area of the curve. This view is consistent with the baroclinic generation of vorticity being tilted onto the vertical by a gradient in sinking motion across the baroclinic zone and then stretched underneath the updraft.

Low-precipitation supercells do not have strong surface cold pools, owing to the lack of or relatively slow rate of evaporation of raindrops, or to only a slow rate of melting of falling hailstones. It would not be expected, then, that they have strong low-level mesocyclones unless there is strong, pre-existing horizontal vorticity in the boundary layer that is advected into the convergent area underneath the updraft and then tilted. Alas, without a downdraft the boundary-layer vorticity vector would become vertically oriented just above the ground—not directly at the ground (Figure 4.29, bottom panel; Figure 4.52). A downdraft is

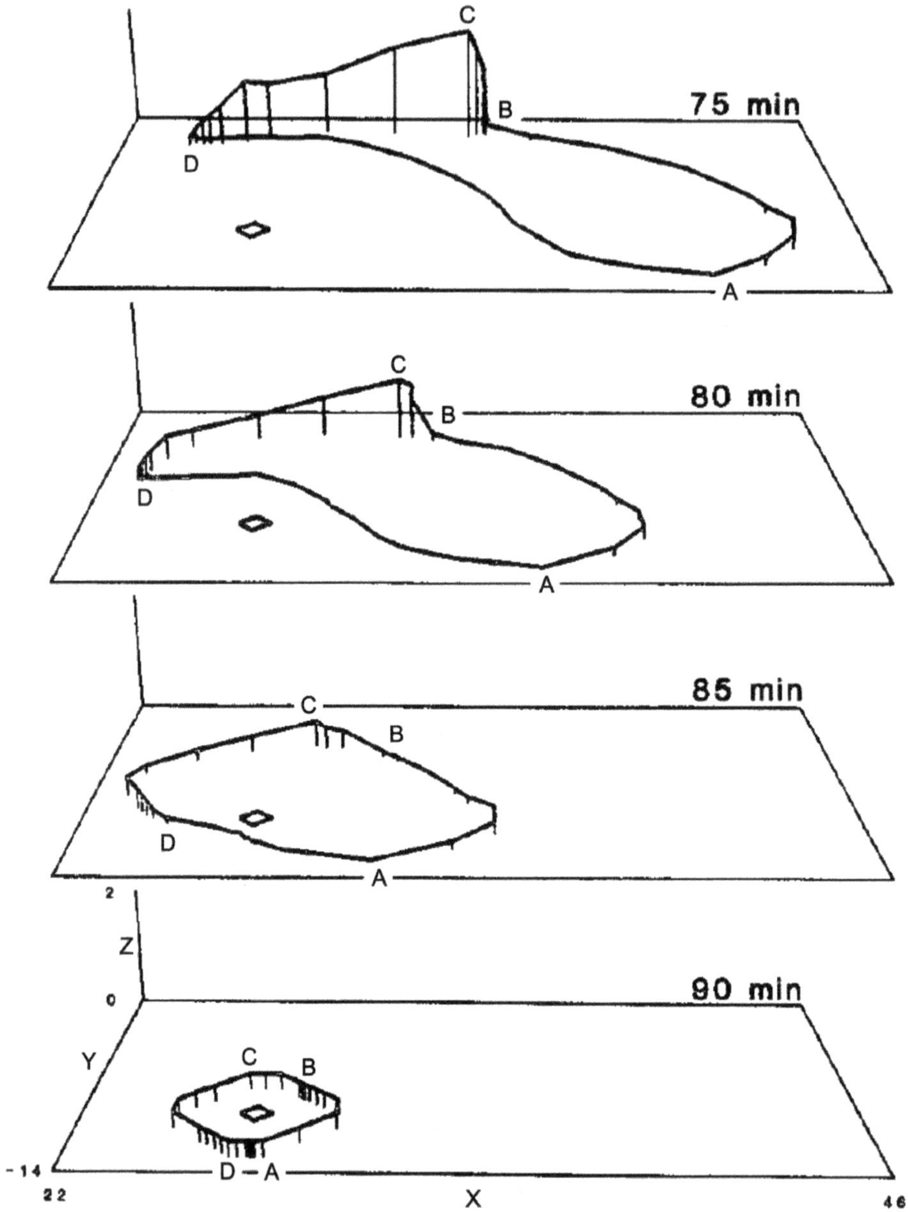

Figure 4.53. Material circuit in the horizontal plane, around a low-level mesocyclone in a numerical simulation of a supercell traced back from when it was well defined at 90 min (from storm initiation) to 15 min earlier. It is seen that some of the circuit, whose enclosed area decreased with time, came from the region outside the storm to the southeast and some came from a vertical plane to the north, within the storm. The circulation north of the storm was produced baroclinically and then tilted downward by a downdraft (from Rotunno and Klemp, 1985).

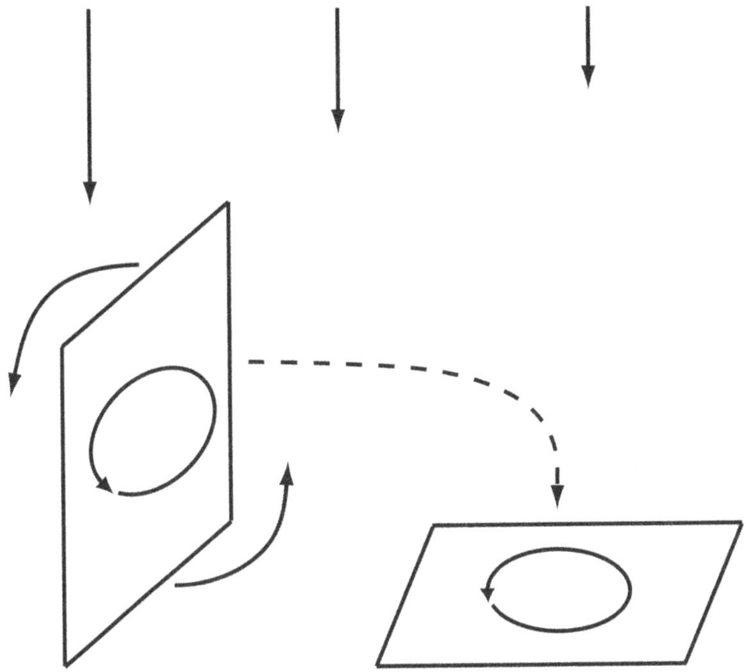

Figure 4.54. Idealized illustration of how circulation about a vertical plane can be tilted onto the horizontal and advected downward to the ground. In this case there is a downdraft everywhere, but the magnitude of the downdraft decreases (vectors at the top) as the material circuit is advected from left to right. The tilting could also occur along a gradient of w in which an updraft is adjacent to a downdraft, but *not* along a gradient in w consisting of only an updraft, because the material circuit could not be brought to the ground in this case.

required so that tilting brings the material curve *down to the ground* (Figure 4.54). It is possible that sinking motion is forced dynamically through a downward-directed perturbation pressure gradient force—not through negative buoyancy—and that this is a mechanism for producing a low-level mesocyclone. To get the downward-directed perturbation pressure gradient force, however, low-level cyclonic vorticity, which decreases with height, is a sufficient, but not necessary condition. So, it appears unlikely that LP storms can produce strong low-level mesocyclones, and that such storms must await a transition to classic supercells, with the appearance of a surface cold pool, for low-level mesocyclogenesis to occur.

Recently, Alex Schenkman at OU has suggested that horizontal vorticity may be produced along the forward flank of a supercell ahead of the FFD, as environmental air rises over the surface cold pool and, under certain conditions, can trigger trapped lee waves that, along with surface friction, are responsible for the production of horizontal rotors: a horizontal vortex sheet associated with surface friction is lifted into the lee wave. Thus, there is a source for horizontal vorticity, which is produced by the convective storm, that is at least partly frictional—not baroclinic. Such vortex lines, however, must be elevated above the cold pool.

4.7.1 The "owl horn" echo

Matthew Kramar and collaborators have found while some supercells are developing that the rear side of the radar reflectivity pattern resembles the profile of the great horned owl. This owl horn echo (Figure 4.55) is evident at low altitude and lasts only for around 5–10 min. The signature is created when horizontal vorticity is tilted onto the vertical as cold/cool outflow at the rear expands through a positive feedback mechanism with a rotational couplet produced inside the cold pool (Figure 4.56).

When the leading edge of the outflow has a head, a raised portion (cf. Figure 3.35), air that approaches the right front-flank side of the outflow head with streamwise vorticity tilts upward and cyclonic vorticity is produced; as the air sinks after passing the head anticyclonic vorticity is produced. The opposite happens on the other side of the cold pool. The cyclonic–anticyclonic couplet produced inside the cold pool acts to increase the flow of cold air farther rearward along the leading edges of the cold pool toward the storm (Figure 4.57). A curved environmental hodograph is necessary to produce the owl horn echo when there is an outflow head.

When an outflow head is not present, as when the hodograph is straight, a vortex couplet is not produced inside the cold pool. The same vortex structure is observed, however, on the outer edge of the cold pool.

4.8 THE LIFE CYCLE OF THE MESOCYCLONE AND CYCLIC MESOCYCLOGENESIS

The degree of steadiness of the main updraft in supercells is an aspect of their behavior that has been analyzed and discussed for many years. It has been found, from both numerical simulation studies and observational studies, that supercell updrafts are not as steady as had first been postulated. Steadiness is not a necessary attribute of supercells. While multicell convective storms exhibit pulses in updrafts, sometimes at quasi-regular intervals, supercells do so also. At one end of the spectrum of supercell behavior are supercells in which the updraft remains intense, but undergoes some slight variations in intensity, along with relatively long-lived mesocyclones, which eventually "occlude" (become detached from the rear-flank gust front as downdraft air wraps completely around the mesocyclone, in an attempted analogy to the life cycle of extratropical cyclones) and then decay. At the other end of the spectrum are supercells in which discrete mesocyclones, particularly at low levels, periodically form and undergo well-defined life cycles. Such behavior is called "cyclic mesocyclogenesis". This process will be discussed again in Chapter 6 regarding smaller scale cyclic tornadogenesis.

During surface mesocyclogenesis, the mesocyclone at low levels may be

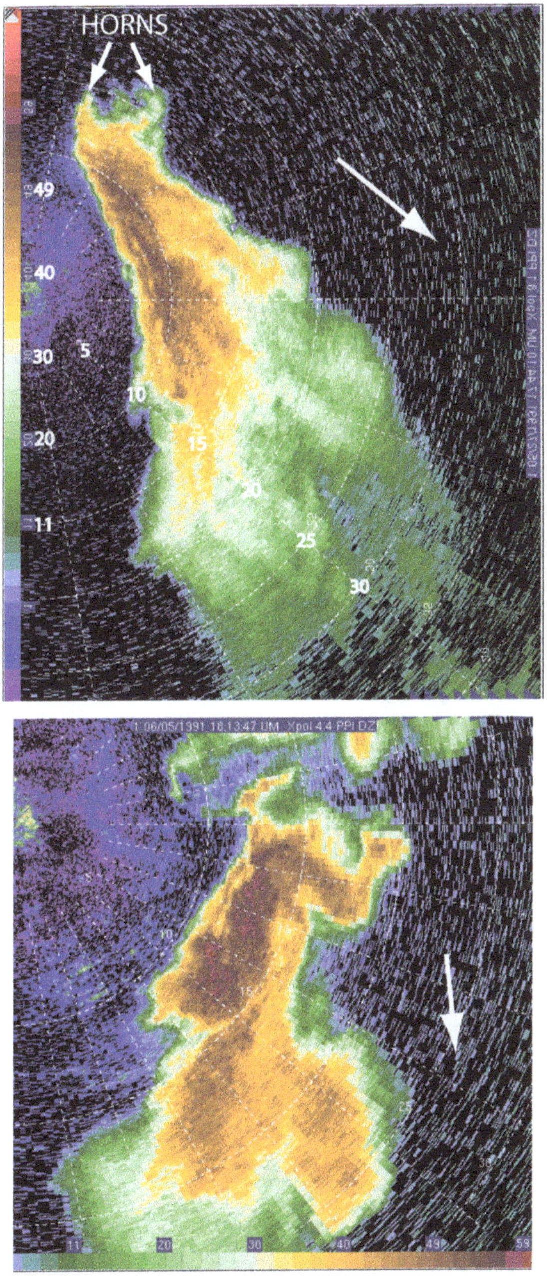

Figure 4.55. "Owl horn" echoes in supercells as detected by the U. Mass. X-Pol. mobile Doppler radar. (Top) In southwest Kansas on May 27, 2001; (bottom) in northwestern Oklahoma on June 5, 2001. Radar reflectivity shown in dBZ_e (range rings shown every 5 km). Arrow indicates storm motion. The radar images have been aligned so that the "horns" at the rear of the storm look like the horns of the great horned owl.

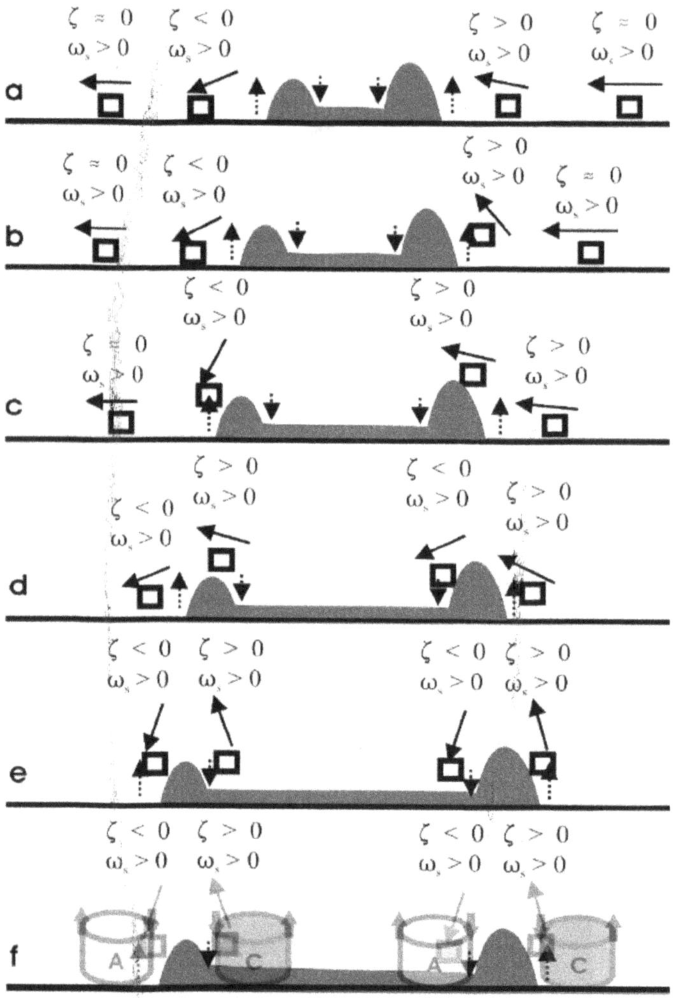

Figure 4.56. Conceptual model of couplets of vorticity produced as streamwise vorticity (ω_s) is tilted upward and downward as it flows from right to left over a head at the edge of a cold pool (vertical cross section shown) at the rear of a supercell and then exits the cold pool at the left in a similar manner. With time (a → f) the cold pool expands and the vorticity couplets become more separated from each other (from Kramar *et al.*, 2005).

stronger than or displaced from the center of the mesocyclone aloft, so that a downward-directed, dynamically driven, perturbation pressure gradient forms and forces a downdraft, the *occlusion downdraft* (Figure 4.58), which is not to be confused with the RFD (which may be driven *both* thermodynamically and dynamically), since it may be located nearby or within the broader RFD. The RFD is thought to be involved in the increase in vorticity near the surface, so that the occlusion downdraft may be a consequence of the RFD. The occlusion down-

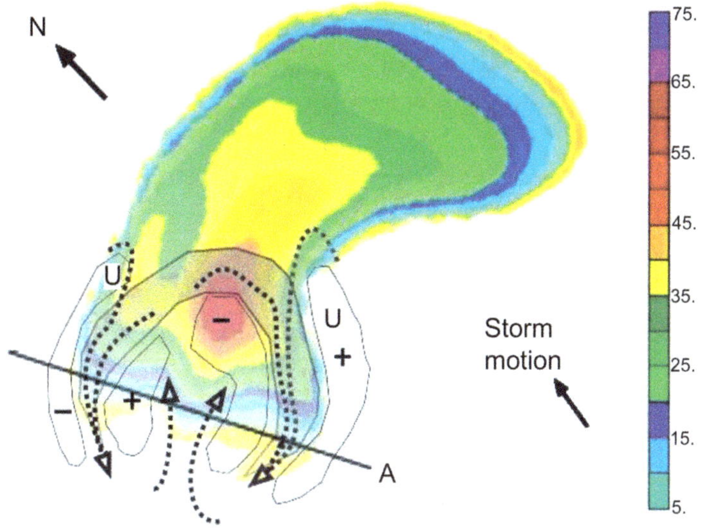

Figure 4.57. Conceptual model of the "owl horn" echo. Radar reflectivity factor color-coded in dBZ at 250 m AGL. Shaded region identifies cold air protrusions. Outer and inner unshaded bounded areas enclosed by thin black contours are regions of cyclonic (+) and anticyclonic (−) vertical vorticity. Updrafts denoted by "U". Stippled streamlines indicate the general low-level storm-relative flow. The line labeled "A" identifies the location of the vertical cross sections seen in Figure 4.56 (from Kramar *et al.*, 2005).

draft, however, is not co-located with the surface vorticity maximum where one would expect the greatest downward force to be. If the surface vorticity maximum is located within an updraft, then a downdraft may be discernible only along the edge of the updraft. Another possible explanation for the lack of exact correspondence between the location of the occlusion downdraft and the vorticity maximum at the surface is that there are comparable contributions to the net vertical force from other dynamic pressure effects or buoyancy. It is still also possible that the vorticity maximum tilts with height.

Following the occlusion, a new mesocyclone may form along the rear-flank gust front where there is a surge in the RFGF, initiating a new updraft separate from the original, previous updraft, and increasing surface vorticity through convergence at the leading edge of the RFGF; subsequently, the new mesocyclone becomes occluded, and this periodic behavior may continue. In a series of influential papers by Ed Adlerman and Kelvin Droegemeier in the early part of the second millennium, results from studies of cyclic mesocyclogenesis simulated numerically have revealed how sensitive it is to environmental conditions (e.g., the vertical shear profile). It is easily seen how the intensity of the cold pool produced by the storm, if any, plays a role in whether or not there is cyclic mesocyclogenesis. Within the category of cyclic mesocyclogenesis, there are instances in which the mesocyclone occludes (OCM, occluding cyclic mesocyclogenesis; Figure

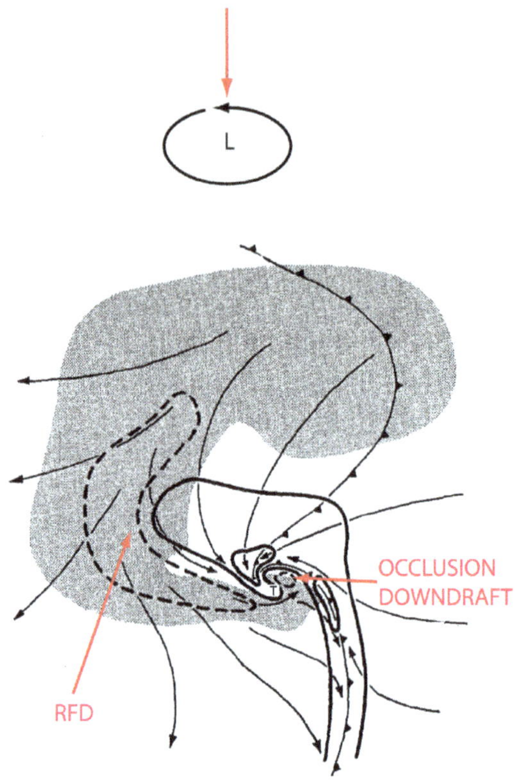

Figure 4.58. Occlusion downdrafts. (Top) Downward-directed dynamic perturbation pressure gradient force (red vector) at the center of a low-level cyclone that decreases in intensity with height; the strongest central pressure deficit is at the surface (L). (Bottom) Summary of features at low levels in a numerical simulation of a supercell. Streamlines indicate storm-relative wind field. Vertical velocity is contoured at $2\,\mathrm{m\,s^{-1}}$ intervals with the zero line omitted; dashed contours indicate downward velocities and solid contours indicate upward velocities. Cold front symbol denotes the leading edge of the rear-flank gust front below and the forward-flank cool outflow boundary above. The "T" denotes the location of maximum vorticity, perhaps where a tornado might be located. Shaded region indicates rainwater content in excess of $0.5\,\mathrm{g\,kg^{-1}}$, which represents roughly the outline of an outer radar reflectivity contour. Note how the occlusion downdraft for the mesocyclone is displaced slightly from the vorticity maximum (from Klemp and Rotunno, 1983).

4.59, top panels) and instances in which the mesocyclone does not occlude (NOCM, non-occluding cyclic mesocyclogenesis). In the latter, low-level mesocyclones do not go through an occlusion process as part of their normal life cycle; instead the mesocyclone propagates away from the updraft along the leading edge of the outflow on the forward flank (Figure 4.59, bottom panels). For vertical shear below some level, results from numerical simulation experiments indicate that OCM is favored by hodographs with some curvature.

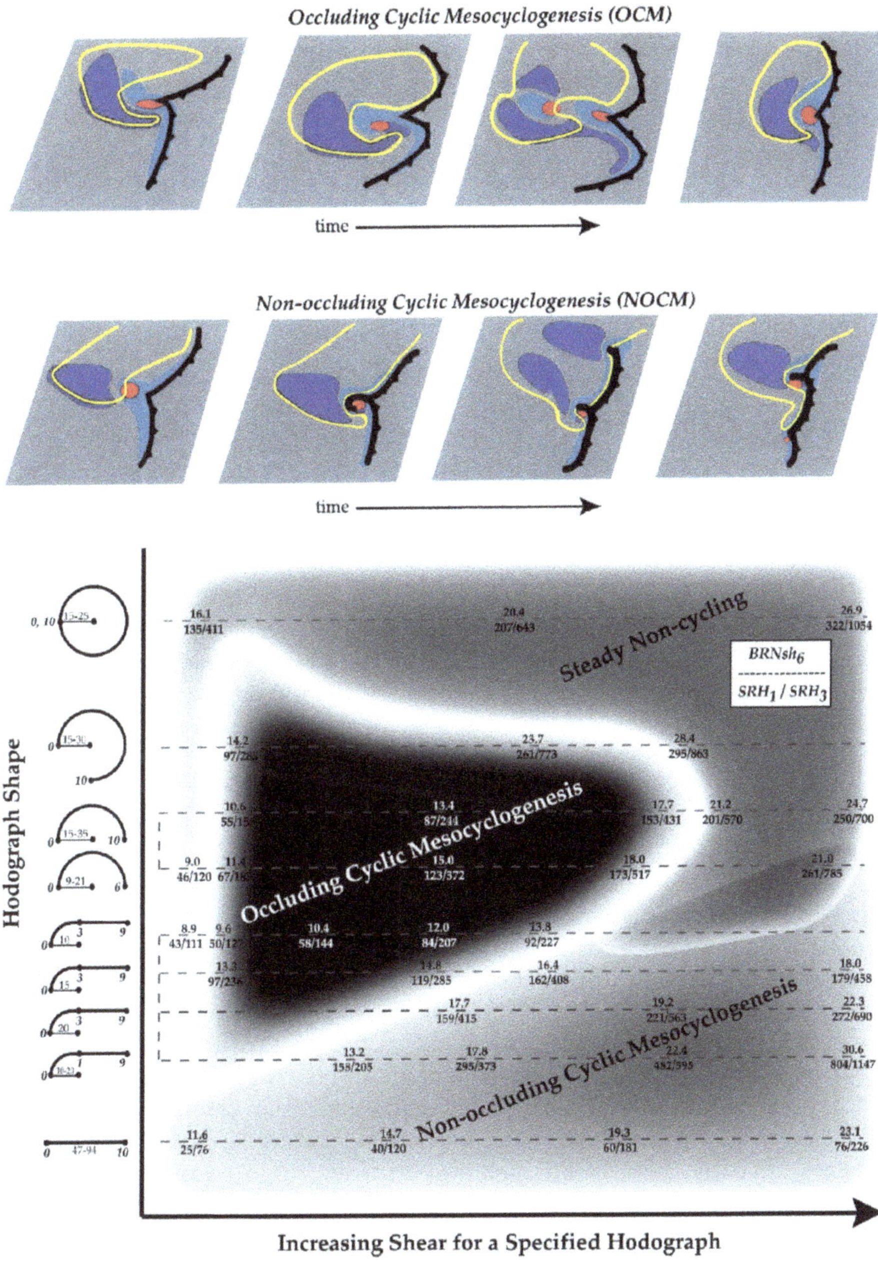

Figure 4.59. Conceptual models of cyclic mesocyclogenesis (top and middle panels) and effects of hodograph shapes and lengths (bottom panel). Cold-pool boundary indicated by cold front symbol. Red area indicates vorticity maximum. Light blue indicates updraft areas; dark blue indicates downdraft areas. Yellow contour marks rain boundary. (Top) Occluding cyclic mesocyclogenesis. (Middle) Non-occluding cyclic mesocyclogenesis (from Adlerman and Droegemeier, 2005).

4.9 SUPERCELL STRUCTURE AND BEHAVIOR IN RELATION TO INHOMOGENEITIES IN THE ENVIRONMENT, AND INTERACTIONS WITH NEIGHBORING STORMS AND SURFACE BOUNDARIES

In most numerical studies of supercell behavior, the simulated supercell is isolated and embedded in a homogeneous environment. In nature, supercells may move from one environment to another or the environment may change with time. For example, storms may move across baroclinic zones, across zones of moisture gradients, and across zones of gradients in vertical shear. Storm environmental shear may be increased substantially as night-time approaches as the low-level jet forms. Yvette Richardson and colleagues, in considering the influence of horizontal variations in vertical shear and low-level moisture on convective storm behavior, have found that they can have significant effects. Variations in low-level moisture can affect the location, timing, and intensity of new-cell development. When vertical shear is strong, a convective storm may survive even if it moves into a region too dry to support convective initiation. Forecasters must anticipate whether a supercell will decay or devolve into an ordinary-cell or multicell complex or whether an ordinary-cell or multicell complex will evolve into a supercell as the environment changes. There are many documented cases of storms changing character as they move through different environments or as the environment changes with time.

There are also some distinctively different behavioral aspects of supercells that are related to the supercell's interaction with neighboring storms and boundaries, and its movement across surface boundaries.

4.9.1 Neighboring cell interaction

Even when the environment is favorable for the formation of a supercell (i.e., when vertical shear and CAPE are matched so that the bulk Richardson number is within the range such that vertical perturbation gradients are significant and vortices form), a supercell may not necessarily evolve from convective storms that are initiated. In some instances, when convective storms are initiated along a surface boundary, neighboring cells may interact with each other so that supercells cannot evolve when cold pools block warm, moist air from entering updrafts.

When relatively widely spaced convective storms are initiated along a boundary, the orientation of the boundary with respect to mean vertical shear in the lower half of the troposphere matters (Figure 4.60). If the boundary is oriented normal to the mean shear vector, then convective storms split into right and left-moving members and adjacent left and right-moving cells collide with each other. Only the right-moving, cyclonically rotating member at the upshear end of the line and the left-moving, anticyclonically rotating member on the downshear end of the line do not interact with their neighbors and can behave like isolated supercells. If the boundary is oriented along the mean shear vector, then left-moving, anticyclonically rotating supercells move across the boundary into cooler or drier

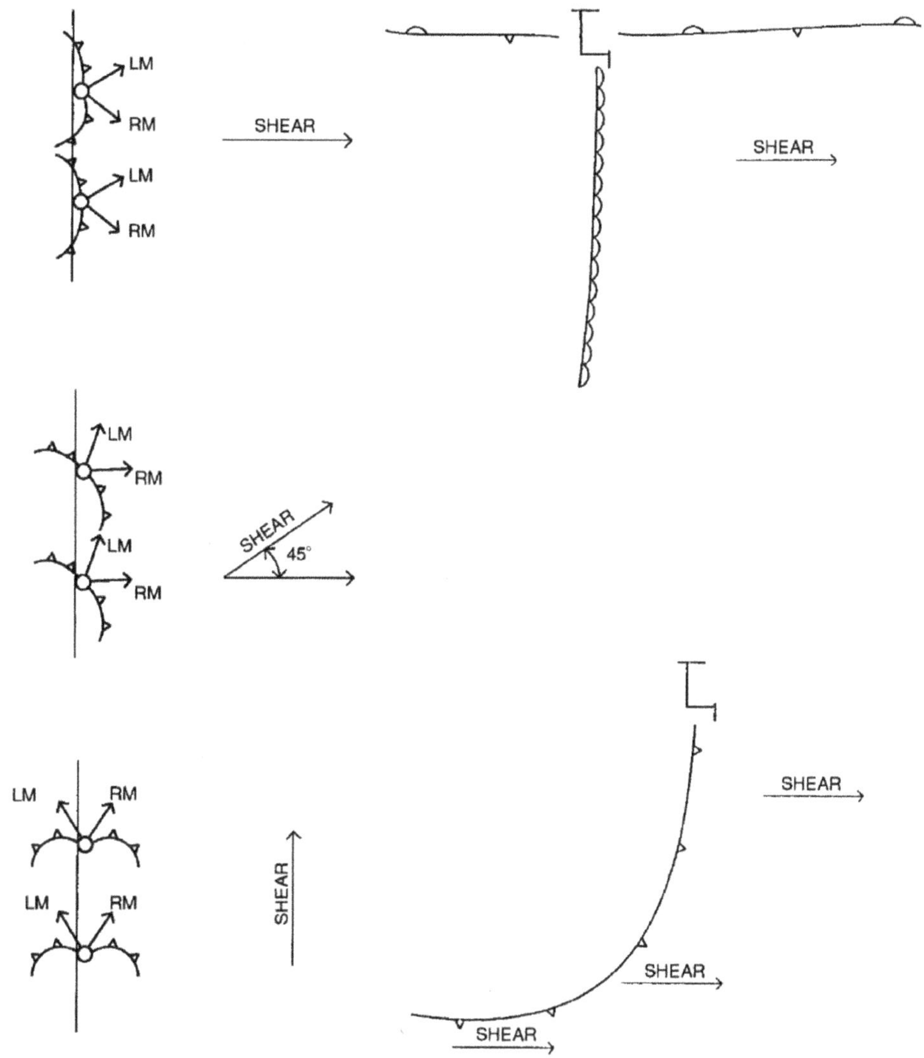

Figure 4.60. Idealized illustration showing how neighboring storms might interact according to how the mean vertical shear vector is oriented with respect to a line along which convective storms are triggered. (Two columns at the left) Initial thermal bubbles shown by open circles. Cold front symbols denote outflow boundaries. "RM" and "LM" denote storm motion (vectors) of right and left-moving cells. Vertical shear vector indicated in the middle. (Right) Typical synoptic-scale features in the Great Plains of the U. S. Note how the orientation of the shear vector may vary along the boundaries where storms may form. In the case in which the vertical shear vector is normal to the initiation line, the storms at the ends of the line in the direction normal and to the right of the shear in the figure are most likely to become an isolated supercell; in the case in which the shear is about 45° from the initiation line, all storms could develop into isolated RM supercells; in the case in which the shear vector is parallel to the initiation line, only the storm at the end of the line in the downshear direction is likely to develop into an isolated RM supercell (from Bluestein and Weisman, 2000).

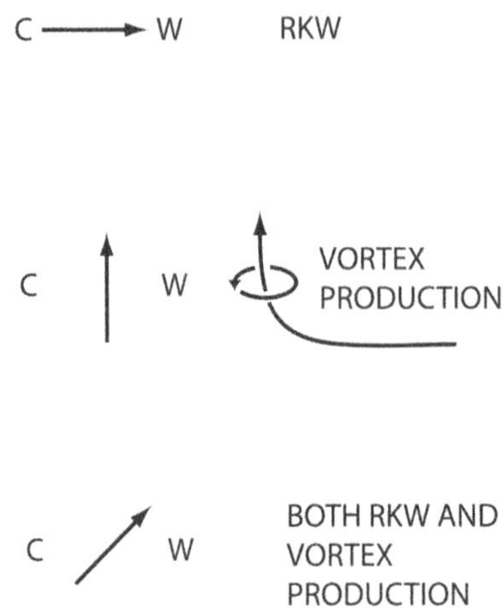

Figure 4.61. Illustration of how the orientation of low-level vertical shear vector (vectors) affects behavior along the leading edge of a cold pool (in between "C" and "W"). (Top) Shear vector points from cold to warm air: RKW effect is dominant. (Middle) Shear vector points along the outflow boundary, with colder air lying to the left: a cyclonic vortex is produced by tilting of environmental horizontal vorticity associated with the shear as air approaches the cold pool from the right and is lifted up and over it (in the reference frame of the cold pool, which moves to the right). (Bottom) Shear vector has components both from cold to warm air and along the outflow boundary, with colder air lying to the left: both RKW effects and vortex production may be significant.

air and decay. With the exception of a cell on the downshear side of the boundary, right-moving, cyclonically rotating cells interact with surface cold pools from adjacent cells. When the boundary is skewed at a 45° angle from mean vertical shear, it is possible that all neighboring right-moving, cyclonically rotating cells and only left-moving, antiyclonically rotating cells on the downshear end of the line do not interact with their neighbors and thus behave like isolated supercells. So, outbreaks of right-moving, cyclonically rotating supercells are most likely to occur when mean vertical shear is oriented at a 45° angle from the line along which they have been initiated.

The effects of the orientation of low-level shear on the behavior of the rear-flank gust front must also be considered (Figure 4.61). For example, according to RKW theory, the potential for longer lasting cells along the flanking line is increased when the shear vector is oriented normal to the gust front and pointing from the cooler side to the warmer side (Figures 4.61, top panel, and 3.40). The cooler the air behind the gust front, the stronger the shear vector must be for new convection to sustain itself. When there is little or no temperature gradient across the rear-flank gust front, vertical shear is not of any such consequence; however, shear along the rear-flank gust front in the absence of a temperature gradient would act to tilt vertical circulation over and reduce the chances for regeneration of new cells (as in the bottom left panel of Figure 3.40). On the other hand, when the low-level shear vector is oriented along the gust front such that cooler air lies to the left of the shear vector, RKW theory is not applicable, but horizontal vorticity is tilted onto the vertical as, relative to the movement of the rear-flank gust front, air parcels are tilted upward at the leading edge of the gust front (Figure 4.61, middle panel). When the low-level shear vector is oriented at an angle of 45° from the gust front, then both RKW theory and the tilting of horizontal vorticity onto the vertical must be considered (Figure 4.61, bottom panel).

4.9.2 Movement across outflow boundaries or fronts

Cells initiated along a surface boundary sometimes cross the boundary. Nolan Atkins and colleagues considered what happens when a numerically simulated supercell interacts with a pre-existing surface boundary in idealized numerical experiments. They found that low-level mesocyclones form earlier, are stronger, and are more long lived when a boundary is present. When a supercell crosses the boundary from the warmer side to the cooler side, the low-level mesocyclone weakens; when a supercell propagates along the boundary or has a component of motion that is toward the warm side, the low-level mesocyclone strengthens.

In a homogeneous environment, the source of vorticity in low-level mesocyclones in supercells is largely from baroclinally generated horizontal vorticity that is associated with the forward-flank outflow boundary and whose baroclinicity is generated by the convective storm itself. When a supercell interacts with a baroclinic boundary such as an outflow boundary or a front, the forward-flank area of the storm is less likely to affect the formation of the low-level mesocyclone, while baroclinically generated horizontal vorticity associated with the pre-existing boundary makes a significant contribution to the formation of the mesocyclone at low levels.

It is thus concluded that, just as the orientation of a boundary along which a broken line of storms is initiated is important in determining whether or not the cells can evolve into long-lived supercells, the orientation of a boundary is also important in determining whether or not a low-level mesocyclone in an isolated supercell will intensify or decay (Figure 4.62). When a supercell crosses a baroclinic boundary and the surface air becomes less potentially buoyant, the supercell and the mid-level mesocyclone may persist, but the low-level mesocyclone weakens. Thus, supercells that cross boundaries may maintain their intensity and rotational characteristics aloft, but it is unlikely that sustained rotation will be produced at low levels. On the other hand, there a number of observational cases in which a tornado forms when a supercell crosses a surface boundary from the warm side to the cooler side. In these cases, winds tend to be backed on the cooler side, so that low-level vertical shear is stronger. Evidently, the effects of increased vertical shear can sometimes be more important than the effects of reduced CAPE.

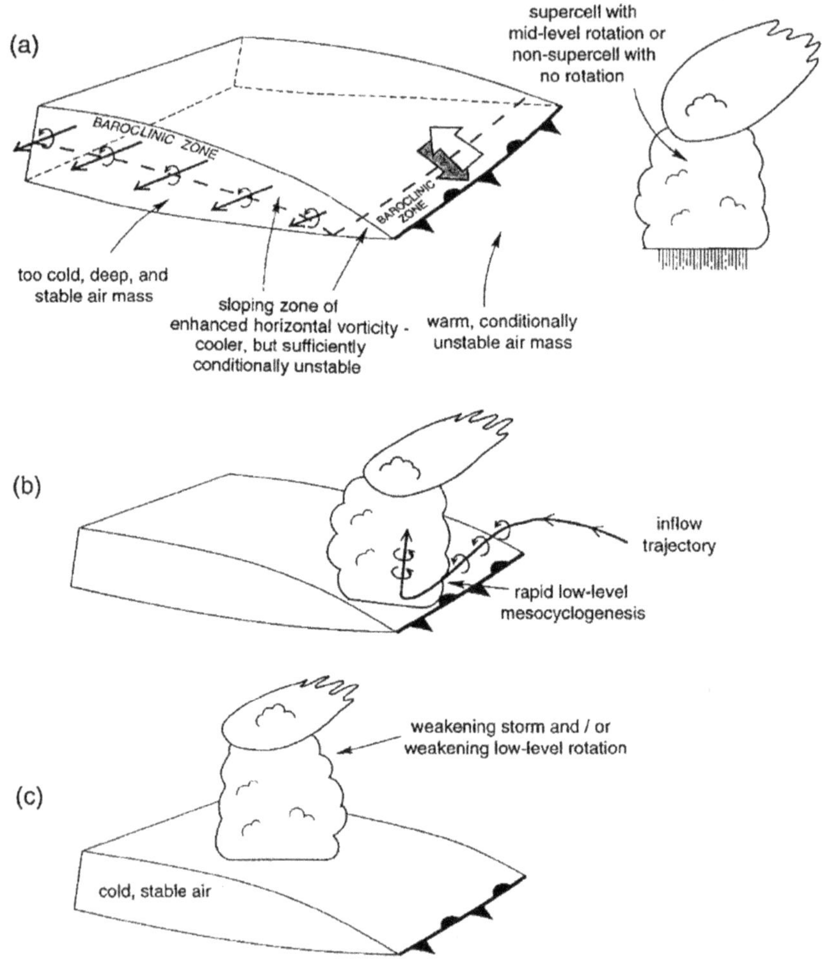

Figure 4.62. Idealized illustration of how a supercell may behave as it crosses a cold pool (from Markowski *et al.*, 1998a).

4.10 ROTATING DOWNDRAFTS IN CONVECTIVE STORMS

Strong vorticity is sometimes observed in downbursts/microbursts and the vortex signature detected by Doppler radar, especially when vorticity is cyclonic, may be mistaken in an ordinary convective storm for a mesocyclone in a supercell. Failure to recognize that the mesocyclone is really a rotating downdraft can lead to inaccurate tornado warnings.

Dave Parsons and Morris Weisman at NCAR demonstrated numerically how in an environment where vertical shear is too weak to support supercells, a strong, cyclonically rotating downdraft can be produced when the vertical shear vector at mid-levels veers with height (the hodograph turns in a clockwise direction with

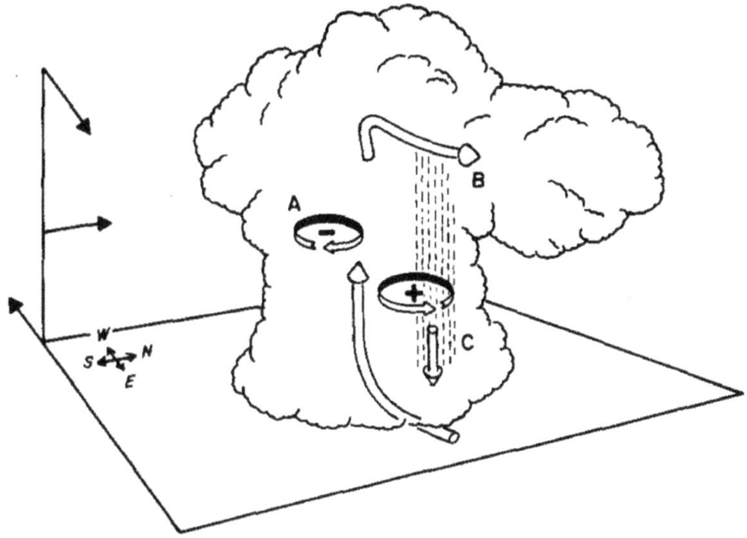

Figure 4.63. Idealized illustration of how a cyclonically rotating downdraft may form in a convective storm when the hodograph curves in a clockwise manner with height. A cyclonic (+) and anticyclonic (–) couplet forms from tilting by the updraft; precipitation falls out at high levels at B; the mid-level vorticity at B is stretched as the downdraft accelerates downward due to evaporative cooling (from Parsons and Weisman, 1993).

height; Figure 4.63). Early on in the life of a convective storm, a cyclonic–anticyclonic vortex couplet is produced at mid-levels as a result of the tilting of environmental horizontal vorticity associated with vertical shear, as in a supercell, with the mesocyclone to the right and the meso-anticyclone to the left of the vertical shear vector at mid-levels where the updraft and shear are strongest. Precipitation is produced in the updraft and then falls out in the downshear direction (with respect to the vertical shear aloft) into the cyclonic member of the vorticity couplet. As the air is dragged downward by falling precipitation and evaporative cooling makes the air negatively buoyant, it accelerates downward, creating convergence that acts to intensify the mesocyclone and advect it downward. When the shear vector backs with height, an anticyclonically rotating downdraft is produced.

It is noted that the mechanism just described does not occur in supercells: when vertical shear is relatively strong, a downdraft produces an anticyclonic–cyclonic vortex couplet at mid-levels through tilting. When the vertical shear vector veers with height, the linear dynamic perturbation pressure term is negative upshear from the downdraft and positive downshear from the downdraft, so that on the anticyclonic side there is a downward-directed perturbation pressure gradient force and on the cyclonic side there is an upward-directed perturbation pressure gradient force. Thus, anticyclonic downdrafts and cyclonic updrafts are produced, which is opposite to what is observed when the vertical shear is weak.

4.11 GENERAL MONOGRAPHS AND BOOKS

The reader is referred to p. 24 for a list of relevant general monographs and books.

4.12 REFERENCES AND BIBLIOGRAPHY

Adlerman, E. J. and K. K. Droegemeier (2005) The dependence of numerically simulated cyclic mesocyclogenesis upon environmental vertical wind shear. *Mon. Wea. Rev.*, **133**, 3595–3623.

Adlerman, E. J., K. K. Droegemeier, and R. P. Davies-Jones (1999) A numerical simulation of cyclic mesocyclogenesis. *J. Atmos. Sci.*, **56**, 2045–2069.

Atkins, N. T., M. L. Weisman, and L. J. Wicker (1999) The influence of preexisting boundaries on supercell evolution. *Mon. Wea. Rev.*, **127**, 2910–2927.

Barnes, S. L. (1968) *On the Source of Thunderstorm Rotation.* National Severe Storms Laboratory, ESSA Tech. Memo. ERLTM-NSSL 38, NSSL, Norman, OK, 28 pp.

Barnes, S. L. (1970) Some aspects of a severe, right-moving thunderstorm deduced from mesonetwork rawinsonde observations. *J. Atmos. Sci.*, **27**, 634–648.

Beck, J. R., J. L. Schroeder, and J. M. Wurman (2006) High-resolution dual-Doppler analyses of the 29 May 2001 Kress, Texas, cyclic supercell. *Mon. Wea. Rev.*, **134**, 3125–3148.

Bluestein, H. (1984) Further examples of low-precipitation severe thunderstorms. *Mon. Wea. Rev.*, **112**, 1885–1888.

Bluestein, H. B. (1986) Visual aspects of the flanking line in severe thunderstorms. *Mon. Wea. Rev.*, **114**, 788–795.

Bluestein, H. B. (1992) *Synoptic-Dynamic Meteorology in Midlatitudes, Vol. 1: Principles of Kinematics and Dynamics.* Oxford University Press, New York, pp. 213, 226.

Bluestein, H. B. (2007) Advances in applications of the physics of fluids to severe weather systems. *Reports on Progress in Physics*, **70**, 1259–1323.

Bluestein, H. B. (2009) The formation and early evolution of the Greensburg, Kansas, tornadic supercell on 4 May 2007. *Wea. Forecasting*, **24**, 899–920.

Bluestein, H. B. and S. G. Gaddy (2001) Airborne pseudo-dual-Doppler analysis of a rear-inflow jet and deep convergence zone within a supercell. *Mon. Wea. Rev.*, **129**, 2270–2289.

Bluestein, H. and C. Parks (1983) A synoptic and photographic climatology of low-precipitation severe thunderstorms in the Southern Plains. *Mon. Wea. Rev.*, **111**, 2034–2046.

Bluestein, H. B. and A. L. Pazmany (2000) Observations of tornadoes and other convective phenomena with a mobile, 3-mm wavelength, Doppler radar: The spring 1999 field experiment. *Bull. Amer. Meteor. Soc.*, **81**, 2939–2951.

Bluestein, H. and C. Sohl (1979) Some observations of a splitting severe thunderstorm. *Mon. Wea. Rev.*, **107**, 861–873.

Bluestein, H. B. and M. L. Weisman (2000) The interaction of numerically simulated supercells initiated along lines. *Mon. Wea. Rev.*, **128**, 3128–3149.

Bluestein, H. B. and G. R. Woodall (1990) Doppler-radar analysis of a low-precipitation (LP) severe storm. *Mon. Wea. Rev.*, **118**, 1640–1664.

Bluestein, H. B., W. P. Unruh, D. C. Dowell, T. A. Hutchinson, T. M. Crawford, and H. Stein (1997) Doppler radar analysis of the Northfield, Texas, tornado of 25 May 1994. *Mon. Wea. Rev.*, **125**, 212–230.

Bluestein, H. B., S. G. Gaddy, D. C. Dowell, A. L. Pazmany, J. C. Galloway, R. E. McIntosh, and H. Stein (1997) Doppler radar observations of substorm-scale vortices in a supercell. *Mon. Wea. Rev.*, **125**, 1046–1059.

Brandes, E. A. (1977a) Flow in severe thunderstorms observed by dual-Doppler radar. *Mon. Wea. Rev.*, **105**, 113–120.

Brandes, E. A. (1977b) Gust front evolution and tornado genesis as viewed by Doppler radar. *J. Appl. Meteor.*, **16**, 333–338.

Brandes, E. A. (1978) Mesocylone evolution and tornadogenesis: Some obervations. *Mon. Wea. Rev.*, **106**, 995–1011.

Brandes, E. A. (1981) Fine structure of the Del City–Edmond tornadic mesovortices. *Mon. Wea. Rev.*, **109**, 635–647.

Brandes, E. A. (1984a) Relationships between radar-derived thermodynamic variables and tornadogenesis. *Mon. Wea. Rev.*, **112**, 1033–1052.

Brandes, E. A. (1984b) Vertical vorticity generation and mesocyclone sustenance in a tornadic thunderstorms: The observational evidence. *Mon. Wea. Rev.*, **112**, 2253–2269.

Brandes, E. A. (1993) Tornadic thunderstorm characteristics determined with Doppler radar. In: C. Church, D. Burgess, C. Doswell, and R. Davies-Jones (Eds.), *The Tornado: Its Structure, Dynamics, Prediction, and Hazards*, AGU Monogr. 79, American Geophysical Union, Washington, D.C., pp. 143–159.

Brandes, E. A., R. P. Davies-Jones, and B. C. Johnson (1988) Streamwise vorticity effects on supercell morphology and persistence. *J. Atmos. Sci.*, **45**, 947–963.

Braun, S. A. and J. P. Monteverdi (1991) An analysis of a mesocylone-induced tornado occurrence in northern California. *Wea. Forecasting*, **6**, 13–31.

Brooks, H. E. and R. B. Wilhelmson (1993) Hodograph curvature and updraft intensity in numerically modeled supercells. *J. Atmos. Sci.*, **50**, 1824–1833.

Brooks, H. E., C. A. Doswell, III, and R. Davies-Jones (1993) Environmental helicity and the maintenance and evolution of low-level mesocyclones. In: C. Church, D. Burgess, C. Doswell, and R. Davies-Jones (Eds.), *The Tornado: Its Structure, Dynamics, Prediction, and Hazards*, AGU Monogr. 79, American Geophysical Union, Washington, D. C., pp. 97–104.

Brooks, H. E., C. A. Doswell, III, and R. B. Wilhelmson (1994) The role of midtropospheric winds in the evolution and maintenance of low-level mesocyclones. *Mon. Wea. Rev.*, **122**, 126–136.

Brown, R. A., D. W. Burgess, J. K. Carter, L. R. Lemon, and D. Sirmans (1975) NSSL dual-Doppler radar measurements in tornadic storms: A preview. *Bull. Amer. Meteor. Soc.*, **56**, 524–526.

Browning, K. A. (1962) Cellular structure of convective storms. *Meteor. Mag.*, **91**, 341–350.

Browning, K. A. (1963) The basis of a general model of the airflow and precipitation trajectories within persistent convective storms. Conference Review, *Third Conference on Severe Local Storms*, American Meteorological Society, Urbana, IL (November 12–14, 1963).

Browning, K. A. (1964) Airflow and precipitation trajectories within severe local storms which travel to the right of the winds. *J. Atmos. Sci.*, **21**, 634–639.

Browning, K. A. (1965a) The evolution of tornadic storms. *J. Atmos. Sci.*, **22**, 664–668.

Browning, K. A. (1965b) Some inferences about the updraft within a severe local storm. *J. Atmos. Sci.*, **22**, 669–677.

Browning, K. A. (1977) The structure and mechanisms of hailstorms. In: G. B. Foote and C. A. Knight (Eds.), *Hail: A Review of Hail Science and Hail Suppression*, Meteor. Monogr. Vol. 16, Issue 38, American Meteorological Society, Boston, pp. 1–43.

Browning, K. A. (1986) Morphology and classification of middle-latitude thunderstorms. Chapter 7 in: E. Kessler (Ed.), *Thunderstorm Morphology and Dynamics*, University of Oklahoma Press, Norman, OK, pp. 133–152.

Browning, K. A. and R. J. Donaldson, Jr. (1963) Airflow and structure of a tornadic storm. *J. Atmos. Sci.*, **20**, 533–545.

Browning, K. A. and C. R. Landry (1963) Airflow within a tornadic storm. Preprint, *Tenth Weather Radar Conference, Washington, D. C.*, American Meteorological Society, Boston, pp. 1116–1122.

Browning, K. A. and F. H. Ludlam (1962) Airflow in convective storms. *Quart. J. Roy. Meteor. Soc.*, **88**, 117–135.

Bunkers, M. J. and J. W. Stoppkotte (2007) Documentation of a rare tornadic left-moving supercell. *Electronic J. Severe Storms Meteor.*, **2**, 1–22.

Bunkers, M. J., B. A. Klimowski, J. W. Zeitler, R. L. Thompson, and M. L. Weisman (2000) Predicting supercell motion using a new hodograph technique. *Wea. Forecasting*, **15**, 61–79.

Burgess, D. W. and R. P. Davies-Jones (1979) Unusual tornadic storms in eastern Oklahoma on 5 December 1975. *Mon. Wea. Rev.*, **107**, 451–457.

Burgess, D. W. and L. R. Lemon (1990) Severe thunderstorm detection by radar. Chapter 30a in: D. Atlas (Ed.), *Radar in Meteorology: Battan Memorial and 40th Anniversary Radar Conference*, American Meteorological Society, Boston, pp. 619–656.

Burgess, D. W., M. A. Magsig, J. Wurman, and D. C. Dowell (2002) Radar observations of the 3 May 1999 Oklahoma City tornado. *Wea. Forecasting*, **17**, 456–471.

Byko, Z., P. Markowski, and Y. Richardson (2009) Descending reflectivity cores in supercell thunderstorms observed by mobile radars and in a high-resolution numerical simulation. *Wea. Forecasting*, **24**, 155–186.

Charba, J. and Y. Sasaki (1971) Structure and movement of the severe thunderstorm of 3 April 1964 as revealed from radar and surface mesonetwork data analysis. *J. Meteor. Society of Japan*, **49**, 191–214.

Davies, J. M. (2006) Tornadoes with cold core 500-mb lows. *Wea. Forecasting*, **21**, 1051–1062.

Davies-Jones, R. P. (1984) Streamwise vorticity: The origin of updraft rotation in supercell storms. *J. Atmos. Sci.*, **41**, 2991–3006.

Davies-Jones, R. (2000) A Lagrangian model for baroclinic genesis of mesoscale vortices, Part I: Theory. *J. Atmos. Sci.*, **57**, 715–736.

Davies-Jones, R. (2002) Linear and nonlinear propagation of supercell storms. *J. Atmos. Sci.*, **59**, 3178–3205.

Davies-Jones, R. P. (2004) Growth of circulation around supercell updrafts. *J. Atmos. Sci.*, **61**, 2863–2876.

Davies-Jones, R. and H. Brooks (1993) Mesocyclogenesis from a theoretical perspective. In: C. Church, D. Burgess, C. Doswell, and R. Davies-Jones (Eds.), *The Tornado: Its Structure, Dynamics, Prediction, and Hazards*, AGU Monogr. 79, American Geophysical Union, Washington, D.C., pp. 105–114.

Davies-Jones, R., R. J. Trapp, and H. B. Bluestein (2001) Tornadoes and tornadic storms. In: C. Doswell, III (Ed.), *Severe Convective Storms*, AMS Monogr. Vol. 28, Issue 50, American Meteorological Society, Boston, pp. 167–221.

Donaldson, R. J., Jr. (1970) Vortex signature recognition by a Doppler radar. *J. Appl. Meteor.*, **9**, 661–670.

Donaldson, R. J. (1990) Foundations of severe storm detection by radar. Chapter 15 in: D. Atlas (Ed.), *Radar in Meteorology: Battan Memorial and 40th Anniversary Radar Conference*, American Meteorological Society, Boston, pp. 115–121.

Doswell, C. A. (1980) Synoptic-scale environments associated with High Plains severe thunderstorms. *Bull. Amer. Meteor. Soc.*, **61**, 1388–1400.

Doswell, C. A., III, H. E. Brooks, and M. P. Kay (2005) Climatological estimates of daily local nontornadic severe thunderstorm probability for the United States. *Wea. Forecasting*, **20**, 577–595.

Dowell, D. C. and H. B. Bluestein (1997) The Arcadia, Oklahoma, storm of 17 May 1981: Analysis of a supercell during tornadogenesis. *Mon. Wea. Rev.*, **125**, 2562–2582.

Dowell, D. C. and H. B. Bluestein (2002a) The 8 June 1995 McLean, Texas, storm, Part I: Observations of cyclic tornadogenesis. *Mon. Wea. Rev.*, **130**, 2626–2648.

Dowell, D. C., and H. B. Bluestein (2002b) The 8 June 1995 McLean, Texas, storm, Part II: Cyclic tornado formation, maintenance, and dissipation. *Mon. Wea. Rev.*, **130**, 2649–2670.

Dowell, D. C., H. B. Bluestein, and D. P. Jorgensen (1997) Airborne Doppler radar analysis of supercells during COPS-91. *Mon. Wea. Rev.*, **125**, 365–383.

Doyle, J. D. and D. R. Durran (2002) The dynamics of mountain-wave-induced rotors. *J. Atmos. Sci.*, **59**, 186–201.

Eastin, M. D. and M. C. Link (2009) Miniature supercells in an offshore outer rainband of Hurricane Ivan (2004). *Mon. Wea. Rev.*, **137**, 2081–2104.

Emanuel, K. A. (1994) *Atmospheric Convection*, Oxford University Press, New York, Chapters 1–6, 9–11.

Fankhauser, J. C. (1971) Thunderstorm–environment interactions determined from aircraft and radar observations. *Mon. Wea. Rev.*, **99**, 171–192.

Forbes, G. S. and H. B. Bluestein (2001) Tornadoes, tornadic thunderstorms, and photogrammetry: A review of the contributions by T. T. Fujita. *Bull. Amer. Soc.*, **82**, 73–96.

French, M. M., H. B. Bluestein, D. C. Dowell, L. J. Wicker, M. R. Kramar, and A. L. Pazmany (2008) High-resolution, mobile Doppler radar observations of cyclic mesocyclogenesis in a supercell. *Mon. Wea. Rev.*, **136**, 4997–5016.

Fujita, T. (1960) *A Detailed Analysis of the Fargo Tornadoes of June 20, 1957*, Research Paper to the U. S. Weather Bureau, No. 42, University of Chicago, Chicago, 67 pp.

Fujita, T. and H. Grandoso (1968) Split of a thunderstorm into anticyclonic and cyclonic storms and their motion as determined from numerical model experiments. *J. Atmos. Sci.*, **25**, 416–439.

Hammond, G. R. (1967) *Study of a Left-moving Thunderstorm of 23 April 1964*, ESSA Tech. Memo. IERTM-NSSL 31, NSSL, Norman, OK, 75 pp.

Heymsfield, G. M. (1978) Kinematic and dynamic aspects of the Harrah Tornadic storm analyzed from dual-Doppler radar data. *Mon. Wea. Rev.*, **106**, 233–254.

Hitschfeld, W. (1960) The motion and erosion of convective storms in severe vertical wind shear. *J. Meteor.*, **17**, 270–282.

Houston, A. L. and R. B. Wilhelmson (2007) Observational analysis of the 27 May 1997 central Texas tornadic event, Part I: Prestorm environment and storm maintenance/propagation. *Mon. Wea. Rev.*, **135**, 701–726.

Houston, A. L. and R. B. Wilhelmson (2011) The dependence of storm longevity on the pattern of deep convection initiation in a low-shear environment. *Mon. Wea. Rev.*, **139**, 3125–3138.

Johns, R. H. (1982) A synoptic climatology of northwest flow severe weather outbreaks, Part I: Nature and significance. *Mon. Wea. Rev.*, **110**, 1653–1663.

Johns, R. H. (1984) A synoptic climatology of northwest flow severe weather outbreaks, Part II: Meteorological parameters and synoptic patterns. *Mon. Wea. Rev.*, **112**, 449–464.

Kennedy, P. C., N. E. Wescott, and R. W. Scott (1993) Single-Doppler radar observations of a mini-supercell tornadic thunderstorm. *Mon. Wea. Rev.*, **121**, 1860–1870.

Klemp, J. B. (1987) Dynamics of tornadic thunderstorms. *Ann. Rev. Fluid Mech.*, **19**, 369–402.

Klemp, J. B. and R. Rotunno (1983) A study of the tornadic region within a supercell thunderstorm. *J. Atmos. Sci.*, **40**, 359–377.

Klemp, J. B., R. B. Wilhelmson, and P. S. Ray (1981) Observed and numerically simulated structure of a mature supercell thunderstorm. *J. Atmos. Sci.*, **38**, 1558–1580.

Kramar, M. R., H. B. Bluestein, A. L. Pazmany, and J. D. Tuttle (2005) The "Owl Horn" radar signature in developing Southern Plains supercells. *Mon. Wea. Rev.*, **133**, 2608–2634.

Kumjian, M. R. (2011) Precipitation properties of supercell hook echoes. *Electronic J. Severe Storms Meteor.*, **6**, 1–21.

Kumjian, M. R. and A. V. Ryzhkov (2008) Polarimetric signatures in supercell thunderstorms. *J. Appl. Meteor. Climatol.*, **47**, 1940–1961.

Kumjian, M. R. and A. V. Ryzhkov (2009) Storm-relative helicity revealed from polarimetric radar measurements. *J. Atmos. Sci.*, **66**, 667–685.

Lehmiller, G. S., H. B. Bluestein, P. J. Neiman, F. M. Ralph, and W. F. Feltz (2001) Wind structure in a supercell thunderstorm as measured by a UHF wind profiler. *Mon. Wea. Rev.*, **129**, 1968–1986.

Lemon, L. R. (1976) The flanking line, a severe thunderstorm intensification source. *J. Atmos. Sci.*, **33**, 686–694.

Lemon, L. R. and C. A. Doswell, III (1979) Thunderstorm evolution and mesocyclone structure as related to tornadogenesis. *Mon. Wea. Rev.*, **107**, 1184–1197.

Lilly, D. K. (1986) The structure, energetics and propagation of rotating convective storms, Part II: Helicity and storm stabilization. *J. Atmos. Sci.*, **43**, 126–140.

Markowski, P. M. (2002) Hook echoes and rear-flank downdrafts: A review. *Mon. Wea. Rev.*, **130**, 852–876.

Markowski, P. M., E. N. Rasmussen, and J. M. Straka (1998a) The occurrence of tornadoes in supercells interacting with boundaries during VORTEX-95. *Wea. Forecasting*, **13**, 852–859.

Markowski, P. M., E. N. Rasmussen, J. M. Straka, and D. C. Dowell (1998b) Observations of low-level baroclinity generated by anvil shadows. *Mon. Wea. Rev.*, **126**, 2942–2958.

Markowski, P. M., J. M. Straka, E. N. Rasmussen, and D. O. Blanchard (1998c) Variability of storm-relative helicity during VORTEX. *Mon. Wea. Rev.*, **126**, 2959–2971.

Markowski, P. M., E. Rasmussen, J. Straka, R. Davies-Jones, Y. Richardson, and R. J. Trapp (2008) Vortex lines within low-level mesocyclones obtained from pseudo-dual Doppler radar observations. *Mon. Wea. Rev.*, **136**, 3513–3535.

Marquis, J., Y. Richardson, J. Wurman, and P. Markowski (2008) Single- and dual-Doppler analysis of a tornadic vortex and surrounding storm-scale flow in the Crowell, Texas, supercell of 30 April 2000. *Mon. Wea. Rev.*, **136**, 5017–5043.

McCaul, E. W., Jr. (1991) Buoyancy and shear characteristics of hurricane–tornado environments. *Mon. Wea. Rev.*, **119**, 1954–1978.

McCaul, E. W., Jr. and M. L. Weisman (1996) Simulation of shallow supercell storms in landfalling hurricane environments. *Mon. Wea. Rev.*, **124**, 408–429.

McCaul, E. W., Jr. and M. L. Weisman (2001) The sensitivity of simulated supercell structure and intensity to variation in the shapes of environmental buoyancy and shear profiles. *Mon. Wea. Rev.*, **129**, 664—687.

Miller, R. C. (1972) *Notes on the Analysis and Severe-storm Forecasting Procedures of the Air Force Global Weather Central*, Tech. Rep. 200 (Rev.), Air Weather Service, Scott AFB, IL, 190 pp.

Moller, A. R. (1978) The improved NWS storm spotters' training program at Ft. Worth, Tex. *Bull. Amer. Meteor. Soc.*, **39**, 1574–1582.

Moller, A. R., C. A. Doswell III, and R. Przybylinski (1990) High-precipitation supercells: A conceptual model and documentation. Preprints, *16th Conference on Severe Local Storms, Kananaskis Park, Alberta, Canada*, American Meteorological Society, Boston, pp. 52–57.

Moller, A. R., C. A. Doswell, III, M. P. Foster, and G. R. Woodall (1994) The operational recognition of supercell thunderstorm environments and storm structure. *Wea. Forecasting*, **9**, 327–347.

Moncrieff, M. W. and J. S. A. Green (1972) The propagation and transfer properties of steady convective overturning in shear. *Quart. J. Roy. Meteor. Soc.*, **98**, 336–352.

Neumann, C. J. (1965) Mesoanalysis of severe South Florida hailstorm. *J. Appl. Meteor.*, **4**, 161–171.

Newton, C. W. and J. C. Fankhauser (1964) On the movements of convective storms, with emphasis on size discrimination in relation to water-budget requirements. *J. Appl. Meteor.*, **3**, 651–668.

Ogura, Y. (1963) *A Review of Numerical Modeling Research on Small Scale Convection in the Atmosphere*, Meteor. Monogr. Vol. 5, Issue 27. American Meteorological Society, Boston, pp. 65–76.

Parsons, D. B. and M. L. Weisman (1993) A numerical study of a rotating downburst. *J. Atmos. Sci.*, **50**, 2369–2385.

Rasmussen, E. N. (2003) Refined supercell and tornado forecast parameters. *Wea. Foreasting*, **18**, 530–535.

Rasmussen, E. N. and D. O. Blanchard (1998) A baseline climatology of sounding-derived supercell and tornado forecast parameters. *Wea. Forecasting*, **13**, 1148–1164.

Rasmussen, E. N. and J. M. Straka (1998) Variations in supercell morphology, Part I: Observations of the role of upper-level storm-relative flow. *Mon. Wea. Rev.*, **126**, 2406–2421.

Rasmussen, E. N., S. Richardson, J. M. Straka, P. M. Markowski, and D. O. Blanchard (2000) The association of significant tornadoes with a baroclinic boundary on 2 June 1995. *Mon. Wea. Rev.*, **128**, 174–191.

Rasmussen, E. N., J. M. Straka, M. S. Gilmore, and R. Davies-Jones (2006) A preliminary survey of rear-flank descending reflectivity cores in supercells. *Wea. Forecasting*, **21**, 923–938.

Ray, P. S., R. J. Doviak, G. B. Walker, D. Sirmans, J. Carter, and B. Bumgarner (1975) Dual-Doppler observation of a tornadic storm. *J. Appl. Meteor.*, **14**, 1521–1530.

Ray, P. S., B. C. Johnson, K. W. Johnson, J. S. Bradberry, J. J. Stephens, K. K. Wagner, R. B. Wilhemson, and J. B. Klemp (1981) The morphology of several tornadic storms on 20 May 1977. *J. Atmos. Sci.*, **38**, 1643–1663.

Richardson, Y. P., K. K. Droegemeier, and R. P. Davies-Jones (2007) The influence of horizontal environmental variability on numerically simulated convective storms, Part I: Variations in vertical shear. *Mon. Wea. Rev.*, **135**, 3429–3455.

Rogers, R. R. (1990) The early years of Doppler radar in meteorology. Chapter 16 in: D. Atlas (Ed.), *Radar in Meteorology: Battan Memorial and 40th Anniversary Radar Conference*, American Meteorological Society, Boston, pp. 122–129.

Rotunno, R. (1981) On the evolution of thunderstorm rotation. *Mon. Wea. Rev.*, **109**, 577–586.

Rotunno, R. (1993) Supercell thunderstorm modeling and theory. In: C. Church, D. Burgess, C. Doswell, and R. Davies-Jones (Eds.), *The Tornado: Its Structure, Dynamics, Prediction, and Hazards*, AGU Monogr. 79, American Geophysical Union, Washington, D.C., pp. 57–73.

Rotunno, R. and J. B. Klemp (1982) The influence of the shear-induced pressure gradient on thunderstorm motion. *Mon. Wea. Rev.*, **110**, 136–151.

Rotunno, R. and J. B. Klemp (1985) On the rotation and propagation of simulated supercell thunderstorms. *J. Atmos. Sci.*, **42**, 271–292.

Schenkman, A., M. Xue, and A. Shapiro (2012) Tornadogenesis in a simulated mesovortex within a mesoscale convective system. *J. Atmos. Sci.* (in press).

Schlesinger, R. E. (1975) A three-dimensional numerical model of an isolated convective cloud: Preliminary results. *J. Atmos. Sci.*, **32**, 934–957.

Schlesinger, R. E. (1978) A three-dimensional numerical model of an isolated thunderstorm, Part I: Comparative experiments for variable ambient wind shear. *J. Atmos. Sci.*, **35**, 690–713.

Schesinger, R. E. (1980) A three-dimensional numerical model of an isolated thunderstorm, Part II: Dynamics of updraft splitting and mesovortex couplet evolution. *J. Atmos. Sci.*, **37**, 395–420.

Schlesinger, R. E. (1984) Effects of the pressure perturbation field in numerical models of unidirectionally sheared thunderstorm convection: Two versus three dimensions. *J. Atmos. Sci.*, **41**, 1571–1587.

Snyder, J. C., H. B. Bluestein, V. Venkatesh, and S. J. Frasier (2012) Observations of polarimetric signatures in supercells by an X-band mobile Doppler radar. *Mon. Wea. Rev.* (in press).

Straka, J. M., E. N. Rasmussen, R. P. Davies-Jones, and P. M. Markowski (2007) An observational and idealized numerical examination of low-level counter-rotating vortices in the rear flank of supercells. *Electronic J. Severe Storms Meteor.*, **2**, 22.

Suzuki, O., N. Hiroshi, O. Hisao, and N. Hiroshi (2000) Tornado-producing mini supercells associated with Typhoon 9019. *Mon. Wea. Rev.*, **128**, 1868–1882.

Tanamachi, R. L., L. J. Wicker, D. C. Dowell, H. B. Bluestein, D. T. Dawson II, and M. Xue (2012) EnKF assimilation of high-resolution, mobile Doppler radar data of the 4 May 2007 Greensburg, Kansas supercell into a numerical cloud model. *Mon. Wea. Rev.* (in press).

Vasiloff, S. V., E. A. Brandes, and R. P. Davies-Jones (1986) An investigation of the transition from multicell to supercell storms. *J. Clim. Appl. Meteor.*, **25**, 1022–1036.

Wakimoto, R. M. and H. Cai (2000) Analysis of a nontornadic storm during VORTEX 95. *Mon. Wea. Rev.*, **128**, 565–592.

Wakimoto, R. M., W.-C. Lee, H. B. Bluestein, C.-H. Liu, and P. H. Hildebrand (1996) ELDORA observations during VORTEX 95. *Bull. Amer. Meteor. Soc.*, **77**, 1465–1481.

Wakimoto, R. M., C. Liu, and H. Cai (1998) The Garden City, Kansas storm during VORTEX-95, Part I: Overview of the storm's life cycle and mesocyclogenesis. *Mon. Wea. Rev.*, **126**, 372–392.

Wakimoto, R. M., H. Cai, and H. V. Murphey (2004) The Superior, Nebraska supercell during BAMEX. *Bull. Amer. Meteor. Soc.*, **85**, 1095–1106.

Weaver, J. F. (1979) Storm motion as related to boundary-layer convergence. *Mon. Wea. Rev.*, **107**, 612–619.

Weisman, M. L. and J. B. Klemp (1982) The dependence of numerically simulated convective storms on vertical wind shear and buoyancy. *Mon. Wea. Rev.*, **110**, 504–520.

Weisman, M. L. and J. B. Klemp (1984) The structure and classification of numerically simulated convective storms in directionally varying wind shears. *Mon. Wea. Rev.*, **112**, 2479–2498.

Weisman, M. L. and J. B. Klemp (1986) Characteristics of isolated convective storms. Chapter 15 in: *Mesoscale Meteorology and Forecasting*, American Meteorological Society, Boston, pp. 331–358.

Weisman, M. L. and R. Rotunno (2000) The use of vertical wind shear versus helicity in interpreting supercell dynamics. *J. Atmos. Sci.*, **57**, 1452–1472.

5

Mesoscale convective systems

"The clouds their backs together laid,
The north begun to push,
The forests galloped till they fell,
The lightning skipped like mice;
The thunder crumbled like a stuff—
How good to be safe in tombs,
Where nature's temper cannot reach,
Nor vengeance ever comes!"
Emily Dickinson—*The Clouds their Backs together laid*

When convective storms are organized on a scale larger than the convective scale, their conglomeration is called a "mesoscale convective system (MCS)." An MCS is composed of a contiguous area of precipitation that is ~100 km or greater across in at least one spatial dimension. It is apparent from radar imagery that much of the area of an MCS seen in satellite imagery[1] is devoid of convective precipitation, especially when the anvil is very large. MCSs include both isolated/amorphous complexes of convective storms and squall lines (*lines* of deep convective cells), some of which are relatively long, but very narrow, perhaps as long as 100 km or more and as narrow as 10 km or less.

MCSs undergo evolution in which both their mode of organization and spatial scale change with time. The individual convective cells that make up at least part of an MCS are considered to be its building blocks. The building blocks may be ordinary cells or supercells. MCSs composed of ordinary cells are sometimes referred to as "multicell" complexes, but the latter term could also include regions

[1] The term "mesoscale convective complex (MCC)" was coined by Bob Maddox in 1980. Based solely on satellite imagery, MCCs were identified as circularly shaped (as opposed to linearly shaped) MCSs.

of precipitation too small to be considered MCSs. Supercells may be embedded within MCSs and also coexist with multicells. Portions of some MCSs during parts of their life may be composed of both a broken line or solid line (Figure 5.1) or a solid line (Figure 5.2) of convective cells and (a broader region of) stratiform precipitation. In parts of the U. S., a relatively large fraction of the annual precipitation falls in MCSs. While MCSs occur in both mid-latitudes and the tropics, the focus of this discussion will be on MCSs in mid-latitudes. The former tend to occur over the land in the presence of vertical shear, mainly westerly.

Since Doppler radar observations and three-dimensional non-hydrostatic numerical cloud models have become available, the major aspects of MCSs have been explained at least qualitatively. The following is a discussion of how MCSs form and a discussion of their two-dimensional and three-dimensional wind structure and thermodynamic structure.

5.1 FORMATION

MCSs frequently begin as squall lines, which at least in the central U. S. form in four major ways (Figure 5.3). (a) When convective cells break out along a line, but the cells are initially discrete and then eventually the spaces between adjacent cells fill in with heavy precipitation, the process is referred to as a "broken line" formation. Such a process often occurs along surface boundaries such as fronts and outflow boundaries, and sometimes along the dryline (a surface boundary separating relatively cool, moist, marine air from relatively, warm, dry, continental air; in the U. S., the marine air is from the Gulf of Mexico and the dry air is from the elevated terrain of the southwest U. S. and Mexico). Less frequently, lines may form in response to boundary-layer rolls that trigger deep convection. (b) When one convective cell forms, and subsequent cells form just upstream with respect to storm motion, the process is called "back-building"; eventually, a line forms as newer and newer cells extend the length of the line. The back-building process may be a result of new convection triggered at a cool/cold outflow boundary on the rear flank of the existing cell, or by upward-directed perturbation pressure gradients on the rear flank. (c) When a region of convective cells conglomerates into a convective line the process is called "broken areal" formation. (d) When a convective line appears within an area of stratiform precipitation, the process is called "embedded areal" formation.

The most common types of squall line formation are probably the broken line and back-building processes. In these situations, narrow lines of deep convection are produced during the early stages of squall line formation. These narrow lines usually form in air masses that are potentially unstable with respect to air based in the boundary layer. The original spacing of the first cells might be a result of the enhanced lift along the rising branch of boundary-layer rolls that intersect the boundary along which the storms are triggered (Figure 2.20), or gravity waves, or some other scale-selecting process. In some instances, a bore (see textbooks on mesoscale meteorology) might play a role in triggering convection (Figure 5.4), or even internal gravity waves, particularly out ahead of a pre-existing squall line

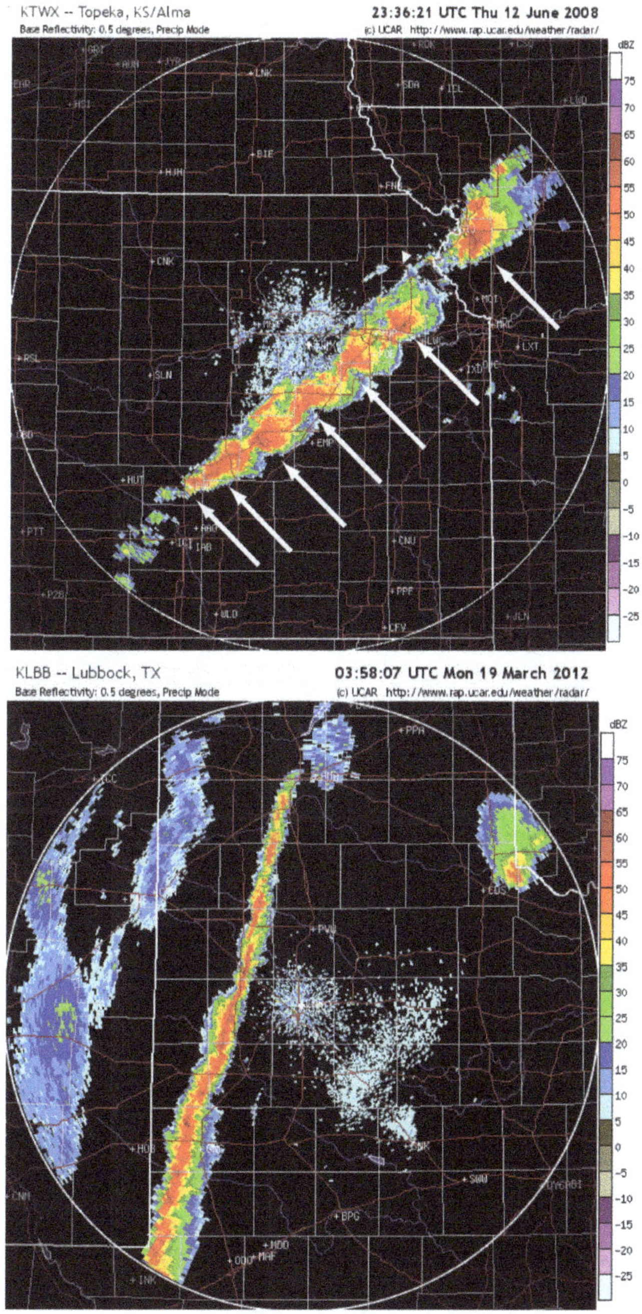

Figure 5.1. Radar reflectivity at low elevation angle coded in dB*Z* for a broken and solid line of convective cells. (Top) Broken line of individual (white arrows) convective cells (in this case, supercells) in southeast Kansas, as depicted by the WSR-88D at Topeka, KS on June 12, 2008. (Bottom) Solid line of convective cells during the early life of an MCS squall line in the Texas Panhandle on March 18, 2012, as depicted by the WSR-88D radar at Lubbock, TX.

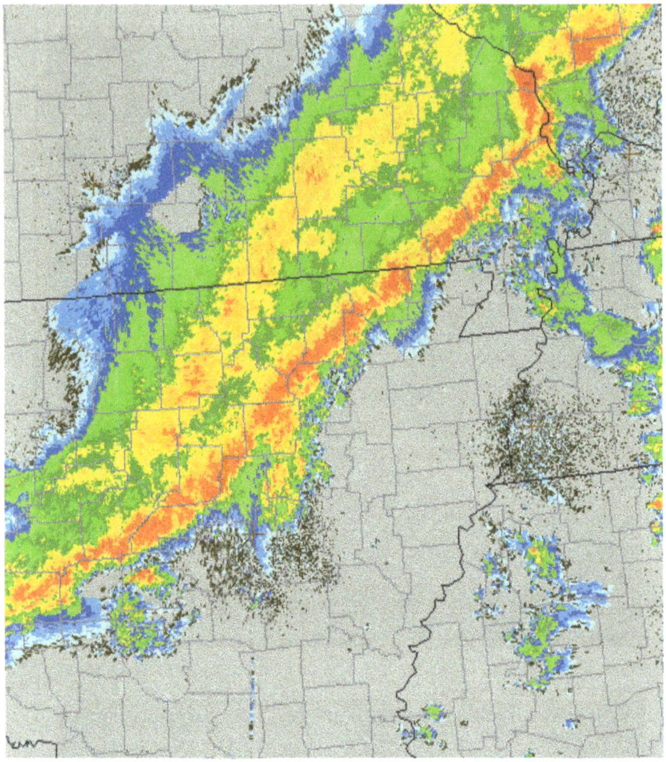

Figure 5.2. Example of a squall line with a leading convective line and a trailing stratiform precipitation area on March 31, 2008 over Arkansas, as seen by WSR-88D radars. Radar reflectivity factor is color-coded such that the warmest colors represent the most intense precipitation.

(Figure 5.5). Bores may be triggered when an outflow boundary from prior convection propagates into a stable air mass. So, when a bore or solitary waves or internal gravity waves trigger a squall line, the squall line may actually be the result of a secondary formation process (one that depends upon an *earlier* squall line).

Broken areal and embedded areal mechanisms involve pre-existing convective cells or pre-existing stratiform precipitation. In the latter case they probably involve convection that is not based at or near the surface in the boundary layer, but rather convection that is "elevated". The term "elevated" convection is used in sharp distinction from boundary-layer based convection. Elevated convection is sometimes referred to as "high-based convection", but is probably not a good name since the boundary layer may be very deep and dry. The air that flows into the cloud base during elevated convection has a recent origin above the boundary layer. Elevated convection is more difficult to study because it often occurs when a *layer* of air is lifted on the mesoscale, such that low, stratiform clouds often hide convective clouds occurring above; in the case of boundary-layer based convection, one sees cumulus or cumulus congestus as a precursor to cumulonimbus clouds. On the other hand, elevated convection is seen initially as altocumulus castellanus clouds (Figure 3.2b), which may be visible only from above if there is an intervening layer of clouds below.

Stable lift above an outflow boundary, particularly enhanced by a low-level jet (LLJ, see textbooks on mesoscale meteorology; Figure 5.6) is a mesoscale mechanism for triggering elevated convection, particularly during the evening when the LLJ is strongest and after earlier convection has produced an outflow boundary. The nature of this lift is similar to that of the lift produced above a density current.

Weaker quasi-geostrophic lift as a result of warm advection or vorticity advection becoming more cyclonic with height may also lift a layer of air covering a broad area, particularly on the cold side of a warm front or stationary front or outflow boundary, or in advance of a baroclinic wave. Moist symmetric instability (see textbooks on mesoscale meteorology) is another possible formation mechanism.

Mesoscale or synoptic-scale lift, which is not driven by buoyancy, acting on an unsaturated, conditionally unstable air mass could lead to a saturated unstable layer, sometimes over 100 hPa deep, called a "MAUL (moist absolutely unstable layer)" (Figure 5.7). MAULs may be 100 km or more wide and persist on time scales longer than that of cumulus clouds (i.e., for as long as 30 min or longer). MAULs first identified on soundings were rather curious since it is usually assumed that they should immediately trigger convection that destroys them, and it was therefore thought that they might be specious or not representative, particularly when a radiosonde leaves the top of a saturated region and evaporation cools the temperature sensor. George Bryan and Mike Fritsch in 2000 argued that many are real and that they represent "slabs of saturated, turbulent flow rather than a collection of discrete cumulonimbus clouds separated by sub-saturated areas." They postulated that the speed of dynamically driven rising air on the mesoscale (e.g., via frontal or other baroclinically driven circulations) can become greater than the speed of buoyancy-induced vertical parcel excursions, so that the rate at which the saturated absolutely unstable layer is maintained by mesoscale ascent is greater than the rate at which it is being removed by cumulus-scale turbulent mixing. They also suggested, in agreement with speculation about there being an analogy between the dry boundary layer and the moist layer in MCSs by Ed Zipser and Peggy LeMone 20 years earlier, that the depth of a MAUL might dictate the size/horizontal scale of moist eddies, just as the depth of a dry, surface mixed layer might dictate the size/horizontal scale of eddies in the dry boundary layer.

Upward forcing of potentially unstable moist air along a line or through the back-building process is not the only way to form an MCS that has a line configuration. Numerically simulated lines can also evolve from initially isolated cells that trigger secondary cells along the gust front of the original cell. As the gust front spreads out, Morris Weisman and Joe Klemp have shown new convective cells can break out along the arc of the outward-expanding cold pool (Figure 5.8). Such a process is a good example of the "upscale growth" of

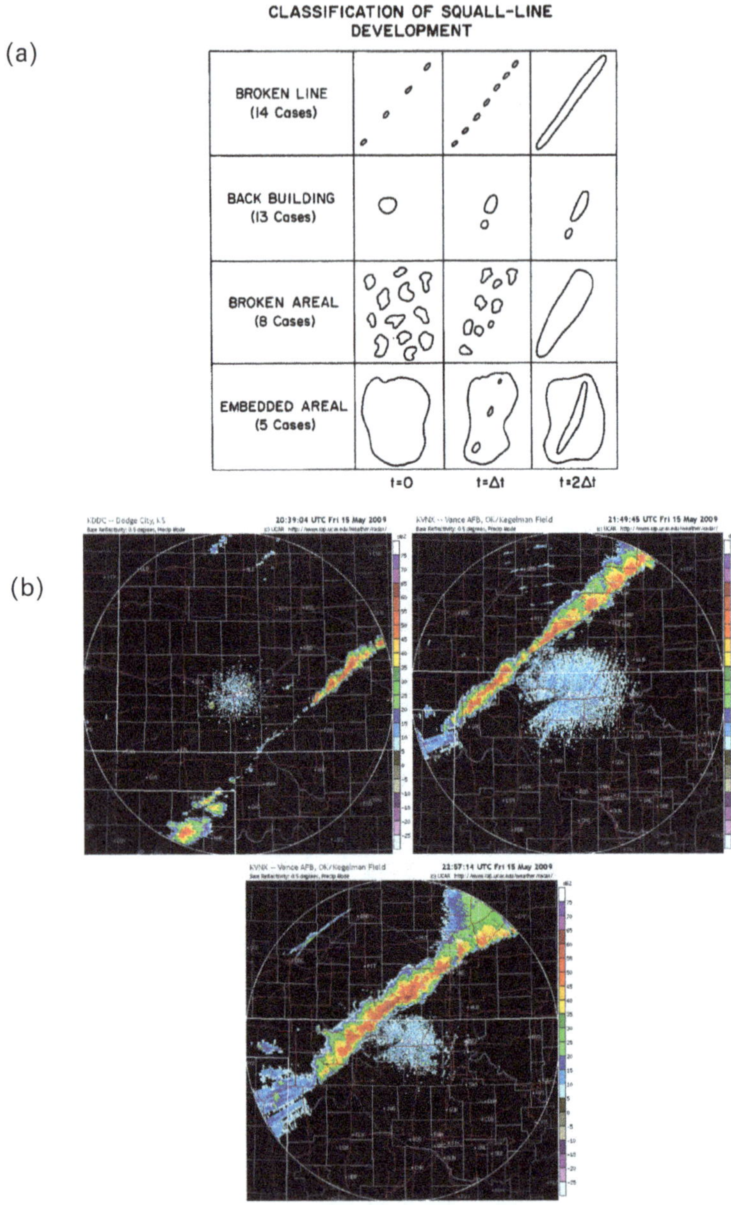

Figure 5.3. (a) Idealized depiction of four types of squall line formation (from Bluestein and Jain, 1985). (b) Example of broken line formation. Sequence of images of radar reflectivity factor in dBZ (color-coded) on May 15, 2009 in southern Kansas and northwest Oklahoma, at approximately 1 h intervals. (c) Idealized representation of a back-building squall line (from Newton and Fankhauser, 1964). (d) Example of an embedded areal squall line on March 30, 2007, as detected by the WSR-88D radar at Oklahoma City, OK. Radar reflectivity factor in dBZ (color-coded).

(c)

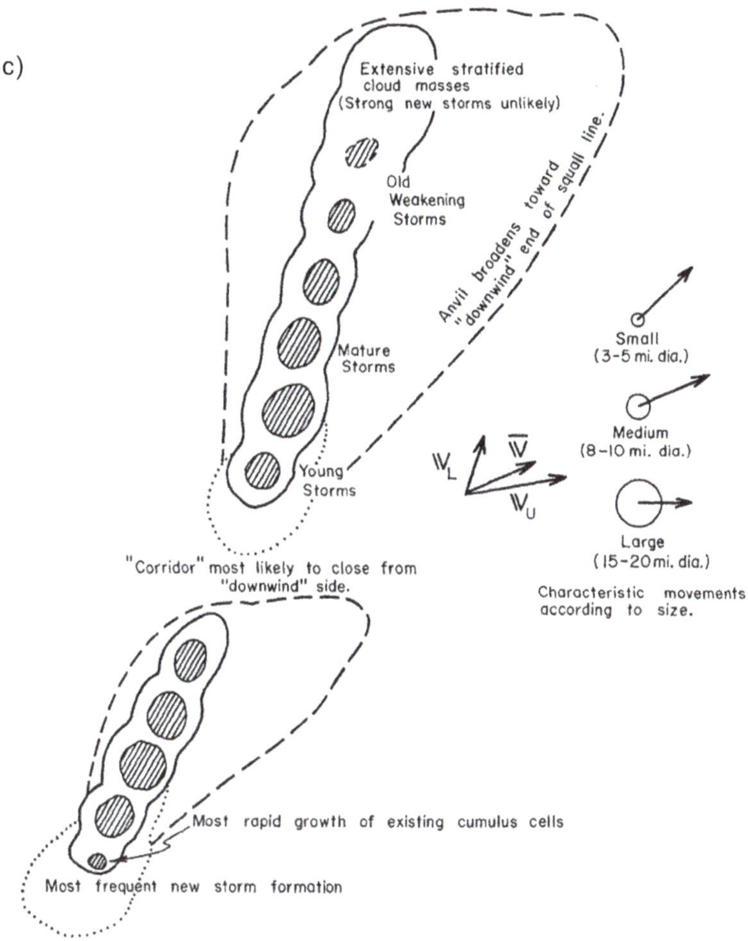
Extensive stratified cloud masses
(Strong new storms unlikely)
Old Weakening Storms
Anvil broadens toward "downwind" end of squall line.
Mature Storms
Young Storms
$\mathbb{V}_L$
$\overline{\mathbb{V}}$
$\mathbb{V}_U$
Small (3-5 mi. dia.)
Medium (8-10 mi. dia.)
Large (15-20 mi. dia.)
Characteristic movements according to size.
"Corridor" most likely to close from "downwind" side.
Most rapid growth of existing cumulus cells
Most frequent new storm formation

(d)

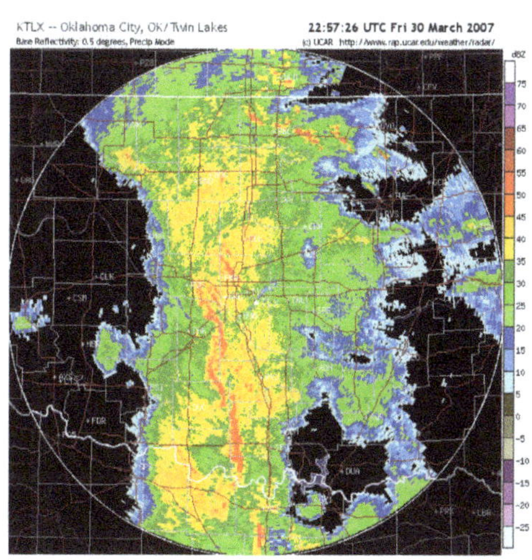
KTLX -- Oklahoma City, OK/Twin Lakes
22:57:26 UTC Fri 30 March 2007

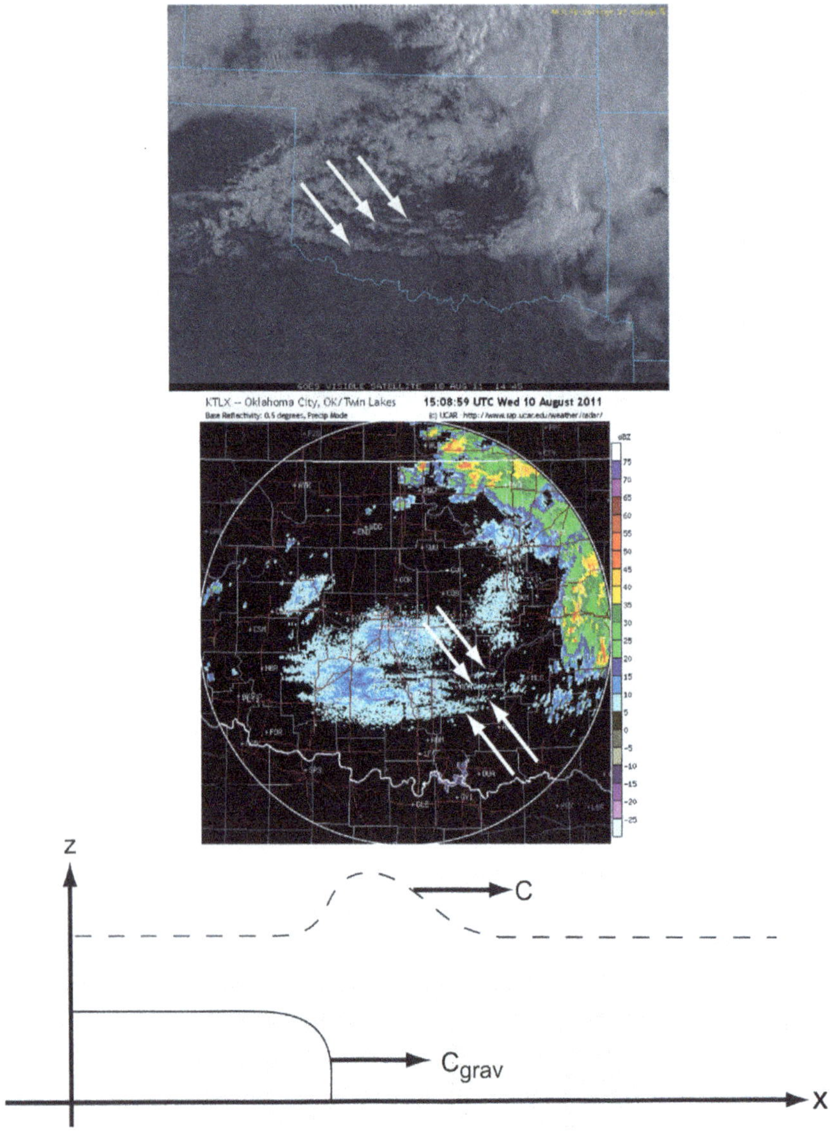

Figure 5.4. Evidence of a bore and/or solitary waves on August 10, 2011 in Oklahoma as seen from satellite (top panel, arrows point to cloud bands) and radar (middle panel, arrows point to fine lines in clear air from WSR-88D at Oklahoma City, OK). A mesoscale convective system is seen in both satellite and radar in the northeasternmost portion of the domain. (Bottom panel) Idealized schematic representation of a vertical cross section showing a density current (lower left, solid outline) impinging on a deeper, relatively dense, but not as dense as the cold pool air mass and triggering a rise in the free surface of ambient stable air (dashed line marks top of air mass of intermediate density between that in the cold pool and that aloft). The phase speed of the bore/solitary waves is c and the speed of the density current is c_{grav}. The fluid far ahead of the density current and bore/solitary waves is resting.

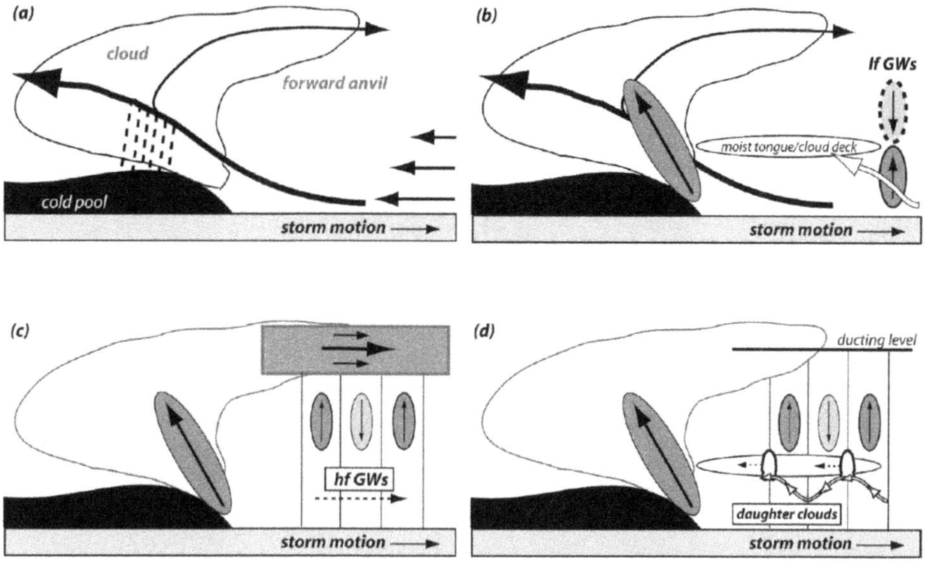

Figure 5.5. Conceptual model of sequence of events that occur when a nocturnal squall line propagates ahead of itself discretely as convection is initiated ahead of the leading convective line, as high-frequency gravity waves (hf GWs) are triggered by the squall line's periodic formation of convective cells (multicellularity) and ducted in the forward anvil, where the Scorer parameter decreases rapidly with height. A moist tongue/cloud deck forms as a result of low-frequency gravity waves (lf GWs) forced by heating and cooling in the squall line's convective region (from Fovell *et al.*, 2006).

convective systems, from an isolated cell to an MCS, without the need for any initial forcing along a line. The leading edge of an approaching MCS cold pool is seen in Figure 5.9.

5.2 MORPHOLOGY

While many MCSs begin as narrow squall lines, they eventually broaden with time. A narrow, leading convective line eventually may develop a trailing region of stratiform precipitation that is much broader than the narrow width of the leading convective line (Figure 5.2). While the rainfall rate in the stratiform precipitation region is less than that in the more intense leading convective line, an observer in the former would experience precipitation for a much longer period of time than an observer in the latter, and so the total (integrated with respect to time) rainfall experienced may be largely due to that from the stratiform region.

In between the intense leading convective line and the stratiform precipitation region there is a region of weaker precipitation rate (a narrow zone of weaker radar reflectivity at low levels) that Brad Smull and Bob Houze named the

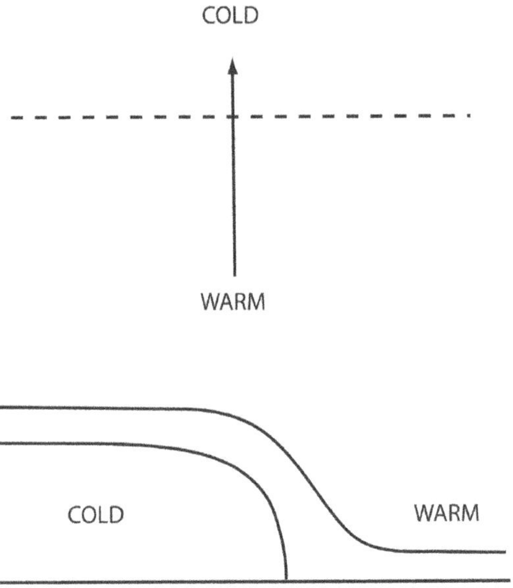

Figure 5.6. Illustration of how a low-level jet (LLJ) oriented normal to a surface boundary such as an outflow boundary or front can lead to locally enhanced lift for triggering convection. (Top) Horizontal view of the LLJ (vector) advecting warm, moist air across a surface boundary towards the cold side. A narrow zone of warm advection results in quasi-geostrophic lift that is concentrated along a portion of the boundary. (Bottom) Vertical cross section showing how the LLJ enhances lift over the cold pool/cold side of the front locally, in addition to the outflow boundary normal flow that is created by its motion from the cold side towards the warm side.

"transition zone" (Figures 5.2 and 5.10).[2] The leading convective line is preceded by a gust front, above which there is a shelf cloud (Figure 5.9), which is formed as environmental air ahead of the MCS is lifted over the cold pool behind the gust front.

When the leading convective line is followed by a stratiform precipitation region that is centered approximately to the rear (with respect to MCS motion) of the MCS, the MCS is said to be "symmetric" (Figure 5.11a). When the leading convective line, however, is centered or is more intense off to the southern, southwestern, or western side (in the Great Plains of the U. S.), the MCS is said to be "asymmetric" (Figure 5.11b).

Symmetric MCSs often evolve into asymmetric MCSs, since with time the Coriolis force becomes significant and convergence at mid-levels above the sinking, evaporatively cooled air and below mesoscale ascending air above in the stratiform precipitation region, acts on Earth's vorticity to produce cyclonic

[2] There is not always a distinct difference between the leading convective line and the stratiform precipitation region. In general, however, the number of convective towers decreases with increasing distance behind the leading convective line.

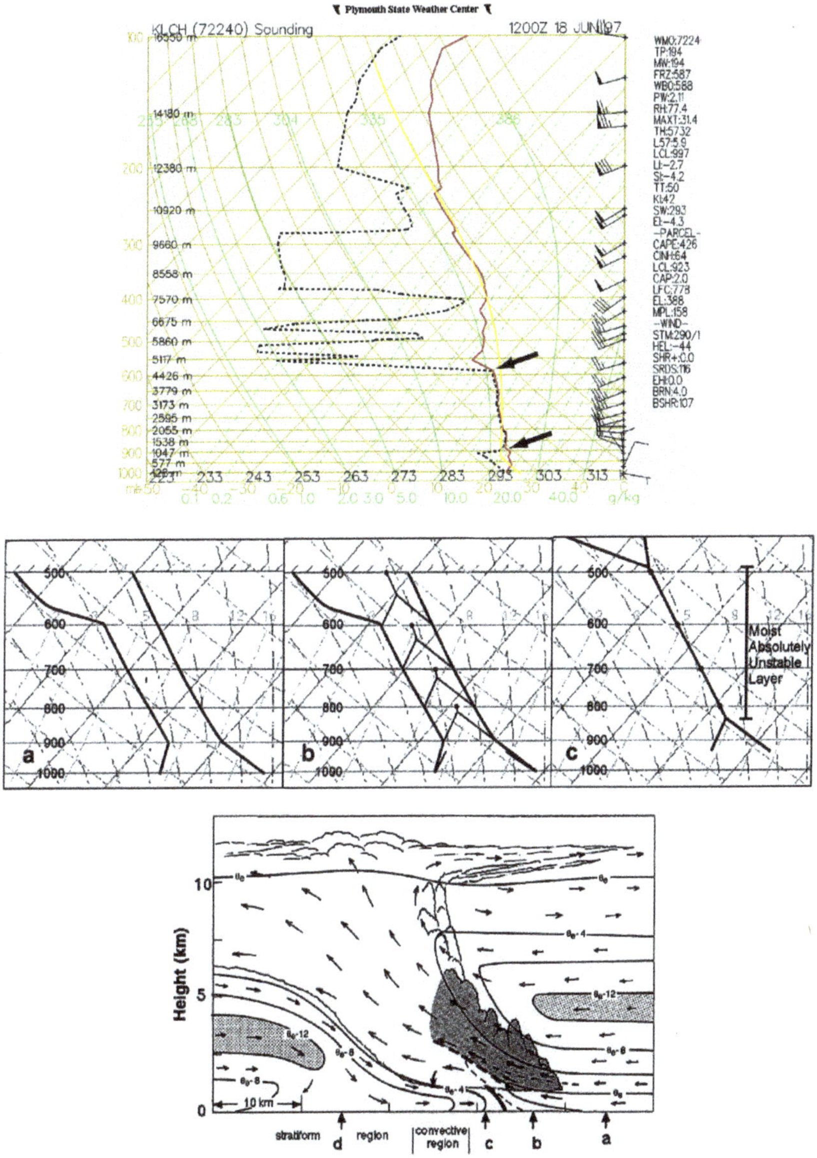

Figure 5.7. The MAUL (moist absolutely unstable layer) in the growing portion of an MCS. (Top) Sounding with a MAUL between 900 and 600 hPa, at Lake Charles, LA at 12:00 UTC on June 18, 1997 (from Plymouth State College archives). (Middle) Initial sounding (a) experiences lift until the LCL is reached for each parcel (four selected for illustration) and brought upward to the location of each dot (b). The result is the MAUL indicated (from Bryan and Fritsch, 2000). (Bottom) Idealized vertical cross section through a "slab" of convection, showing the MAUL as a dark, shaded area. Flow vectors are system relative. Heavy solid line indicates the leading edge of the cold pool. Soundings with a MAUL can be found at points labeled "b" and "c" (from Bryan and Fritsch, 2000)

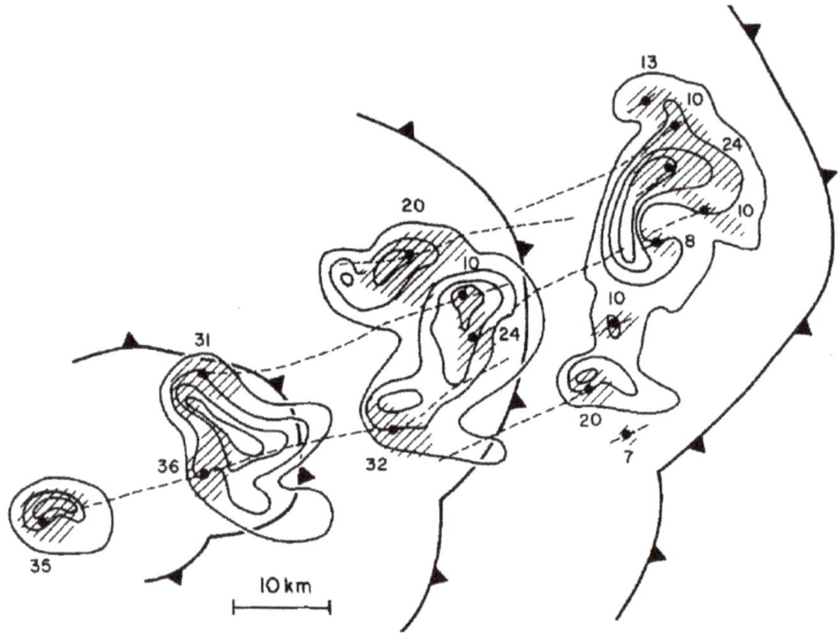

Figure 5.8. Example of the upscale growth of a convective system from an initial isolated convective storm from a numerical simulation. Evolution of the storm structure at mid-levels (4.6 km AGL) at 30, 60, 90, and 120 min after initiation. Rainwater mixing ratio is contoured at $2\,\mathrm{g\,kg^{-1}}$ intervals, beginning at $0.1\,\mathrm{g\,kg^{-1}}$. Regions of updraft in excess of $5\,\mathrm{m\,s^{-1}}$ are hatched and updraft maxima are located by dots and labeled in $\mathrm{m\,s^{-1}}$. Surface gust front (isotherm of $-0.5°\mathrm{C}$ deviation from domain average) marked by a cold front symbol. Dashed lines track path of updrafts (from Weisman and Klemp, 1984).

Figure 5.9. The leading edge of an MCS on June 4, 2002 in the Texas Panhandle (composite photograph by the author).

vorticity—cf. (2.50) with the inclusion of the Coriolis force in the horizontal equation of motion (2.13), so that the term "$-\delta f$" appears on the RHS of (2.50)—which is air in the ascending front-to-rear branch of the vertical circulation that turns to the right, which advects hydrometeors in that direction. In addition, the

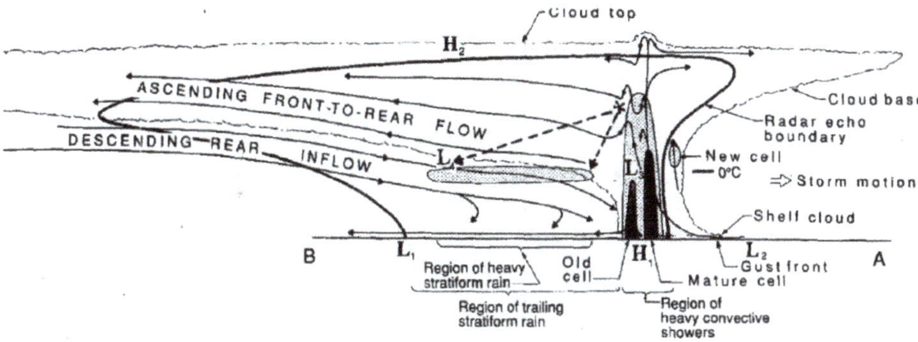

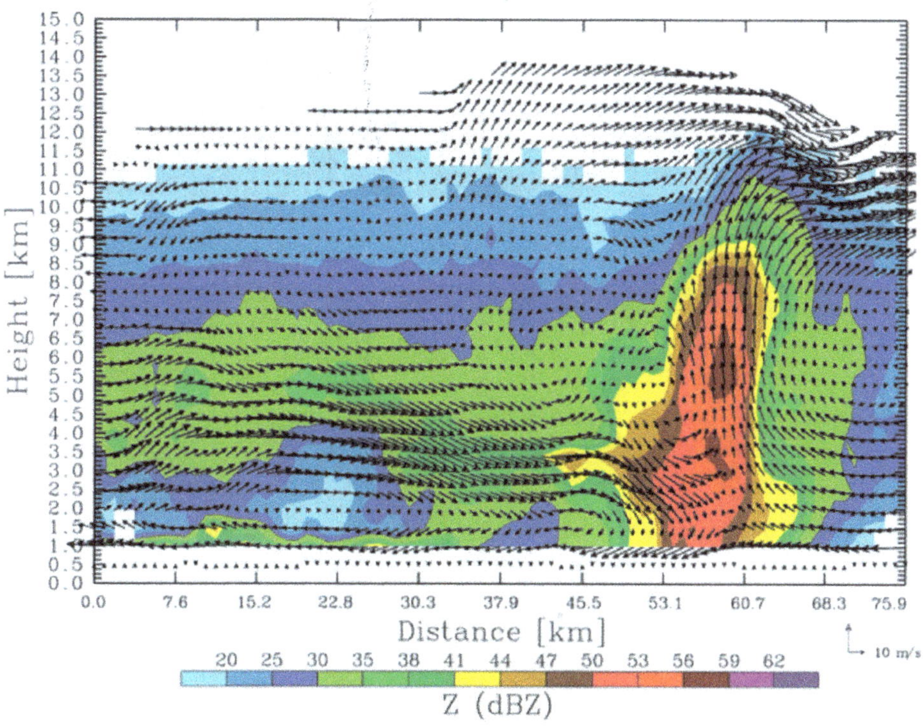

Figure 5.10. Vertical cross section through a squall line MCS. (Top) Idealized model (from Houze *et al.*, 1989). (Bottom) From quad-Doppler analysis of a bow echo on June 10, 2003 near St. Louis, MO during BAMEX. System-relative winds (vectors) and radar reflectivity factor in d*BZ* (from Davis *et al.*, 2004).

Coriolis force turns the rear-inflow jet to the right, which advects the cold pool in that direction, so that new cells are generated along this edge of the cold pool. When symmetric MCSs evolve into asymmetric MCSs, the stratiform precipitation region (area) shifts location (usually poleward when the leading convective line is oriented mainly in a meridional direction).

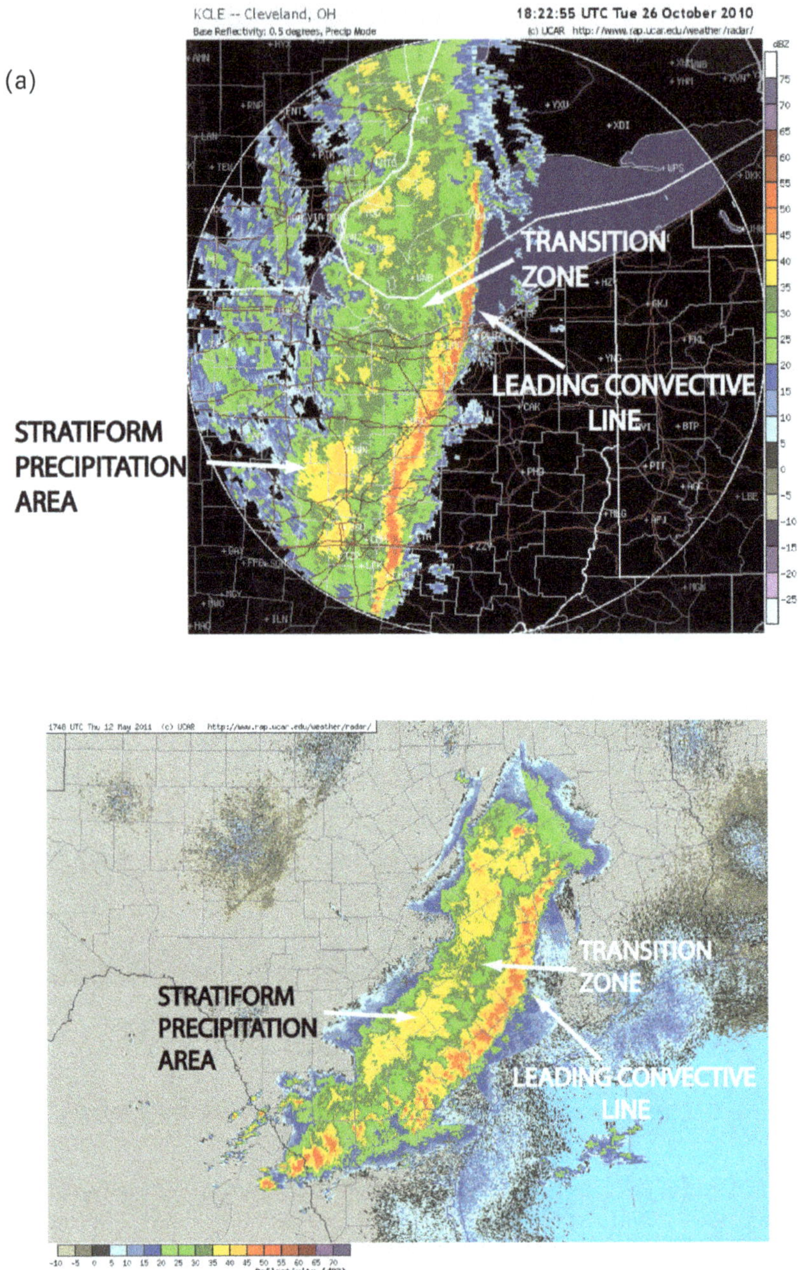

Figure 5.11. (a) Symmetric mesoscale convective systems (top) on October 26, 2010, as depicted by the Cleveland, OH WSR-88D radar and (bottom) on May 12, 2011 in south Texas, as depicted by WSR-88D radars. (b) Asymmetric mesoscale convective system on June 18, 2010, as depicted by the WSR-88D radar at Quad Cities, IA. (c) Idealized depiction of symmetric and asymmetric mesoscale convective systems (from Houze *et al.*, 1990).

(b)

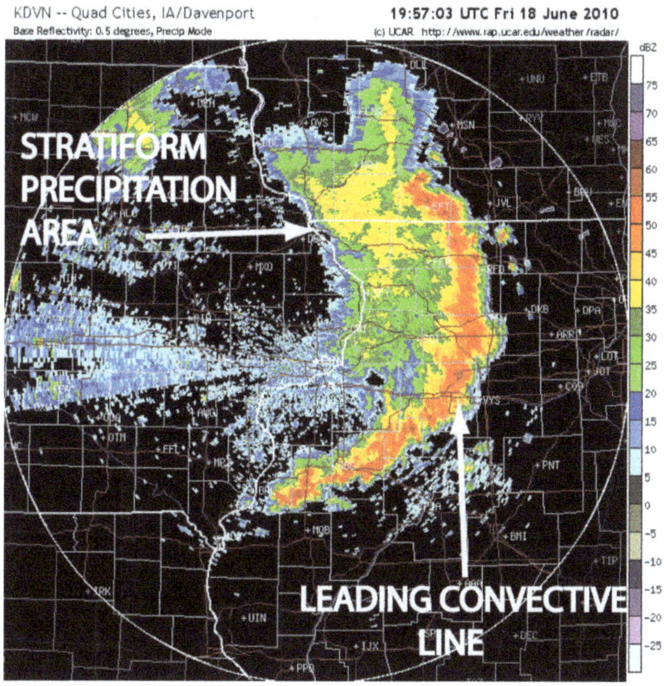
KDVN -- Quad Cities, IA/Davenport
19:57:03 UTC Fri 18 June 2010
Base Reflectivity: 0.5 degrees, Precip Mode
(c) UCAR http://www.rap.ucar.edu/weather/radar/
STRATIFORM PRECIPITATION AREA
LEADING CONVECTIVE LINE
dBZ

(c)

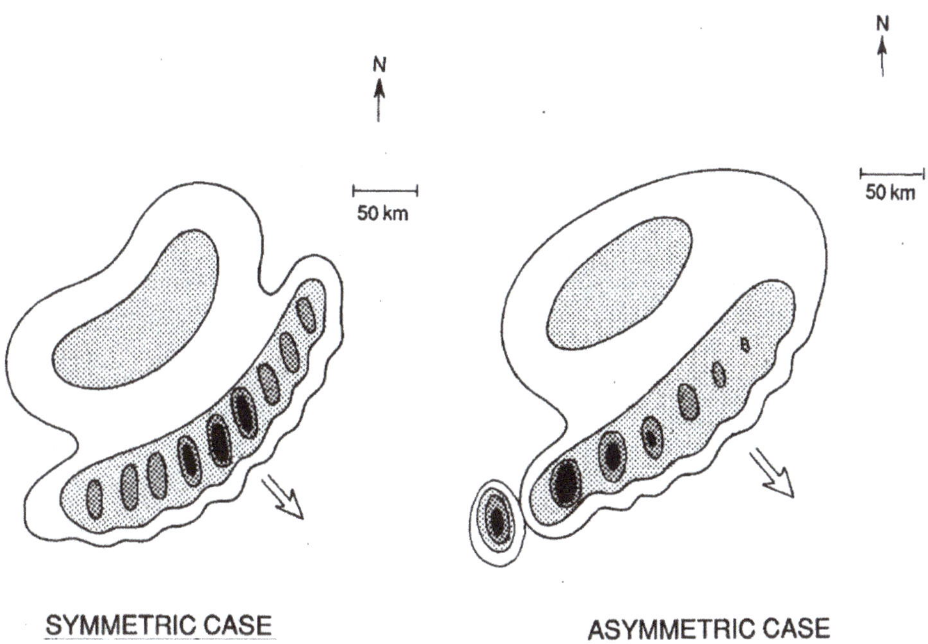
N
50 km
N
50 km
SYMMETRIC CASE
ASYMMETRIC CASE

It was found in the 1980s from Doppler radar studies that a jet of unsaturated air from the environment enters the stratiform precipitation region from the rear side of the MCS. Brad Smull and Bob Houze in the late 1980s named this feature the "rear-inflow" jet (Figure 5.10, top panel); it aids in the production of a cold pool at the surface as stratiform precipitation falls into the unsaturated air and cools evaporatively. However, it will be noted in the next section that the very presence of a cold pool aids in the production of the rear-inflow jet, so there is a positive feedback mechanism between the rear-inflow jet and the cold pool. Ascending front-to-rear air motion is found above the rear-inflow jet, in the anvil cloud region (Figure 5.10), which is composed of condensed/frozen water substance formed in the leading convective line.

New convective cells form ahead of the leading convective line and eventually become the leading convective line, while the dissipating rear edge of the leading convective line is absorbed into the trailing stratiform precipitation area. A vertical cross section normal to an MCS composed of a leading convective line with a trailing stratiform precipitation area (Figure 5.10) is like looking at the temporal history of air parcels in the MCS, and is perhaps similar to tree rings, which depict spatially the annual growth cycle in trees. An area of enhanced radar reflectivity is observed in the stratiform precipitation area at the freezing level; this *bright band* is a result of ice particles coated with water, which have higher radar reflectivity than the snow above and the rain below.

The classic MCS squall line with a leading convective line and stratiform precipitation area, symmetric and asymmetric, may take on various appearances. While the archetype structure is that of a linear leading convective line, sometimes the leading convective line has the appearance more of blobs of intense echo cores (Figure 5.11a, bottom panel) or individual cells that are canted with respect to the leading line (Figure 5.12). The leading line itself may contain wiggles and bow-shaped line segments (Figure 5.13).

The pressure field underneath MCS squall lines is characterized by a mesohigh underneath the main area of the leading convection and a "wake low" to the rear, just at the rear edge of the enhanced stratiform precipitation area (Figure 5.14). These mesoscale, surface pressure features were first mentioned in the literature in the mid-1950s by Ted Fujita and were discovered through mesoanalysis of a mesoscale network of observations then and until a decade or two ago, during special field programs, but now routinely observed in some states such as Oklahoma and Texas (in a subsection of the state) which maintain operational mesonetworks. Mesohighs were first seen in data from the Thunderstorm Project. Dick Johnson at CSU has found that when the rearward-directed pressure gradient force is relatively weak the rear-inflow jet continues toward the leading convective line; when the rearward-directed pressure gradient force is relatively strong the rear-inflow jet may be blocked (Figure 5.15). A much weaker "pre-squall low" may also be found just ahead of the leading gust front (Figure 5.14). In asymmetric MCSs, the mesohigh is still found underneath the stratiform precipitation region (Figure 5.16).

A mesohigh is produced in large part by the hydrostatic pressure excess of the

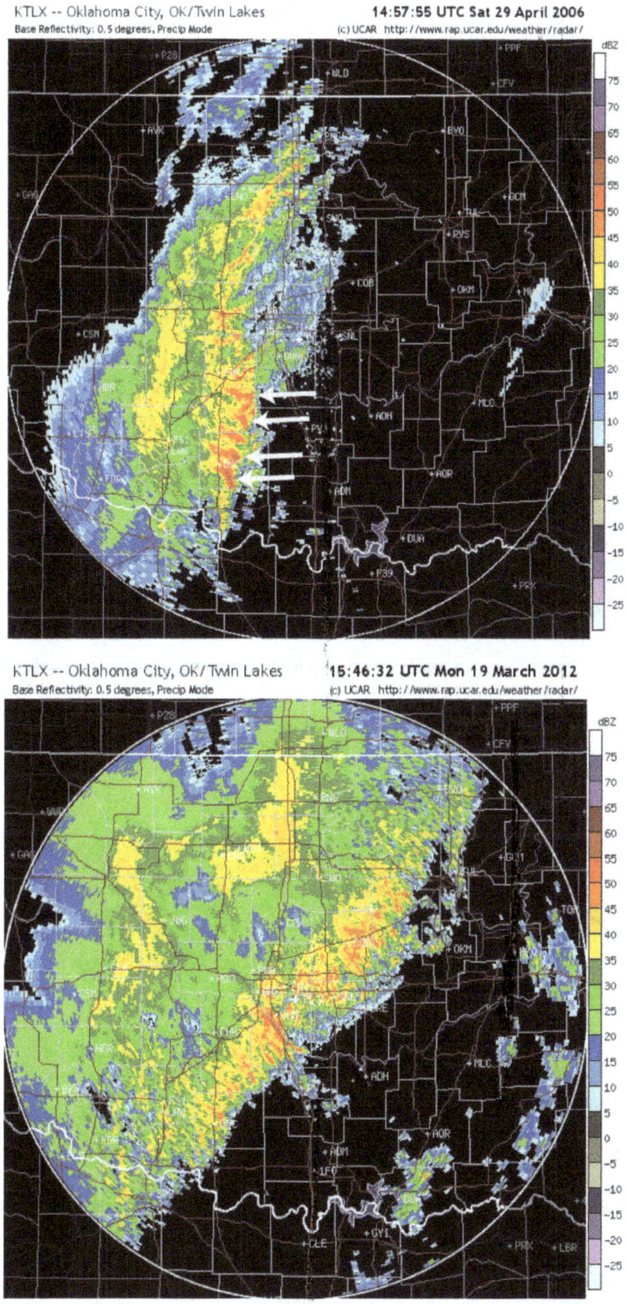

Figure 5.12. Mesoscale convective systems whose leading convective line consists of individual cells (arrows) whose major axes are oriented approximately normal to the leading line (top) on April 29, 2006 and (bottom) on March 19, 2012, as depicted by the Oklahoma City, OK WSR-88D radar.

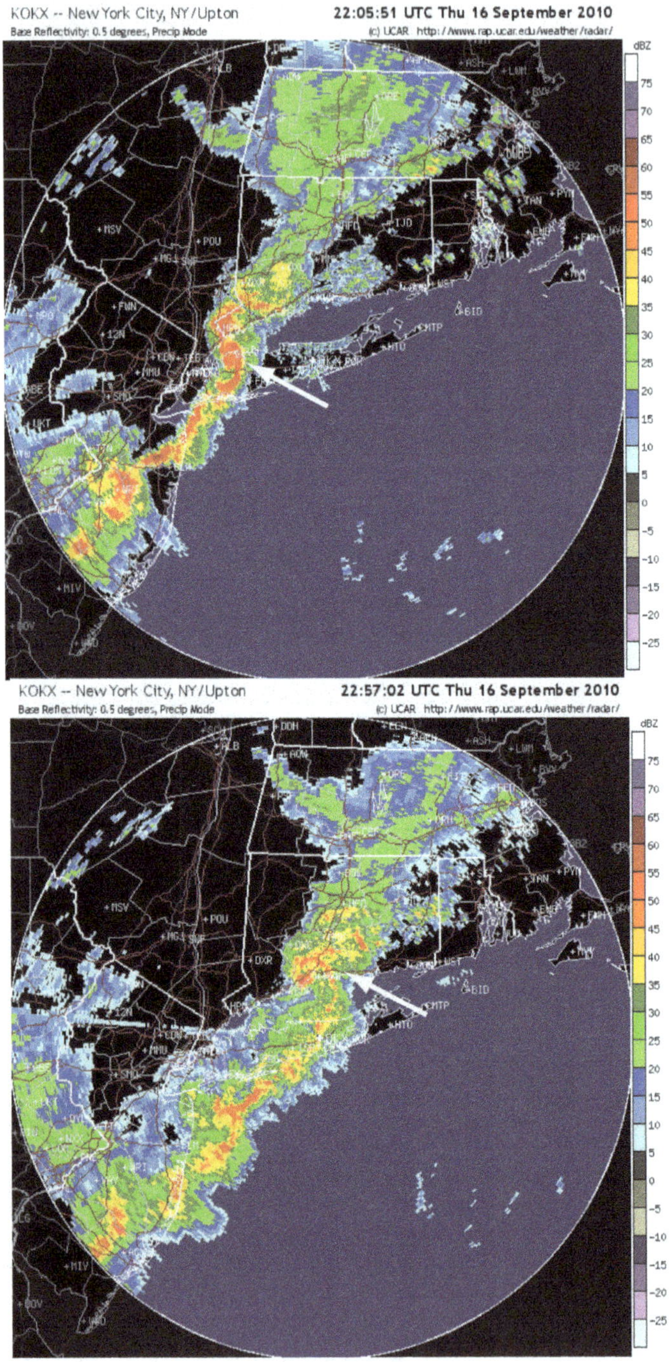

Figure 5.13a. Squall lines with mesoscale waves (white arrow) along the leading convective line on September 16, 2010, as detected by reflectivity from the New York City WSR-88D radar.

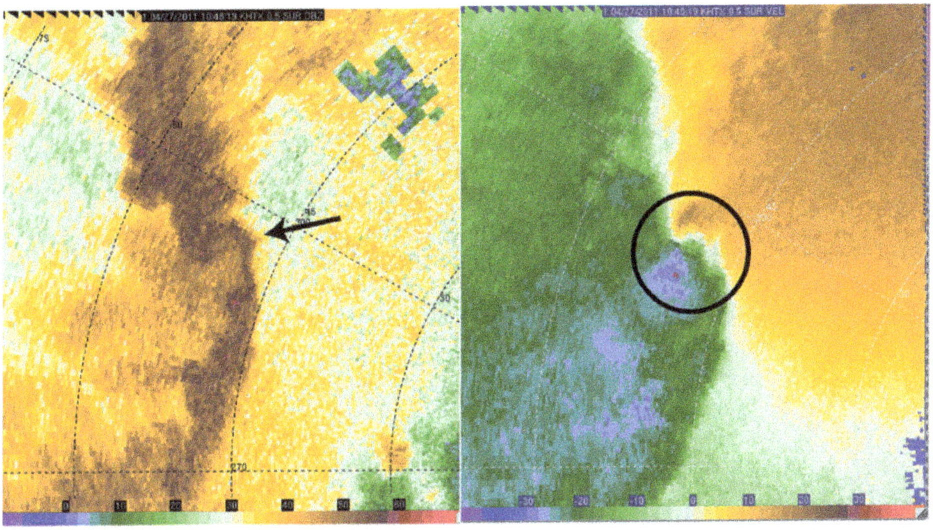

Figure 5.13b. Squall line with wave (arrow) and mesoscale vortex (signature is circled) on May 24, 2011, as depicted by the Huntsville, AL WSR-88D radar; (left) radar reflectivity in dBZ; (right) Doppler velocity in m s^{-1}.

evaporatively cooled cold pool underneath the storm. Melting or sublimation may increase the amount of cooling. The pressure at the leading edge of the cold pool/gust front also has a non-hydrostatic, dynamic component: air approaching the leading edge must slow down and rise, which is consistent with an adverse horizontal pressure gradient. Beneath the downdraft, there is also a non-hydrostatic, dynamic component: air descending to the ground must spread out laterally, which is consistent with a local maximum in pressure and diverging surface pressure gradient force. Hydrometeor loading can also increase surface pressure. When a symmetric MCS evolves into an asymmetric MCS, the mesohigh shifts location along with the stratiform precipitation area.

A wake low is produced as dry air subsides, warming the air and adiabatically producing a hydrostatic deficit of pressure. It is thought that a descending rear-inflow jet warms the air adiabatically more than the evaporation of raindrops (or sublimation of ice crystals or melting of ice crystals) cools it. When there is very strong warming, as there is in a heat burst, descending air parcels may overshoot their equilibrium level as they lose their negative buoyancy, but have enough kinetic energy to continue to descend and warm even more (cf. Section 3.2.1.2). Wake lows may contain transient decaying convective cells associated with stronger, narrower low-pressure areas. A necessary condition for wake low formation is the existence of a stratiform precipitation area; the intensification of wake lows occurs during the latter stages of an MCS.

A pre-squall mesolow is associated with subsiding air aloft ahead of squall lines. This subsiding air warms adiabatically, and hydrostatically reduces the surface pressure.

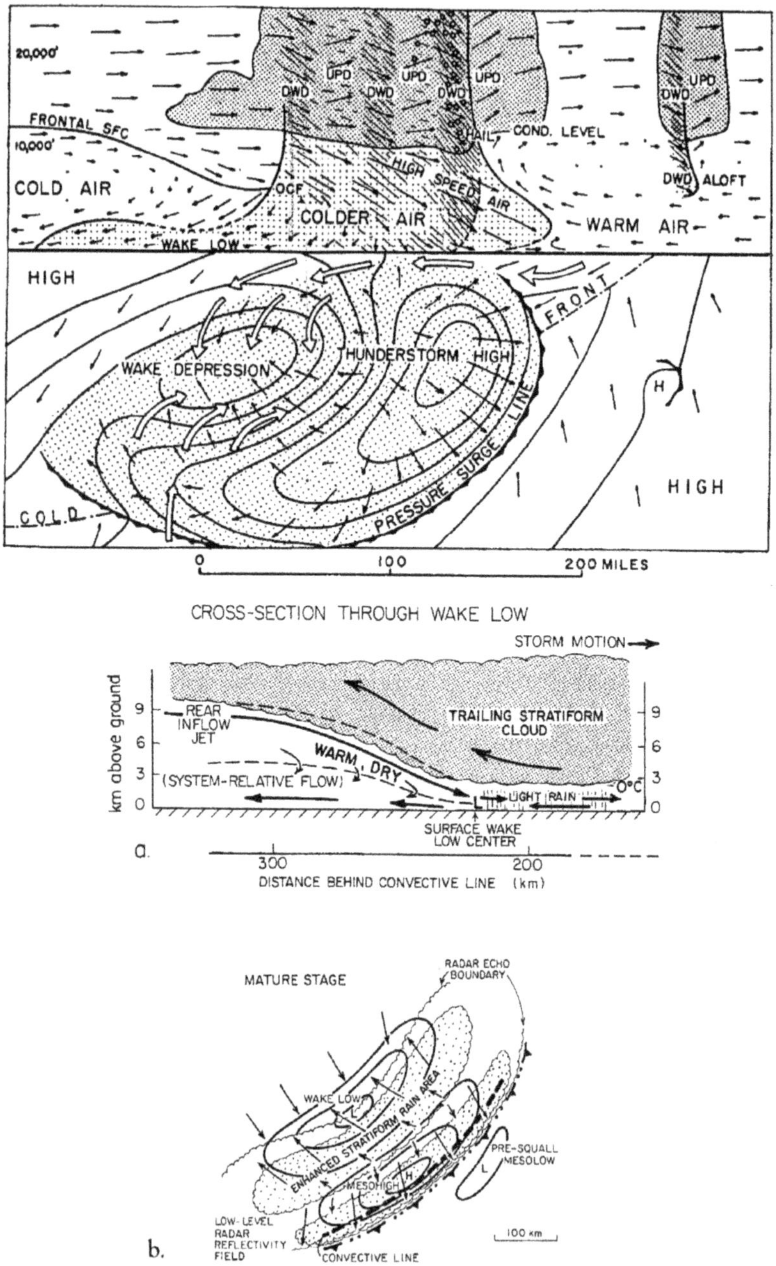

Figure 5.14. Conceptual model of a mesohigh and wake low in an MCS. (Top) Fujita's model of a mesohigh ("thunderstorm high") and wake low ("wake depression') from 1955. Updrafts and downdrafts are noted by "UPD" and "DWD", respectively. (Bottom) (a) Vertical cross section through the stratiform precipitation area of a mature MCS. (b) Horizontal cross section through a squall line MCS (from Johnson, 2001).

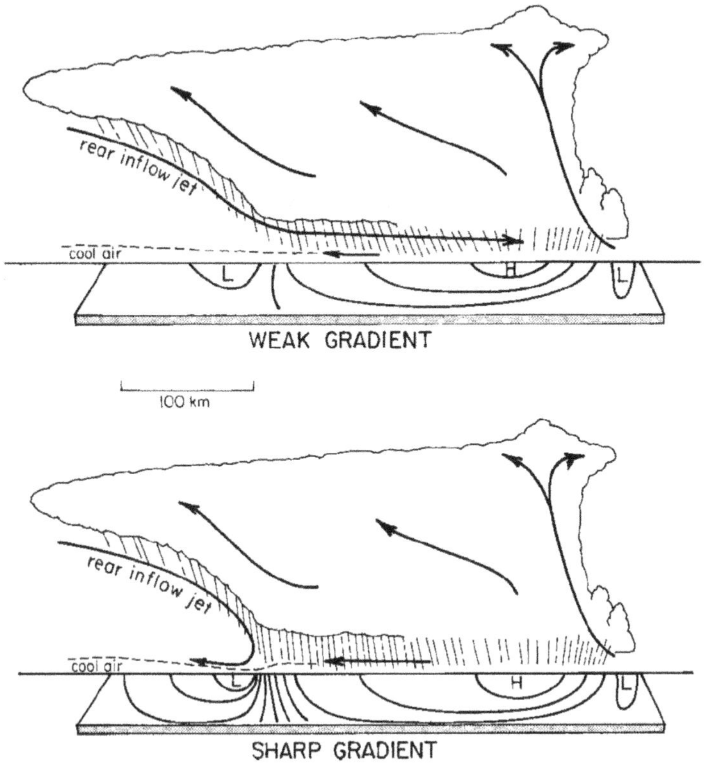

Figure 5.15. Idealized depiction of (top) weak and (bottom) strong surface horizontal pressure gradients in association with rear-inflow jets that continue forward toward the leading convective line or are blocked, respectively (from Johnson, 2001).

There is also a theory in which low-level convergence associated with gravity waves drives convection when the atmosphere is conditionally unstable, while the diabatic heating–cooling couplet induced by latent heat release–evaporation (or sublimation or melting) excites gravity waves: many modes of gravity waves are possible, but there could be constructive coupling between gravity waves and the convective system. Such a mechanism is called "Wave CISK", where CISK stands for "conditional instability of the second kind"; it was first postulated to explain tropical convection by Dick Lindzen back in the 1970s, and demonstrated in an idealized model by his student Dave Raymond. It has been suggested that Wave CISK may explain the surface mesoscale pressure features in MCSs. On a variation of the Wave CISK theme, it has been suggested that gravity waves excited in a linearized dynamical framework by the cooling that takes place underneath a stratiform precipitation area (without the need for diabatic heating above) may explain the mesohigh–wake low couplet and its propagation.

While we have described the more common "trailing stratiform" linear MCS (Figure 5.17a), there are also MCSs in which the stratiform area develops ahead of

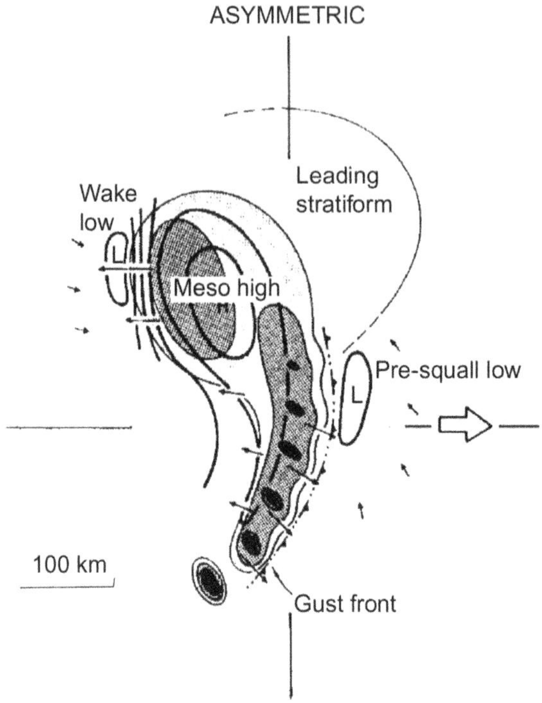

Figure 5.16. Conceptual model of the surface pressure, wind, and precipitation field associated with an asymmetric MCS. Levels of shading denote increasing radar reflectivity factor, with darkest shading representing convective cores. Pressure is given in increments of 1 hPa. Small vectors represent the wind field at the surface. Large arrows represent storm motion (from Johnson, 2001 and Loehrer and Johnson, 1995).

the MCS. This type of MCS is called a "leading stratiform" MCS (Figure 5.17b). When the stratiform area develops along the line of convection, it is called a "parallel stratiform" MCS (Figure 5.17c). The storm-relative winds prior to MCS formation vary among trailing, leading, and parallel stratiform MCSs (Figure 5.18). The leading stratiform vertical profile has a component of wind aloft that advects cloud material and precipitation to the right of the line (when viewed towards the north), while the trailing stratiform profile has a component of wind that advects cloud material and precipitation to the left. In the parallel stratiform vertical profile, cloud material and precipitation is advected parallel to the line. The most salient characteristics of the common trailing stratiform profile are that the leading line propagates quickly in the direction of the inflow, while at high altitudes in the storm the storm-relative wind is weak.

In the conceptual model of a trailing stratiform linear MCS, a two-dimensional flow pattern is characterized by three separate flow trajectories (Figure 5.10, top panel): at low levels air approaches the gust front, is lifted upwards, and eventually flows rearward aloft (the "ascending front-to-rear flow").

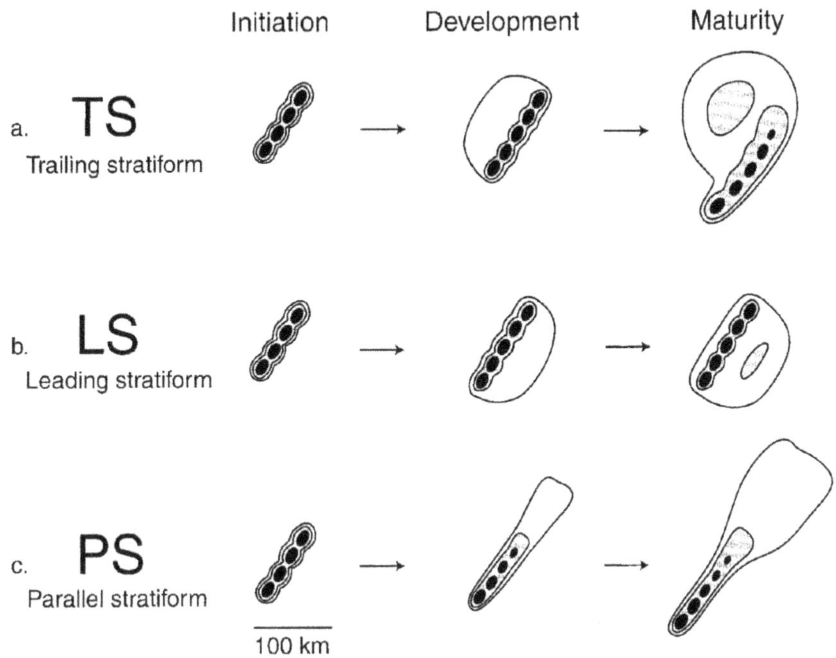

Figure 5.17. Linear MCS archetypes—(a) trailing stratiform (TS), (b) leading stratiform (LS), (c) parallel stratiform (PS)—from initiation to maturity. Low-level horizontal radar reflecivity factor levels shaded at approximately 20 (no shading), 40, and 50 dB*Z*. The approximate time intervals between phases are 3–4 h for the TS, 2–3 h for the LS, and 2–3 h for the PS archetypes (from Parker and Johnson, 2000).

At mid-levels, air enters the MCS underneath the cloud and reverses direction, flowing rearward, and out of the MCS near the surface. At mid-levels, air approaches the MCS, but is reversed aloft. The last branch is not always found, especially if deep-layer shear is not too strong. An example of a snapshot of the flow, derived from a multi-Doppler wind analysis, in a vertical plane cutting across an MCS, is shown in Figure 5.10, bottom panel; this real-life example resembles the idealized model depicted in Figure 5.10, top panel. A simplified model is therefore one in which air approaches at low levels, rises along the leading convective line, and then flows toward the rear in the stratiform precipitation area (Figure 5.19). A second branch of airflow enters from the rear at mid-levels, the rear-inflow jet, and then reverses direction near the surface (Figure 5.19). These are the two fundamental branches of airflow in a typical trailing stratiform linear MCS. This type of analysis is similar to that in synoptic meteorology of the conveyor belt in idealized models of extratropical cyclones.

In this simple, idealized conceptual model of a trailing stratiform linear MCS (squall line) over a depth H, described by Kerry Emanuel in his 1994 textbook on convection (the basic technique was first used by Mitch Moncrieff and collaborators in the 1970s, with inspiration from the earlier observational work done

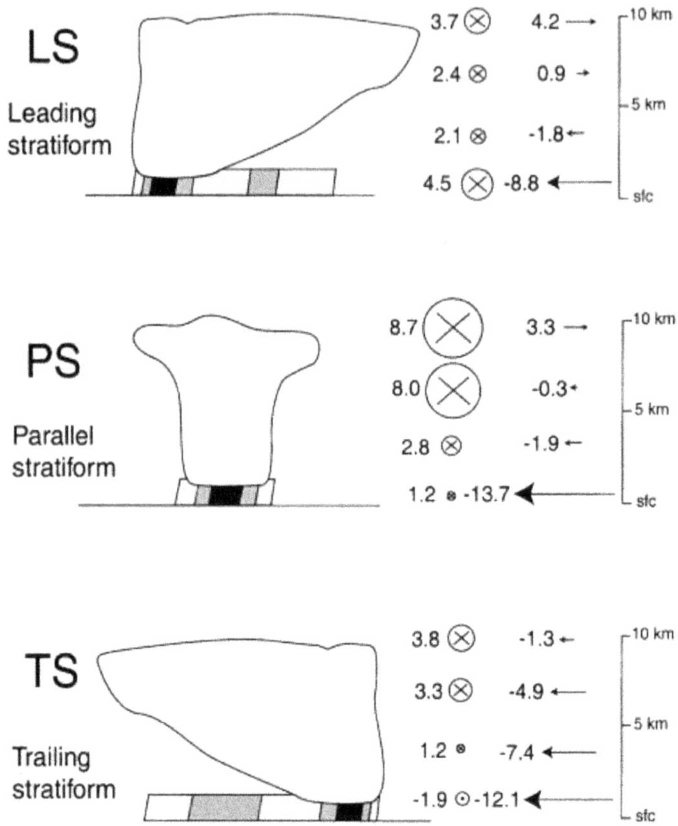

Figure 5.18. Vertical profiles of layer mean storm-relative pre-MCS winds for linear MCS archetypes. Wind vectors depicted as line parallel ($\otimes, \odot$) and line perpendicular ($\rightarrow, \leftarrow$) components in m s^{-1} (a $\otimes$ sign indicates flow into the page; a $\odot$ indicates flow out from the page). Idealized reflectivity factor (shadings) and cloud outlines (solid line outline) are shown. Leading edges of MCSs are to the right (from Parker and Johnson 2000).

by Keith Browning and collaborators), air flows toward the MCS in a moist boundary layer of depth h_0 at a speed of U_0. This branch of the MCS (the rearward-leaning, ascending branch) flows up and rearward, departing at the rear with a speed of U_b. Rear inflow has a speed of U_c over a depth of h_c and turns around just to the rear of the leading edge of the cold pool; the retreating branch also has a depth of h_c and exits at low levels with a speed of U_c. It follows that the rearward-leaning, ascending branch has a depth of $H - 2h_c$ at the left. At the rear of the MCS (to the left in Figure 5.19), the rearward-ascending branch is characterized by positive buoyancy B_u above $z = 2h_c$ (Emanuel in his text let $B = B_u$ above $z = h_c$ at the rear (left), but the air in the rear inflow in nature is *not* positively buoyant and may even be negatively buoyant). At the top of the rear inflow layer (to the left in Figure 5.19), there is no buoyancy; at the rear in low levels, the cold pool has negative buoyancy, B_d. This simple model, while nonlinear, is restricted

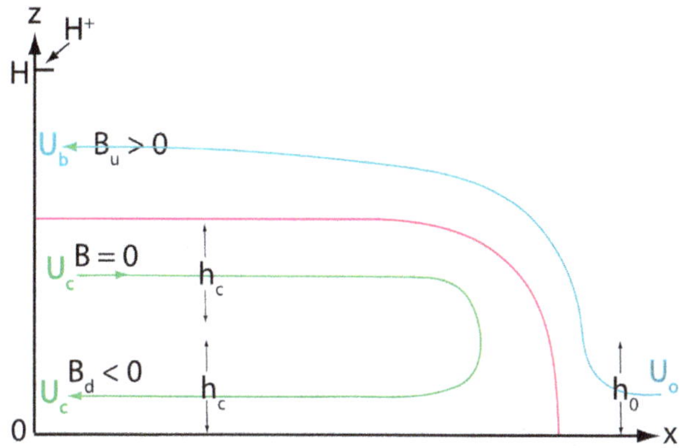

Figure 5.19. Idealized model of two-dimensional vertical circulation in a squall line. Buoyant air stream, ascending front-to-rear flow, (blue streamline) with speed in the x-direction noted in the lower-right to upper-left direction; rear-inflow jet (green streamline) at mid-levels at left without buoyancy at speed in the x-direction noted at the beginning and end of the streamline; negatively buoyant return flow at lower left. Red line separates two separate branches of the flow (modified from Emanuel, 1994).

to be steady state and two dimensional,[3] and therefore of limited application to real MCSs which have some degree of two dimensionality and some three dimensionality.

The object of this exercise is to find out what values of U_0, U_c, and h_c are dynamically and thermodynamically consistent with the model and if they are reasonably close to what is observed. The acceleration from the pressure gradient force at $x = -\infty$ is found from the hydrostatic equation—expressed as the vertical component of (4.11) with $Dw/Dt = 0$

$$\alpha_0\, \partial p'/\partial z = \begin{cases} B_u & z \geq 2h_c \\ 0 & h_c \leq z < 2h_c \\ B_d & z < h_c \end{cases} \tag{5.1}$$

We define

$$p'(x = -\infty, z = H) = p_t \tag{5.2}$$

Integrating (5.1) with respect to pressure at $x = -\infty$, we find that

$$\alpha_0\, p' = \begin{cases} \alpha_0\, p_t - B_u(H - z) & z \geq 2h_c \\ \alpha_0\, p_t - B_u(H - 2h_c) & h_c \leq z < 2h_c \\ \alpha_0\, p_t - B_u(H - 2h_c) - B_d(h_c - z) & z < h_c \end{cases} \tag{5.3}$$

Consider now the x-equation of motion (2.13) subject to the simplifications of no friction (inviscid flow), two dimensionality, steady flow, and a Boussinesq atmosphere, with $1/\bar{\rho} = \alpha_0$

$$u\, \partial u/\partial x + w\, \partial u/\partial z = -\alpha_0\, \partial p'/\partial x \tag{5.4}$$

[3] Two-dimensional circulations in squall lines are sometimes referred to as examples of "slab overturning".

At $z = 0$, $w = 0$ (the kinematic lower-boundary condition for a level surface), so the integral of (5.4) from $-\infty$ to $+\infty$ yields the following:

$$\tfrac{1}{2}U_0^2 = \alpha_0 p_s + \tfrac{1}{2}U_c^2 \tag{5.5}$$

where $p'(x = \infty, z = 0) = 0$ (i.e., the atmosphere at the surface is hydrostatic far upstream from the squall line leading edge); and $p'(x = -\infty, z = 0) = p_s$ (the pressure far downstream at the surface). From (5.3) evaluated at $z = 0$ and from (5.5) we find that upon eliminating p_s we get the relation

$$\tfrac{1}{2}U_0^2 = \tfrac{1}{2}U_c^2 + \alpha_0 p_t - B_u(H - 2h_c) - B_d h_c \tag{5.6}$$

which is valid at $z = 0$. We will use this equation shortly to solve for p_t. Now consider the continuity equation

$$\partial u/\partial x + \partial w/\partial z = 0 \tag{5.7}$$

The first equation in our model is derived from the two-dimensional, horizontal vorticity equation for the y-component of vorticity (a technique used previously by Mitch Moncrieff). The steady-state, frictionless form of the y-component of the vorticity ($\eta = \partial u/\partial z - \partial w/\partial x$) equation in flux form is based on (2.51) (the divergence term is zero because the atmosphere is Boussinesq and the tilting term is zero because the flow is two dimensional in the x–z-plane) and the continuity equation (5.7):

$$\partial/\partial x(u\eta) + \partial/\partial z(w\eta) = -\partial B/\partial x \tag{5.8}$$

This equation is integrated over the domain ($x = -\infty$ to $x = +\infty$, and $z = 0$ to $z = H_+$, where H_+ is just above the tropopause, in the lower stratosphere). By intruding into the stratosphere, we are able to assume that u is zero at $z = H_+$. The result is the following:

$$\tfrac{1}{2}U_0^2 = \tfrac{1}{2}U_c^2 - B_u(H - 2h_c) - B_d h_c \tag{5.9}$$

So, we now have a relation among the inflow velocity U_0, the geometry of the problem in terms of H, h_0, and h_c, rear inflow U_c, and updraft and downdraft buoyancies B_u and B_d: Since there are three dependent variables (h_c, U_0, and U_c), we need three independent equations to find solutions. Equation (5.9) is one such equation. The quantities H, h_0, B_u, B_d are specified constants. H is the depth of the troposphere, B_u and h_0 are determined from an environmental sounding (lapse rate of temperature and depth of the boundary layer), and B_d is determined from both an environmental sounding and an estimate of liquid water content in the storm.

We now compare (5.9) with (5.6) and see that p_t must be zero. From (5.4) and (5.7) we derive the flux form of the x-component of the equation of motion as

$$\partial u^2/\partial x + \partial(uw)/\partial z = -\alpha_0\, \partial p'/\partial x \tag{5.10}$$

Integrating (5.10) over the domain in the x–z-plane from 0 to H in z and $-\infty$ to $+\infty$ in x, we find that

$$U_0^2 h_0 = 2U_c^2 h_c + (H - 2h_c)U_b^2 + \int_0^H \alpha_0 p' \Big|_{x=-\infty}^{x=+\infty} dz \tag{5.11}$$

The continuity equation (5.7) is integrated over the entire x–z-plane and the result is:

$$U_b = U_0 h_0/(H - 2h_c) \tag{5.12}$$

Substituting (5.12) into (5.11) to eliminate U_b and integrating p' in (5.11) using (5.3) with p_t set to zero, we get the following:

$$U_0^2[h_0 - h_0^2/(H - 2h_c)] = 2U_c^2 h_c - \tfrac{1}{2}B_u(H^2 - 4h_c^2) - \tfrac{1}{2}B_d h_c^2 \tag{5.13}$$

This is the second equation in our model.

Rear inflow U_c is driven by evaporative cooling, so that the diabatic form of the thermodynamic equation

$$dQ/T = ds = C_p\, d\ln\theta \tag{5.14}$$

is used, where Q is the diabatic heating rate and s is the specific entropy. The rate of cooling in the downdraft is given approximately by integrating (5.14) from θ_m, the potential temperature of rear inflow (elevated above the ground), to θ_s, the potential temperature at the surface to the rear of the squall line (Fig. 5.19), to get $C_p T_D \ln(\theta_s/\theta_m)U_c h_c$, where T_D is the mean temperature in the downdraft. The product $U_c h_c$ is the downdraft mass flux. Similarly, the rate of heating in the updraft due to latent heat release is $C_p T_U \ln(\theta_t/\theta_0)U_c h_c$, where T_U is the mean temperature in the updraft, θ_t is the potential temperature at the top of the updraft, and θ_0 is the potential temperature at the bottom of the updraft.

Kerry Emanuel assumed, based on numerical simulations, that the ratio of the *magnitude* of the cooling to the magnitude of the heating is a constant given by $1 - \varepsilon_p$, where ε_p is the bulk precipitation efficiency. It follows that the third and last equation in the model is

$$U_c = U_0(1 - \varepsilon_p)(h_0/h_c)(T_U/T_D)[\ln(\theta_t/\theta_0)/\ln(\theta_m/\theta_s)] \tag{5.15}$$

New parameters ε_p, T_U, T_D, θ_t, θ_0, θ_s, and θ_m now appear, but H and the buoyancies do not. The three equations in the model are therefore (5.9), (5.13), and (5.15).

For $H = 9$ km, $h_0 = 2$ km, updraft temperature excess of 4°C, downdraft temperature deficit of 4°C, $\varepsilon_p = 0.6$, $T_U = 250$ K, $T_D = 280$ K, $\theta_t = 400$ K, $\theta_0 = 300$ K, $\theta_s = 290$ K, and $\theta_m = 330$ K, two sets of solutions are mathematically possible: $U_0 = 24.6\,\mathrm{m\,s^{-1}}$ or $49.9\,\mathrm{m\,s^{-1}}$, $h_c = 3.62$ km or 5.95 km, and $U_c = 10.8\,\mathrm{m\,s^{-1}}$ or $13.3\,\mathrm{m\,s^{-1}}$. We will reject the second set of solutions because such strong inflow is not observed. If U_0 is a measure of vertical shear in the inflow layer, so that Δu (over the depth of the inflow layer h_0) $\approx 2U_0$ (i.e., the wind goes from $-2U_0$ at the surface and increases to 0 at $z = h_0$), then vertical shear for the first set of parameters ($\Delta u = 49.2\,\mathrm{m\,s^{-1}}$ over the lowest 3.62 km) is much stronger than what is typically observed. As H is decreased, U_0 and consequently Δu is also decreased, but so is h_c. This exercise demonstrates the difficulty in producing a dynamically and thermodynamically consistent *steady-state*, *two-dimensional* squall line with sloping and ascending front-to-rear inflow in the presence of strong vertical shear.

In contrast to the model in Figure 5.19, the conceptual model of a *rear*-fed convective line with leading precipitation (Figure 5.17), as is found on the cool side of a stationary front or warm front or outflow boundary, air flows from the boundary layer up and over the front/outflow boundary, producing a convective line, and an ascending rear-to-front flow branch carries cloud material and precipitation to the front of the MCS, while a "leading inflow jet" enters the MCS at mid-levels.

5.3 THE DYNAMICS AND THERMODYNAMICS OF MATURE MCS SQUALL LINES

The time-dependent behavior of MCS squall lines can be understood through an extension of RKW theory, discussed earlier (Section 3.2.2) in connection with the behavior of gust fronts in the presence of vertical shear. As a brief review, the two-dimensional aspects of the evolution of a convective line summarized by Morris Weisman (cf. Figure 3.40) are considered.

First, before precipitation falls, when deep convection is initiated, the convective cloud leans in the downshear direction: baroclinically generated horizontal vorticity is produced along the edges of the cloud in response to the latent heat release from condensation; this vorticity is augmented by the import of low-level environmental horizontal vorticity associated with vertical shear on the downshear side (Figure 5.20a). (In the case of supercell formation when vertical shear is deep and very strong, the initial rate of production of horizontal vorticity baroclinically is much less than the rate of vorticity advected in from the environment, so that the circulation of the cloud leans so far over that the magnitude of the updraft is reduced significantly.)

Later, after precipitation has begun to fall (on the downshear side) and an evaporatively produced cold pool is produced, the rate of generation of circulation induced at the leading edge of the cold pool may become counterbalanced by the rate of circulation produced by the import of environmental horizontal vorticity (of opposite sign) associated with environmental vertical shear at low levels (Figure 5.20b). In this case, air feeding into the convective cloud at the leading edge of the cold pool follows a completely vertical trajectory, thus maximizing the chances for new-cell growth as air is lifted to its LFC. This is referred to as the "optimal state".

As the convective system evolves, the cold pool may build up in intensity and deepen if more and more precipitation evaporates, so that the rate of generation of horizontal vorticity baroclinically at the leading edge of the cold pool is no longer balanced by the advection of environmental vorticity, but instead overwhelms it. The resulting circulation produced at the leading edge of the convection system now leans in the upshear direction and air flowing into the convective system moves rearward with respect to the convective system, carrying cloud particles and precipitation with it. Thus, a stratiform precipitation region forms to the rear of the leading convective line (Figure 5.20c). This stage requires that precipitation continue and that unsaturated air from the environment continues to be advected underneath the rear of the convective storm.

The development of the transition zone just to the rear of the leading convective line is likely a result of microphysical processes, as has been suggested

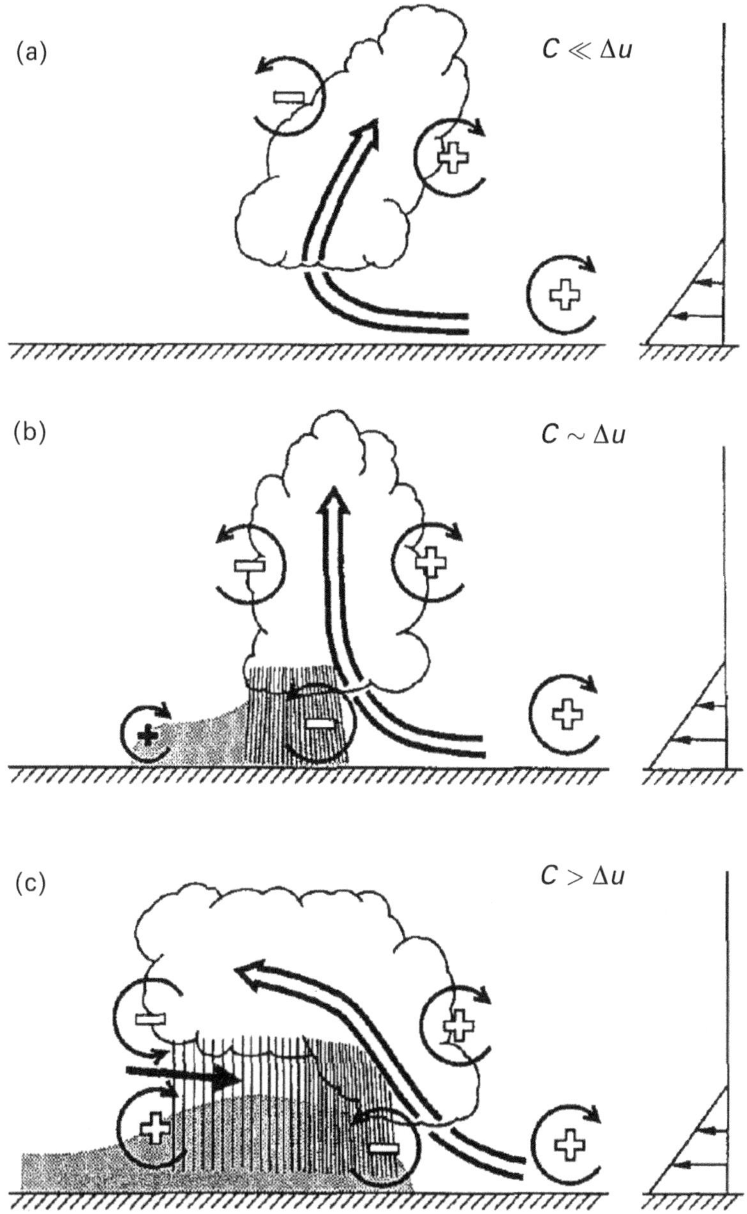

Figure 5.20. Idealized depiction of the three stages in the evolution of an MCS. (a) Downshear-leaning updraft. (b) Upright updraft. (c) Upshear-leaning updraft, development of stratiform precipitation region, and development of rear-inflow current (vector). Vertical profile of low-level system-relative wind shown at the lower right. Sense of horizontal vorticity indicated by curved arrows, with a positive sign indicating a horizontal vorticity vector into the page and a negative sign indicating a horizontal vorticity vector out from the page. Precipitation is indicated by hatched lines; cold pool is indicated by shaded area near the ground. Scalloped line indicates the cloud edge. "C" qualitatively indicates the intensity of the cold pool and Δu indicates the strength of low-level vertical shear (from Weisman, 1992).

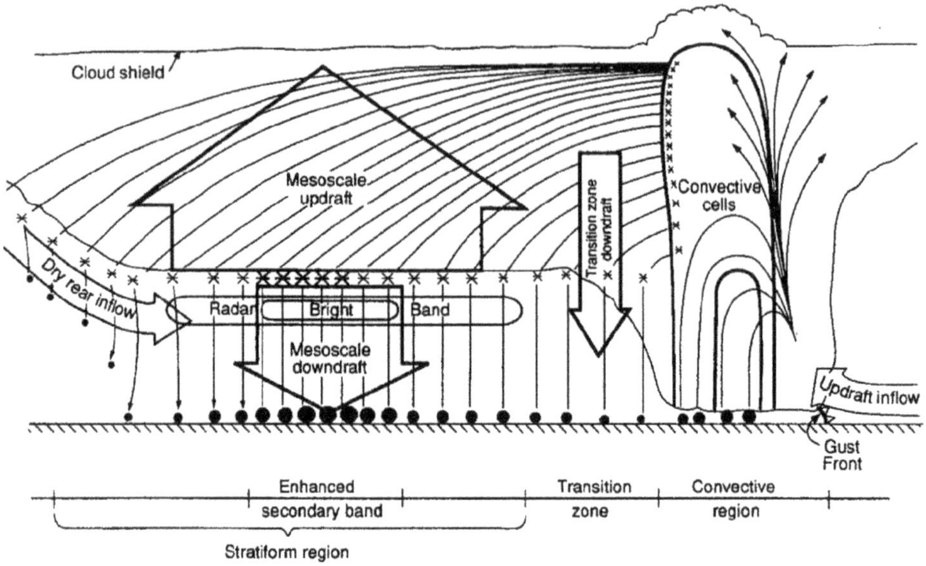

Figure 5.21. Conceptual model of precipitation particle trajectories and mean vertical motions in a trailing stratiform MCS (from Biggerstaff and Houze, 1993).

by Bob Houze and his students Steve Rutledge, Mike Biggerstaff, and Scott Braun: they have presented evidence that there is less aggregation (see texts on cloud microphysics) above the melting layer just to the rear of the leading convective line than farther to the rear of the convective system at the same altitude, because mid-level subsidence just to the rear of the leading convective line reduces the availability of small ice crystals; at the same time a mesoscale updraft farther to the rear enhances the growth of particles there (a mesoscale downdraft is found at lower levels) (Figure 5.21). In addition, precipitation particles falling out just to the rear of the leading convective line originate at low levels and thus have less time to grow than particles originating at higher levels, which travel farther back and have a much longer period of time over which to grow. The largest particles fall out quickly near the leading convective line, while the smaller particles with slower fall speeds are advected farther rearward before they fall out and are available for precipitation formation.

At the rear edge of the convective system, baroclinically generated horizontal vorticity is produced aloft as a result of latent heat release in the cloud, and below as a result of the cold pool. The horizontal vorticity produced aloft is opposite in direction to that produced below, so that a rear-inflow jet develops (Figure 5.20c) in between the vorticity couplet. The strength of the rear-inflow jet is proportional to CAPE, since the higher the CAPE, the greater the temperature excess in the cloud above and the lower the hydrostatic pressure deficit underneath the cloud; the stronger the low-pressure deficit under the cloud, the greater the rear-to-front pressure gradient force. The rear-inflow jet advects unsaturated environmental air into the convective system, thus enhancing the cold pool even more through continuing evaporative cooling. So, in addition, the cooler the cold pool, the stronger the rear-inflow jet. If the air becomes saturated beneath the stratiform cloud, then no more evaporational cooling can occur. There may also be some cooling due to melting of frozen precipitation and sublimation of ice crystals.

The most important consequence of the rear-inflow jet, however, is that there is horizontal vorticity associated with it of opposite sign just above and below it. Eventually, horizontal vorticity associated with vertical shear just *underneath* the rear-inflow jet, which is the same sign as that of the low-level environmental shear ahead of the system, is advected forward through the convective system until it reaches the leading edge, where it now can re-establish a balance between the baroclinically generated horizontal vorticity at the leading edge of the cold pool and the horizontal vorticity advected into it; as a consequence, the rising branch of the circulation at the leading edge is stronger, deeper, and more erect (Figure 5.22, bottom panel). Thus, the dynamics of the convective system itself result in a re-invigoration of the system and promote longevity. The rear-inflow jet of the

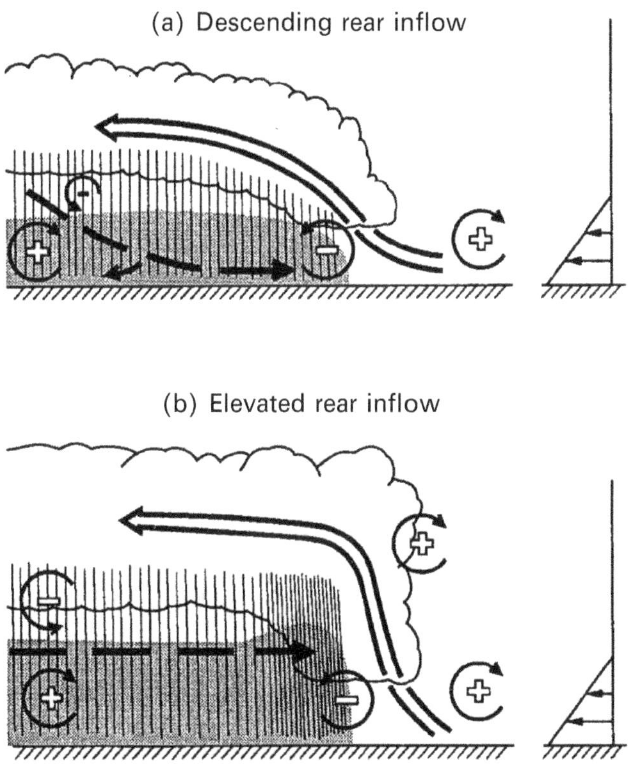

Figure 5.22. A conceptual model of the mature structure of a long-lived squall line MCS for a descending rear-inflow jet (top) and an elevated (bottom) rear-inflow jet. Vertical cross section showing the updraft branch (thick, double-lined, storm-relative streamline), the rear-inflow jet/current (dashed streamline). All other features as in Figure 5.20 (from Weisman, 1992).

type seen in Figure 5.22 (bottom panel) is called, more specifically, an "elevated rear-inflow jet".

When the rate of generation of horizontal vorticity in the convective system is greater at the rear edge of the surface cold pool than it is above at the rear edge of the cloud, the rear-inflow jet descends as it passes through the convective system (Figure 5.22, top panel), owing to the overwhelming effect of horizontal vorticity associated with the rear, cold-pool edge. In this case, this feature is called a "descending rear-inflow jet", and horizontal vorticity just below it, which is opposite in sign to that of baroclinically generated horizontal vorticity at the leading edge of the cold pool, is not advected all the way to the leading edge as it is by the elevated rear-inflow jet. An important consequence of the descent of the rear-inflow jet is that the rate at which horizontal vorticity is generated baroclinically at the leading edge of the cold pool is *not* balanced by the advection of horizontal vorticity (into it), so that the MCS is not as long lived as it would have been if the rear-inflow jet were elevated (and the ascending branch of inflow would not have been less intense).

In summary, the main factor determining whether or not a rear-inflow jet is elevated or descending (i.e., whether or not the convective system can be long lived or not) depends on the relative horizontal buoyancy gradients at the rear edge associated with the warm cloud above and the cold pool below. Morris Weisman found in numerical simulations that, in general, when CAPE is "low to medium" and low-level vertical shear is "weak to moderate" the rear-inflow jet descends to the surface behind the leading edge of the gust front. When CAPE is "high" and vertical shear "strong", the rear-inflow jet is elevated. Further refinement of estimates of how rapidly horizontal vorticity is generated at the leading edge of the cold pool and especially at the rear edge of the cold pool depends on cloud microphysics and the consequent melting and release of latent heat of condensation and fusion and on water loading, which reduces buoyancy inside clouds.

5.4 THE PRODUCTION OF VORTICES IN MCSs

In the preceding discussions of the dynamics of MCSs, the two-dimensional aspects of squall line MCSs were emphasized. It has been found observationally that many MCSs are fully three dimensional (but "quasi-linear" in shape) and several types of vortices can occur at low or mid-levels. For example, Ted Fujita in the late 1970s identified and named the "bow echo" (Figure 5.23), in which a 40–100 km long convective line segment bulges outward and is associated with damaging straight line winds at the surface. In addition, bow echoes sometimes produce counter-rotating vortices at either end of the line at ~2–3 km AGL: in the Northern Hemisphere, an anticyclonic (cyclonic) vortex is produced on the right (left) side of the end of the line with respect to the mean vertical shear vector. Morris Weisman in the early 1990s named these features "bookend vortices" (Figure 5.24) (also referred to as "line end vortices"). Not only may bow echoes that comprise the scale of the entire system form, but smaller, bowed line segments may also occur (Figures 5.13a, 5.23a, bottom panel).

Bow echoes early in their life tend to be symmetrical, but later in their life the cyclonic bookend vortex becomes stronger than the anticyclonic bookend vortex. The bow echoes develop into a comma shape (Figures 5.11b, 5.23b) as an asymmetric MCS. The cyclonic vortex is favored over the anticyclonic vortex because, as noted earlier, in a stratiform precipitation region, convergence at mid-levels, above a low-level downdraft and below an upper-level updraft, acts on the Earth's vorticity to produce cyclonic vorticity. The cyclonic bookend vortex turns into a "mesoscale convective vortex (MCV)". It is possible that over a warm ocean surface, under the proper environmental conditions, a tropical cyclone can develop from an MCV. MCVs, unlike mesocyclones (and meso-anticyclones) in supercells, are larger in scale (as large as hundreds of kilometers) and can last much longer (up to days).

Weisman and others have found in numerical simulation experiments that the production of bow echoes with bookend vortices and strong surface winds is most likely in environments of high CAPE and strong low-level vertical wind shear (15–20 $\mathrm{m\,s^{-1}}$ or higher over the lowest 2–5 km). Less well-organized (with weaker, shallower, and shorter lived vortices) bow echoes can occur when vertical shear is only 10–15 $\mathrm{m\,s^{-1}}$ over the lowest 2–5 km. When a rear-inflow jet forms as described in Section 5.3, it is enhanced by the development of bookend vortices, which can focus the rear-inflow jet into the center of the convective system (Figure 5.25).

Some vortices originate as the cyclonic and anticyclonic vortices along the right and left flanks of a pair of cells that originated from the splitting of a parent supercell (when deep-layer shear, 0–6 km or deeper, is high); the radar echo expands as cold outflow in between the two members of the splitting cell forces new convective cells to grow in between them. The vortices in effect are produced when a downdraft or updraft acts on strong or deep westerly vertical shear and occur very early in the life of an MCS when there may be isolated supercells. Or, later on in the life of the MCS, they may form preferentially at the ends of the linear MCS where there is minimal interference from neighboring cells. They may make the MCS leading convective line break up into shorter line segments, which may evolve into short bowed line segments.

However, since most bookend vortices from along the ends of a pre-existing line segment or bowing segment within a much longer line, Chris Davis and Morris Weisman suggested that most *subsystem-scale* (~5–10 km) bookend vortices are produced when a localized downdraft, most likely initiated by precipitation loading and enhanced by evaporation, sublimation, and melting, tilts the horizontal vorticity associated with environmental vertical wind shear (Figure 5.26, bottom panel) early on in the evolution of the MCS before a substantial cold pool has been formed.

On the other hand, system-scale (tens of kilometers) bookend vortices are produced when an updraft acts on vorticity that is baroclinically generated at the leading edge of the (mature) gust front as air is lifted over the cold pool and the

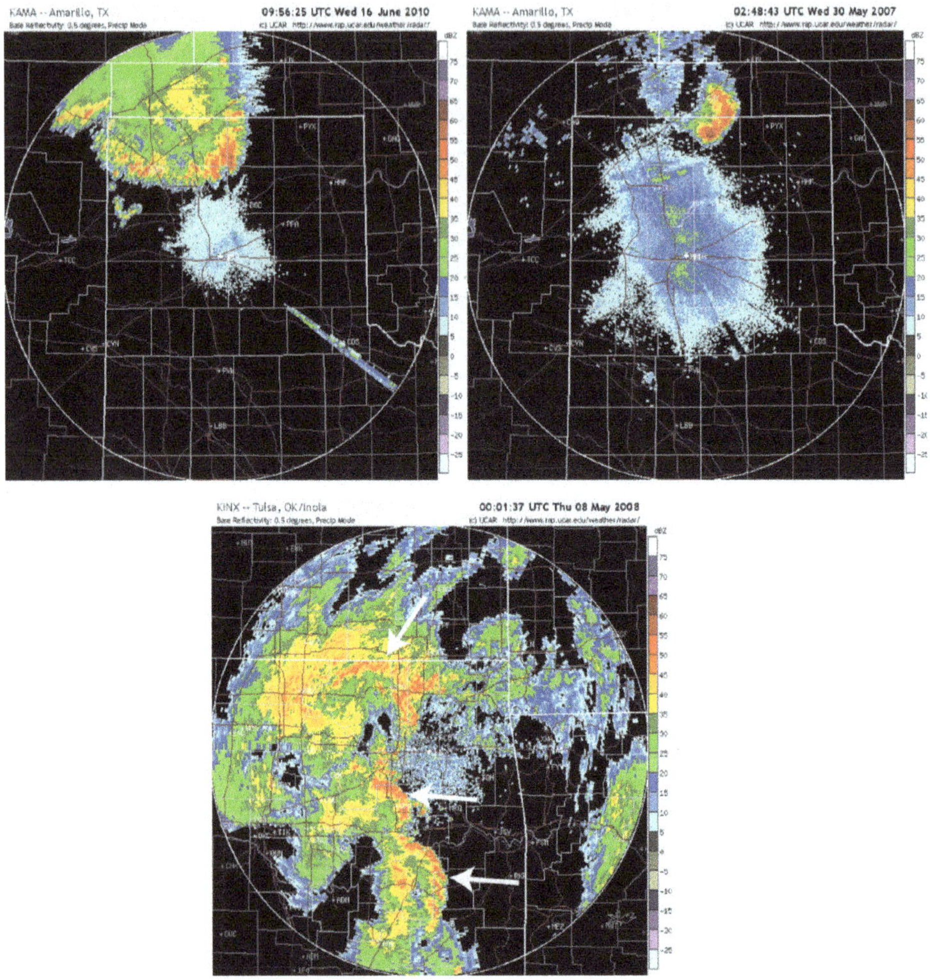

Figure 5.23a. Bow echoes as depicted by WSR-88D radars. Radar reflectivity color-coded in dBZ. (Top, left) A symmetric, narrow bow echo on June 16, 2010 entering the northwestern Texas Panhandle. (Top, right) An isolated short bow echo line segment on May 29, 2007 over the Oklahoma and northern Texas Panhandle. (Bottom) Short bowed line segments (indicated by white arrows) in Oklahoma on May 7, 2008.

LFC is reached (Figure 5.26, top panel). In this scenario, which occurs later on in the lifetime of an MCS, after a substantial cold pool at the surface has been built up, and in an environment of only weak-to-moderate vertical wind shear, tilting results in an anticyclone to the right of baroclinically generated, low-level easterly vertical shear and a cyclone to the left of the easterly vertical shear (to the south and north, respectively, in the Northern Hemisphere, when the MCS leading line has a meridional orientation). These vortices are lifted and advected rearward by

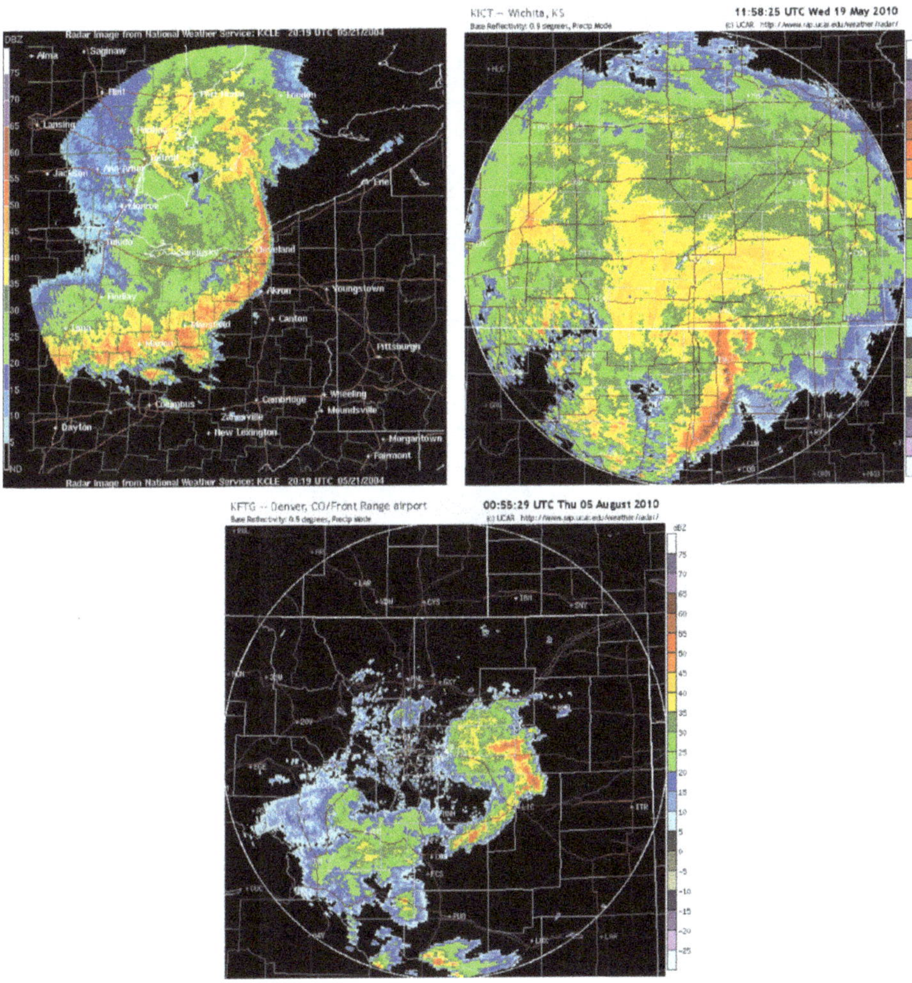

Figure 5.23b. Bow echoes as depicted by WSR-88D radars. Radar reflectivity color-coded in dBZ. Asymmetric bow echoes with varyingly strong mesoscale convective vortices in the stratiform precipitation regions on (top, left) May 21, 2004 over Ohio and Michigan, on (top, right) May 19, 2010 over southern Kansas and northern Oklahoma, and on (bottom) August 4, 2010 over northeastern Colorado.

the front-to-rear relative flow. According to RKW theory, when low-level vertical shear is relatively weak, the front-to-rear ascending flow is highly tilted, so that bookend vortices are advected relatively far to the rear; when vertical shear is stronger, the front-to-rear ascending flow is more upright, so that bookend vortices are located closer to, but still behind, the leading edge of the cold pool. Bookend vortex formation is delayed in the case of stronger low-level shear because it takes longer for the baroclinic generation of horizontal vorticity along

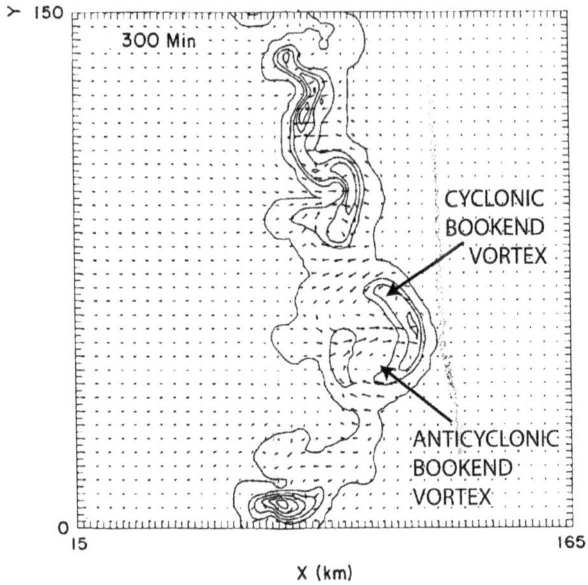

Figure 5.24. Bookend vortices in a simulated squall line MCS. Horizontal cross section at 5 h after initiation of storm-relative flow (vectors) at 2.5 km AGL in an environment of strong low-level shear. Vectors plotted every other grid point, corresponding to a distance of $25\,\mathrm{m\,s^{-1}}$ every two grid points). Rainwater field (solid lines) is contoured every $2\,\mathrm{g\,kg^{-1}}$ (from Weisman, 1993).

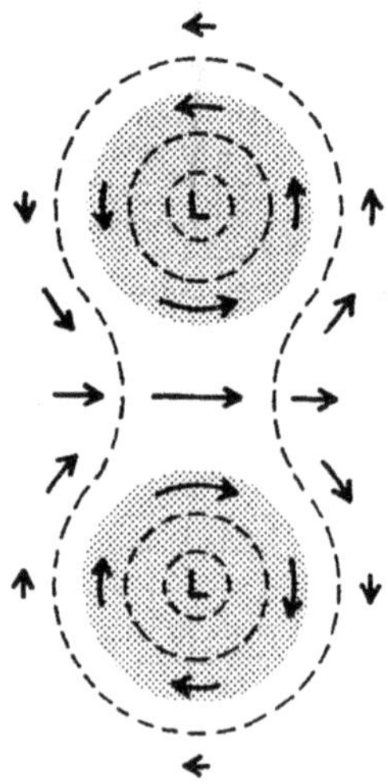

Figure 5.25. Schematic representation of an idealized two-dimensional counter-rotating vortex couplet showing the stronger flow (wind field depicted by vectors) induced in between the vortices. Shading marks regions of constant vertical vorticity and unshaded areas have zero vorticity. Dashed contours represent an approximate negative perturbation pressure field consistent with the wind field. Locations of lowest pressure are denoted by an "L" (from Weisman, 1993).

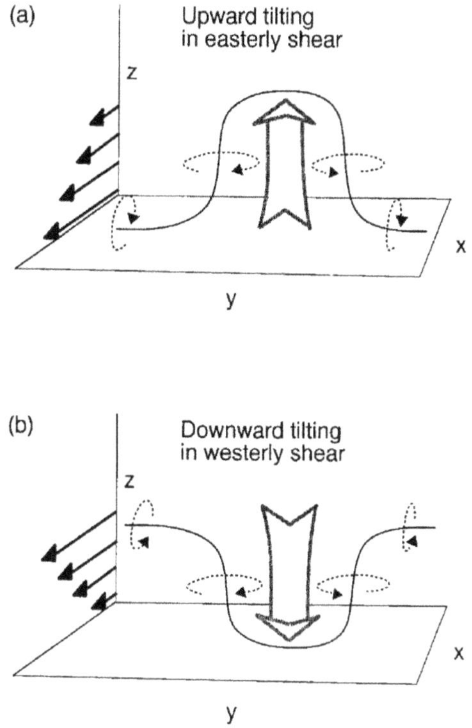

Figure 5.26. Idealized illustration of how (a) upward tilting of vorticity in the $-y$-direction associated with easterly vertical shear produces an anticyclonic vortex to the south and cyclonic vortex to the north and how (b) downward tilting of vorticity in the $+y$-direction associated with westerly shear produces the same kind of vortex couplet. In (a) easterly shear is produced baroclinically at the leading edge of the cold pool while in (b) westerly shear is associated with environmental vertical shear (from Weisman and Davis, 1998).

the leading edge of the cold pool to overwhelm the rate at which environmental horizontal vorticity of the opposite sign (associated with low-level shear) is advected to the leading edge of the cold pool.

When bookend vortices are relatively close together, the rear-inflow jet is enhanced as rear-to-front flow is strengthened, which can increase the intensity of the updraft at the leading edge of the cold pool; as a consequence, the tilting of system-generated horizontal vorticity (baroclinically generated along the leading edge of the cold pool) is stronger. In effect, the shorter the length of the convective system, the less time it takes for bookend vortices to form.

Late in the evolution of MCSs, small-scale vortices (both cyclonic and anti-cyclonic vortex pairs) within the MCS, not just at the ends, may be produced by the tilting of baroclinic vorticity produced at the leading edge of the gust front by either localized updrafts *or* downdrafts (Figure 5.27).

Since vortices are associated with dynamically induced regions of relatively low pressure, there is a downward-directed pressure gradient force that tends to

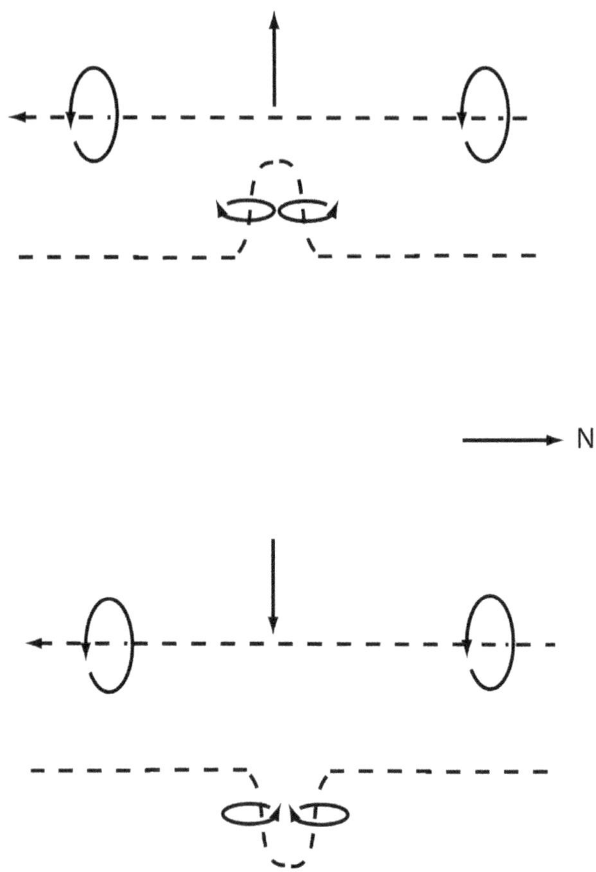

Figure 5.27. Idealized illustration of how small-scale counter-rotating vortices may be produced later in the life of a squall line MCS by the tilting of baroclinically generated horizontal vorticity in the $-y$-direction (y points toward the north; the cold pool lies to the west) by (top) an updraft and (bottom) a downdraft.

act against any updraft above the vortex. This suppression of updrafts near low-level vortices acts to break up an existing continuous line into segments when there are vortices produced at low levels (Figure 5.28). Bookend vortices then develop along the new line ends, and the process may be repeated.

The vortices in MCSs at low levels are unlike mesocyclones in supercells in that they are not characterized by upward-directed, dynamically forced accelerations and rising motion at mid-levels. In addition, they tend to move along with the mean wind and do not propagate normal to vertical shear, so that the helicity paradigm is not applicable to their dynamics. Furthermore, these "mesovortices" are produced from the tilting of crosswise vorticity—not streamwise vorticity. Forecasters and nowcasters must be careful not to assume that mesocyclones/

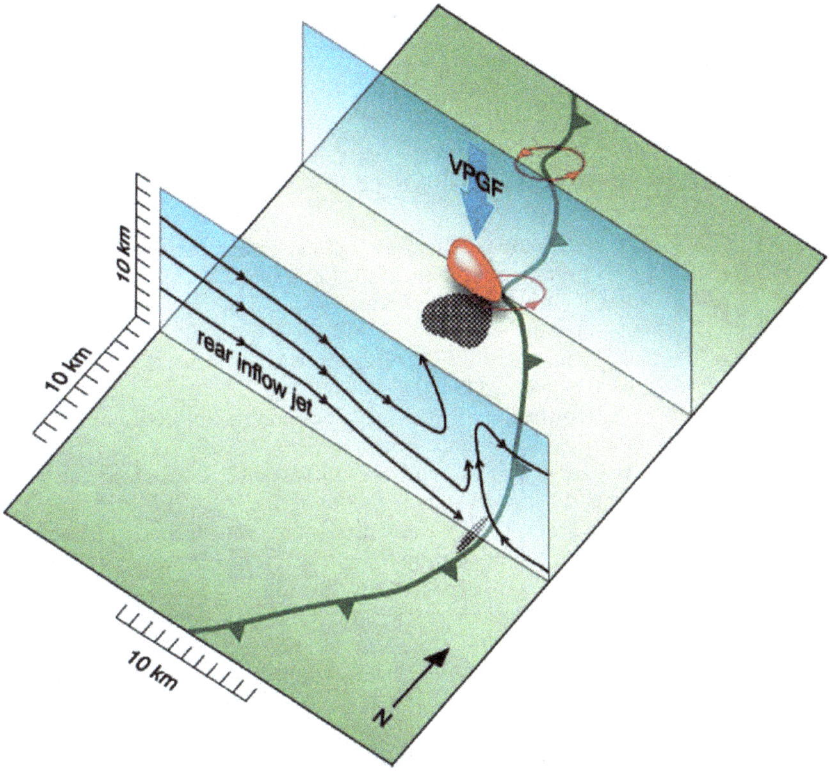

Figure 5.28. Idealized illustration showing how a downward-directed dynamic pressure gradient force (blue arrow) is present above a mesovortex (red surface), which suppresses new updrafts and fractures the convective line. The gust front/leading edge of the cold pool is denoted by a cold front symbol. A rear-inflow jet is shown descending to the surface south of the mesovortex and producing an area of enhanced surface wind (stippled). Enhanced surface winds are also indicated just southwest of the mesovortex (from Trapp and Weisman, 2003).

MCVs in linear (or "quasi-linear") MCSs are supercell mesocyclones and produce supercell-associated severe weather phenomena.

When environmental vertical shear is relatively strong, there may be high straight line winds at the surface, which may inflict damage at the apex of the bow echo, in between two counter-rotating vortices, perhaps closer to the cyclonic member of the pair as it has become stronger than the anticyclonic member in response to the Coriolis force. On the other hand, when vertical shear is weaker, well-defined mesovortices may not form and strong straight line winds occur from a rear-inflow jet that has descended to the ground some distance to the rear of the leading edge of the cold pool. Long-lived MCSs that move rapidly and produce widespread strong straight-line winds at the surface are called *derechos* (from the Spanish word for "straight ahead"), as opposed to tornadoes (from the Spanish word for "to turn").

5.5 GENERAL MONOGRAPHS AND BOOKS

The reader is referred to p. 24 for a list of relevant general monographs and books.

5.6 REFERENCES AND BIBLIOGRAPHY

Adams-Selin, R. D. and R. H. Johnson (2010) Mesoscale surface pressure and temperature features associated with bow echoes. *Mon. Wea. Rev.*, **138**, 212–227.

Biggerstaff, M. I. and R. A. Houze, Jr. (1991a) Kinematic and precipitation structure of the 10–11 June 1985 squall line. *Mon. Wea. Rev.*, **119**, 3034–3065.

Biggerstaff, M. I. and R. A. Houze, Jr. (1991b) Midlevel vorticity structure of the 10–11 June 1985 squall line. *Mon. Wea. Rev.*, **119**, 3066–3079.

Biggerstaff, M. I. and R. A. Houze, Jr. (1993) Kinematics and microphysics of the transition zone of the 10–11 June 1985 squall line. *J. Atmos. Sci.*, **50**, 3091–3110.

Bluestein, H. B. and M. H. Jain (1985) Formation of mesoscale lines of precipitation: Severe squall lines in Oklahoma during the spring. *J. Atmos. Sci.*, **42**, 1711–1732.

Bluestein, H. B. and G. T. Marx (1987) Formation of mesoscale lines of precipitation: Nonsevere squall lines in Oklahoma during the spring. *Mon. Wea. Rev.*, **115**, 2719–2727.

Braun, S. A. and R. A. Houze, Jr. (1994) The transition zone and secondary maximum of radar reflectivity behind a midlatitude squall line: Results retrieved from Doppler radar data. *J. Atmos. Sci.*, **51**, 2733–2755.

Braun, S. A. and R. A. Houze, Jr. (1996) The heat budget of a midlatitude squall line and implications for potential vorticity production. *J. Atmos. Sci.*, **53**, 1217–1240.

Braun, S. A. and R. A. Houze, Jr. (1997) The evolution of the 10–11 June 1985 PRE-STORM squall line: Initiation, development of rear inflow, and dissipation. *Mon. Wea. Rev.*, **125**, 478–504.

Browning, K. A. and F. H. Ludlam (1962) Airflow in convective storms. *Quart. J. Roy. Meteor. Soc.*, **88**, 117–135.

Bryan, G. H. and J. M. Fritsch (2000) Moist absolute instability: The sixth static stability state. *Bull. Amer. Meteor. Soc.*, **81**, 1207–1230.

Coniglio, M. C., D. J. Strensrud, and L. J. Wicker (2006) Effects of upper-level shear on the structure and maintenance of strong quasi-linear mesoscale convective systems. *J. Atmos. Sci.*, **63**, 1231–1252.

Davis, C. A. and M. L. Weisman (1994) Mesoscale vortices produced in simulated convective systems. *J. Atmos. Sci.*, **51**, 2005–2030.

Davis, C. A., N. Atkins, D. Bartels, L. Bosart, M. Coniglio, G. Bryan, W. Cotton, D. Dowell, B. Jewett, R. Johns *et al.* (2004) The Bow Echo and MCV Experiment. *Bull. Amer. Meteor. Soc.*, **85**, 1075–1093.

Emanuel, K. A. (1994) *Atmospheric Convection*, Oxford University Press, New York, Chapters 1–6, 9–11.

Fovell, R. G. and Y. Ogura (1988) Numerical simulation of a midlatitude squall line in two dimensions. *J. Atmos. Sci.*, **45**, 3846–3879.

Fovell. R. G., G. L. Mullendore, and S.-H. Kim (2006) Discrete propagation in numerically simulated nocturnal squall lines. *Mon. Wea. Rev.*, **134**, 3735–3752.

Fritsch, J. M. and G. S. Forbes (2001) Mesoscale convective systems. Chapter 9 in: C. Doswell, III (Ed.), *Severe Convective Storms*, AMS Monogr. Vol. 28, Issue 50, American Meteorological Society, Boston, pp. 323–357.

Fujita, T. (1955) Results of detailed synoptic studies of squall lines. *Tellus*, **4**, 405–436.

Gallus, W. A., Jr. and R. H. Johnson (1995a) The dynamics of circulations within the trailing stratiform regions of squall lines, Part I: The 10–11 June PRE-STORM system. *J. Atmos. Sci.*, **52**, 2161–2187.

Gallus, W. A., Jr. and R. H. Johnson (1995b) The dynamics of circulations within the trailing stratiform regions of squall lines, Part II: Influence of the convective line and ambient environment. *J. Atmos. Sci.*, **52**, 2188–2211.

Haertel, P. T. and R. H. Johnson (2000) The linear dynamics of squall line mesohighs and wake lows. *J. Atmos. Sci.*, **57**, 93–107.

Hane, C. E. and D. P. Jorgensen (1995) Dynamic aspects of a distinctly three-dimensional mesoscale convective system. *Mon. Wea. Rev.*, **123**, 3194–3214.

Houze, R. A., Jr. (2004) Mesoscale convective systems. *Rev. Geophys.*, **42**, RG4003, doi: 10.1029/2004RG000150.

Houze, R. A., Jr., S. A. Rutledge, M. I. Biggerstaff, and B. F. Smull (1989) Interpretation of Doppler weather radar displays of midlatitude mesoscale convective systems. *Bull. Amer. Meteor. Soc.*, **70**, 608–619.

Houze, R. A., Jr., B. F. Smull, and P. Dodge (1990) Mesoscale organization of springtime rainstorms in Oklahoma. *Mon. Wea. Rev.*, **118**, 613–654.

James, E. P. and R. H. Johnson (2010) Patterns of precipitation and mesolow evolution in midlatitude mesoscale convective vortices. *Mon. Wea. Rev.*, **138**, 909–931.

Johns, R. H. and W. D. Hirt (1987) Derechos: Widespread convectively induced windstorms. *Wea. Forecasting*, **2**, 32–49.

Johnson, R. H. (2001) Mesohighs and mesolows. *Bull. Amer. Meteor. Soc.*, **82**, 13–31.

Johnson, R. H. and P. J. Hamilton (1988) The relationship of surface pressure features to the precipitation and airflow structure of an intense midlatitude squall line. *Mon. Wea. Rev.*, **116**, 1444–1472.

Johnson, R. H., S. Chen, and J. J. Toth (1989) Circulations associated with a mature-to-decaying midlatitude mesoscale convective system, Part I: Surface features—heat bursts and mesolow development. *Mon. Wea. Rev.*, **117**, 942–958.

Jorgensen, D. P. and B. F. Smull (1993) Mesovortex circulations seen by airborne Doppler radar within a bow-echo mesoscale convective system. *Bull. Amer. Meteor. Soc.*, **74**, 2146–2157.

Jorgensen, D. P. and T. M. Weckwerth (2003) Forcing and organization of convective systems. Chapter 4 in: R. M. Wakimoto and R. C. Srivastava (Eds.), *Radar and Atmospheric Science: A Collection of Essays in Honor of David Atlas*, AMS Monogr. Vol. 30, Issue 52, American Meteorological Society, Boston, pp. 75–103.

Knievel, J. C. and R. H. Johnson (1998) Pressure transients within MCS mesohighs and wake lows. *Mon. Wea. Rev.*, **126**, 1907–1930.

Lindzen, R. S. (1974) Wave-CISK in the tropics. *J. Atmos. Sci.*, **31**, 156–179.

Loehrer, S. M. and R. H. Johnson (1995) Surface pressure and precipitation life cycle characteristics of PRE-STORM mesoscale convective systems. *Mon. Wea. Rev.*, **123**, 600–621.

Maddox, R. A. (1980) Mesoscale convective complexes. *Bull. Amer. Meteor. Soc.*, **11**, 1374–1387.

Moncrieff, M. W. (1978) The dynamical structure of two-dimensional steady convection in constant vertical shear. *Quart. J. Roy. Meteor. Soc.*, **104**, 543–567.

Moncrieff, M. W. and J. S. A. Green (1972) The propagation and transfer properties of steady convective overturning in shear. *Quart. J. Roy. Meteor. Soc.*, **98**, 336–352.

Newton, C. W. and J. C. Fankhauser (1964) On the movements of convective storms, with emphasis on size discrimination in relation to water-budget requirements. *J. Appl. Meteor.*, **3**, 651–668.

Nolen, R. H. (1959) A radar pattern associated with tornadoes. *Bull. Amer. Meteor. Soc.*, **40**, 277–279.

Ogura, Y. and M.-T. Liou (1980) The structure of a midlatitude squall line: A case study. *J. Atmos. Sci.*, **37**, 553–567.

Parker, M. D. and R. H. Johnson (2000) Organizational modes of midlatitude mesoscale convective systems. *Mon. Wea. Rev.*, **128**, 3413–3436.

Parker, M. D. and R. H. Johnson (2004a) Structures and dynamics of quasi-2D mesoscale convective systems. *J. Atmos. Sci.*, **61**, 545–567.

Parker, M. D. and R. H. Johnson (2004b) Simulated convective lines with leading precipitation, Part I: Governing dynamics. *J. Atmos. Sci.*, **61**, 1637–1655.

Parker, M. D. and R. H. Johnson (2004c) Simulated convective lines with leading precipitation, Part II: Evolution and maintenance. *J. Atmos. Sci.*, **61**, 1656–1673.

Raymond, D. J. (1975) A model for predicting the movement of continuously propagating convective storms. *J. Atmos. Sci.*, **32**, 1308–1317.

Rutledge, S. A., R. A. Houze, Jr., M. I. Biggerstaff, and T. Majejka (1988) The Oklahoma–Kansas mesoscale convective system of 10–11 June 1985: Precipitation structure and single-Doppler radar analysis. *Mon. Wea. Rev.*, **116**, 1409–1430.

Skamarock, W. C., M. L. Weisman, and J. B. Klemp (1994) Three-dimensional evolution of simulated long-lived squall lines. *J. Atmos. Sci.*, **51**, 2563–2584.

Smull, B. F. and R. A. Houze, Jr. (1985) A midlatitude squall line with a trailing region of stratiform rain: Radar and satellite observations. *Mon. Wea. Rev.*, **113**, 117–133.

Smull, B. F. and R. A. Houze, Jr. (1987) Rear inflow in squall lines with trailing stratiform precipitation. *Mon. Wea. Rev.*, **115**, 2869–2889.

Stensrud, D. J., M. C. Coniglio, R. P. Davies-Jones, and J. S. Evans (2005) Comments on "A theory for strong long-lived squall lines" revisited. *J. Atmos. Sci.*, **62**, 2989–2996.

Storm, B. A., M. D. Parker, and D. P. Jorgensen (2007) A convective line with leading stratiform precipitation from BAMEX. *Mon. Wea. Rev.*, **135**, 1769–1785.

Trapp, R. J. and M. L. Weisman (2003) Low-level mesovortices within squall lines and bow echoes, Part II: Their genesis and implications. *Mon. Wea. Rev.*, **131**, 2804–2823.

Verlinde, J. and W. R. Cotton (1990) A mesoscale vortex couplet observed in the trailing anvil of a multicellular convective complex. *Mon. Wea. Rev.*, **118**, 993–1010.

Vescio, M. D. and R. H. Johnson (1992) The surface-wind response to transient mesoscale pressure fields associated with squall lines. *Mon. Wea. Rev.*, **120**, 1837–1850.

Wakimoto, R. M. (2001) Convectively driven high winds. Chapter 7 in: C. Doswell, III (Ed.), *Severe Convective Storms*, AMS Monogr. Vol. 28, Issue 50, American Meteorological Society, Boston, pp. 255–298.

Wakimoto, R. M., H. V. Murphey, A. Nester, D. P. Jorgensen, and N. T. Atkins (2006) High winds generaged by bow echoes, Part I: Overview of the Omaha bow echo 5 July 2003 storm during BAMEX. *Mon. Wea. Rev.*, **134**, 2793–2812.

Weisman, M. L. (1992) Convectively generated rear-inflow jets in the evolution of long-lived mesoconvective systems. *J. Atmos. Sci.*, **49**, 1826–1847.

Weisman, M. L. (1993) The genesis of severe, long-lived bow echoes. *J. Atmos. Sci.*, **50**, 645–670.

Weisman, M. L. (2001) Bow echoes: A tribute to T. T. Fujita. *Bull. Amer. Meteor. Soc.*, **82**, 97–116.

Weisman, M. L. and C. A. Davis (1998) The generation of mesoscale vortices within quasi-linear convective systems. *J. Atmos. Sci.*, **55**, 2603–2622.

Weisman, M. L. and J. B. Klemp (1984) The structure and classification of numerically simulated convective storms in directionally varying wind shears. *Mon. Wea. Rev.*, **112**, 2479–2498.

Weisman, M. L. and R. J. Trapp (2003) Low-level mesovortices within squall lines and bow echoes, Part I: Overview and dependence on environmental shear. *Mon. Wea. Rev.*, **131**, 2779–2802.

Weisman, M. L., J. B. Klemp, and R. Rotunno (1988) Structure and evolution of numerically simulated squall lines. *J. Atmos. Sci.*, **45**, 1990–2013.

Yang, M.-J. and R. A. Houze, Jr. (1995) Multicell squall-line structure as a manifestation of vertically trapped gravity waves. *Mon. Wea. Rev.*, **123**, 641–661.

Zipser, E. J. and M. A. LeMone (1980) Cumulonimbus vertical velocity events in GATE, Part II: Synthesis and model core structure. *J. Atmos. Sci.*, **37**, 2458–2469.

6

Tornadoes

"In this review no consideration has been given to the tornado, which energetically is only a detail in the severe storm. However, its importance as a hazard and the interest of the problems which it poses make it desirable to indicate its probable place in the cumulonimbus model."

F. H. Ludlam (1963)

6.1 BASIC OBSERVATIONAL ASPECTS OF TORNADOES

Tornadoes are rapidly rotating columns of air that make contact with the ground. In general we consider them to be intense, columnar vortices in contact with the ground that are capable of inflicting damage. They are either connected to or are situated underneath a cumuliform, buoyant convective cloud above (Figure 6.1). Tornadoes are sometimes, but not always, visible as a funnel cloud when the pressure deficit in them is low enough that water vapor condenses and cloud particles form (Figure 6.2). When the air is too dry or the pressure not low enough for a condensation funnel to form or to extend down to the ground, the tornado may be visualized as a column of rotating dust or debris (Figure 6.3) or not visualized at all if there are no visible particles available to be lofted. In some instances they are encased in precipitation ("rain-wrapped") and either not visible at all or visible only from a restricted viewing angle (Figure 6.4).

Lasting anywhere from just a few seconds to an hour or more, most tornadoes last only ~10 min. The diameter of a tornado is usually ~200 m, but can vary from as narrow as ~10 m to as wide as ~1.5–2 km. Most tornadoes are distinctly narrower than mesocyclones, which are ~2–5 km in diameter. Tornadoes are the most violent and destructive of all the severe weather phenomena that localized convective storms produce.

Figure 6.1. Early drawing of condensation funnels, underneath cumuliform clouds, associated with tornadoes, in the U. S. in 1882 (courtesy of History of Science Collections, University of Oklahoma Libraries).

Figure 6.2a. Old drawing (or photograph) of a tornado in the U. S. in 1884 (courtesy of History of Science Collections, University of Oklahoma Libraries).

Figure 6.2b. Tornadoes with condensation funnels: May 12, 2004, near Attica, KS, silhouetted (top left); June 10, 2004, southwestern Nebraska, fully illuminated (top right); May 22, 1981, near Binger, OK (middle left); May 22, 1981, near Binger, OK, dust obscuring center of tornado (middle right); May 13, 1989, near Hodges, TX, dissipating stage, when gust front from parent storm makes condensation funnel lean with height (bottom left); May 13, 1989, near Hodges, TX, dissipating stage when condensation assumes a narrow, ropelike appearance (bottom right) (photographs by the author).

Wind speeds in tornadoes range from $\sim$20 to $\sim$140 m s^{-1}, although it is not generally agreed what the lower cutoff wind speeds are, and some rather weak vortices (producing only rotating dust whirls) have been reported as tornadoes. The translation speed of a tornado is important in determining the lower speed limit: A rapidly moving (e.g., at 20 m s^{-1}) very weak vortex (e.g., with azimuthal wind speeds of only 20 m s^{-1}) can inflict significant damage on the side where the ground-relative wind is highest. Efforts have been made to estimate wind speeds in tornadoes based on the nature and extent of the damage they inflict (Figure 6.5). The Fujita "F-scale", devised by Ted Fujita in the 1970s, ranges from F0

Figure 6.2c. As Figure 6.2b, but near Canadian, TX, May 7, 1986, dissipating stage, condensation aloft separating from condensation just above the ground (top); as in previous image, but slightly later when condensation funnel aloft has completely disappeared (bottom) (photographs by the author).

(18–32 m s^{-1}; 40–72 mph) to F5 (117–142 m s^{-1}; 261–318 mph), but is not calibrated (i.e., the relationship between wind speed and the nature of the damage inflicted has not actually been measured under controlled conditions). The Fujita scale was replaced by the "Enhanced Fujita (EF) scale" in 2007; unlike the F-scale, it is calibrated to some extent or at least made more consistent (Table 6.1). When tornadoes strike sparsely populated rural areas with little vegetation, it becomes very difficult, if not impossible, to estimate tornado wind speeds based on damage alone.

Damage from tornadic winds comes from the pressure exerted by the wind (proportional to the square of the wind speed), by transient responses to wind

Figure 6.3. Tornadoes with debris cloud at and just above the ground and a small funnel cloud at cloud base. (Top) June 10, 2004, in southwest Nebraska; (middle) April 12, 1991, in north central Oklahoma; and (bottom) May 12, 2004, near Attica, KS (photographs by the author).

Figure 6.4. Tornadoes surrounded by precipitation and partially obscured, when viewed looking to the south or southwest. (Top) May 14, 1986, near Snyder, OK; (bottom) June 13, 2010, in the eastern Oklahoma Panhandle (photographs by the author).

gusts, by aerodynamic effects produced as the airflow interacts with structures (e.g., lift may develop and rip roofs from buildings), from the impact of flying debris, and from structures tipping over and even being rolled over repeatedly. Engineers have been investigating the nature of tornado wind damage under controlled conditions in the laboratory, particularly at Texas Tech University in Lubbock. The results of such studies are beyond the scope of this text.

Damage debris may be lofted directly by upward wind currents or by aerodynamic effects. Debris may be detected by polarimetric Doppler radar as a region of relatively low co-polar cross-correlation coefficient ρ_{HV} (Figure 6.6) because debris tends to tumble so that the longest axis of radar cross sections of the debris tend not to be favored along any direction. To be detected by radar at long range, the debris must obviously be lofted high enough so that it is above the radar horizon. While ground teams often survey damage from tornadoes, swaths of damage are often best seen from aircraft or helicopters or from satellites (Figure 6.7).

Figure 6.5. Tornado damage in Moore, OK from a large tornado on May 3, 1999. (Top panel) A pickup truck and sheet metal wrapped against a tree. (Bottom panel) The foundation of a house swept nearly clean (photographs by the author).

Most tornadoes in supercells rotate cyclonically, while anticyclonic tornadoes are also observed but much more rarely, alone, but more often in tandem with nearby cyclonic tornadoes or with cyclonic circulations at the surface that are the remnants of cyclonic tornadoes (Figure 6.8). When counter-rotating tornado pairs occur, the anticyclonic tornadoes are located at the opposite end (usually the equatorward side) of the rear-flank gust front (Figure 6.9).

Some tornadoes contain even smaller-scale vortices within them that rotate about the main axis of the tornado (Figure 6.10). These subtornado-scale vortices have been referred to as "suction vortices" (named so by Ted Fujita) or "satellite vortices" or "secondary vortices". Tornadoes that have subtornado-scale vortices are frequently called "multiple-vortex" tornadoes. It has been postulated that these secondary vortices are responsible for extreme damage and local variability of damage. Ted Fujita first showed that over open fields they can produce

Table 6.1. A comparison of the Fujita scale with the enhanced Fujita scale.

Fujita Scale			*EF Scale*	
Fujita Scale	*Fastest $\frac{1}{4}$-mile wind speeds* (mph)	*3-second gust speed* (mph)	*EF Scale*	*3-second gust speed* (mph)
F0	40–72	45–78	EF0	65–85
F1	73–112	79–117	EF1	86–109
F2	113–157	118–161	EF2	110–137
F3	158–207	162–209	EF3	138–167
F4	208–260	210–261	EF4	168–199
F5	261–318	262–317	EF5	200–234

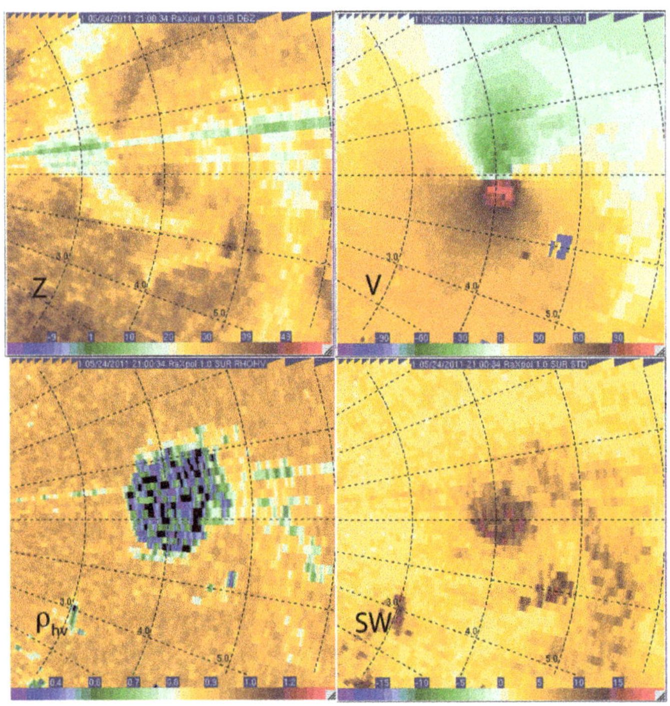

Figure 6.6. Polarimetric radar debris signature in an EF5 tornado, as detected by the mobile radar RaXPol, in Oklahoma, on May 24, 2011 (data processing courtesy of Jana Houser and Jeff Snyder). (Top left) Radar reflectivity factor (Z) in dBZ_e; (top right) Doppler velocity (V) in $m\,s^{-1}$; (bottom left) co-polar cross-correlation coefficient (ρ_{HV}); (bottom right) spectrum width in $m\,s^{-1}$. The region of $\rho_{HV} < 0.8$ is characteristic of debris, which encloses the "debris ball" of moderate reflectivity (>30–35 dBZ_e), the cyclonic vortex shear signature (red–green couplet), and the region of relatively high spectrum widths. Spectrum width is a measure of the variation in wind speeds in a radar volume. The maximum wind speeds indicated are $125\,m\,s^{-1}$.

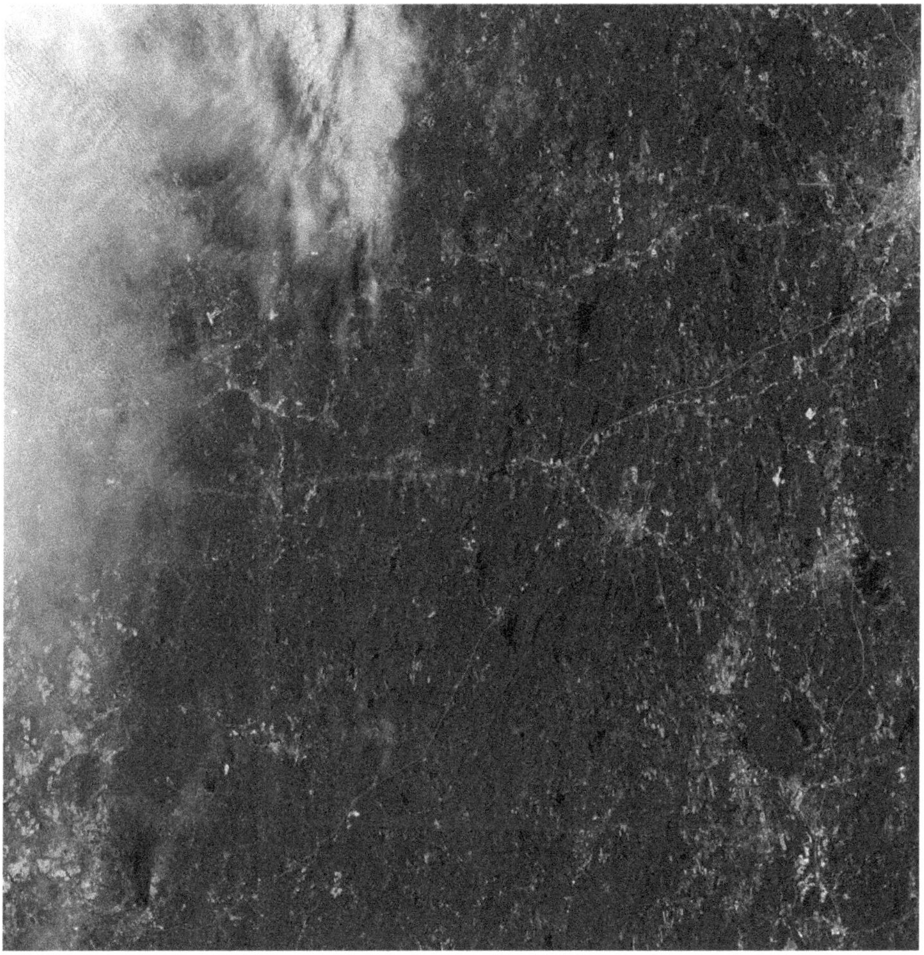

Figure 6.7. Tornado damage track (horizontal brown swath surrounded by greenery from vegetation) over Massachusetts from a tornado on June 1, 2011, as seen by the Landsat 5 satellite.

cycloidal damage marks indicative of smaller scale vortices rotating around a common axis (Figure 6.11). They typically do revolve all the way around the common axis, but rather appear in one location with respect to the larger vortex, rotate part way around, and dissipate. They often lean outwards with height from the central axis of rotation (Figure 6.10a).

When a rotating column of air is visualized as a condensation funnel that does not appear to be making contact with the ground (i.e., it visibly terminates well above the ground and no surface debris cloud is apparent), then the feature is referred to as a "funnel cloud" (Figure 6.12). It is not always clear whether or not a funnel cloud is associated with a strong circulation at the ground, especially if the ground surface does not contain material that is easily swept airborne.

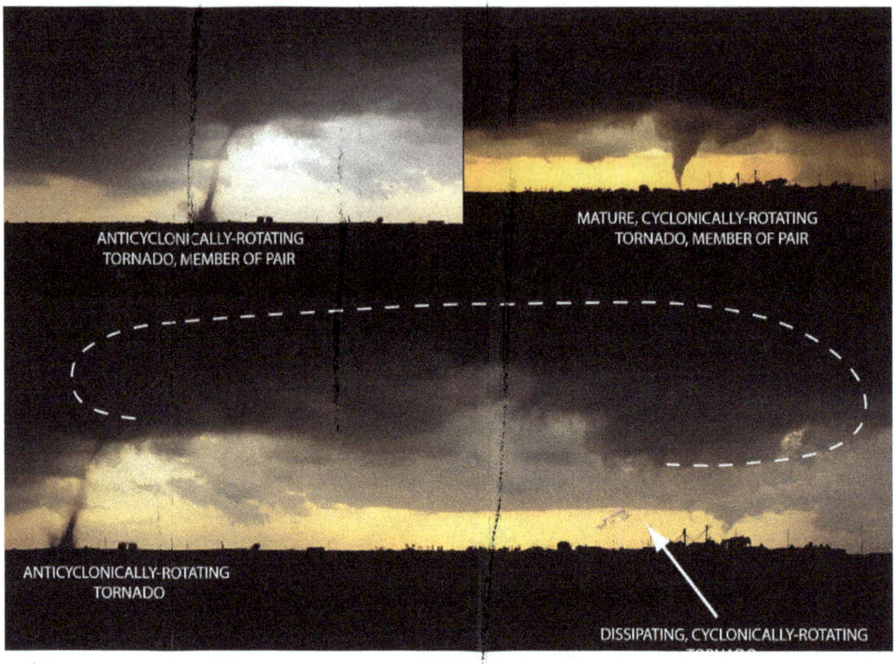

Figure 6.8. Cyclonic–anticyclonic tornado pair, April 24, 2006, near El Reno, OK. (Top left) Anticyclonically rotating tornado, part of pair; (top right) cyclonically rotating tornado, part of pair, before anticyclonically rotating member had formed; (bottom) anticyclonic tornado to the south (left), dissipating, cyclonic tornado with ropelike condensation funnel to the north (right). Dashed white line marks approximate location of cloud base associated with the rear-flank gust front (photographs by the author).

Tornadoes that appear over the water are called "waterspouts" (Figure 6.13). Waterspouts sometimes make landfall and by definition become tornadoes, while tornadoes sometimes pass over bodies of water and by definition become waterspouts. There is nothing fundamentally different dynamically about tornadoes and waterspouts, except perhaps for the characteristics of surface friction and their effect on the character of the vortex in the boundary layer. Tornadoes have been observed over all types of terrain, including mountainous areas (Figure 6.14). Tornadoes do not avoid complex terrain, but certainly high-elevation locations over mountainous terrain typically are drier than low-elevation locations, especially those sites that have access to air that has had contact with warm water surfaces. Also, the effects of surface friction are different over complex terrain and these might affect the nature of the tornadoes that form over them. Channeling and other terrain-induced features could affect vertical wind shear locally and increase or decrease vertical shear, thus possibly increasing or decreasing, respectively, the chances for tornado formation.

Related to tornadoes and waterspouts are dust devils (Figure 6.15), steam devils (Figure 6.16), and fire whirls, which are all driven by surface heating, which

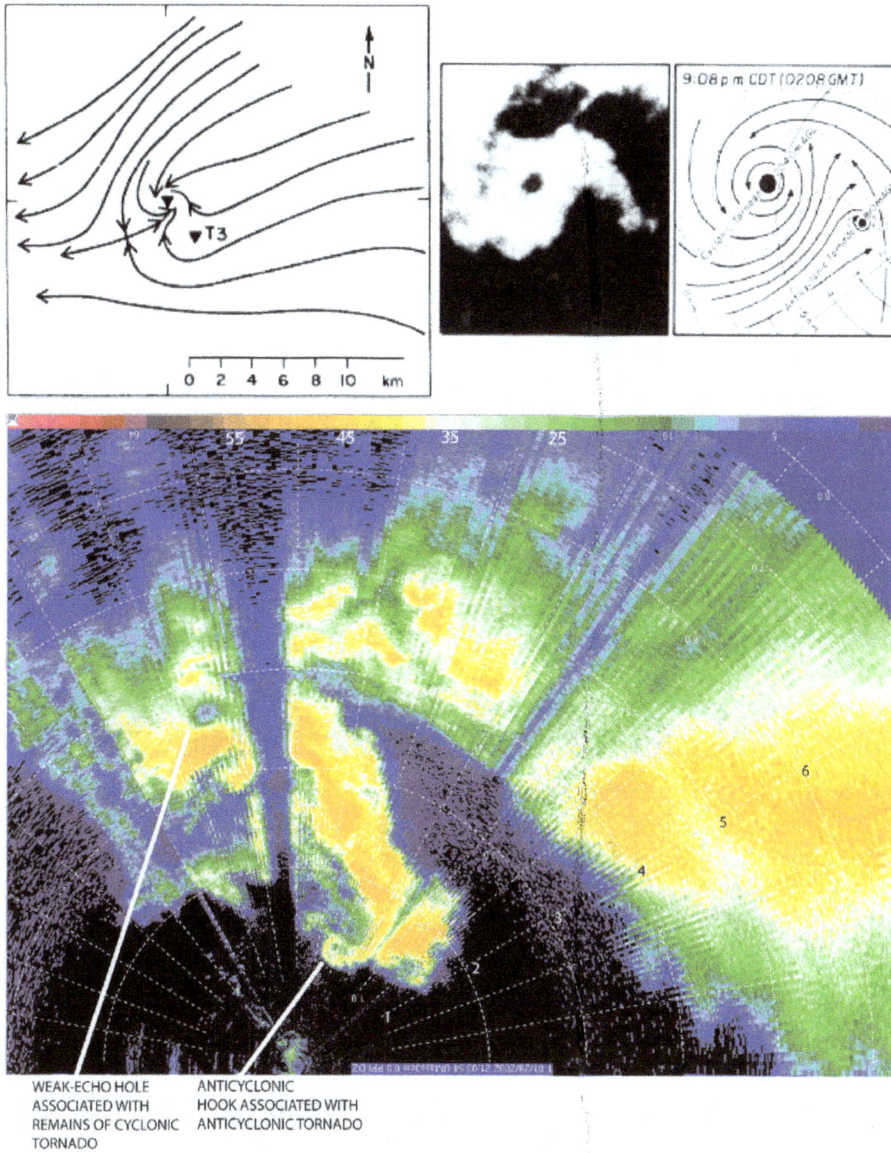

Figure 6.9. Examples of cyclonic–anticyclonic tornado pairs in cyclonically rotating, right-moving supercells. (Top left) Smoothed streamlimes of the surface wind field relative to the motion of the cyclonically rotating tornado, indicated by the upside-down triangle without a label; the anticyclonic tornado is indicated by the upside-down triangle labeled "T3". For tornadoes in central Iowa on June 13, 1976 (from Brown and Knupp, 1980); (top right) WSR-57 radar reflectivity image of a cyclonic–anticyclonic tornado pair near Grand Island, NE, on June 3, 1980 (from Fujita, 1981); (bottom) color-coded radar reflectivity factor (dBZ_e) for a cyclonic–anticyclonic tornado pair near Calumet, OK on May 29, 2004; from the U. Mass. X-Pol mobile Doppler radar. Range rings shown in km.

Figure 6.10a. Photographs of multiple-vortex (i.e., having smaller, secondary, satellite tornadoes rotating around a common axis) tornadoes. (Top) May 11, 1982, Friendship, OK; (middle panels) May 3, 1999, near Verden, OK; (bottom) close-up view of a "suction vortex", May 3, 1999, near Verden, OK. The parent tornado circulation went on to strike parts of Oklahoma City and Moore as its parent supercell tracked northeastward (photographs by the author).

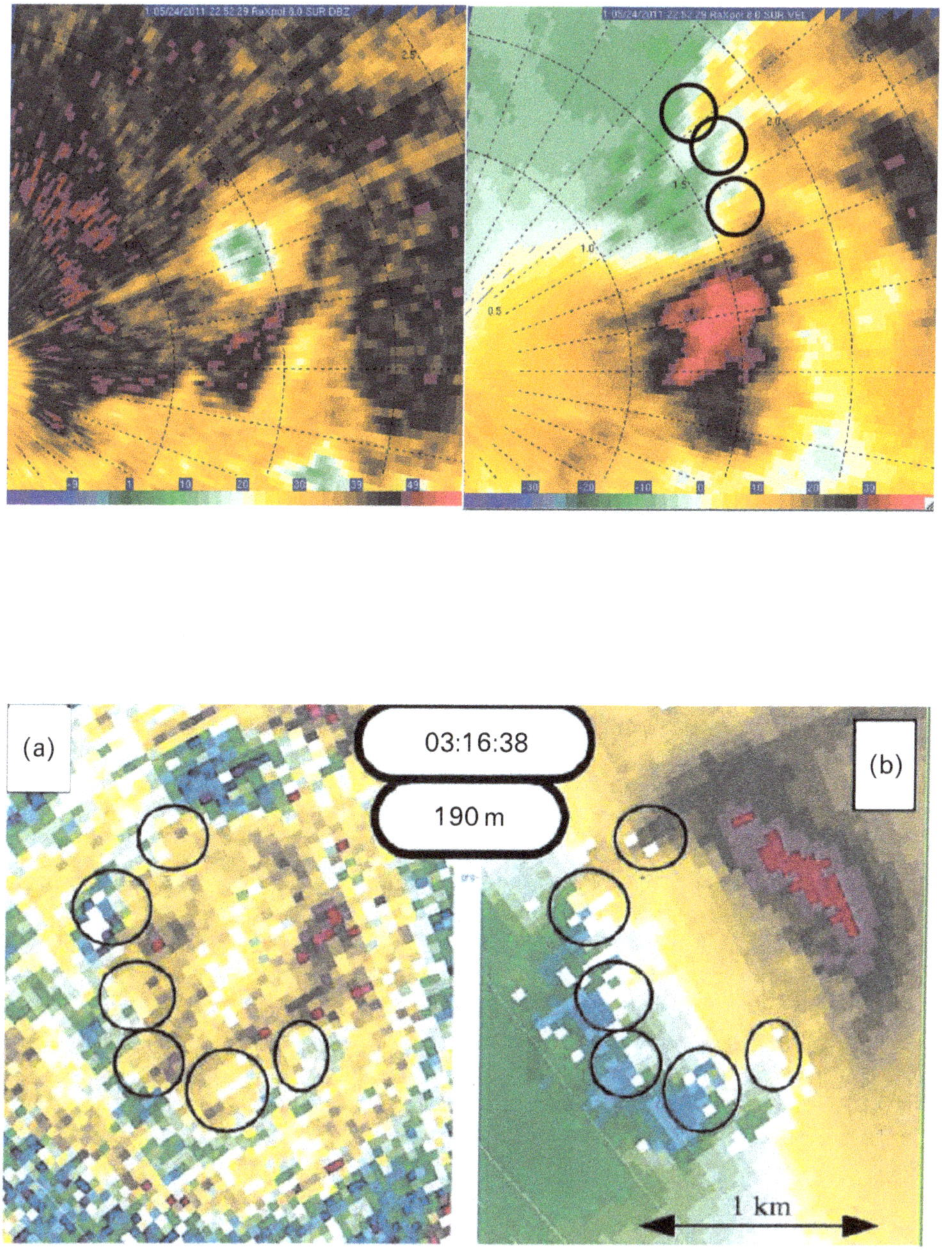

Figure 6.10b. Doppler radar imagery of multiple-vortex tornadoes. (top) Tornado east of Newcastle, OK on May 24, 2011, as detected by RaXPol, (left) radar reflectivity in dBZ_e, (right) Doppler velocity in $m\,s^{-1}$, range markers given every 500 m, circles mark vortex couplet signatures of multiple vortices, at 8° elevation angle, near 180 m ARL; (bottom) tornado near Mulhall, OK on May 3, 1999, at 190 m ARL, (left) power returned to DOW radar, (right) vortex couplet signatures of multiple vortices marked by circles (from Lee and Wurman, 2005).

Figure 6.11. Cycloidal marks (two of which are indicated by arrows) in a field created by small-scale vortices in a tornado rotating around an axis of rotation associated with a larger vortex (photograph by Ted Fujita, circa mid-1960s; courtesy of History of Science Collections, University of Oklahoma Libraries).

leads to superadiabatic lapse rates near the ground. Fire whirls owe their existence entirely to intense surface heat sources. Volcanic eruptions are sometimes responsible for producing tornadoes and waterspouts, which are associated with "pyro"-cumulonimbus clouds. Steam devils occur when very cold air is advected over a relatively warm water surface. Dust devils are driven by solar insolation at the ground, but may depend on frictionally induced vertical shear near the surface, or clear air cellular convection and the tilting of horizontal shear due to the vertical circulation of cells, or baroclinically generated horizontal vorticity. Dust devils typically occur in an unsaturated atmosphere and may not be visible when there is no dust to act as tracers. While the causes of each of these vortices are somewhat different, they all involve intense vortices making contact with the ground.

There are also funnel clouds that do not become tornadoes and are pendant from high-based cumuliform clouds—not cumulonimbus clouds—which most likely are not fed by boundary-layer air because their bases are so high (Figure 6.17). The dynamics of these vortices are not well known, though funnel clouds are observed relatively frequently, particularly underneath ragged cloud bases and when the cumuliform tower above begins to dissipate.

Figure 6.12. (Top) Funnel cloud in a supercell over northwestern Oklahoma, April 14, 2012. (Bottom) Funnel cloud (arrow) underneath a line of moderate cumulus, Ft. Lauderdale, FL, July 3, 1971 (photographs by the author).

Figure 6.13. Waterspouts. (Top) May 28, 1975, Key Biscayne, FL; (bottom) August 24, 1993, near Key West, from a NOAA helicopter (photographs by the author).

Figure 6.14. Tornadoes or funnel clouds over mountainous terrain. (Top left) South of Niwot Ridge, Indian Peaks, CO, July 12, 1996; (top right) below LaPlata Peak, CO, August 9, 1994; (middle left) July 14, 2011, southeast of Lake Isabelle, Indian Peaks, CO; (middle right) July 31, 2010, north of Niwot Ridge, Indian Peaks, CO (photographs by the author). (Bottom center) Tornado just east of Mt. Audubon, Indian Peaks, CO, August 9, 2010 (courtesy of Emily Riley and Josi Taylor).

Figure 6.15. Dust devils. (Top) May 31, 2002, Arizona; (bottom) August 7, 1978, Nevada (photographs by the author).

Figure 6.16. Steam devil over Lake Thunderbird, Norman, OK, December 22, 1989 (photograph by the author).

Figure 6.17a. Photographs of high-based funnel clouds. (Top, left) August 20, 2011, near the summit of Pawnee Peak, Indian Peaks, CO; (top right) August 24, 2002, northeastern Colorado; (bottom left) May 15, 1991, underneath the cloud base of tilted cumulus congestus just east of the dryline, northeastern Texas Panhandle; (bottom right) under a dissipating, narrow cumulus congestus that formed near the dryline, in southwest Oklahoma, on May 28, 1985 (photographs by the author).

Figure 6.17b. U-shaped funnel underneath the anvil of a tornadic supercell in eastern Oklahoma on May 27, 1997 (photograph by the author)

6.2 TORNADO CLIMATOLOGY

Tornadoes occur all over the world, but are most common in the Great Plains of the U. S. and just to the east during the spring (Figures 6.18–6.20), when there is a ready supply of moisture from the relatively warm Gulf of Mexico and climatologically there is a jet of air aloft that enables there to be vertical wind shear strong enough to support supercells, the most prolific parent storms of the strongest tornadoes. As air subsides in the lee of the Rocky Mountains, it warms and produces a trough of low pressure (see textbooks on synoptic meteorology for more details). In response to this surface trough, surface geostrophic winds become southerly and, under the influence of surface friction, the winds become southeasterly over the states of the Great Plains. This southeasterly flow advects moisture

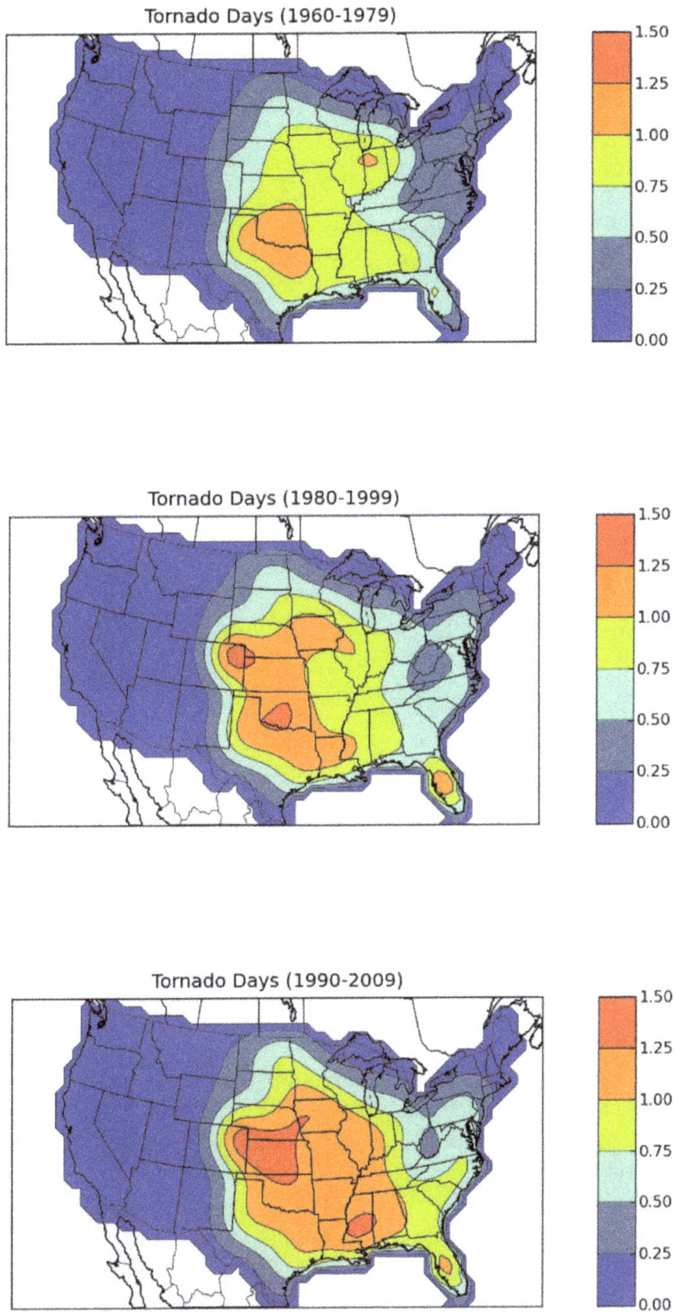

Figure 6.18. Tornado climatology information (U. S.). Annual tornado climatology (mean number of days/year with one or more tornadoes within 25 miles of a point; purple area represents values ≥ 0.2 and the interval between successive colors is 0.2) in the U. S. for (top) 1960–1979, (middle) 1980–1999, (bottom) 1990–2009 (courtesy of NOAA).

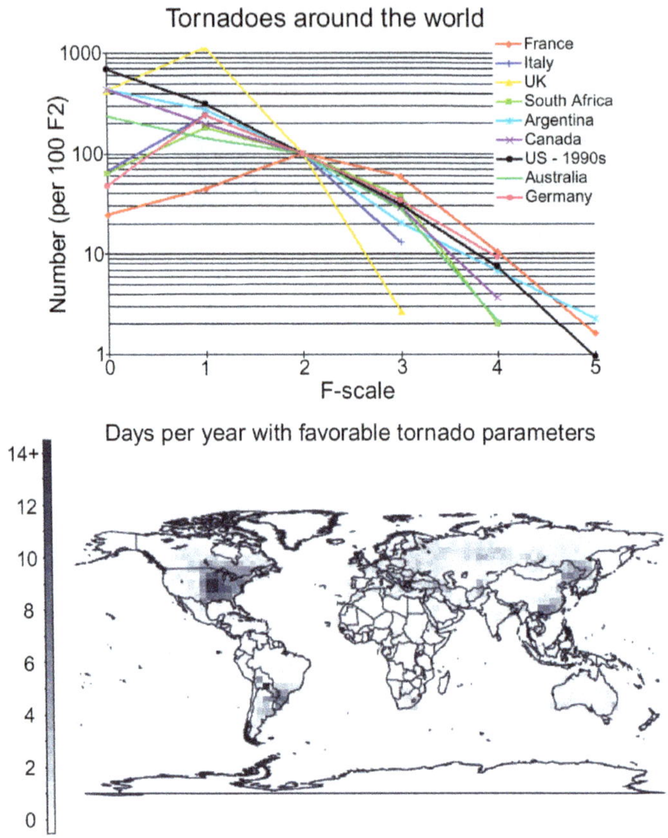

Figure 6.19. Tornado climatology information (worldwide). (Top) Distribution of number of tornadoes/(100 F2 tornadoes) for selected countries (from Brooks and Doswell, 2001); (bottom) days/year with "favorable" tornado parameters (from Brooks *et al.*, 2003a). In lieu of a good, consistent worldwide climatology of tornadoes, it is useful to estimate this climatology from environmental data.

from the Gulf of Mexico, which is situated upstream, into the Great Plains and nearby environs, an area sometimes colloquially referred to as "Tornado Alley", and also adds to the vertical shear associated with strong westerlies, southwesterlies, or northwesterlies aloft. Periodic incursions of cold air behind frontal zones, however, act to remove the potential instability needed for convective storm development. The frequency of occurrence of tornadoes varies from decade to decade (Figure 6.18), but the location of Tornado Alley is consistently from Texas, northward through Oklahoma, Kansas, Nebraska, and into Iowa. The dryline, a boundary separating moist, relatively cool Gulf of Mexico air from dry, relatively warm continental air that has had contact with higher terrain to the west, is a frequent locus of storm initiation, since just east of it the equivalent potential temperature is greatest; thus, the potential for reaching convective tem-

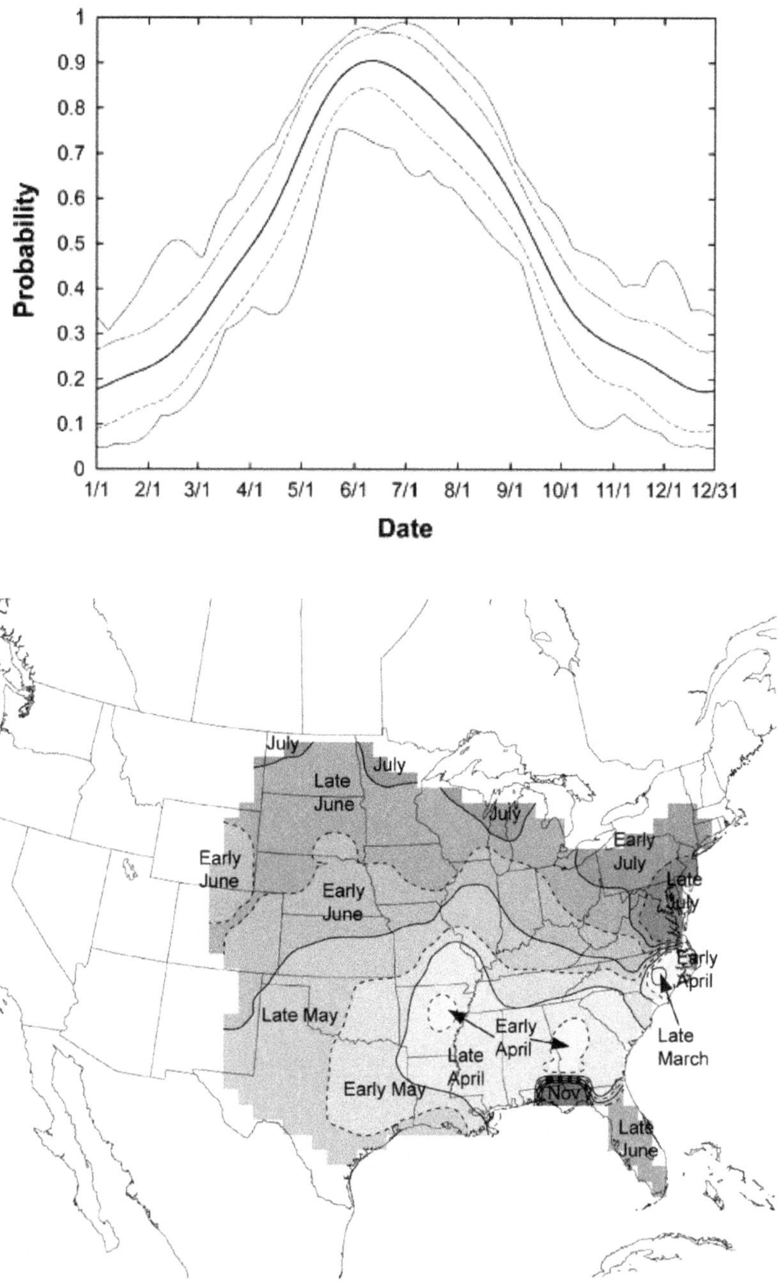

Figure 6.20. Tornado threat in the U. S. by time of year. (Top) Probability of a tornado day anywhere in the U. S. The heavy black line marks the mean value, the dashed lines indicate the mean ±1 standard deviation value, and the thin black lines mark the maximum and miniumum values, all for the 1980–1999 period; (bottom) month (or part of month) of maximum tornado threat in the U. S. for locations with at least 0.25 tornado days/year (from Brooks *et al.*, 2003b).

perature is maximized at and just east of the dryline and CAPE is greatest there also, if the conditions aloft are uniform. Storms are also initiated along fronts and outflow boundaries where there is upward motion driven by mesoscale vertical circulations at their leading edges.

Other regions that have supercell tornadoes are associated with synoptic patterns consistent with the necessary environmental shear and CAPE for supercells, but are different from those responsible for supercells in the Great Plains of the U. S. in the spring. For example, over eastern Colorado and Wyoming (Figure 6.21) it is common for the air to the rear of (north of) cold fronts to be moister than air ahead of (south of) cold fronts. In addition, the upslope flow of surface easterly winds enhances the vertical wind shear underneath westerly (or southwesterly or westerly or northwesterly) flow aloft. Furthermore, as air that has had contact with the elevated mountainous region has heated up, it flows downstream, producing a cap that inhibits storm formation. But near the western edge of the low-level upslope regime, convective temperature may be attained and storms may form there first, before moving eastward onto lower terrain.

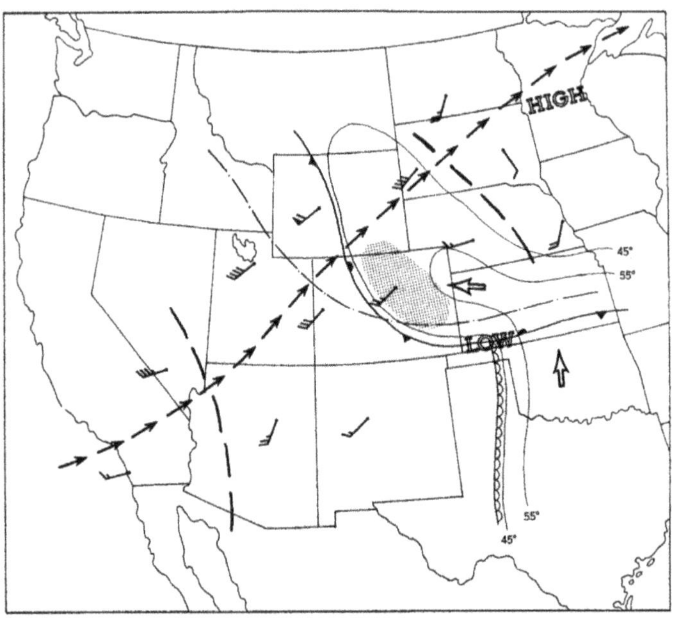

Figure 6.21. Composite High Plains severe convective storm parameter chart. Fronts indicated by conventional symbols; surface isodrosotherms (°F) denoted by fine lines; dryline at the surface indicated by scalloped line; surface flow indicated by large arrows; 700 hPa thermal ridge indicated by dash–dot line; wind barbs show winds at 500 hPa (half barb represents $2.5\,\mathrm{m\,s^{-1}}$; full barb represents $5\,\mathrm{m\,s^{-1}}$; flag represents $25\,\mathrm{m\,s^{-1}}$); heavy dashed lines indicate locations of shortwave troughs at 500 hPa; chain of arrows marks the core of strong winds at high altitudes, above 500 hPa; stippling denotes region of severe convective storm threat (from Doswell, 1980).

California tends to get tornadoes when cold upper-level troughs/cyclones from the Pacific enhance vertical shear and onshore flow from the Pacific provides the necessary moisture, which in tandem with the relatively cool air aloft associated with the upper-level troughs/cyclones, provides the necessary CAPE for supercells (cf. Figure 4.49). These supercells are usually shallower in depth than most tornadoes in Tornado Alley because the tropopause is relatively low. Such conditions tend to occur more frequently during the late winter months. It is beyond the scope of this text to enumerate the synoptic conditions necessary for tornadic supercells everywhere on Earth. The preceding examples of highly different synoptic patterns were highlighted in order to make the point that there are wide variations in synoptic patterns that can support tornadic supercells.

In the U. S., tornado activity begins along the Gulf states in late winter and activity peaks first at low latitudes east of the Rockies and in much of the southeastern U. S., except for Florida, and peaks progressively later in the season with latitude along the Canadian border in the Great Plains and Midwest in July (Figure 6.20), when the overall probability of there being a tornado somewhere in the U. S. is greatest.

Tornadoes have been well documented in China, Japan (where a tornado is called a *tatsu maki*), all over Europe (including especially the U. K., Spain, France, the Netherlands, Romania, Italy, and Germany), Russia, Australia, New Zealand, Argentina, Brazil, Bangladesh, South Africa, and Canada, just to name some, even though they are not as common as they are in the U. S. In most of the mid-latitudes over land areas conditions are sufficient for the formation of supercells on some days.

Tornadoes can strike at any time of day or night, but in "Tornado Alley" in the U. S., part of the Great Plains, they are most common late in the afternoon and early evening, which indicates that there is a connection between most tornadoes and the diurnal heating cycle: convective storms are most likely to form late in the day after maximum surface temperatures have been reached and shortly thereafter; any supercells that form may go on to persist for hours into the early evening. Other factors must also be taken into account: during the early evening, as surface heating decreases and ceases altogether by sunset, the boundary layer becomes decoupled from the free atmosphere above as vertical mixing decreases. As a result, surface winds back (mixing down of westerly momentum from aloft, when the winds have a substantial westerly component aloft, which is typical in mid-latitudes, ceases) and low-level vertical shear increases. When this happens, inertia gravity waves may be excited and may induce additional upward motion. Other storm-chasers and I have colloquially referred to the apparent increase in the likelihood of storm initiation and the formation of tornadoes beginning at 6 PM local time as "6 o'clock magic". (I recently learned from Jim Wilson at NCAR that the same term has also been used to describe the behavior of convective storms in eastern Colorado.)

In addition to tornadoes that are spawned by supercells, tornadoes also occur in other types of storms. Some of these convective storms are tied to the diurnal cycle, which varies from location to location. In mountainous terrain, the convective temperature may be reached fairly early in the day, around noon; along coastal areas, the sea breeze front may initiate convective storms by midday; the offshore land breeze, on the other hand, may initiate convection early in the morning. Tornadoes in MCSs, which develop from more isolated convection late in the afternoon, but mature after dark when the cold pool from the cells has built up in strength and areal extent, occur during the overnight hours. In addition, supercell tornadoes have been documented overnight when storm formation is not tied at all to the diurnal heating cycle, but occurs, for example, when a front finally reaches a region of enough moisture and CAPE for storm initiation.

6.3 TORNADO RESEARCH

The two main scientific problems related to tornadoes are (1) explaining their structure, particularly their three-dimensional wind distribution, and (2) why they form (and why at other times they do not form). The former involves an understanding of boundary-layer dynamics and how the tornado vortex in the boundary layer is coupled to a vortex produced aloft, which in supercells is usually independent of boundary-layer processes (just the vortex aloft). The latter involves storm-scale dynamics and how vorticity is produced and amplified on the storm scale. The two problems may be related, however, for example, when boundary-layer behavior results in the amplification/diminution of vorticity independent of storm-scale processes or when changes in storm-scale flow affect the character and behavior of the boundary layer. The first problem involves a vortex of air rubbing against the ground and forcing from above by positive buoyancy (or an upward-directed dynamic pressure gradient force) or, in the case of a laboratory model, an exhaust fan. The second problem has been studied using idealized numerical simulations, in which the practical limits of spatial resolution and the realism of subgrid-scale parameterizations are the main issues affecting the utility of simulations. Single and dual-Doppler radar analyses have been helpful, as has the analysis of photographs, movies, and videos.

To date, it is not an understatement to say that whether or not a parent convective storm will go on to produce a tornado cannot be predicted accurately. Owing to the difficulties in collecting detailed observations in the right place for a sufficiently long period of time, and the difficulty in numerically simulating both a convective storm and a tornado together (simulation of tornadoes requires extremely high spatial resolution), tornadoes are arguably the least well understood of all phenomena associated with convective storms. However, there are some instances when there is an "outbreak" of tornadic supercells over a broad area, and one can therefore be confident that there should be a "signal" in the environment that points to why they occur.

The first studies of tornadoes included finding exact analytic solutions to highly idealized conditions, which have yielded considerable physical insight; these studies were absolutely necessary before powerful computers became available. Early research (some of the following is given in more detail in Chapter 1) on

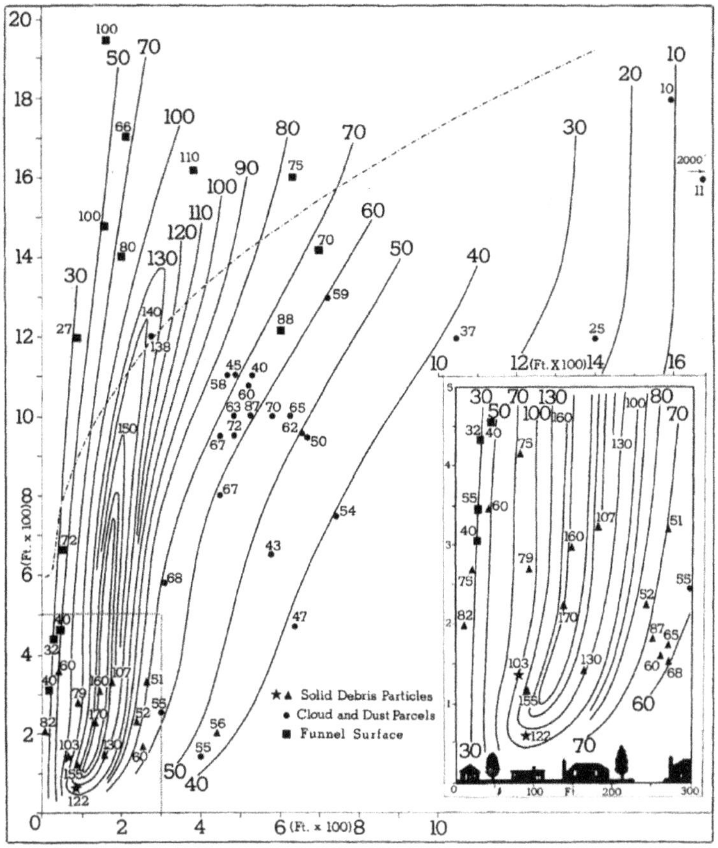

Figure 6.22. Vertical cross section of estimated azimuthal wind speeds (mph) in a tornado from photogrammetric analysis debris and cloud tags in movies of the Dallas, TX tornado of April 2, 1957 (from Hoecker, 1960).

tornadoes consisted mostly of serendipitous in situ measurements and photographs, studies with conventional radar, and photogrammetric analyses of debris movies. Analysis of movies and photographs and, in the last 30 years or so, of videos, showing tornadoes and tornadic debris, have been valuable (Figure 6.22), though debris cannot be seen inside the condensation funnel and only two-dimensional analyses are possible without a second view from another viewing angle. Fixed site Doppler radar networks in the 1970s and early 1980s afforded a look at the wind field in storms with spatial resolution on the scale of ~500 m–1 km and temporal resolution ~2–5 min. "Storm-chasers" began to document storm features in a systematic way beginning in the 1970s and with in situ instrumentation in the 1980s. In situ measurements are difficult to obtain while measurements are valid only at a point or several points. Laboratory experiments in vortex chambers were carried out in the 1970s and 1980s at Purdue University and at the University of Oklahoma/NSSL. Vortex chamber experiments yielded the first quantitative measurements of vortex characteristics independent of radar and computer technology. Numerical simulations of convective storms on the storm scale, with nested grids used to simulate substorm vortices, were also first carried out then. Airborne Doppler radars were first used to probe severe convective storms in the early 1990s and especially during VORTEX in 1994 and 1995. Radars mounted on aircraft allowed storms to be followed and documented, with 300 m spatial resolution, for longer time durations, but the time between aircraft passes was ~5 min and features near the ground could not be detected very well, owing to ground clutter contamination. Mobile, ground-based radars mounted in vans were first used in the late 1980s; while it was difficult to follow storms as well as in an aircraft, data in tornadoes near the ground could be obtained with even higher spatial resolution and at much shorter time intervals and much more frequently than possible at a fixed site radar. Mobile Doppler radar observations have been vital in obtaining data from real tornadoes, but they are difficult to obtain, have limited spatial resolution, and suffer from ground clutter contamination at low levels. High-frequency radars require only modest-size antennas, but are very susceptible to attenuation; low-frequency radars are much less susceptible to attenuation, but require larger antennas to yield the same high azimuthal resolution and are therefore too large for mobile work. Large-eddy simulations (LES) of tornadoes (vortices interacting with the ground and isolated from their parent storms) were first carried out in the late 1990s using grid spacing as short as ~1–3 m in some places, so that the turbulent aspects of tornadoes could be better represented. LES models have yielded interesting measurements, but in the absence of interaction with a parent storm. Studying tornadoes using just an LES model of the vortex is like studying how a finger works in the absence of the arm, the brain, etc. On the other hand, studying tornadoes using a model simulating the entire parent storm is inadequate because boundary-layer processes cannot be simulated faithfully without higher spatial resolution.

The state of the art for tornado research at the time of this writing in 2011 and 2012, just a year or so after VORTEX2, involves observing tornadoes with ground-based, rapid-scan (both electronic and mechanical) mobile, Doppler radars; mobile, rapid-scan (mechanically scanning—not electronically scanning), polarimetric Doppler radar; mobile, pulsed Doppler lidar, instrumented surface probes; UAVs; and three-dimensional numerical simulation experiments with grid spacing down to tens of meters in non-hydrostatic cloud models and 1 m LES models.

6.4 TYPES OF TORNADOES AND TORNADO-LIKE VORTICES

Tornadoes have been classified according to whether they are associated with a pre-existing, larger-scale circulation that existed *before* the parent storm harboring the tornado had formed and the vorticity associated with that circulation then became the source of vorticity for the tornado, or whether the tornadoes are associated with a mesocyclone produced in a supercell. The largest and most intense, and consequently the most damaging, tornadoes are associated with the latter, a mesocyclone produced in a parent supercell. Bob Davies-Jones at NSSL named this type of tornado a "Type I tornado", also known as a "supercell tornado". By "associated with" we mean that a mesocyclone preceded the tornado and the tornado is the mesocyclone contracted in size or is a separate vortex located within the mesocyclone. A surface vortex on the scale intermediate between that of the mesocyclone and that of the tornado has been referred to as a "tornado cyclone". It got its name in 1949 from Ed Brooks, who found that tornadoes were embedded in regions of relatively low pressure whose horizontal scale was greater than that of the tornado. Perhaps there really is just one vortex, but it can be resolved on several spatial scales. There is, however, some evidence from mobile Doppler radars that a tornado cyclone or mesocyclone in some instances may be broader than a tornado and coexist with a tornado, so that the tornado cyclone does not necessarily always contract in scale radially down to that of a tornado (Figure 6.23). Other tornadoes that are not associated with a mesocyclone produced by a supercell were named "Type II tornadoes" by Bob Davies-Jones, but are therefore sometimes also referred to as "non-supercell tornadoes (NSTs)". These tornadoes appear in a number of different situations:

(a) Some are found under growing cumuliform towers (Figure 6.24), often even before precipitation has formed or once it has formed but has not reached the ground yet or just about to reach the ground. Since these tornadoes frequently look like waterspouts that are pendant from lines of cumulus congestus, they are colloquially known as "landspouts" (the origin of this term has been attributed to the author who used the term in a conference publication, but the term may have already been in use by other storm-chasers).[1] When the air is very dry, landspouts appear as rotating whirls of dust near the ground, possibly without any condensation funnel overhead. In some instances, rotating dust whirls precede the appearance of a funnel cloud aloft. They begin near the ground and expand upwards toward cloud base. Although this type of tornado is found in ordinary-cell convective storms, it is also found in supercells, but not in or near the mesocyclone. When "non-supercell" tornadoes are found in supercells, they are found along the rear-flank gust front. The term "non-supercell tornado" can therefore be misleading. The term "Type II tornado" is therefore probably preferable. From visual observations and radar observations we know that some of the Type II landspout tornadoes must form in the absence of a downdraft, especially when no radar echo has yet been detected and the parent cloud is still growing, as the cloud top has not reached the tropopause.

It is thought that pre-existing vorticity along a surface boundary characterized as a "vortex sheet"[2] may become barotropically unstable, breaking up into small vortices, and that vorticity is stretched by an updraft in a convective

[1] The term "landspout" appeared in the literature many years ago. Ironically, a tornado in the early 20th century was once referred to as "a violent landspout".

[2] A vortex sheet is the band of vorticity associated with a discontinuity across a line of the component of the wind along the line.

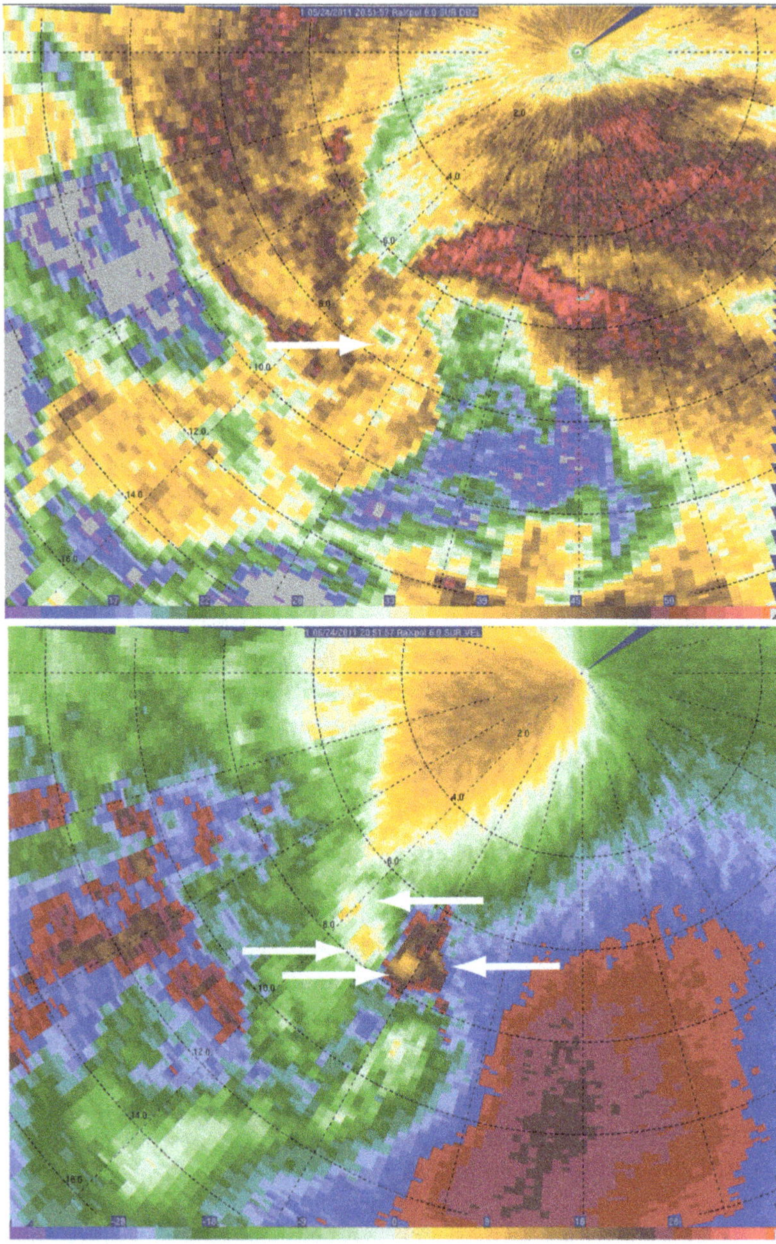

Figure 6.23a. Examples of vortices of different scales in and around a tornado, as seen by a mobile Doppler radar. (Top) Radar reflectivity from RaXPol in central Oklahoma on May 24, 2011; arrow points to the weak-echo hole at the center of the tornado. (Bottom) As in the top, but for aliased Doppler velocity; velocity data are not corrected for folding to highlight the Doppler velocity extrema (in yellow); left-pointing arrows highlight extrema for the mesocyclone, while right-pointing arrows highlight extrema for the smaller scale tornado.

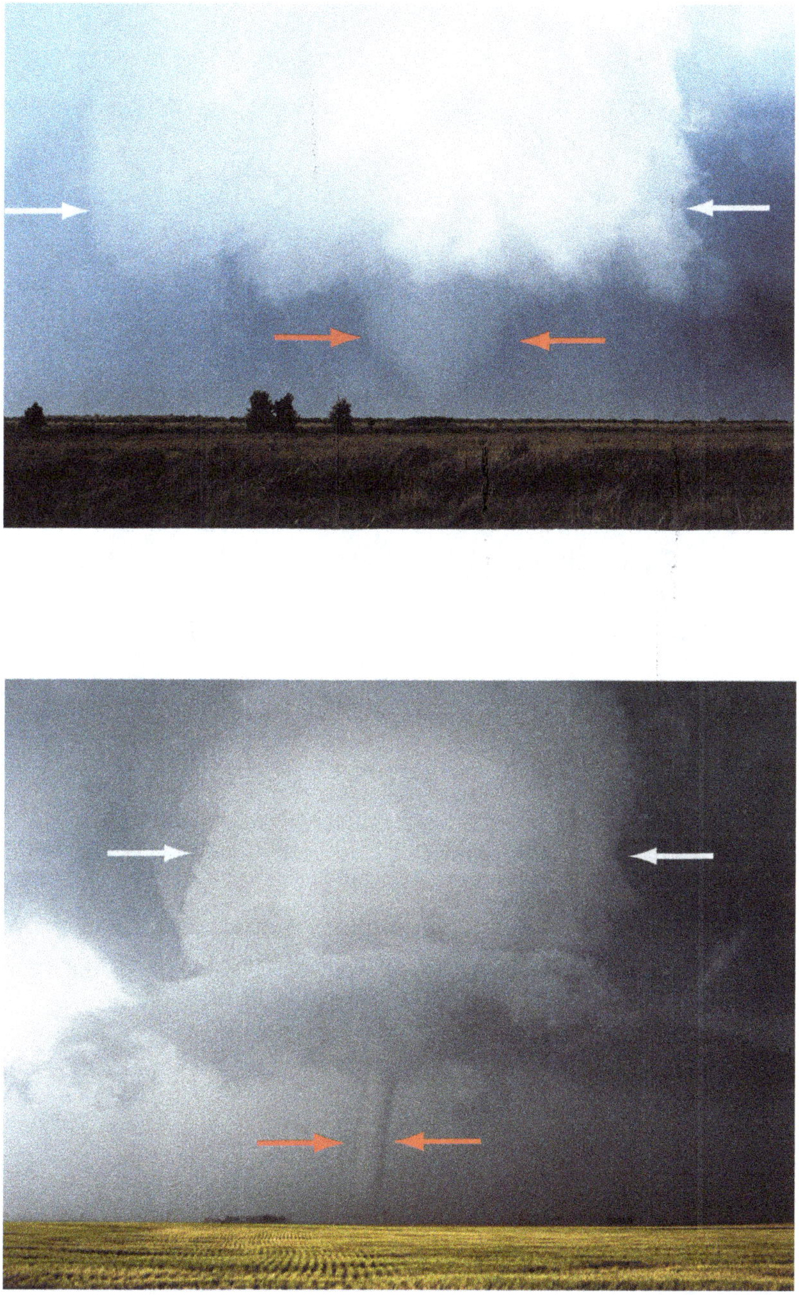

Figure 6.23b. (Top) Photograph of a tornado (red arrows show the condensation funnel) embedded within a broader, cylindrically shaped cloud (white arrows), which approximately marks the larger scale mesocyclone in a supercell in northwest Texas on May 25, 1994 and (bottom) in Nebraska on June 10, 2004 (photographs by the author).

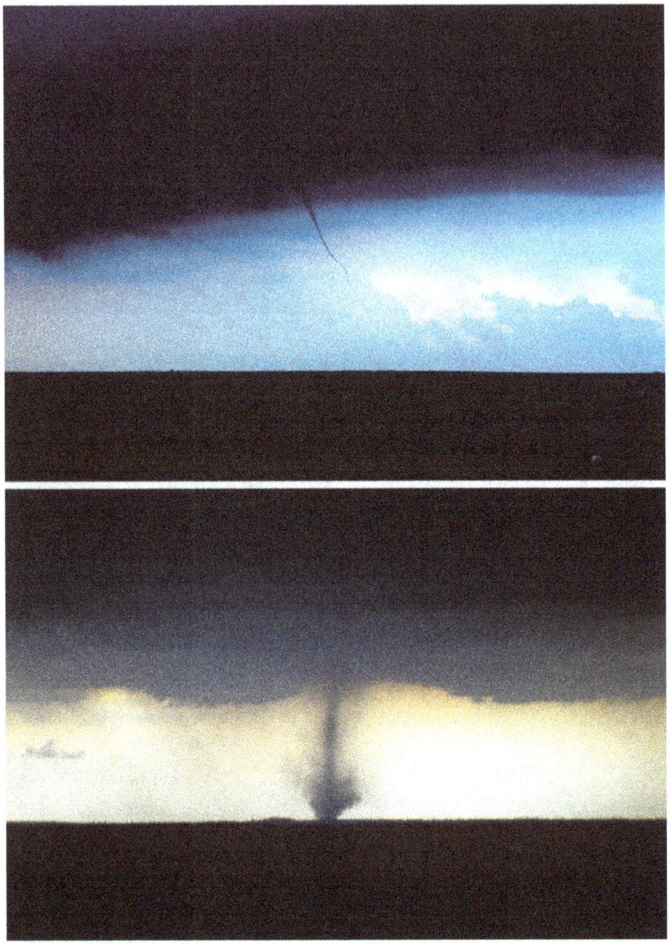

Figure 6.24. Photographs of landspouts. (Top) April 22, 1981, south central Oklahoma; (bottom) August 9, 2004, west of Limon, CO (photographs by the author).

Figure 6.25. Conceptual model of the life cycle of a non-supercell tornado. The solid line marks the radar-detectable convergence boundary. Low-level vortices are assigned letters. The three-dimensional wind field is indicated by arrows. In the last stage (rightmost panel) a boundary-layer vortex ("C") is stretched underneath an updraft in a buoyant cloud aloft (from Wakimoto and Wilson, 1989).

Figure 6.26. Conceptual model of the evolution of non-supercell tornadoes along a weak outflow boundary. In (I) a vortex sheet develops along the leading edge of the outflow when there is a component of motion in the environment, ahead of the outflow boundary, along (parallel to) the leading edge. The vortex sheet becomes unstable and small-scale vertical vortices develop (II). Vortices interact and merge (III) through pairing of similar vortices and the "mining" of weaker neighboring vortices by stronger vortices, leading to a wider separation between neighboring vortices. Vorticity is advected upward out of the boundary layer into growing convective cells and stretched. Frictionally induced radial inflow near the ground (discussed in more detail in Section 6.6.1) stretches vorticity to tornadic strength (IV) and environmental horizontal vorticity is tilted upward and also stretched. During the late mature stage (V) cold pools produced by new convective storms increase low-level convergence at the leading edge of the outflow boundary, increasing the stretching, and wrap relatively cold air around the vortices. Downward-directed dynamic perturbation pressure gradient forces develop above the vortices and force downdrafts, as does the negative buoyancy associated with the strengthening cold pool (VI). The outflow boundary then accelerates and tilts the vortex so that much of it is in descending air and the tornadoes dissipate. The tilting of baroclinically generated horizontal vorticity along the outflow boundary does not appear to be important in NST tornadogenesis. (from Lee and Wilhelmson, 1997).

cloud that is triggered by vertical motion associated with low-level convergence along the boundary (Figure 6.25). Bruce Lee at the University of Illinois Urbana-Champaign and his collaborators have done seminal numerical studies of this type of tornadoes. Baroclinicity along the surface boundary could also play a role, especially as a source for horizontal vorticity that is tilted onto the vertical along the edges of updrafts or downdrafts (Figure 6.26), but it appears it is not of primary importance.

(b) Some, which appear along the edge of gust fronts as rotating dust whirls (Figure 6.27), are colloquially known as "gustnadoes". Since gust fronts are found in

Figure 6.27a. Gustnadoes/non-supercell tornadoes along an outflow boundary of a non-supercell convective storm, on June 14, 2010 during VORTEX2, in southwest Texas. (Top) Two vortices visualized as rotating dust columns (white arrows) along the leading edge of the cloud base associated with the outflow; (bottom) close-up view of one of the rotating dust columns (photographs by the author).

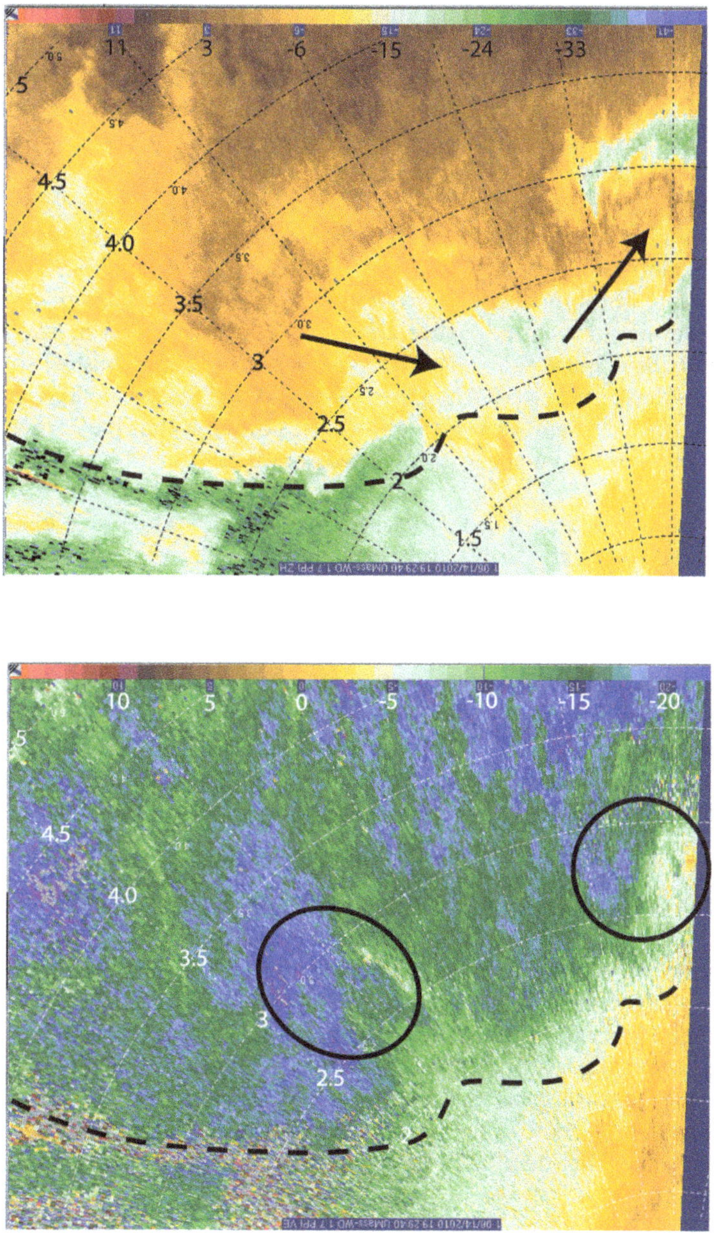

Figure 6.27b. Doppler radar data from the U. Mass. mobile, W-band, Doppler radar of the storm shown in Figure 6.27a. Dashed line indicates the approximate leading edge of the outflow boundary. (Top) Radar reflectivity factor (dBZ_e); arrows point to two spiral/appendage-like echoes; range shown in km. (Bottom) Doppler velocity in m s^{-1} (color-coded at the top of each panel); vortex shear couplets are circled/marked by an ellipse (data courtesy of Robin Tanamachi and Krzysztof Orzel).

all types of convective storms, gustnadoes are found in both supercells and ordinary-cell convective storms. Gustnadoes, like landspouts, are probably the result of a barotropic shearing instability along a gust front boundary. Gustnadoes do not require an updraft above and typically are very short lived. In a sense, gustnadoes are non-supercell tornadoes in the absence of buoyant updrafts aloft or with buoyant updrafts that get left behind as the convergence boundary underneath the convective storm moves away from the convective towers; landspouts, on the other hand, are non-supercell tornadoes with buoyant updrafts aloft.

(c) Some are typically not readily visible, forming within regions of precipitation along lines of convection, particularly along cold fronts and in bow echoes or within MCSs (Figure 6.28). Rit Carbone of NCAR has published some classic studies of the former, while Greg Forbes and Roger Wakimoto have published a classic study of tornadoes in bow echoes. It is thought that the vorticity in these tornadoes comes from shear instability as in (a) or (b).

In the spirit of the so-called "forecasting funnel" in which one considers first the influence of the large scale, and then the influences of progressively smaller scales, we will first consider the formation of the parent vortex, and then the smaller scale tornado vortex. Simply put, *tornadogenesis associated with mesocyclones in supercells involves formation of a storm-scale vortex, a mesocyclone aloft (i.e., above the boundary layer), and the interaction of the storm-scale vortex with the surface in the boundary layer and its contraction in scale or the production of a smaller scale vortex and its contraction in scale, and increase in intensity. Tornadogenesis not associated with a mesocyclone aloft is associated with a boundary-layer vortex that is intensified and advected upward.*

6.5 TORNADO VORTEX FORMATION: TORNADOGENESIS

Rich Rotunno at NCAR has noted "the tornado does not fit a simple model like the spin-up that skaters experience when they pull in their arms." It may be inferred from observations that the proximity of tornado formation to surface boundaries separating warm, ambient air from evaporatively cooled outflow, and of very strong updrafts near the ground, "that complex boundary-layer interactions are (also) important." To understand how tornadoes form, the source of their vorticity must be identified and the mechanisms for its rapid increase must be accounted for.

6.5.1 Tornado-like vortices in a vortex chamber

If a tornado is barotropic (as it is in a vortex chamber), then one may use vortex line analysis to visualize tornado–vorticity dynamics. Rich Rotunno has pointed out that vortex lines in a vortex chamber for steady-state flow with no swirling motion (no azimuthal velocity) consist of a ring of clockwise-turning vortex lines

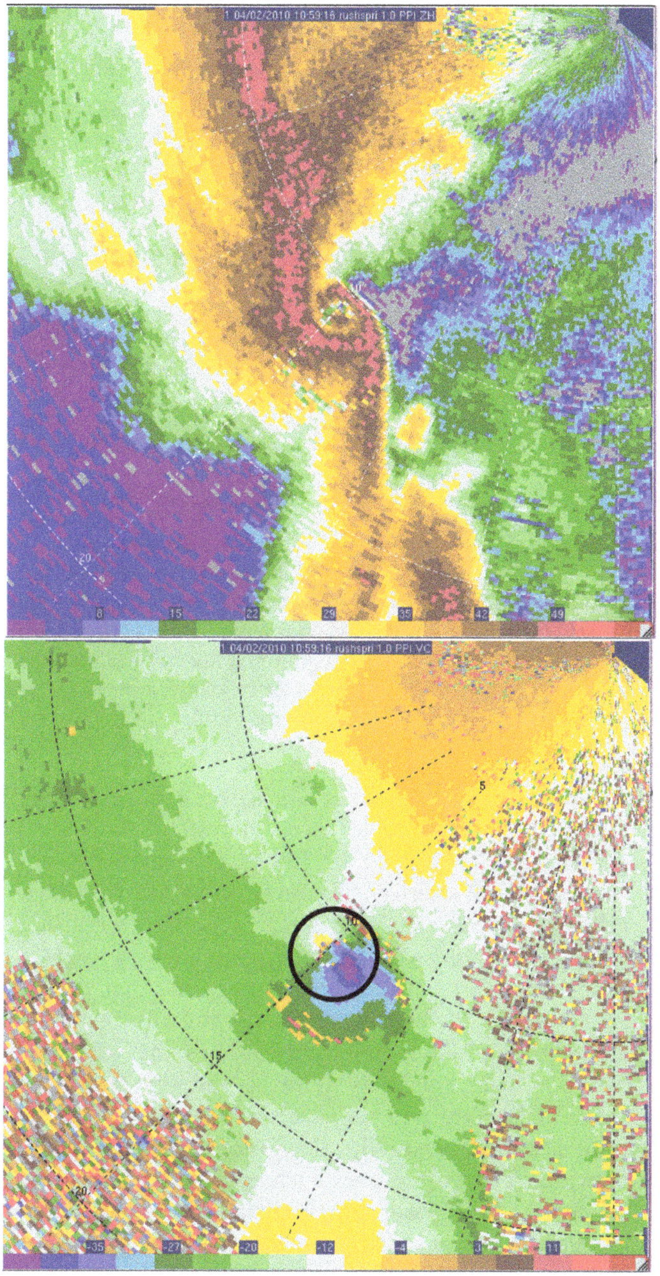

Figure 6.28. Tornado along a QLCS in central Oklahoma as detected by a CASA (Collaborative Adaptive Sensing of the Atmosphere) Doppler radar (which operates at X-band) on April 2, 2010. (Top) Radar reflectivity factor in dBZ_e; (bottom) Doppler velocity in $m\,s^{-1}$. Range rings shown every 5 km. Vortex signature circled (data courtesy of Vivek Mahale and Jerry Brotzge).

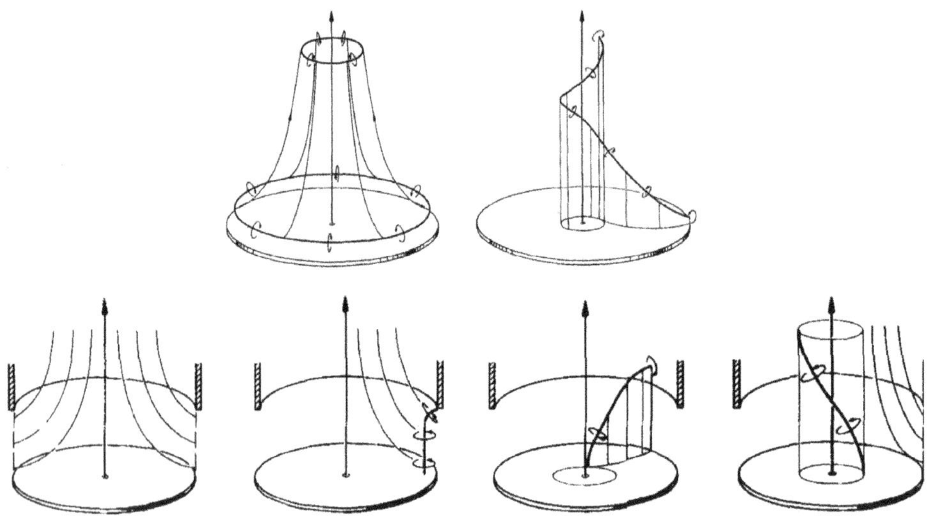

Figure 6.29. Vortex lines in a tornado simulator. (Top) For steady-state flow: (left) when there is no azimuthal velocity (no swirl), but only radial and vertical velocity; (right) when there is swirl. (Bottom) From left to right, time-dependent flow characteristics: air converges toward the center and exhausted outward at the top, without any surface friction; a vertically oriented vortex line is introduced at the outer edge; it is advected inward and becomes tilted; sinking motion develops at the center and the vortex stops getting any closer to the center, along the axis where there is no vertical motion (from Rotunno, 1980).

(Figure 6.29, top left); these vortex lines are associated with radial inflow that increases with height (as a result of a forced updraft and surface drag). Radial convergence associated with the updraft brings the rings radially inward and upward. With the addition of swirling motion in the vortex chamber, vortex lines do not form rings, but instead enter the domain of the tornado near the surface and then spiral counterclockwise with height (Figure 6.29, top right).

To spin up a tornado, vertically oriented vortex lines that point upward (for a cyclonic tornado) (Figure 6.29, bottom) converge radially inward from the "environment", and then are tilted radially outward, as the vortex line encounters decreasing radial inflow with height as it travels radially inward. The azimuthal vorticity vector near the surface points in the clockwise direction and radial vorticity points radially inward. Eventually the vortex line cannot be advected any farther radially inward. The vortex lines then curl in a clockwise direction with height around the central axis. Tornado-like vortices in a vortex chamber are produced with no pre-existing downdrafts.

6.5.2 Stretching of pre-existing vertical vorticity

We first consider how pre-existing vorticity can be increased to tornado intensity. Vorticity in a tornado is $\sim 1\,\mathrm{s}^{-1}$, since the radial shear of the azimuthal wind ~ 50–$100\,\mathrm{m\,s}^{-1}/100\,\mathrm{m}$. Another way of estimating vorticity in a tornado is to use circulation: if the maximum azimuthal wind speed is $\sim 50\,\mathrm{m\,s}^{-1}$ at a "core"

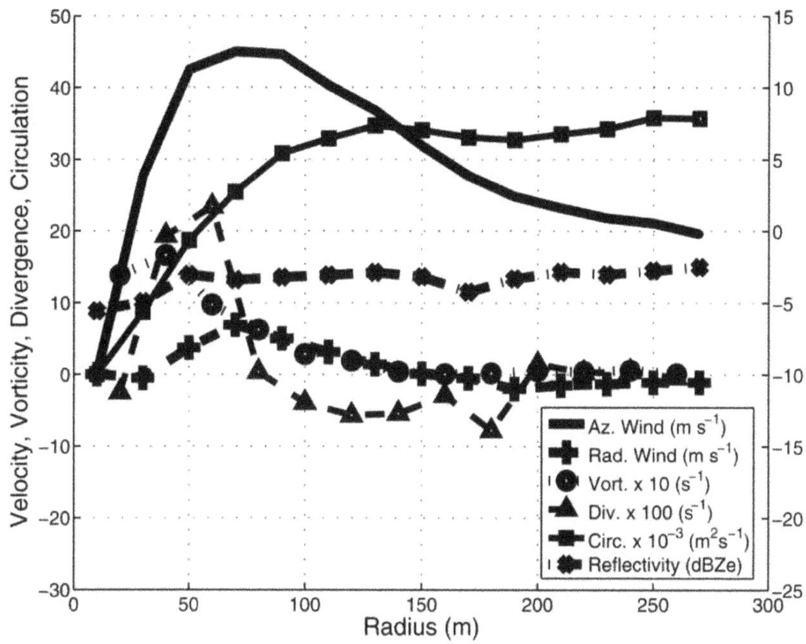

Figure 6.30. Radial profiles of azimuthally averaged azimuthal velocity ($\mathrm{m\,s^{-1}}$), radial velocity ($\mathrm{m\,s^{-1}}$), vertical vorticity ($\times\ 10\,\mathrm{s^{-1}}$), divergence ($\times\ 100\,\mathrm{s^{-1}}$), circulation ($\times\ 10^{-3}\,\mathrm{m^2\,s^{-1}}$), and radar reflectivity factor (dBZ_e), from analyses of data collected at low elevation angle in a tornado in northern Kansas on May 15, 1999 by the U. Mass. mobile W-band Doppler radar. Circulation increases radially outward until about just a little more than 100 m, beyond which it levels off at about $35{,}000\,\mathrm{m^2\,s^{-1}}$ (from Tanamachi *et al.*, 2007).

radius—(to be discussed later); not necessarily the radius of maximum azimuthal wind, but usually close to it (the radius beyond which the angular momentum of the vortex does not vary with radius)—of 100 m, then the circulation of the core is $2\pi(100\,\mathrm{m})\ (50\,\mathrm{m\,s^{-1}}) = \pi \times 10^4\,\mathrm{m^2\,s^{-1}}$ ($\sim$30,000 $\mathrm{m^2\,s^{-1}}$). This magnitude of circulation seems to be, based on mobile Doppler radar measurements, characteristic of the near environment of tornadoes (Figure 6.30). Since vorticity is circulation/area, vorticity $\sim\pi \times 10^4\,\mathrm{m^2\,s^{-1}}/\pi(100\,\mathrm{m})^2 = 1\,\mathrm{s^{-1}}$.

Background vorticity associated with a mesocyclone is $\sim\geq 10^{-3}\,\mathrm{s^{-1}}$, since radial shear of the azimuthal wind $\sim\geq 10\,\mathrm{m\,s^{-1}}/5\,\mathrm{km}$. The mesocyclone may be a "low-level mesocyclone" that was produced by the parent supercell or a pre-existing one along a surface boundary, perhaps enhanced by shearing instability. In the case of the latter, the pre-existing vortex may also be called a "miso-cyclone" which, according to Ted Fujita's classification scheme, has a shorter (average) space scale than a mesocyclone ($\sim$40 m–4 km in diameter for a miso-cyclone, as compared with $\sim$2–10 km for a mesocyclone). As in the previous paragraph, another way of estimating vorticity in a mesocyclone is to use circulation. Since circulation is conserved in a frictionless, barotropic atmosphere, the mesoscale circulation from which the tornado develops (neglecting friction and baroclinic effects) must have vorticity $\sim\pi \times 10^4\,\mathrm{m^2\,s^{-1}}$/area of the mesocyclone; for a mesocyclone of radius 2.5 km, vorticity is $\sim 1.5 \times 10^{-3}\,\mathrm{s^{-1}}$.

Considering only the divergence term ($-\delta\zeta$) in the vorticity equation (2.50), it is seen that under the effect of a field of constant convergence at the base of a convective storm of $\sim 10\,\mathrm{m\,s^{-1}}/1\,\mathrm{km} = 10^{-2}\,\mathrm{s^{-1}}$, which is consistent with a buoyant updraft in a convective cloud above ($\partial w/\partial z \sim \partial u/\partial x$ and $\partial v/\partial y \sim 10\,\mathrm{m\,s^{-1}}/1\,\mathrm{km}$), the e-folding time for vorticity is 1/convergence $\sim 100\,\mathrm{s}$, or on the order of a minute. Considering the divergence term in (2.50) alone

$$t = -\ln[\zeta|_f/\zeta|_0]/\delta \tag{6.1}$$

where the subscripts f and 0 represent the final and initial values. In $\sim$10 min, the vorticity of tornado intensity ($1\,\mathrm{s^{-1}}$) can be "grown" from mesocyclone vorticity ($\sim 1.5 \times 10^{-3}\,\mathrm{s^{-1}}$) with $10^{-2}\,\mathrm{s^{-1}}$ convergence. In $\sim$10 min, air flowing radially inward at $\sim 10\,\mathrm{m\,s^{-1}}$ from the outer edge of a mesocylone at the ground at 2.5 km radius would have enough time to make it to near the center of the tornado and then be transported upward.

Since vorticity is scale dependent, it is possible, for example, to identify vortices such as dust devils, whose wind speeds can vary $\sim 10\,\mathrm{m\,s^{-1}}$ over only 10 m, having vorticity as high as that of tornadoes. *Circulation is therefore a better indicator of the intensity of tornadic vortices than vorticity because circulation is less scale dependent. In addition, circulation is conserved if friction and baroclinic effects are neglected, just like vorticity.* This assertion must be accepted with extreme caution, however, since friction does play a role and tornadoes are often embedded in baroclinic environments, in which circulation may be further enhanced, but not necessarily exponentially with time. The circulation enclosing the core of a strong tornado, as noted earlier, is $\sim 30{,}000\,\mathrm{m^2\,s^{-1}}$. In constrast, the circulation of a dust devil is only $\sim 2\pi(10\,\mathrm{m})(10\,\mathrm{m\,s^{-1}}) \sim 600\,\mathrm{m^2\,s^{-1}}$. A 5 km wide mesocyclone having a circulation of $30{,}000\,\mathrm{m^2\,s^{-1}}$, with vorticity averaging $1.5 \times 10^{-3}\,\mathrm{s^{-1}}$, could be shrunken to 100 m radius in $\sim$4.5 min by radial inflow of just $10\,\mathrm{m\,s^{-1}}$. However, while circulation is a better measure of tornado intensity than vorticity, circulation also is deficient because it depends on the precise location of the material curve about which it is being computed. Since circulation is a spatially averaged quantity, the wind speeds may be much higher inside the circuit than they are outside the circuit and vice versa. Thus, depending on the radial profile of the wind, much higher and more damaging winds may exist inside the circuit or outside the circuit. To be a true measure of damage potential, the circuit should be taken at the radius of maximum wind.

In addition, one must consider that tornado damage estimates and Doppler wind measurements include both the wind speed associated with the tornado vortex and that of the translation speed of the vortex. A weak vortex moving very quickly can inflict strong damage, even though the azimuthal wind speeds in the reference frame of the vortex are relatively low. It could be argued that one should subtract out the translation speed to assess the "true" intensity of a tornado.

Earth's vorticity probably does not contribute *directly* to tornadogenesis. Convergence of $10^{-2}\,\mathrm{s^{-1}}$ (e.g., if convergence acts at cloud base, above which there is an updraft 5 km in diameter and if the wind velocity varies linearly from zero at the center of the cloud base to $20\,\mathrm{m\,s^{-1}}$ radially inward at the perimeter, then convergence is $\sim 8 \times 10^{-3}\,\mathrm{s^{-1}}$) would have to act on Earth's vorticity ($\sim 10^{-4}\,\mathrm{s^{-1}}$)

for about 15 min, to produce a tornado, which would require a strong, very wide, and persistent updraft. Suppose that a typical cloud base is $\sim$5 km in diameter; in 15 min, air moving radially inward at an average of 10 m s^{-1} would travel $\sim$9 km, which is much farther than the radius of the updraft. Thus, air in that time period would have turned upward in the updraft, advecting vorticity upward and moving away from the region of low-level convergence. The cloud base diameter would have to be $\sim$>18 km wide in order for tornado intensity vorticity to be produced from Earth's vorticity if the horizontally convergent wind speed averages $\sim$10 m s^{-1}.

In just 4.5 min, however, a convergence of 10^{-2} s^{-1} acting on Earth's vorticity could produce mesocyclone intensity vorticity of 1.5×10^{-3} s^{-1}, and air would have moved radially inward (at 10 m s^{-1}) $\sim$2.7 km, which is about the width of a mesocyclone; so, in just 2.25 min air would have turned upward in the updraft. In any event, numerical cloud models produce a realistic-looking supercell structure and tornado-like vortices without Earth's vorticity present in the model. However, in nature, the vertical wind shear necessary to produce a supercell is associated with Earth's rotation via the thermal wind relation (4.23), so in a sense Earth's rotation is necessary for the production of supercell tornadoes at mid-latitudes, even though it is not itself the source of vorticity. It is likely, though, that Earth's vorticity enhances whatever relative vorticity there already is and can in effect speed up tornadogenesis.

The nature of the pre-existing vorticity, as we just noted, may be from the Earth's rotation about its axis, or along surface boundaries that preceded the parent storm's formation, or were produced by the storm itself (e.g, a gust front). In the case of surface boundaries along which there is horizontal shear, a vertical "vortex sheet"—vertically oriented surfaces along which there is intense shear vorticity created when the flow parallel to the boundary changes abruptly across it—can be found along gust fronts or other sharp surface boundaries and these vortex sheets may be barotropically unstable, allowing for the growth of a series of vortices. Many non-supercell tornadoes occur as periodically spaced vortices along a line (Figure 6.26). However, vortices strung out in a line along the rear-flank gust front in supercells have also been noted (Figure 6.31). The merging of these vortices by larger scale convergence may also play a role in producing tornadoes, though in some instances the vortices do not seem to play any role because they remain along the gust front.

6.5.3 Tilting of horizontal vorticity into the vertical, followed by stretching underneath an updraft

Other possible sources of cyclonic vorticity for tornadoes include horizontal vorticity generated baroclinically along the edge of an outflow boundary (such as near the edge of the FFD or RFD or a gust front from a neighboring storm) or internally (i.e., not along a sharp baroclinic boundary, but instead along a broad baroclinic zone; Figures 4.51 and 6.32), resulting from gradients in the phase change of different types of hydrometeors, such as large raindrops next to small raindrops (stronger evaporative cooling in the case of the latter) or raindrops of

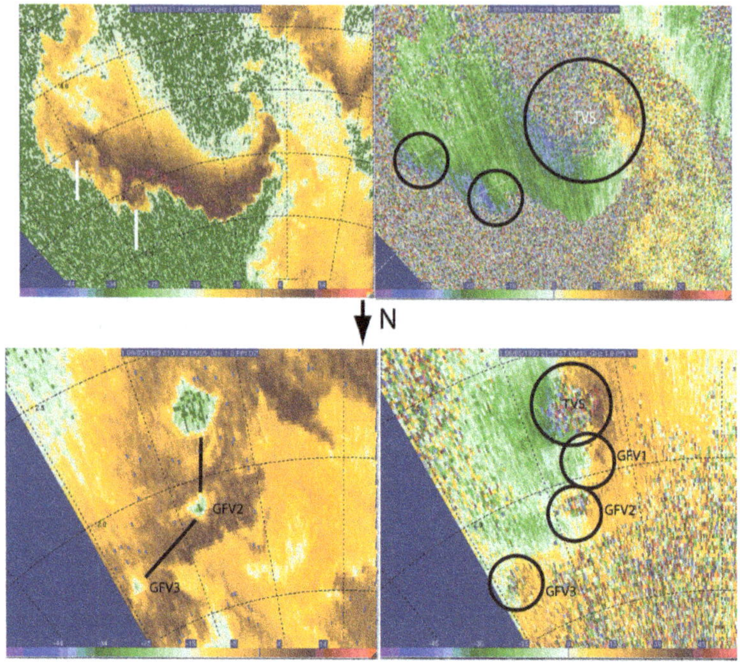

Figure 6.31. Small-scale vortices along the rear-flank gust front as detected by the U. Mass. W-band mobile Doppler radar on June 5, 1999 in north central Nebraska. Range rings shown every 500 m. (Left) Radar reflectivity in dBZ_e; (right) Doppler velocity in $m\,s^{-1}$. (Top) Two spiral echoes (white lines) and associated vortex shear signatures (small circles); large circle encloses vortex couplet associated with the tornado vortex signature (TVS). (Bottom) Valid about 3 min later showing weak-echo holes (WEHs) for the tornado and gust front vortices (GFV) 2 and 3; small circles show vortex signatures for GFV1, GFV2, and GFV3, the line along which they are connected curves along the gust front; no WEH is evident for GFV1.

any size next to hailstones (stronger evaporative cooling in the case of the former) or along the edge of water-loaded air or pre-existing horizontal vorticity associated with boundary-layer vertical shear, which is subsequently tilted onto the vertical as air parcels pass through the gradient in vertical velocity as they enter an updraft or pass in between a downdraft and an updraft or as they exit a downdraft and then enter an updraft (Figures 4.52–4.54 and 6.32). Polarimetric radar observations can determine if there are gradients in hydrometeor type. An additional complication is that the humidity of the air into which hydrometeors are falling may vary in space, particularly in the case in which the convective storm is situated just behind a frontal boundary; in some instances it may be drier farther behind a front, while in others it may actually get moister. Another complication is that the height at which hydrometeors begin their descent can affect surface temperature: descent from high aloft can yield much higher net cooling than descent from a much lower altitude. If θ_e decreases with height as (by definition) it does in a convectively unstable environment, then low values of θ_e

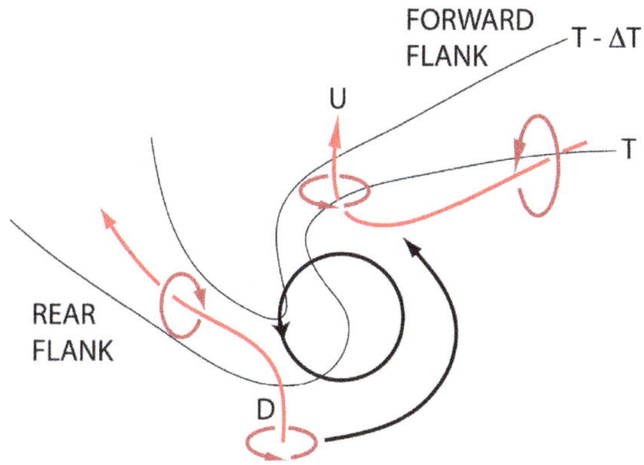

Figure 6.32. Idealized illustration showing how baroclinically generated horizontal streamwise vorticity in the forward flank of a supercell can be tilted in the updraft and stretched underneath the updraft, and how baroclinically generated horizontal anti-streamwise vorticity in the rear flank can be tilted by the downdraft and advected around the mesocyclone at the surface and stretched underneath the updraft. Isotherms denoted by thin solid lines and labeled at T and $T - \Delta T$; streamlines denoted by heavy lines with arrows; three-dimensional vortex lines shown in red, with sense of rotation indicated by curved red lines with arrows. This diagram is shown for illustrative purposes only and does not necessarily represent what is found in every supercell. See also Figure 4.15.

may be brought down to the surface; if θ_e does not change with height (i.e., if the lapse rate is moist adiabatic), then the temperature at the surface in saturated air cannot be changed by descending air parcels.

The baroclinic generation of horizontal vorticity depends not only on the trajectories of air parcels passing along baroclinic zones, but also on the length of time the air parcels pass over baroclinic zones. If the wind speed is too strong, then air parcels do not reside in the baroclinic zone long enough to acquire high values of horizontal vorticity; if the wind speed is too weak, then air parcels may take so long to acquire horizontal vorticity that disruption from new surges in outflow created when new pulses of updraft and the subsequent production of precipitation occur, and modify the nature of the baroclinic zone and disrupt the updraft. George Bryan at NCAR and Leigh Orf at Central Michigan University have recently found evidence in idealized numerical experiments that low-level mesocyclogenesis is favored when the low-level shear vector is relatively large in magnitude and oriented *normal* to the shear vector at higher altitudes (see insets at upper left-hand corner of Figure 4.14g). It is possible that stronger low-level mesocyclones are produced because air parcels tend to produce horizontal vorticity baroclinically under the FFD for a longer period of time than possible when the low-level shear is not normal to the shear aloft. Or, it might be that such a configuration of shear (i.e., low-level hodograph that bends at 90°) favors an increase in the upward-directed perturbation pressure gradient force. The dynamic and thermodynamic reasons for the bent hodograph effect as of 2011 are still not well understood.

It is also possible that horizontal vorticity is tilted onto the vertical as air parcels encounter an outflow boundary and turn upward as they pass over the cold dome. Unless air parcels turn very sharply upward, tilting along the gradient of an updraft alone is insufficient to create a tornado because vertical vorticity is rapidly advected upward, away from the ground. While gust fronts may present opportunities for air parcels to be turned sharply upward (Figure 4.5), in many instances air parcels begin to turn upward ahead of the gust front not so sharply (Figures 3.37 and 3.39a).

When a downdraft forms adjacent to an updraft (underneath which there is stretching), the vertical vorticity created through tilting remains closer to the ground, as noted in Section 4.7, because material curves are brought to the ground by the downdraft—not by the updraft. In many tornadoes, the tornado is indeed located along a gradient in vertical motion, in between the rear-flank downdraft (RFD) and the tip of the horseshoe-shaped updraft along the leading edge of the RFD (Figure 4.15). The role of baroclinically generated horizontal vorticity in the FFD region of a supercell has been called into question recently, owing to observations that temperature gradients at the surface near strong tornadoes seem to be much weaker than those in the absence of tornadoes. On the other hand, upstream of the tornado, in some direction, air parcels may have passed through regions of stronger temperature gradients or there is strong low-level shear in the environment, such that environmental horizontal vorticity is the main source for vorticity in the tornado. In recent field campaigns such as VORTEX2 efforts have been made to map out the temperature at the surface using probes and the complete results are forthcoming at the time of this writing. Evidence from observational and numerical work so far suggests that the baroclinic generation of horizontal vorticity is more important than the import of existing horizontal vorticity associated with vertical shear in the environment. Since surface temperatures realized in numerical simulations of supercells are very sensitive to the type of microphysical parameterization employed, results from numerical simulations, however, are to be viewed with extreme caution.

While Rotunno and Klemp showed using circulation analysis how cyclonic circulation is produced in a low-level mesocyclone (cf. Chapter 4) as a result of part of the material curve being pushed downward by the downdraft in the FFD and then the material curve in the convergent area shrinks under the main updraft, Bob Davies-Jones and Harold Brooks have more recently offered, also based upon numerical experiments, an explanation for the appearance of strong low-level cyclonic vorticity in terms of both the main updraft and the rear-flank downdraft using (local/individual) vortex line analysis rather than an area-averaged (non-local/group of vortex lines) circulation analysis: consider a downdraft in which (1) there is streamwise vorticity; (2) storm-relative flow through the downdraft is parallel to isentropic surfaces, with the colder air lying to the right, as it would in the RFD, west/northwest/north of a low-level mesocyclone (in the Northern Hemisphere); and (3) that environmental flow has a component of motion that decreases with height, as there would be if there were a density current (gust front, surge of cool air) approaching from the north (Figure 6.33).

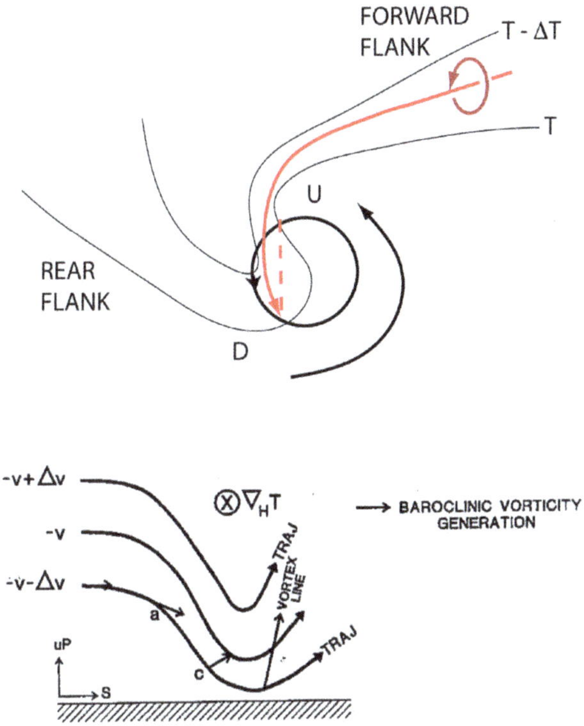

Figure 6.33. Schematic showing how cyclonic vorticity may be generated, as a combination of tilting and baroclinic generation causes the vorticity of parcels to change from anticyclonic to cyclonic while descending in a downdraft. (Top) Consider the vertical plane indicated by the dashed line. The orange steamline represents streamwise vorticity in the forward flank—not as it approaches the downdraft, but where it represents the trajectory. Otherwise, as in Figure 6.32. (Bottom) Suppose that (a) just inside the cold side of an RFD that is wrapping around a mesocyclone, the temperature gradient vector points into the page, to the left of the flow, so that baroclinic generation of vorticity is in the direction of the arrow at the right. As the parcel sinks baroclinic vorticity continues to be generated, while the vorticity vector, which was tilted downward by the downdraft (accompanied by anticyclonic vorticity) becomes tilted upward (accompanied by cyclonic vorticity) as it is advected faster southward below than it is aloft, as happens at the "foot" of a density current (where the flow of the density current is toward the ambient air to the south, but much weaker above the density current) or in the presence of strong low-level environmental vertical shear in the southerly direction (as in Figure 4.14g, upper-left inset) and (c) finally enters the base of the updraft where it is stretched. Trajectories in the vertical plane are denoted by solid curved lines with arrows; the three-dimensional vorticity vector is indicated by the vectors. (The reader is cautioned that in Figure 6.32 vorticity generated in the downdraft is on the opposite side of the RFD, so that the horizontal temperature gradient is reversed and vorticity is anti-streamwise.) Figure 6.33 effectively shows how an air parcel coming from the forward flank may wrap around the mesocyclone and then enter the downdraft, but become tilted *upward* by vertical shear and then pass underneath the updraft. Figure 6.32, on the other hand, shows how anti-streamwise vorticity can be tilted *downward* just as it enters the downdraft and be advected to the ground where no more tilting is possible, and then pass underneath the updraft.

If the air to the rear (west or northwest) of the RFD is evaporatively cooled and advected southward and southeastward, then northerly (i.e., from the north) horizontal vorticity is generated baroclinically as air parcels descend and turn cyclonically around the low-level mesocyclone (see also Figure 6.32). Vortex lines do not simply follow storm-relative trajectories, because northerly vorticity is being generated baroclinically. In addition, there is tilting as a result of baroclinic boundary parallel wind speeds decreasing with height. Vortex lines therefore do not initially tilt downward as much as trajectories, such that anticyclonic vorticity that is produced through tilting is lessened. Let us follow an air parcel descending along a streamline intermediate between the ones above and below. While air parcels are descending, vortex lines eventually gain a component of tilt upward as a result of vertical wind shear (and their moving from the downdraft to the main updraft) (Figure 6.33). *Thus, vortex lines change their orientation from anticyclonic (having a downward component) to cyclonic (having an upward component) even in the presence of a downdraft and finally cyclonic vorticity is enhanced as vortex lines encounter the convergent, lower portion of the updraft.* The important point here is that cyclonic vorticity is produced in a downdraft and can be brought to the surface and then stretched in the updraft. Vertical shear is essential for tilting vortex lines upward even before air parcels reach the main updraft. In effect, the analysis of individual vortex lines is here in agreement with circulation analysis and is a special case, representing only a very particular trajectory. If we choose a trajectory from the warm side of the mesocyclone, we would not be able to generate cyclonic vorticity near the surface. Arguably, circulation analysis is more useful for explaining low-level mesocyclogenesis, while vortex line analysis may be more useful for explaining the intensification phase of low-level mesocyclogenesis just prior to tornadogenesis, though one might disagree on when mesocyclogenesis ends and tornadogenesis begins.

6.5.4 The dynamic pipe effect and the vertical propagation of vortices

As was shown in Section 4.4, tilting of environmental horizontal vorticity along the edges of an updraft may produce vertical vorticity that is elevated. Mid-level mesocyclones are elevated and can be produced solely by an updraft. Roger Smith and Lance Leslie suggested in the late 1970s that a vortex localized in height propagates downward (and upward) through the "dynamic pipe effect" (DPE). Jeff Trapp and Bob Davies-Jones in the late 1990s further studied the DPE in mid-level mesocyclones from a theoretical standpoint. When a vortex forms at mid-levels, air is drawn up into it by an upward-directed, dynamic perturbation pressure gradient force (associated with the increase in vorticity with height; consider the vorticity term in (4.48)). The resulting convergence increases vorticity below the original vortex as it acts on any vorticity there might be at that level, creating a mechanism for the downward propagation of the vortex (above the vortex vorticity is also increased by convergence; at the center of the vortex, there must be compensating divergence; so, vorticity increases above and below the vortex, but decreases at the level of vorticity maximum; Figure 6.34). How fast

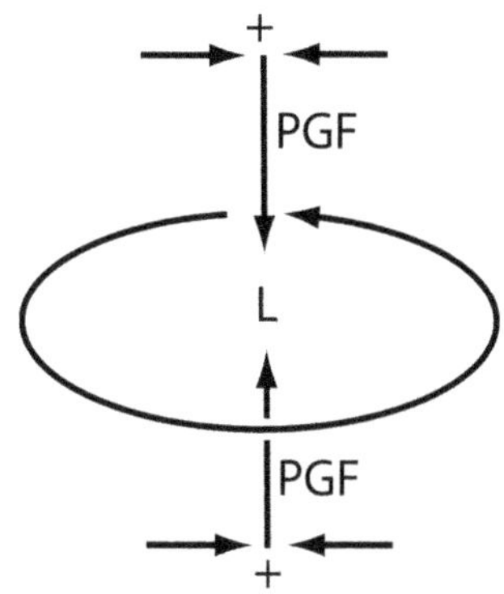

Figure 6.34. Illustration of the dynamic pipe effect. A steady vortex in an inviscid fluid (curved streamline) is associated with a dynamic perturbation central pressure deficit "L" (it does not matter if the vortex is cyclonic or anticyclonic). There are therefore downward and upward-directed vertical dynamic peturbation pressure gradient forces (PGFs) above and below the vortex, respectively. Fluid converges inward and vorticity is increased both above and below (+) (while divergence (not shown) occurs at the center of the vortex, where vorticity is decreased). When a stable vortex is squeezed inward locally, centrifugal waves are generated.

this occurs probably depends on the vertical gradient in vorticity or, in other words, how localized in height the mesocyclone is: the shorter the vertical scale of the mesocyclone, the sharper the vertical gradient in vorticity, the more intense the vertical perturbation pressure gradient force, and the faster the propagation speed. However, the propagation speed of the mid-level vortex could depend on the speed of vertically propagating centrifugal waves, which for solid body rotation (solid body rotation and centrifugal waves are discussed in more detail later in this chapter) that is height dependent and with no mean updraft or downdraft superimposed, the vertical phase speed as shown by Alan Shapiro depends upon the rate of rotation divided by the wavenumber: the fastest vertically propagating centrifugal waves are those with the longest vertical wavelength. For a vertical wavelength of 10 km and vorticity of $10^{-2}\,s^{-1}$, the phase speed is around $50\,m\,s^{-1}$, which is comparable with vertical velocity in supercell updrafts; at this propagation speed, waves can travel half the depth of the troposphere in just a few minutes. In nature, vortices may not be in solid body rotation, may be confined to a layer, and there may be an updraft or downdraft embedded, so that the problem of determining the centrifugal wave speed is more complicated.

It has been found from Doppler radar observations on time scales of minutes that many "tornadic vortex signatures" (and hence tornadoes) begin at mid-levels and then propagate downward. Numerical experiments have demonstrated how when the buoyancy driving the updraft in a supercell is greatest at mid-levels, there is the highest potential for the DPE to propagate mid-level vortices downward; on the other hand, when buoyancy is greatest at low levels, vorticity increases throughout the vertical column simultaneously.

More recent rapid scan Doppler radar observations of the formation and subsequent behavior of a few mid-level mesocyclones on time scales of 10 s suggest that mesocyclone vertical propagation speed is nearly infinite (i.e., mesocyclones/TVSs instantly form in a column); earlier radar observations on time scales of minutes simply could not resolve the true time scale of the vertical propagation of mesocyclones or tornadoes. The DPE has therefore been called into question, owing to a lack of observations of it actually occurring when the time resolution of observations is such that observations are capable of resolving it.

The DPE is a process by which vortices produced aloft propagate downward, so that the origin of the tornado vortex is aloft. On the other hand, in vortex chambers tornado-like vortices develop as a result of frictional convergence near the surface driven by an updraft and vortex aloft. It is therefore thought that there might be two different types of tornado formation: one in which the parent vortex begins aloft and builds downward and intensifies via either the DPE (if it actually exists) or builds downward through transport by a downdraft, and one in which the tornado forms at low levels through frictional convergence and a strong updraft aloft and is advected upward so that it appears to build upward from the ground. The distinction between these two types of tornado formation may be artificial: any vortex that gets near enough to the ground can be intensified as will be discussed subsequently. Any vortex that forms aloft, away from the effects of surface friction, may descend via the DPE process or be advected downward. When it gets near enough to the surface, friction may take over. In a sense, *the DPE process and friction are similar, because they both act to draw air radially inward below the vortex.*

How can we explain how seemingly different processes through which a mid-level mesocyclone and a low-level mesocyclone form can result in a tornado? Tornadogenesis probably owes its existence more to the intensification of a low-level mesocyclone rather than the intensification of a mid-level mesocyclone. *The role of a mid-level mesocyclone may be to promote the sustenance and intensification of a low-level mesocyclone.* For example, a mid-level mesocyclone promotes supercell updraft propagation normal to vertical shear, which may allow surface air parcels to pass into the nearby downdraft so that streamwise vorticity is tilted onto the vertical and then can be stretched underneath the main updraft. In a sense, mid-level and low-level mesocyclones may be like the upper-level and surface isentropic potential vorticity (IPV) (or potential temperature) anomalies in the analysis of baroclinic instability of synoptic-scale flows. Should a mid-level mesocyclone not be situated above a low-level mesocyclone, the low-level mesocyclone might not be able to intensify. This hypothesis probably requires further examination.

6.5.5 Role of downdrafts in enhancing and transporting vorticity

Bob Davies-Jones at NSSL has considered the possibility that the hook echo in supercells, which has been thought for decades to be formed passively as precipitation is transported in a curved trajectory by a low-level mesocyclone, actually plays an active role in tornadogenesis. Raindrops may be advected around just outside the updraft and fall as a "rain curtain". In many instances the rain curtain may be very narrow (Figure 6.35); I have named these narrow curtains "umbilical cords". Narrow curtains of rain should produce narrow corridors of evaporative cooling, which lead to a region where horizontal vorticity is generated baroclinically. This horizontal vorticity can be tilted onto the vertical by nearby updrafts or downdrafts. Why narrow rain curtains are observed in supercells is not well understood, but horizontal deformation and sharp updraft gradients are possible relevant processes.

There is some observational evidence that hook echoes can also form as precipitation falls, so that advection is not the only mechanism responsible for the hook shape. Davies-Jones postulated that precipitation-induced drag just outside the center of a strong updraft drives a downdraft that can transport high angular momentum downward, as first suggested by Ted Fujita in the 1970s, and can increase surface convergence, owing to surface friction. It is not the baroclinically generated azimuthal vorticity (due to the radial gradient in precipitation loading or evaporative cooling) that becomes the tornado, but rather the radially inward-directed frictionally generated vorticity that is tilted upward and stretched.

Localized, transient areas of convergence may act to spin up tornadoes. When the RFD descends to the ground, it spreads out and wraps around a low-level mesocyclone. As it hits the leading edge of the rear-flank gust front, it can produce localized strong convergence (Figure 6.36). This localized convergence may act to intensify vorticity. Microbursts might therefore be able to trigger tornadoes. Some researchers such as Erik Rasmussen have considered "descending reflectivity cores" (DRC) in supercells as being associated with wet microbursts

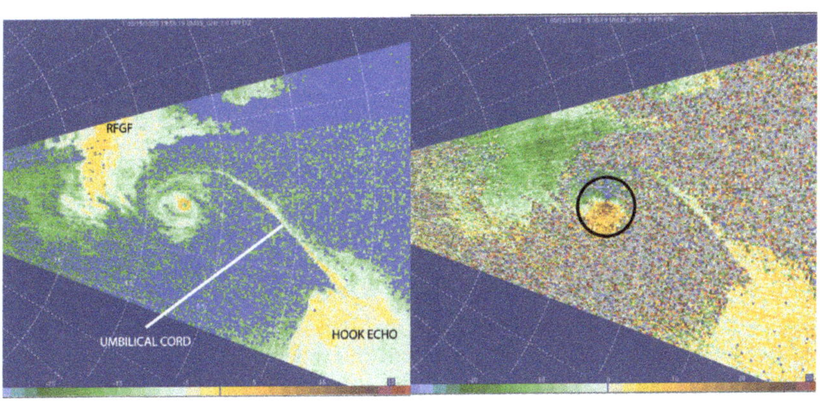

Figure 6.35. Radar reflectivity in dBZ_e (left) and Doppler velocity in $m\,s^{-1}$ (right) for a tornado in Kansas on May 15, 1999, as detected by the U. Mass. W-band, mobile Doppler radar. The tornado (vortex signature circled at the right) is marked by a weak-echo hole surrounded by a ring, and connects the hook echo to the bulge in the rear-flank gust front (RFGF) at an inflection point in the reflectivity field. The hook echo is connected to the ring by a very thin line of reflectivity marked as an "umbilical cord", which appears to connect the tornado to the parent storm. Range rings shown every 1 km.

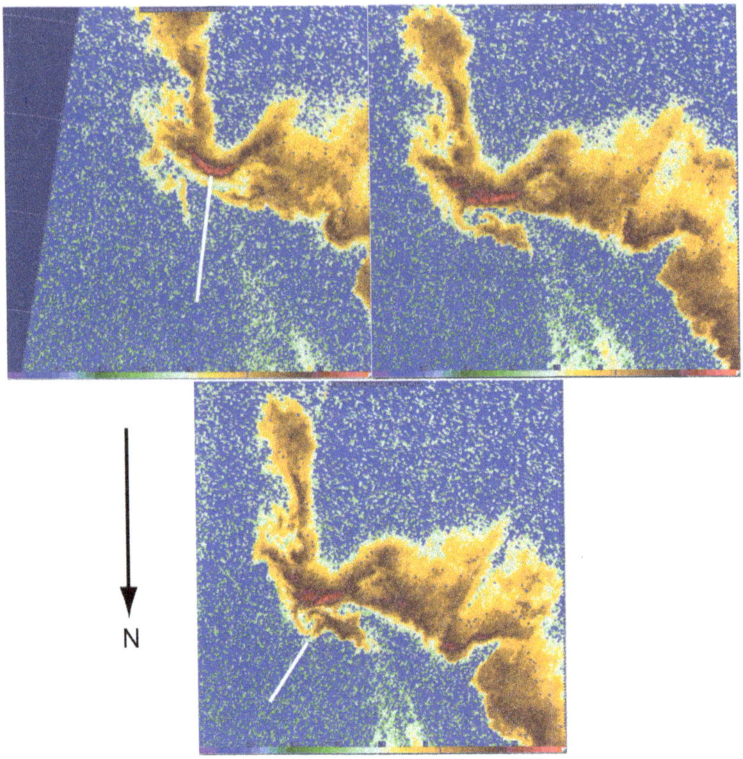

Figure 6.36. Example of tornadogenesis at the leading edge of a bulge in the rear-flank gust front. Radar reflectivity (dBZ_e) from the U. Mass. W-band, mobile Doppler radar in north central Nebraska on June 5, 1999. (Top left) White line points to bulge in rear-flank gust front (U-shaped echo); (top right) about 30 s after the image on the top left, showing a spiral echo just ahead of the bulge; (bottom) about 30 s after the previous image and showing a tightly curled spiral echo just ahead of the bulge. The "emerging" tornado was headed in the general direction of the radar. Range rings shown every 1 km.

capable of initiating tornadoes (Figure 4.16). Recent observations do not show that DRCs are always associated with tornadoes and that tornadoes are not always preceded by DRCs. There is also recent observational evidence from a tornadic supercell during VORTEX2 that a DRC may alter the way circulation is enhanced or vorticity increased as air enters the mesocyclone.

6.5.6 Negative viscosity

The mechanisms hitherto considered involve mean azimuthal velocities. In tornadoes that have eddies, such as secondary vortices or vortex Rossby waves (these waves are analogous to Rossby waves in large-scale flow, except that the radial gradient of the centrifugal force plays the same role as the latitudinal gradient of Earth's vorticity) or Ekman instabilities, etc., it is possible that the eddies are tilted (with respect to the radial direction) so as to convert eddy angular momentum into mean angular momentum, as happens, for example, in some synoptic-scale waves in baroclinic westerlies. Victor Starr at MIT in 1968 termed such a process "negative viscosity" and Doug Lilly suggested it could be responsible for affecting the intensity of some tornadoes.

A process similar to that of negative visocity might be the conglomeration of pre-existing, smaller scale vortices into a larger one: the author and his co-workers found evidence of tornado formation when smaller vortices along a gust front seemed to interact to produce a larger scale, tornadic vortex; similar behavior has been noted in some numerical simulations.

6.5.7 Two-celled mesocyclones and shear instabilities

Roger Wakimoto and his student Chinghwang Liu have suggested that some tornadoes may be initiated when an occlusion/RFD downdraft forms in a mesocyclone (as a result of vorticity becoming less cyclonic with height or precipitation loading or evaporative cooling), leading to an annulus of strong shear and resulting barotropically unstable air within it, which then breaks down into multiple vortices, each of which could become a tornado. This process is like the formation of multiple vortices in a tornado, but on the scale of the mesocyclone. (It is important to realize that vortex breakdown, to be discussed later, cannot occur in mesocyclones because a supercritical end-wall vortex does not form.) It is the descent of air in the middle of the vortex that is responsible for producing the barotropically unstable shear zone. However, it is not clear how each unstable vortex can be stretched up to tornado intensity unless each vortex is co-located with a region of convergence under the updraft.

6.5.8 Cyclic tornadogenesis

Just as mesocyclones sometimes form and decay in a cyclical fashion (Figure 4.59), so do tornadoes in a process Don Burgess at NSSL has referred to as "cyclic tornadogenesis". David Dowell and the author, using airborne Doppler radar data from VORTEX in 1995, further refined the conceptual model of cyclic tornadogenesis proposed by Don Burgess *et al.* in the early 1980s (Figure 6.37). An incipient vortex forms along the rear-flank gust front and propagates along the horseshoe-shaped updraft associated with flanking line towers, until it reaches the tip of the end of the horseshoe-shaped updraft, near the RFD. When it reaches this location of strong horizontal gradient in vertical motion, the tornado is mature. The tornado then propagates away from the updraft entirely and dissipates in the downdraft region. A new tornado may then form along the bulge in the rear-flank gust front, and the process of tornadogenesis is repeated. In some rare instances, a tornado may become locked into such a position that it does not propagate away from the tip of the updraft region and a long-lived (and possibly very intense) tornado may result. It is not known why in these rare instances the tornado remains locked into its mature phase, but observational evidence suggests

Figure 6.37. Cyclic tornadogenesis. (Top left) Conceptual model from Burgess *et al.* (1982) for a succession of parent mesocyclones; (top right) refined conceptual model from Dowell and Bluestein (2002). In left panel, numbered circles identify vortices and thick lines indicate wind-shift lines. Tornado tracks are shaded. In the inserts on the three panels to the right the shaded area indicates the updraft and the spotted area indicates the downdraft; dashed outlines indicate regions of production of cyclonic vertical vorticity by tilting of horizontal vorticity, solid outlines indicate the stretching of vertical vorticity. The arrows denote vortex-relative trajectories. The time between successive tornadoes is about 20 min. (Bottom) Two tornadoes in the northern Texas Panhandle on May 7, 1986; the thin one to the left is the previous tornado in a series about to dissipate, the wider one to the right is the next in a series, near its mature stage (photo by the author).

that interaction with a gust front from a neighboring convective storm might play an important role.

6.5.9 Counter-rotating tornado pairs

It was noted earlier that while most tornadoes in supercells rotate cyclonically, there are a number of well-documented cases of anticyclonic tornadoes forming in cyclonically rotating, right-moving (RM) supercells. Anticyclonic tornadoes have been documented extremely rarely in anticyclonically rotating, left-moving supercells (Figure 6.38). When anticyclonically rotating tornadoes are observed in cyclonically rotating supercells, though, they are found near or along the edge of the rear-flank gust front, ~5–10 km from a surface mesocyclone or cyclonic tornado (i.e., anticyclonic tornadoes in supercells are found paired with nearby cyclonic tornadoes or the remnants of them; Figures 6.8 and 6.9). Anticyclonic vortices of larger scale and weaker vorticity than that associated with tornadoes are sometimes observed in the region where anticyclonic tornadoes have been found (Figure 4.24). These anticyclonic vortices may be associated with the tilting of baroclinically generated horizontal vorticity along the edge of the rear-flank gust front at the end of the updraft associated with the flanking line and the main updraft or the tilting of environmental horizontal vorticity (associated with vertical shear) along the edge of the rear-flank downdraft (Figures 6.39). (Vertical vorticity is thus produced in a manner similar to that of bookend vortices in MCSs; cf. Figure 5.26.) Perhaps tornadoes can be formed when low-level convergence in the vicinity of such an anticyclonic vortex appears, in response to a rapidly growing updraft along the flanking line, above the rear-flank gust front, as for landspouts. It is not known whether anticyclonic tornadoes are related dynamically to the cyclonic member of the pair of vortices, or whether they are independent of it.

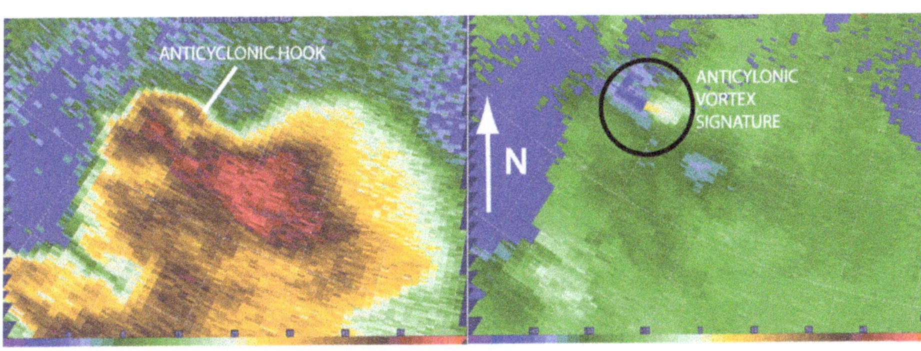

Figure 6.38. Left-moving, anticyclonically rotating supercell containing a meso-anticyclone and an anticyclonic tornado. WSR-88D radar imagery from KTLX on May 10, 2010, southwest of Oklahoma City, OK. Range rings are shown every 10 km. (Left) Radar reflectivity factor in dBZ; (right) Doppler velocity in m s^{-1}.

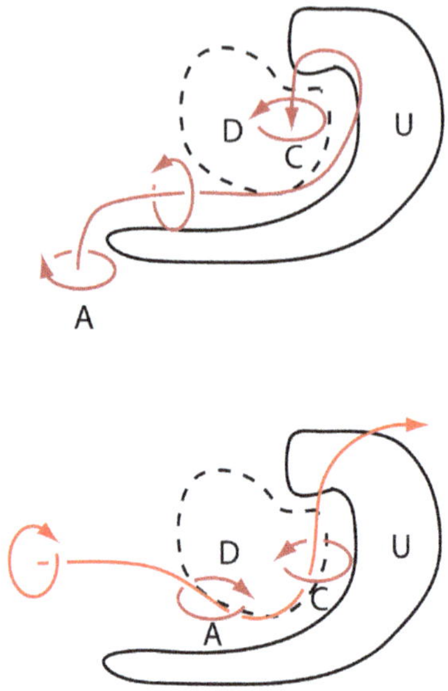

Figure 6.39. Idealized illustration of how counter-rotating vortices (anticyclonic member A, cyclonic member C) can form in a supercell at the ends of an updraft associated with the flanking line and main updraft. Horizontal cross-section shows solid line enclosing an area of updraft (U); dashed line encloses a region of downdraft (D). Red line with arrow shows vorticity streamline (vortex line). Curved red lines with arrows indicate sense of rotation. (Top) Tilting at the ends of the updraft of horizontal vorticity generated baroclinically near the flanking line updraft and main updraft, at the edge of the rear-flank downdraft cold pool, if there is one. (Bottom) Tilting of horizontal vorticity from the environment at the edge of the rear-flank downdraft and between the rear-flank downdraft and main updraft. In the top case, the vortex line is arched upward by the updraft; in the bottom the vortex line is arched downward. In the top illustration, the vortex line may curve back around the other side of the cold downdraft to form a vortex ring (not shown).

6.6 VORTEX DYNAMICS

The problems of what determines the character of the wind field in a tornado and what determines whether or not a tornado will form have been addressed using laboratory models of vortices and numerical models of vortices under highly idealized, controlled, laboratory model-like conditions. While these idealized models do not include the effects of external convective storm features producing and interacting with the vortex and translating along it and do not include the asymmetric effects present in nature, such as the advection of cooler/warmer and moister/drier air in specific quadrants of the vortex, they have been very useful in elucidating many of the observed characteristics of tornadoes in nature.

Axisymmetric, tornado-like vortices in laboratory models (or "simulators") are driven from above by an exhaust fan (Figure 6.40); air is drawn into a rotating lower section with vanes that channel flow so that some of the airflow is in

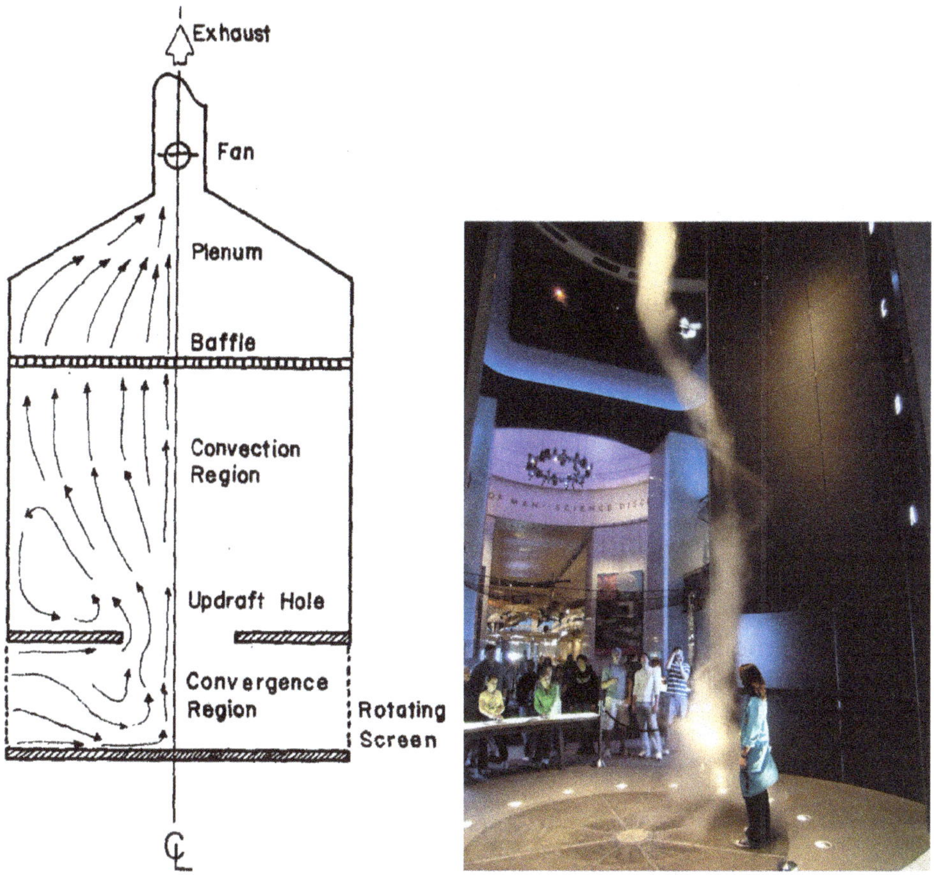

Figure 6.40. (Left) Schematic of the "Ward" tornado simulator (from Church *et al.*, 1979). (Right) Large tornado simulator at the Museum of Science and Industry in Chicago, IL (photograph by the author).

the azimuthal direction, where it acquires vertical vorticity. Neil Ward at NSSL pioneered the use of such laboratory models. While there are a number of parameters that are not allowed to vary (by design, for geometric similarity with a convective storm), such as the size of the opening into which horizontally converging air at the bottom turns into the updraft aloft, measurements have been made that bear good resemblance to observations of some real tornadoes.

While laboratory models allow us to conduct controlled experiments with tornado-like, columnar vortices that make contact with the ground, we do have the problem of measuring wind variables and thermodynamic variables without disturbing the flow so that the vortex structure is not altered by the measurements. In effect, we have a problem similar to (but not directly analogous to) that in quantum physics in which the better one knows the momentum of a particle, the less well one knows the position of the particle: The observer affects the phenomenon being measured. To avoid observing problems laser velocimeters have been used to map the wind field unobtrusively, but measurements of thermodynamic variables still require in situ probes. Ironically, some researchers have conducted numerical simulations of laboratory experiments. Rich Rotunno at NCAR exploited this approach in the late 1970 and early 1980s.

In a "Ward"-type vortex chamber, the tornado is separated from the "storm" above and the air coming into the bottom of the chamber: the boundaries of the tornado (exhaust fan aloft and inflow below) are open. Inflow below and outflow aloft are specified as boundary conditions and can significantly affect the behavior of the flow. Brian Fiedler introduced an alternative with a closed domain, in which an updraft is driven by a bubble of positive buoyancy—not by an exhaust fan. The domain extends upward to the top of the storm at the tropopause. The Ward chamber is one in which air enters from somewhere else, goes up, and then out; the Fiedler model is one in which air is forced radially inward below, upward, and radially outward aloft by localized buoyancy, and then re-cycled. The buoyancy bubble emulates a buoyant updraft in the parent convective storm. In addition to the Ward chamber and the Fiedler model, there are also models which use water as the fluid and one in which a propeller is used to force air upward.

An obvious problem with interpreting laboratory model simulations or idealized numerical simulations of laboratory vortices is that of the difference in spatial scales: in a simulator, the characteristic wind scale

$$U \sim 1\,\mathrm{m\,s^{-1}} \tag{6.2}$$

and the characteristic horizontal length (L) and depth (H) scales are

$$L \sim H \sim 0.1\,\mathrm{m} \tag{6.3}$$

Thus, the Reynolds number (the ratio of the acceleration due to inertial forces $D\mathbf{v}/Dt$ to that due to friction)

$$\mathrm{Re} \sim UL/\nu_m \sim (1\,\mathrm{m\,s^{-1}})(0.1\,\mathrm{m})/(2 \times 10^{-5}\,\mathrm{m^2\,s^{-1}}) \sim 10^4 \tag{6.4}$$

where the kinematic coefficient for molecular viscosity in the atmosphere

$$\nu_m = 2 \times 10^{-5}\,\mathrm{m^2\,s^{-1}} \tag{6.5}$$

has been used. In the case of a real tornado, however,

$$U \sim 75\,\mathrm{m\,s^{-1}} \tag{6.6}$$

$$L \sim H \sim 100\,\mathrm{m} \tag{6.7}$$

so that the Reynolds number is

$$\mathrm{Re} \sim (75\,\mathrm{m\,s^{-1}})(100\,\mathrm{m})/(2 \times 10^{-5}\,\mathrm{m^2\,s^{-1}}) \sim 10^8 \tag{6.8}$$

Thus, the characteristics of turbulent, subgrid-scale diffusion may be quite different in vortices in the vortex chamber and in nature. Recently, Brian Fiedler and Gabe Garfield at OU (Oklahoma University) have shown how the behavior of modeled vortices is sensitive to the type of subgrid-scale parameterization scheme used.

To simulate idealized laboratory vortices, there is the problem of what boundary conditions to use. No-slip lower-boundary conditions are appropriate when the effects of surface friction need to be taken into account, while free-slip lower-boundary conditions are used to isolate the behavior of the vortex when surface friction plays no role. What are the boundary conditions at the top of the domain if the parent storm is not explicitly represented? The reader is referred to Rich Rotunno's early papers and the review papers by Davies-Jones *et al.* (2001) and Rotunno (2012), and elsewhere for a summary of the more technical issues.

6.6.1 Vortex structure

The idealized vortex that is produced may be thought of as an intense ("primary") vortex that is intensified by storm-updraft-associated convergence acting on pre-existing vorticity. In the meantime, the intensifying vortex rubs against the surface, where friction slows it down. The surface boundary-layer flow may be laminar or turbulent, depending on the Reynolds number. The degree of smoothness of the surface underneath the vortex can therefore play a role in the nature of the flow. The Reynolds number, as we have shown, is extremely high and therefore indicative of turbulence. It is therefore thought that in nature tornado boundary layers are turbulent, especially when the surface of the Earth is relatively rough.

The radial and azimuthal components of the equations of motion for axisymmetric ($\partial/\partial\phi = 0$, where ϕ is the azimuthal angle in cylindrical coordinates) motions in a non-rotating atmosphere, including turbulent friction, are given in cylindrical coordinates as follows:

$$\begin{aligned} Du/Dt &= \partial u/\partial t + u\,\partial u/\partial r + w\,\partial u/\partial z - v^2/r \\ &= -\alpha_0\,\partial p'/\partial r + \nu(\partial^2 u/\partial^2 r + 1/r\,\partial u/\partial r - u/r^2 + \partial^2 u/\partial z^2) \end{aligned} \tag{6.9}$$

$$\begin{aligned} Dv/Dt &= \partial v/\partial t + u\,\partial v/\partial r + w\,\partial v/\partial z + uv/r \\ &= \nu(\partial^2 v/\partial r^2 + 1/r\,\partial v/\partial r - v/r^2 + \partial^2 v/\partial z^2) \end{aligned} \tag{6.10}$$

where u is the radial wind component; v is the azimuthal wind component; r is the radial coordinate; z is the vertical coordinate; α_0 is the specific density at the surface; and ν is the kinematic coefficient of viscosity for turbulent eddies—not for molecules. The $-v^2/r$ term on the LHS of (6.9) is the centripetal acceleration; if moved to the RHS so that the reference frame changes from that of an air parcel to that of the rotating reference frame of the vortex, it (i.e., v^2/r) is the centrifugal acceleration. The $-\alpha_0\,\partial p'/\partial r$ term represents the acceleration due to the radial pressure gradient force. The last term on the RHS of (6.9) is the turbulent friction term. The most significant contribution to the friction term comes from the vertical term, $\nu\,\partial^2 u/\partial z^2$. The equation for the azimuthal wind component (6.10) is sometimes expressed in terms of angular momentum $\Gamma = rv$. Since the

Coriolis force is much less than the centrifugal force in a tornado, the Coriolis force is neglected.

The vertical equation of motion is

$$Dw/Dt = \partial w/\partial t + u\,\partial w/\partial r + w\,\partial w/\partial z$$
$$= -\alpha_0\,\partial p'/\partial z + B + \nu(\partial^2 w/\partial r^2 + 1/r\,\partial w/\partial r + \partial^2 w/\partial^2 z) \quad (6.11)$$

where $B = gT'/\bar{T}$. The reader is reminded that $B = 0$ in laboratory models, which are driven by exhaust fans—not by positive buoyancy in a cloud overhead—as a result of latent heat release.

The components of vorticity in an axisymmetric vortex in the radial, azimuthal, and vertical directions respectively are

$$\xi = -\partial v/\partial z \quad (6.12)$$

$$\eta = \partial u/\partial z - \partial w/\partial r \quad (6.13)$$

$$\zeta = 1/r\,\partial/\partial r(rv) \quad (6.14)$$

The reader should note that these three components of vorticity in cylindrical coordinates are given the same names (Greek letters) as the components of vorticity in Cartesian coordinates ((4.36)–(4.38)), and should not be confused with them. Vorticity in cylindrical coordinates will be used later but are provided now for reference.

The equation of continuity in a Boussinesq atmosphere for axisymmetric motions is

$$1/r\,\partial/\partial r(ru) + \partial w/\partial z = 0 \quad (6.15)$$

The adiabatic form of the thermodynamic equation is

$$DT'/Dt = (\partial/\partial t + u\,\partial/\partial r + w\,\partial/\partial z)T' + w\,\partial\bar{T}/\partial z$$
$$= \kappa(\partial^2/\partial^2 r^2 + 1/r\,\partial/\partial r + \partial^2/\partial z^2)T' \quad (6.16)$$

where κ is the eddy coefficient of turbulent diffusivity. The simplest models of tornadoes are those in an atmosphere that is adiabatic and one in which density is constant. It follows that no thermodynamic equation is needed to determine vortex structure in these models.

Beyond a certain radius, the effects of surface friction are felt in the boundary layer (Figure 6.41). Al Barcilon in 1967 was perhaps the first to analyze the problem of having a tornado-like vortex interact with the ground and infer that there are several dynamically different regions. The boundary layer is divided up into the "friction layer" (2b) and the "inertial layer" (2a). Above the boundary layer, in the free atmosphere, away from the axis of rotation (beyond $r = r_c$), the region is typically referred to as the "outer flow" (1). In the outer flow region, a radially inward-directed pressure gradient force (the acceleration is $-\alpha_0\,\partial p/\partial r$) is counterbalanced by a radially outward-directed centrifugal force (the acceleration is v^2/r), the condition of "cyclostrophic balance", if the flow is steady state. It is assumed that the outer flow is characterized by "constant angular momentum" Γ,

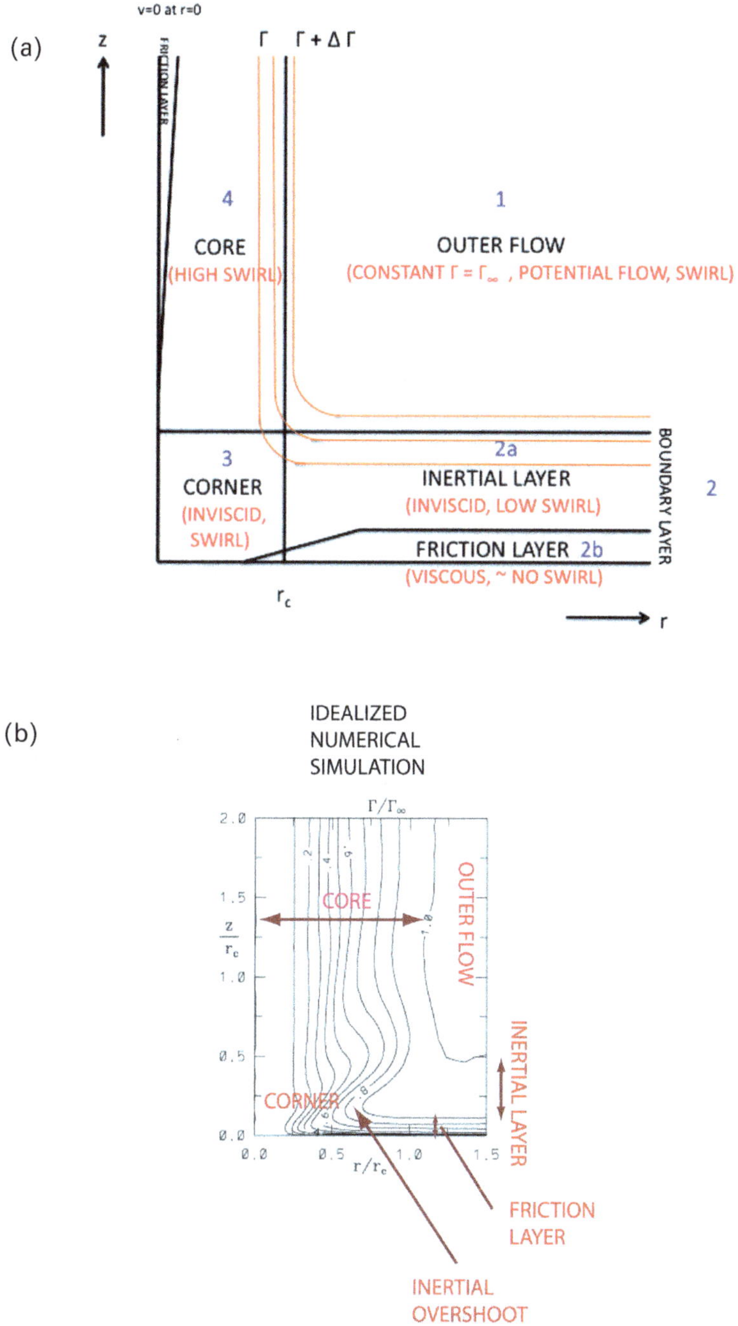

Figure 6.41. The four characteristic regions of a tornado and their properties (vertical cross sections): (a) idealized model; contours of constant angular momentum (solid orange lines); (b) LES; lines of constant angular momentum (solid black lines) (from Lewellen *et al.*, 2000).

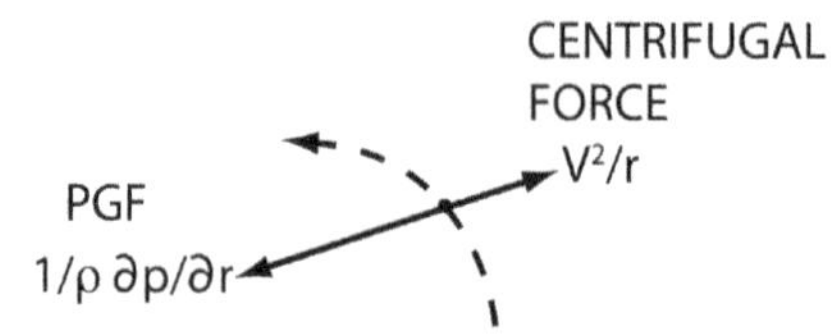

Figure 6.42. Forces in the inertial layer, within the boundary layer. The path of an air parcel is indicated by the dashed streamline. The pressure gradient force (PGF) acts radially inward and is the same as it is above the boundary layer. The centrifugal force is reduced by friction from what it is above the boundary layer so that there is a net force (the forces are not balanced) acting radially inward.

or circulation $C = 2\pi rv = 2\pi\Gamma$, as a "potential vortex". It is assumed that this angular momentum is produced by the parent storm (baroclinically or from environmental vorticity or via frictional generation). *It is the pressure gradient force associated with this vortex that drives the flow in the boundary layer below.* For simplicity, we consider a vortex in which the radial pressure gradient force is independent of height. This assumption is probably pretty good at low levels, but not aloft, where the intensity of the vortex that drives boundary-layer flow decreases with height above some altitude.

It is assumed that the azimuthal wind speed in the boundary layer is less than it is above the boundary layer owing to surface drag, and that the radial pressure gradient force in the boundary layer is the same as it is at the top of the boundary layer, as just noted. As a result of the departure from cyclostrophic balance near the ground, there is radial inflow (Figure 6.42). Such conditions have been famously illustrated through the tempest-in-a-teacup analogue: stir a cup of tea into a vortex and the tea leaves will quickly collect at the center of the bottom of the cup as a result of frictionally induced convergence.

In the section of the boundary layer closest to the surface (i.e., the "friction layer" for steady-state, axisymmetric flow), the radial equation of motion from (6.9) is—based on solutions to the equations of motion with a no-slip lower-boundary condition and by laser velocitometer measurements in the Purdue vortex simulator—given approximately by

$$0 = -\alpha_0 \, \partial p'/\partial r + \nu \, \partial^2 u/\partial z^2 \tag{6.17}$$

The pressure gradient force acts radially inward (<0) and the friction force acts radially outward (>0). In other words, air is accelerated radially inward, but as it rubs against the ground it is retarded, so that it experiences a force that acts radially outward: there is a turbulent transport of zero radial momentum upward and non-zero radial momentum downward, so that radially inward-flowing air is slowed down (i.e., there is radial acceleration outward). The two forces just about cancel each other (Figure 6.43). Horizontal turbulent transports of momentum are much smaller than vertical transports. Equation (6.17) may be rewritten without

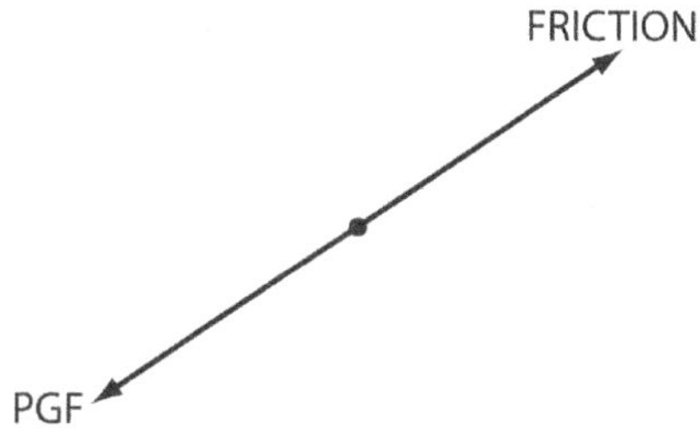

Figure 6.43. Balance of forces in the friction layer.

pressure explicitly expressed as

$$0 = -V^2/r + \nu\, \partial^2 u/\partial z^2 \tag{6.18}$$

where V is the azimuthal wind speed if there were cyclostrophic balance ($V^2/r = \alpha_0\, \partial p'/\partial r$). In the friction layer, the flow is sub-cyclostrophic.

Since the turbulent friction term on the RHS of (6.18) is the same order of magnitude as the $-V^2/r$ term, the kinematic coefficient of turbulent viscosity

$$\begin{aligned}\nu &\sim (V^2/r)/\partial^2 u/\partial z^2 \sim [(10\ \mathrm{m\ s^{-1}})^2/(100\ \mathrm{m})]/[10\ \mathrm{m\ s^{-1}}/(100\ \mathrm{m})^2] \\ &\sim 10^3\ \mathrm{m^2\ s^{-1}}\end{aligned} \tag{6.19}$$

In the friction layer, the vertical derivatives of wind components are much greater than radial gradients.

Just above the friction layer, where the effects of friction are negligible, it follows from (6.9) that

$$u\, \partial u/\partial r + w\, \partial u/\partial z \approx (v^2 - V^2)/r = -\alpha_0\, \partial p'/\partial r + v^2/r \tag{6.20}$$

(Figure 6.42). This layer (just above the friction layer) is called the "inertial layer" because inertial accelerations—the LHS of (6.20)—are significant: the flow in the reference frame of the vortex is unbalanced. It is noteworthy that while turbulent mixing does not appear explicitly in this equation, the effects of surface friction have been communicated to this layer via deviation of the azimuthal component of the wind from its cyclostrophic value. Since the flow is sub-cyclostrophic, the term on the RHS of (6.20) is <0. In the reference frame of a rotating air parcel, the LHS of (6.20) is just the parcel acceleration in the radial direction, so that air parcels accelerate radially inward toward the center. Air parcels become less sub-cyclostrophic with height until they are exactly cyclostrophic at the top of the inertial layer. The vertical derivatives of wind components are much larger than radial derivatives, as they are in the friction layer. At the surface, where $w = 0$, $u\, \partial u/\partial r < 0$, since $u < 0$ and $\partial u/\partial r > 0$; this pattern is consistent with the sign of the RHS of (6.20).

We now consider the depth of the friction and inertial layers. From analytic solutions O. Burggraf and co-authors in 1971 showed in a seminal paper that the depth of the friction layer for a vortex having potential flow is $\sim(\nu/\Gamma)^{1/2}r$: In other words, the depth of the friction layer decreases as one goes radially inward toward the center of the vortex if the angular momentum is constant. The reader should appreciate the difficulties in finding analytic solutions for the wind field in a tornado boundary layer since as $r \to 0$ there are singularities when r is the denominator of terms and asymptotic solutions must therefore be found. One can also appreciate how interesting a tornado boundary layer might be dynamically owing to the circular symmetry of a vortex, something that will be made more explicit soon. For a kinematic coefficient of molecular viscosity of $2 \times 10^{-5}\,\mathrm{m}^2\,\mathrm{s}^{-1}$ and $\Gamma \sim (75\,\mathrm{m\,s}^{-1})\,(100\,\mathrm{m})$, the depth at $r_c \sim 100\,\mathrm{m}$ is only $\sim$0.5 cm. The depth of the friction layer for the kinematic coefficient of turbulent viscosity of $10^3\,\mathrm{m}^2\,\mathrm{s}^{-1}$ (6.19) at $r_c \sim 100\,\mathrm{m}$ is $\sim$10 m, which is much deeper than the depth of the friction layer when not taking turbulent eddies into account. Most of the flow in the friction layer is in the radial direction.

From both vortex chamber studies and analytic solutions for vortex chamber flow, the depth of the tornado boundary layer (friction layer + inertial layer) is $\sim(\nu/\Gamma)^{1/2} r_b$, where r_b is approximately the radius of the rotating bottom of a vortex chamber, or in the atmosphere the radius at which a boundary layer for rotating ("swirling") begins to form. To estimate r_b for the real atmosphere, we make use of vortex chamber data from which it is inferred that the *depth of the inertial layer is approximately the same as the "core" radius*

$$r_c \sim (\nu/\Gamma)^{1/2} r_b \tag{6.21}$$

So, for $r_c \sim 100\,\mathrm{m}$, $\nu \sim 10^3\,\mathrm{m}^2\,\mathrm{s}^{-1}$, and $\Gamma \sim (75\,\mathrm{m\,s}^{-1})(100\,\mathrm{m})$, $r_b \sim 300\,\mathrm{m}$. It is noted that unlike the depth of the friction layer the depth of the inertial layer does not vary significantly with radius. Note that the depth of the friction layer at the core radius ($\sim$10 m) is much shallower than the depth of the inertial layer at the core radius ($\sim$100 m).

In the inertial layer air parcels accelerate radially inward, so the time they rub against the ground in the friction layer decreases as they move radially inward, limiting the effects of turbulent friction (vertical exchange of turbulent eddies from aloft with eddies from the surface). Also, it will be shown later that there is a dynamically induced downdraft that can also act to suppress the height of the friction layer: thus, the depth of the friction layer decreases radially inward.

Now, consider what happens at low levels inside the core radius. Owing to circular symmetry, accelerating radial inflow in the inertial region *must* decelerate near the origin, since u must vanish at $r = 0$. Air rushing in from all directions abruptly slows down and, from continuity considerations, turns and flows upward. This is perhaps the most interesting aspect of boundary-layer flow in a tornado alluded to earlier. Paraphrasing comments Rich Rotunno once made to the author, the radially inward accelerating flow is on a direct "collision course" with air coming in from the opposite direction (and others) and something "catastrophic" must happen. This region, where the flow abruptly "turns the corner", is called the "corner" region (Figure 6.41). In the corner region, *both* vertical and radial variations are significant. Owing to the mass continuity constraint, there is an intense frictionally induced updraft jet. When the vortex is accompanied by this frictionally induced, intense updraft it is called an "end-wall" vortex, where "end-wall" refers to the solid lower boundary.

It is thought that light debris is lofted high in the parent storm from the corner region's upward-flowing jet and then caught up in the storm's updraft; subsequently the debris may be deposited downstream from the storm, sometimes at great distances from its source. There have been many cases in which personal banking data and family photographs have been carried and dropped as much as 100 km downstream. John Snow and his students at OU have conducted studies of lofted debris in powerful tornadoes. Personal effects have been found and re-united with their owners whose houses have been demolished by strong tornadoes.

The equation of motion for radial flow in the corner region from (6.9) is

$$u\,\partial u/\partial r + w\,\partial u/\partial z - v^2/r = -\alpha_0\,\partial p'/\partial r \tag{6.22}$$

Based on numerical experiments, Tim Wilson and Rich Rotunno, in a 1986 paper, showed that corner flow is considered inviscid because the turbulent friction term is not significant in comparison with the other terms in (6.22). This finding is in accord with Burggraf *et al.*'s finding that the depth of the friction layer varies as the radius: in the corner region the depth of the viscous layer $\rightarrow 0$ as $r \rightarrow 0$. As the flow turns upward, as noted earlier, there is less rubbing of the horizontal wind against the ground. It is ironic that where tornadoes are the most violent (their wind speeds are the highest), they are characterized by laminar flow.

Doppler radar measurements in the friction and corner regions are difficult to obtain, owing to ground clutter contamination from trees, utility poles, houses, etc. and because the friction layer is so shallow that the vertical resolution needed to discern vertical variations in wind speed is difficult to achieve in practice. Nevertheless, Bluestein *et al.* in a paper published in 2007a have provided some relatively high-resolution Doppler radar measurements that may indicate that the wind speed can increase by more than 25% in the surface friction layer.

The resulting flow pattern in the vertical plane (i.e., for vertical and radial motions) of the friction layer, the inertial layer, and the corner region, driven in this case by friction, is called the "secondary" circulation. (Synoptic meteorologists also refer to circulation in the vertical plane that is an instantaneous response to and counteracts quasi-geostrophic forcing from the horizontal wind field acting on the temperature field as the "secondary circulation".) Vertical motions are coupled with horizontal motions via the Boussinesq equation of continuity (6.15). To understand the behavior of tornado boundary layers better and to put in perspective what we already know, we first consider what happens to synoptic-scale vortices in the boundary layer. In synoptic meteorology, one considers what happens when vortices such as synoptic-scale cyclones and anticyclones rub against a rotating surface (rotating at the speed of the local rotation rate of the Earth about its axis projected onto the local horizontal plane). Traditional analysis of a steady-state boundary layer yields the familiar Ekman spiral hodograph, for which there is convergence and divergence in the boundary layer of cyclones and anticyclones, respectively. The former produces Ekman pumping out of the boundary layer into the free atmosphere above and the latter produces

Ekman suction into the boundary layer from the free atmosphere (see textbooks on synoptic meteorology: e.g., Bluestein, 1992). The depth of the Ekman layer is approximately $(2\nu/f)^{1/2}$, where f is the Coriolis parameter; this is the height at which the ageostrophic component of the wind falls to a factor of $1/e$ of the ageostrophic wind component at the anemometer level. For a flat Earth, $f = 2\Omega$, so that the depth of the Ekman layer is $(\nu/\Omega)^{1/2}$. In the case of an Ekman layer, we have a vortex having much less vorticity ($\sim 10^{-5}\ \mathrm{s}^{-1}$) than that of the rotating surface below ($10^{-4}\ \mathrm{s}^{-1}$). We remind the reader, for future reference, that the wind actually overshoots its geostrophic value at and just above the "gradient wind level", the height at which the wind direction first becomes identical to that of the geostrophic wind.

The behavior of the secondary circulation in a tornado boundary layer depends on the radial profile of the azimuthal wind. Two extremes are represented by solid body rotation for which vorticity is constant, and for a potential vortex for which there is no vorticity at all. So far, we have considered what happens when there is a potential vortex. In general

$$v \sim r^{\beta} \tag{6.23}$$

where for solid body rotation $\beta = 1$ and for potential flow $\beta = -1$.

Von Bödewadt in 1940 found analytically that the secondary circulation for solid body rotation is similar to that described by Ekman theory. To see why, consider the equations of motion ((6.9) and (6.10)) for steady-state flow in which $v = \Omega r$ and for which vertical eddy transports of momentum are much greater than horizontal eddy transports

$$u\,\partial u/\partial r + w\,\partial u/\partial z = -\alpha_0\,\partial p'/\partial r + \Omega v + \nu\,\partial^2 u/\partial z^2 \tag{6.24}$$

$$u\,\partial v/\partial r + w\,\partial v/\partial z = -\Omega u + \nu\,\partial^2 v/\partial z^2 \tag{6.25}$$

Note that the pressure gradient term does not appear in (6.25) as a result of axisymmetry. When the advective terms are neglected on the LHS, equations (6.24) and (6.25) are similar in form to the Ekman-layer equations for synoptic-scale flow, except that the Coriolis parameter is replaced by Ω.

There is convergence at low levels and divergence just above, which is accompanied by rising motion in the layer of convergence and just above it. The depth of the friction layer in a tornado vortex characterized by solid body rotation is $(\nu/\Omega)^{1/2}$, where 2Ω is the vorticity of the vortex. For $\nu \sim 10^3\ \mathrm{m^2\,s^{-1}}$ and $\Omega \sim 0.5\ \mathrm{s^{-1}}$ (i.e., vorticity $\sim 1\ \mathrm{s^{-1}}$), the depth of the friction layer is $\sim$40 m. The reader is reminded of the similarity of the formula for the depth of the tornado friction layer to that of the Ekman layer, in which 2Ω is the vorticity of the rotating surface. *Just as in an Ekman boundary layer the wind speed overshoots the geostrophic value in the layer aloft, the azimuthal wind speed overshoots the cyclostrophic value in the layer aloft*, in this case by as much as $\sim$20%. The reason for overshooting is that as air parcels converge radially inward, they spin up, but are also spun down through frictional dissipation; the effect of the former is greater than that of the latter, however, and enhanced vorticity is advected upward. When air parcels get higher, they encounter divergence, and therefore spin down. The result of these processes is that there is a level at which the azimuthal wind is greatest and this level is elevated at a height just above the depth of the boundary layer. So, we would expect to find maximum wind speeds in tornadoes just above the friction layer—not at the surface—if there is solid body rotation.

Alas, evidence from numerical simulations is that where solid body rotation is found in a tornado-like vortex *the turbulent friction term is relatively small compared with the other terms.* It therefore appears as if the boundary layer under solid body rotation is not applicable to tornadoes; it seems, on the other hand, to be applicable to the hurricane/typhoon/tropical cyclone boundary layer. However, that a rotating boundary layer for solid-body rotation is inviscid may not be a correct inference if one takes into account that the effect of turbulent viscosity itself might be to reduce the friction term to zero, so that just because the friction term is negligible does not mean that it is not important in bringing the boundary layer into a steady state. (One can say the same for a convective boundary layer in which stratification is dry adiabatic. One might incorrectly conclude that since stratification is neutral there are no turbulent eddies; in reality local, short-lived episodes of super-adiabatic flow occur and the vertical exchange of eddies stabilizes the atmosphere so that it appears as if the atmosphere is inviscid.)

Burggraf, Stewartson, and Belcher, in their 1971 paper, showed that for potential flow there is a layer of radial inflow, whose intensity is a maximum above the surface, but then decreases with height; there is no layer of radial outflow. As r gets smaller and smaller, the level at which the intensity of radial inflow begins to decrease with height lowers. Most importantly, it is found that there is no overshooting of azimuthal velocity with respect to the cyclostrophic value of the azimuthal velocity as there was when the radial profile of azimuthal wind is that of potential flow; the azimuthal wind is always less than the cyclostrophic value.

The flow in tornadoes in nature and in idealized simulations contains aspects of *both* solid body rotation and potential flow. In the inertial region air flows radially inward and angular momentum is conserved (there is no turbulent friction). So

$$v(r)r = \Gamma \tag{6.26}$$

where Γ is a constant given by the angular momentum, and air parcels therefore spin up as they approach the center of the vortex. It follows that when Γ is spatially uniform outside of some radius the radial profile of azimuthal wind is

$$v(r) = \Gamma/r \tag{6.27}$$

In the outer flow and inertial regions, then, there is potential flow (no vorticity), and the azimuthal wind weakens rapidly with distance from the center of the vortex. The curvature vorticity of the vortex is counterbalanced by shear vorticity of the opposite sign (Figure 6.44). Formally, the potential vortex is a solution to the inviscid forms of (6.9) and (6.10) when $w = u = 0$.

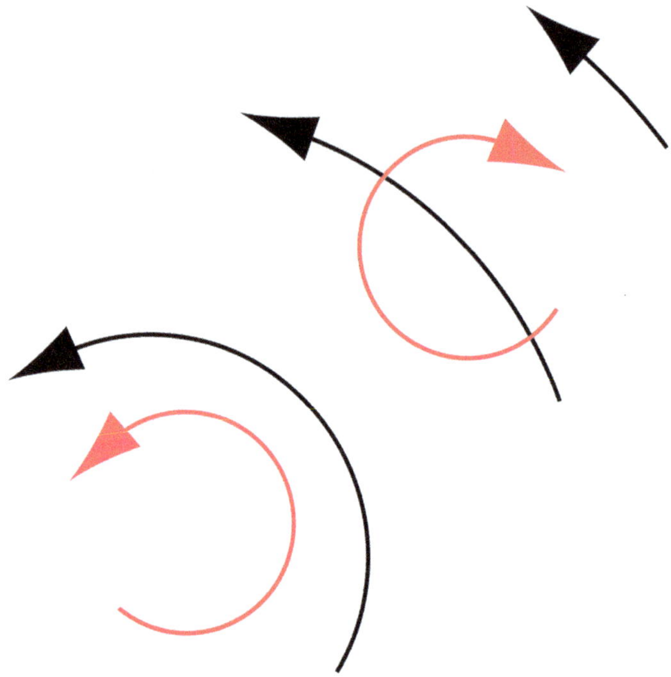

Figure 6.44. Cancellation of shear and curvature vorticity in a potential vortex. The rotation (red curved streamline) induced by shear associated with the decrease in azimuthal wind speed with radius from the center of the vortex (black curved streamlines; wind speed is proportional to the length of each curved streamline) is equal and opposite to the rotation induced by the curvature of the flow.

The circulation (2.52) about a potential vortex (6.27) is finite, since $r > 0$ and $v(\Gamma)$ is nonzero everywhere beyond the origin ($r = 0$). However, the vorticity in a potential vortex is zero. Since vorticity is circulation divided by the area of the material curve about which circulation is computed, there must be infinite vorticity at $r = 0$ (as a point source or line source), and this vorticity is averaged along with the zero vorticity for $r_a > r > 0$, where r_a is the radius of the material curve when circulation is computed. So, if we compute circulation using (2.52) and take the limit as the area of the material curve approaches zero, we find that circulation is due entirely to the point or line source of vorticity at the origin. If, on the other hand, we choose as our material curve one that excludes the center of the vortex, then circulation in the potential vortex is zero (Figure 6.45).

At the center of the tornado, by symmetry there must be zero azimuthal velocity and angular momentum. Beyond the "core" of the tornado the angular momentum is a constant (Figure 6.41a) fixed set by environmental flow (cf. (6.26)). If air parcels were brought towards the center of the vortex, without any turbulent mixing, the azimuthal velocity would approach infinity and so would its radial gradient. So, an air parcel transported radially inward toward the center

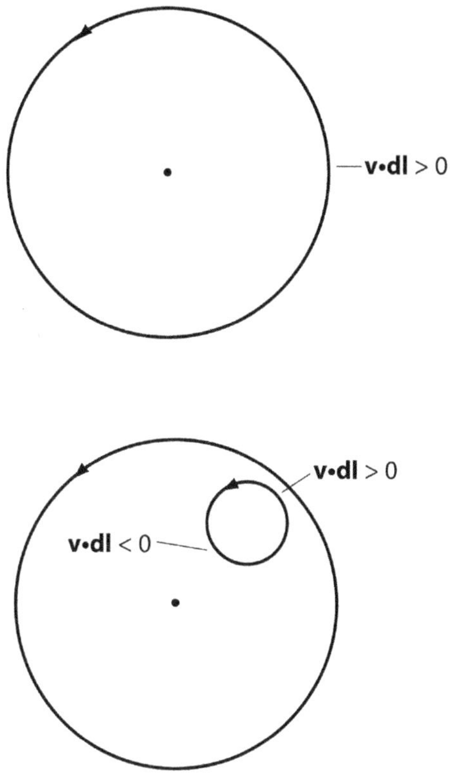

Figure 6.45. Circulation in a potential vortex. (Top) In a circuit around the vortex (circle with arrow indicating the direction of the flow and bold dot indicating the center of the vortex) the circulation is positive because $\mathbf{v} \cdot \mathbf{dl} > 0$ all the way around the curve. (Bottom) Around a circuit inside the previous circuit (curve indicated by smaller circle with arrow indicating the direction of integration) *not including the center of the vortex* the circulation is zero because $\mathbf{v} \cdot \mathbf{dl} > 0$ on the far side of the circuit, but $\mathbf{v} \cdot \mathbf{dl} < 0$ on the near side of the circuit; the contribution at the far side is at larger radius, so the wind speed is relatively low, while the contribution at the near side is at smaller radius, so the wind speed is relatively high. But, the length of arc containing the lower wind speeds at far radius is greater than the length of the arc containing the higher wind speeds at smaller radius: the net contributions around the circuit add up to zero.

would therefore have to lose all its angular momentum as a consequence of turbulent mixing in either the radial or vertical direction (if v vanishes at the ground). How should the azimuthal velocity vary as a function of radius from where it has zero angular momentum at the origin to the core radius, r_c, beyond which angular momentum is a constant? Also, how far radially inward can an air parcel be brought before turbulent mixing becomes significant and dominates? We will address the first question now.

The simplest solution for $v(r)$ is that in the core,

$$v(r) = \Omega r \tag{6.28}$$

where Ω is a constant, which is the rotation rate of a solid body. It is noteworthy that radial or vertical turbulent diffusion may be responsible for bringing about a linear azimuthal wind profile close to the axis of the tornado, which itself is associated *without* any radial turbulent diffusion (i.e., diffusion produces a profile that reduces diffusion to zero; numerical model simulations show nearly solid body rotation close to the center of the axis of rotation and the friction term is negligible). The vortex model in (6.28) is attributed to William John Macquorn Rankine in the late 19th century. In effect, the Rankine vortex is a solution to the inviscid equations of motion for which $w = 0$, $u = 0$, and v is a function of r only. We will now address the second question posed at the end of the previous paragraph.

The Rayleigh criterion for stability in an axisymmetric, inviscid vortex of azimuthal flow only, which is independent of height, is that

$$d^2/dr^2[rv(r)]^2 > 0 \tag{6.29}$$

that is, that the square of the angular momentum increases with increasing radius (squared). In the core of the tornado vortex, the square of the angular momentum is Ωr^2, whose square increases with increasing r^2; the more rapid the rotation, the more stable the vortex. If the vortex is stable with respect to lateral displacements, then it takes work to bring air in closer to the axis of rotation.

Howard and Gupta in 1962 showed that an axisymmetric, inviscid vortex with both azimuthal and vertical motions is stable with respect to axisymmetric perturbations if the "Richardson number" defined as

$$\mathrm{Ri}(r) = \partial(v^2r^2)/\partial r/[r^3(\partial w/\partial r)^2] > \tfrac{1}{4} \tag{6.30}$$

In the corner region, where there are prominent vertical motions, the flow is even more resistant to axisymmetric radial perturbations (i.e., is more stable) than the core region because there is large radial shear in vertical velocity, which appears in the denominator of (6.30)—compare (6.30) with (6.29).

Let us compute the work needed to bring a ring of air radially inward from the core radius in the inertial layer by forcing it against the restoring force on the ring. In effect, we are finding out how much work is needed to bring the outer edge of the core radially inward. Neglecting turbulent diffusion and vertical motion (or simply at $z = 0$ where $w = 0$), we find from (6.9) that for a steady state a vortex in solid body rotation that is in cyclostrophic balance (and for which angular momentum is conserved) follows the following equation of motion in the radial direction:

$$u\,\partial u/\partial r = -\Omega^2 r + v_c^2 r_c^2/r^3 \tag{6.31}$$

To compute the work needed to bring a ring of air radially inward from the core radius r_c to an arbitrary radius r', we integrate (6.31) from r_c to r' and find that

$$\tfrac{1}{2}u_{r'}^2 - \tfrac{1}{2}u_{r_c}^2 = (\Omega^2/2)(r'^2 - r_c^2) + (v_c^2/2)[1 - (r_c/r')^2] \tag{6.32}$$

where v_c is the azimuthal velocity at the core radius, the radius at which solid body flow changes to potential flow. First, note from the RHS that since $r' < r_c$

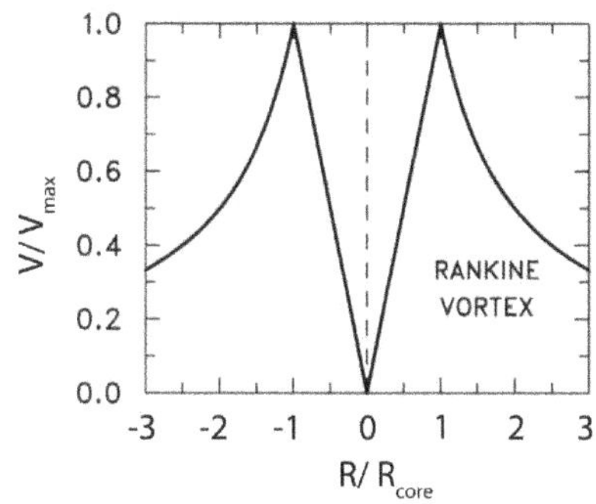

Figure 6.46. Radial (R) profile of azimuthal (tangential) wind (V) in a Rankine combined vortex. R_{core} is the core radius; $V_{\max}$ is the maximum azimuthal wind speed (adapted from Brown and Wood, 2012).

the work done is negative (i.e., work must be done on the system). Also, as $r' \to 0$, the work required $\to \infty$: it becomes more and more difficult to bring in rings of air the closer one gets to the center of the vortex.

The simplest model of a tornado is the "Rankine combined vortex", which is a core of solid body rotation surrounded by (whence the adjective "combined" is used) a region of potential flow, with no vertical motion in either region (Figure 6.46). This type of vortex wind profile satisfies the axisymmetric equations of motion subject to the approximations of steady-state azimuthal flow ((6.9) and (6.10)).

Outside the core radius, where angular momentum is constant, $d^2/dr^2[rv(r)]^2 = 0$, so it takes no work to bring a ring of fluid radially inward, but *only* down to the core radius—not within it. One may verify this claim simply by rewriting (6.31) for the case when a cyclostrophic wind balance is imposed on a potential flow vortex. The work needed to bring a ring of air radially inward from the core radius or within must be related via the continuity equation to the intensity of the updraft forced from above (in the parent storm or vortex chamber), as the radial gradient of u is related to the vertical gradient in w (6.15) and $w = 0$ at the surface.

Recent mobile Doppler radar measurements of the wind field in tornadoes exhibit radial profiles of azimuthal wind that are similar to the Rankine combined vortex, except that there is a smooth transition from solid body rotation to potential-like flow near the radius of maximum wind (RMW) (Figure 6.47a) instead of an abrupt transition. The reader should note that in the Rankine combined vortex, the RMW is identical to the core radius. However, for other radial profiles of azimuthal wind, this may not necessarily be the case: *we define the core radius as that beyond which angular momentum is nearly constant* (at least until we are outside the parent vortex). The formal definition of the core radius is given by the following:

$$r_c = \Gamma_\infty / v_c \tag{6.33}$$

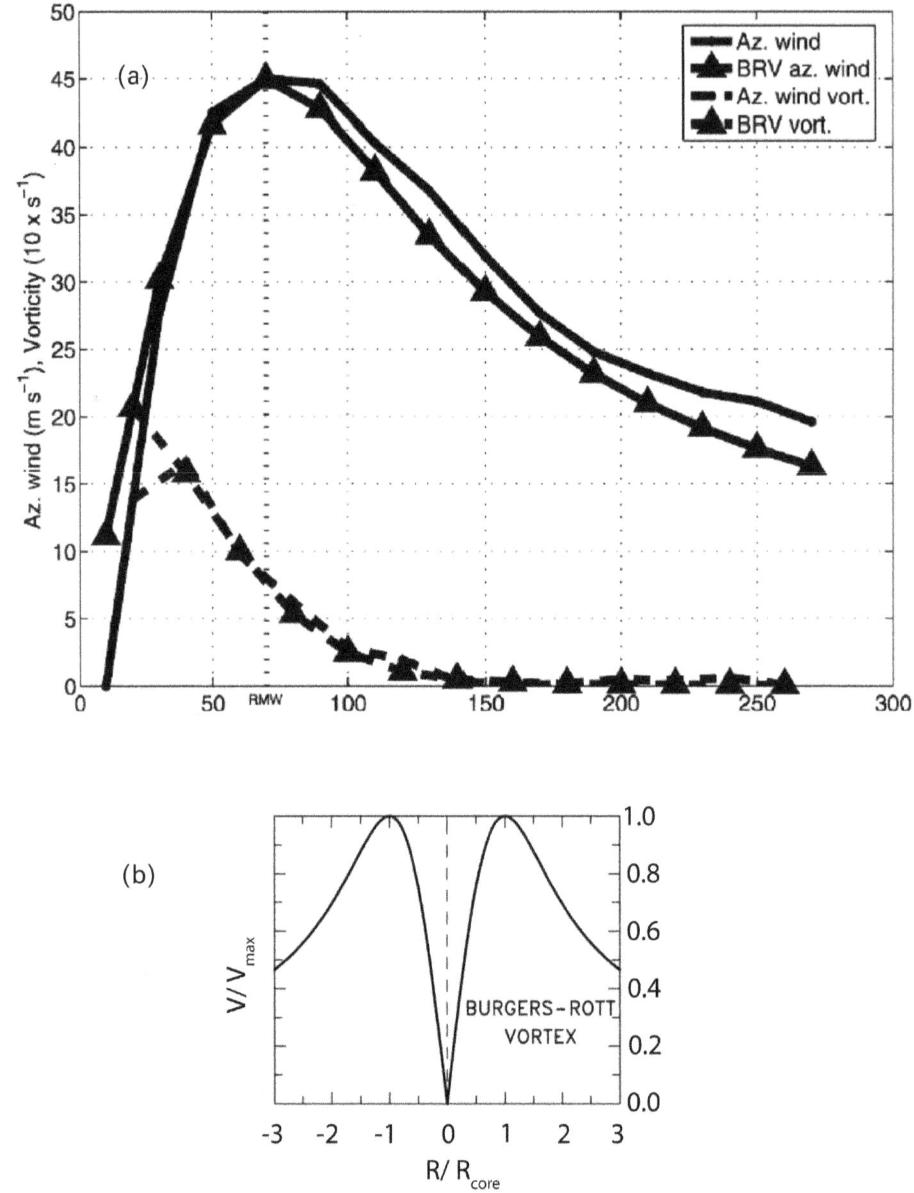

Figure 6.47. (a) Radial profile of azimuthally averaged azimuthal wind component in a tornado in Kansas on May 15, 1999, as estimated from data collected by the U. Mass. W-band, mobile Doppler radar. Also plotted are the radial profile of the azimuthal velocity for a Burgers–Rott vortex having the same maximum wind speed and radius of maximum wind, the vorticity of the tornado, and the vorticity of the Burgers–Rott vortex. The key to each profile is shown in the inset at the upper right (adapted from Tanamachi *et al.*, 2007). (b) Radial (R) profile of azimuthal (tangential) wind (V) in a Burgers–Rott vortex. R_{core} is the core radius; V_{max} is the maximum azimuthal wind speed (adapted from Brown and Wood, 2012).

where Γ_∞ is the angular momentum in the outer, potential flow region; and v_c is the azimuthal velocity at the core radius. It is thought that the smooth transition in the data is not an artifact and that diffusion at the interface between the core and the outer flow region must be responsible for the smooth transition. The velocity field in the presence of diffusion, such that the radial and azimuthal wind components are functions of radius only and the vertical wind component is a function of height only, is called a "Burgers–Rott vortex" (Figure 6.47b). In a Burgers–Rott vortex, the steady-state solutions are

$$u(r) = -ar \tag{6.34}$$

$$v(r) = [\Gamma/2\pi r][1 - \exp(-ar^2/2\nu)] \tag{6.35}$$

$$w(z) = 2az \tag{6.36}$$

where a is a positive constant; ν is the kinematic coefficient of viscosity; and Γ is the angular momentum as in (6.26). For the Burgers–Rott solutions, the azimuthal momentum is diffused in the radial direction only, and radial and vertical momentum are not diffused at all.

Both the Rankine combined vortex and the Burgers–Rott vortex are limited because they assume a steady state and are axisymmetric: real tornadoes change intensity quite rapidly and are not necessarily axisymmetric. A more serious problem with the Rankine combined vortex is that it ignores radial and vertical motions, which are responsible for vortex intensification and are necessary by-products of surface friction. Vertical velocity in the Burgers–Rott model increases with height and is unbounded, which is not realistic: the Burgers-Rott vortex does not take into account boundary conditions. Furthermore, neither the Rankine combined vortex nor the Burgers–Rott vortex allow for any sinking motion.

The "Sullivan" vortex, *which allows for both rising and sinking motion*, is also a solution to the axisymmetric, steady-state equations of motion with diffusion. One of its solutions is (Figure 6.48)

$$u(r) = -ar + 6\nu/r[1 - \exp(-ar^2/2\nu)] \tag{6.37}$$

$$v(r) = A/r\, H(ar^2/2\nu)/H(\infty) \tag{6.38}$$

where

$$H(x) = \int_0^x e^{f(t)}\, dt \tag{6.39}$$

and where

$$f(t) = -t + 3\int_0^t (1 - e^{-y})/y\, dy \tag{6.40}$$

As $x \to \infty$, $H(x)/H(\infty) \to 1$, so that the azimuthal wind component varies as in a potential vortex (6.27). The Sullivan vortex is defined in terms of integrals of functions, which makes it quite complicated, but is easily used nowadays with sophisticated mathematical software that is readily available. The vertical and

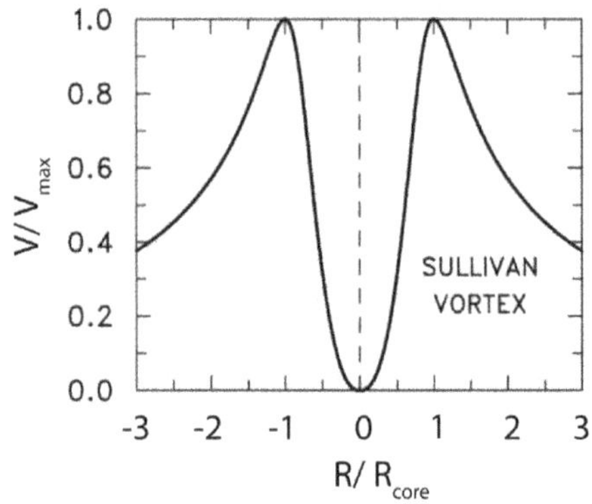

Figure 6.48. Radial (R) profile of azimuthal (tangential) wind (V) in a Sullivan vortex. R_{core} is the core radius; V_{max} is the maximum azimuthal wind speed (adapted from Brown and Wood, 2012).

radial distribution of vertical velocity is given by

$$w(z,r) = 2az[1 - 3\exp(-ar^2/2\nu)] \tag{6.41}$$

So, unlike the Burgers–Rott vortex, vertical velocity is a function of both height and radius. It can be seen that the azimuthal component of vorticity in a Sullivan vortex is

$$\eta = (\nabla \times \mathbf{v})_\varphi = -6a^2rz/\nu \exp(-ar^2/2\nu) \tag{6.42}$$

which is greatest at

$$r = (\nu/a)^{0.5} \tag{6.43}$$

so that there is a circular ring of horizontal vorticity in the clockwise direction associated with sinking motion at the center and rising motion at greater distance from the center. Finding steady-state analytic solutions is an art and may be pursued usefully up to a point, but it then becomes more worthwhile to integrate the equations of motion numerically to find solutions. Because the Sullivan vortex also includes vertical motions, it is the most realistic of the analytic vortex solutions when there is surface friction.

While solutions for the behavior of boundary layers for solid-body rotation and potential flow have been solved for separately, as noted earlier, it was not until H. L. Kuo, in a mathematically complicated 1971 paper, described the solutions to the problem for a Rankine combined vortex, for which motions in both regimes are coupled to each other; for turbulent flow, Kuo used the boundary condition

$$\mathbf{v} = K\, d\mathbf{v}/dz \qquad \text{at } z = 0 \tag{6.44}$$

where $K = 0$, for no-slip boundary conditions; and $K > 0$ for some degree of "slip". He found that in the solid-body rotation region, inside the core, the boundary layer is Ekman-like as we have discussed earlier, but with the caveat that in tornado-like vortices the friction term is actually negligible. Air rises most rapidly just inside the RMW and sinks relatively slowly outside the core.

The reader is reminded that the idealized solutions presented for the Rankine, Burgers–Rott, and Sullivan vortices are for free-slip lower-boundary conditions. Kuo included the effects of "surface stress". It is more complicated to find solutions for free-slip boundary conditions or for those intermediate between no slip and free slip. It might be that "partial" no-slip boundary conditions are most realistic, but there is no assurance that this is the case. In the case of partial slip (chunks of earth may be hurled), we have some semblance of a boundary layer.

We now come back to discuss the corner region in more detail. At the surface, where $w = 0$, we find from (6.22) that

$$u\,\partial u/\partial r - v^2/r = -\alpha_0\,\partial p'/\partial r \tag{6.45}$$

In the corner region, since radial inflow must decelerate (unlike in the inertial layer, in which the flow accelerates radially inward), $u\,\partial u/\partial r > 0$, since $u < 0$ and $\partial u/\partial r < 0$. The second term on the LHS of (6.45), the centripetal term $(-v^2/r)$ is always < 0. It follows that acceleration due to the radial pressure gradient force (the RHS of (6.45)) is radially outward (acting to decelerate radial inflow; p' decreases with radius, so that there is an "adverse" pressure gradient) for relatively large radius within the corner region or when v is small (i.e., when there is "low swirl"). For small radius or when v is large (i.e., when there is "high swirl") the pressure gradient force may vanish (there is a stagnation point) or reverse, becoming negative (acting to accelerate radial inflow; p' increases with radius). For increasing swirl (value of v), a reversal occurs at greater radius. *There are a number of possible configurations of the radial velocity field that vary as a function of swirl* (v) *and radial pressure gradient. If the radial pressure gradient is held fixed, then the radial velocity field depends on the swirl and distance from the axis of rotation.* The behavior of the wind field in the corner region using (6.45) will be discussed in more detail subsequently.

In the corner region, there can be "inertial overshoot", which is manifested by a bulge toward the axis of rotation in the lines of constant angular momentum, such that there is a layer in which the angular momentum is greater than the angular momentum aloft in the core region at the same radii (Figure 6.41b). The inertial overshoot in the corner region is associated with the highest azimuthal velocities (swirl) in a tornado. There will be a further discussion on the effects of swirl in the corner region later when we try to find the condition(s) under which tornado intensity is optimized.

A climatology of the core radius in tornadoes (and other characteristics) based on mobile Doppler radar data from Doppler on Wheels (DOW) radars has been compiled by Curtis Alexander for many tornadoes. The author and his mobile radar group at OU, using mobile Doppler radars from the University of Massachusetts, have also made measurements of core radius. The median core radius for tornadoes in the Great Plains of the U. S. is ~150 m, while the core radius may be as narrow as ~100 m and as wide as ~500 m.

The core radius, as explained earlier, is in part controlled by the amount of work needed to bring a ring of air in towards the center of the tornado; the amount of work available is due in large part to buoyancy above in the updraft: The stronger the buoyancy, the closer in air can be brought. Thermodynamic buoyancy, however, is not the only significant vertical force. Lou Wicker and Bob Wilhelmson showed numerically how non-hydrostatic, upward-directed, dynamic pressure gradient forces could also be significant below the level of free convection if vorticity increases with height below the cloud base (cf. (4.48)): it is natural for vorticity to increase with height below the cloud base as a result of frictional pumping. Within the core radius, where angular momentum increases from zero at the origin, there is approximate solid body rotation and cyclostrophic balance. In this region, the "core region" (Figure 6.41a), radial gradients are much greater than vertical gradients and the core of solid body rotation is advected upward by air flowing upward from the corner region.

Beginning at lower levels in the core region, near the center of the vortex, there is also a friction layer that Wilson and Rotunno have referred to as a "viscous subcore" (Figure 6.41a). This type of friction layer, which may be found when air flows past an aircraft wing and is associated with the core of a "leading edge vortex", was discussed by M. G. Hall in 1961, but is not of any great consequence for us. This viscous core forms about the central axis of the core region so that the azimuthal component of vorticity η (6.13) does not tend to infinity as $r \to 0$. This viscous subcore is unlike the boundary layer beyond the core radius in that there is no solid surface slowing down the flow. However, u and v must be zero at $r = 0$ owing to axisymmetry.

To summarize (Figure 6.41), we re-state that there are four main regions in and just surrounding a tornado. (1) The outer flow, which lies beyond the core radius and above the boundary layer, is inviscid and characterized by constant angular momentum, potential flow, and non-zero swirl (there is an azimuthal component to the wind). (2) The boundary layer is composed of an inertial layer and a friction layer. The former is inviscid, has low swirl, and angular momentum surfaces are horizontally oriented. The friction layer, which is in contact with the surface and is much thinner than the inertial layer, has almost no swirl. (3) The corner region, which encircles the axis of rotation of the tornado, but lies within the core radius, is inviscid and has high swirl. Angular momentum surfaces turn from horizontal at the interface with the boundary layer to vertical at the top of the layer. (4) The core is inviscid, except for a thin, conical region centered on the axis of the tornado; angular momentum surfaces are vertically oriented.

Since dual-Doppler analyses of tornadoes are very difficult to obtain, but many tornadoes have been probed at close range by mobile Doppler radars, techniques have been developed to make good use of these single-Doppler datasets. To measure the characteristics of real tornadoes using just one Doppler radar, one can deduce the structure based on a simple, circularly symmetric model and fitting the observations to the model. Observational data using mobile Doppler radars have been used to estimate the dimensions and characteristics of tornado vortices. In particular, Wen-Chau Lee at NCAR and collaborators have shown how to fit

Doppler wind data to an idealized, circularly symmetric model of a vortex, so that the average radial and azimuthal components are estimated and the continuity equation can be used to estimate vertical motion. The first version of Wen-Chau Lee's technique was first applied to tropical cyclones and is known as GBVTD (ground-based velocity track display), and is essentially a VAD (velocity azimuth display) applied to a vortex whose center is not coincident with the location of the radar. This technique fits a Fourier series to azimuthal and radial wind fields. It is a difficult problem to find a closed solution, but it has been solved subject to some simple restrictions.

Radar-observed tornadoes (and dust devils and other vortices too weak to be classified as tornadoes) frequently have a central weak-echo "eye" or hole (WEH) (Figure 6.49). Ted Fujita discussed early measurements of this radar signature based on conventional radars, while many others have shown more recent examples of the WEH with higher resolution, more sophisticated Doppler radars. It is thought that by centrifuging precipitation particles and other debris radially outward into sheaths or rings, the center of a tornado/small-scale vortex is left devoid of large scatterers, leaving only scatterers too small to be detected by radar, or few if at all. John Snow was one of the first to explain this phenomenon and David Dowell and collaborators described it more quantitatively based on numerical experiments. It has been demonstrated more recently, using polarimetric Doppler radars, that the rings of enhanced radar echoes in tornadoes are indeed most likely composed of debris (Figure 6.6; see the $\rho_{h\nu}$ panel at the lower left).

The vertical structure of tornadoes observed by radar sometimes exhibits a bowl-shaped or otherwise closed-off bottom in the radar reflectivity factor and an open, weak-echo or echo-free eye/hole above (Figure 6.50). The closing off of radar reflectivity near the ground is evidence of the frictionally induced radial inflow of scatterers in the friction layer. Above the friction layer, centrifuging removes the largest pieces of debris and scatterers, while in the friction layer radial inflow may more than compensate for outward centrifuging.

The weak-echo hole may extend all the way to the storm top as a weak-echo column (WEC) (Figure 6.51). In this case is it possible that there is subsidence originating at the storm top? Or, is it possible that there is rising motion, but the air mass is composed of only a few, small scatterers, which cannot be detected by radar, because the largest scatterers have already been centrifuged radially outward beyond the eye down below? Unfortunately, it is difficult to find radar evidence for either an updraft or downdraft on the scale of the eye (evidenced by divergence/convergence) because radar reflectivity is too weak to detect Doppler velocities accurately there. Because whether or not there is a deep updraft or downdraft has implications for the potential intensity of the tornado, we will return to this issue later. Does the WEC exist because there is centrifuging of larger scatterers all the way to the top of the storm? This is unlikely because the highest azimuthal wind speeds in a tornado are confined to relatively low levels. If much or even part of a tornado is embedded within an updraft connected to the parent storm above, then the updraft advects vorticity upwards and it would not be surprising if tornadoes extended rather high up in the troposphere, especially

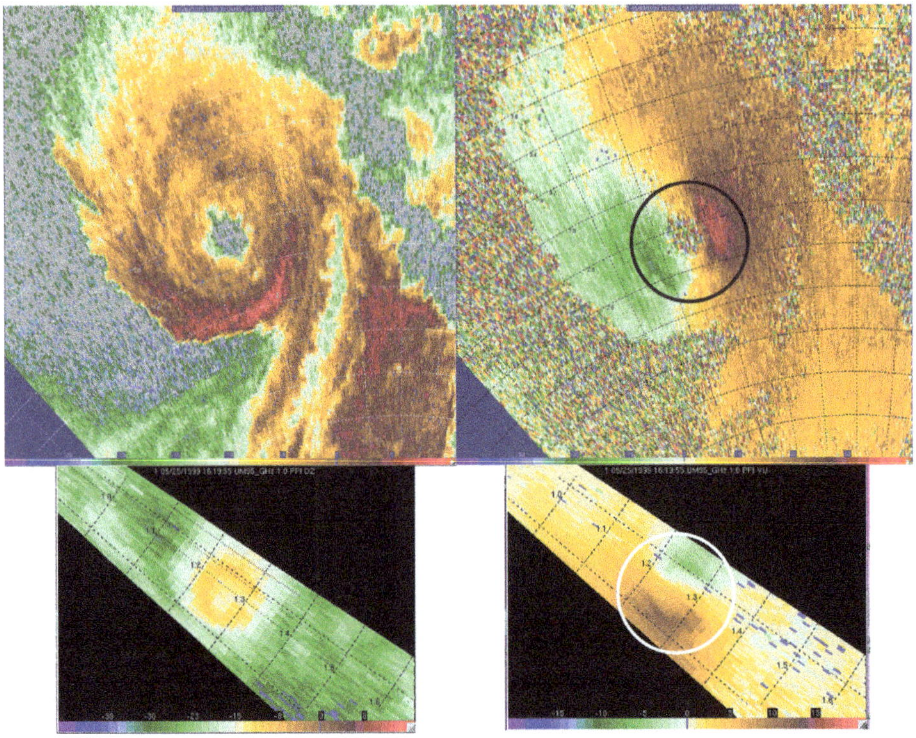

Figure 6.49. Weak-echo holes (WEHs) in (top) a tornado and (bottom) a dust devil. (Top) The WEH, 200 m in diameter, in the center of a tornado in central Oklahoma, southwest of Oklahoma City, on May 3, 1999, as detected by the U. Mass. W-band, mobile Doppler radar. Range rings shown every 200 m. (Left) Radar reflectivity in dBZ_e. The WEH has six undulations along its inner interface, perhaps the result of an instability in vortical flow. (Right) Doppler velocity in $m\,s^{-1}$. The vortex signature is marked by a black circle; the distance between the couplet extrema is ~400 m, which is approximately twice the core radius. (Bottom) The WEH in a dust devil in northwest Texas on May 25, 1999, as detected by the U. Mass. W-band radar. (Left) Radar reflectivity factor in dBZ_e. (Right) Doppler velocity in $m\,s^{-1}$. The vortex signature is marked by a white circle; the distance between the couplet extrema is ~120 m. Range rings shown every 100 m.

when the updraft extends up to the tropopause. However, as we will see later, there are dynamical reasons a strong tornado should not extend too high up in the parent storm.

Ekman-like instability is possible in the boundary layer, owing to friction and rotation. Alan Faller of Woods Hole Oceanographic Institute demonstrated in the early 1960s that horizontal roll vortices may form at an angle of ~15° to the left of the flow above the boundary layer as a consequence of Ekman instability, as realized in spiral bands in tropical cyclones. Doppler radar evidence has been found of regularly spaced spiral bands around tornadoes (Figure 6.52), which may represent boundary-layer rolls.

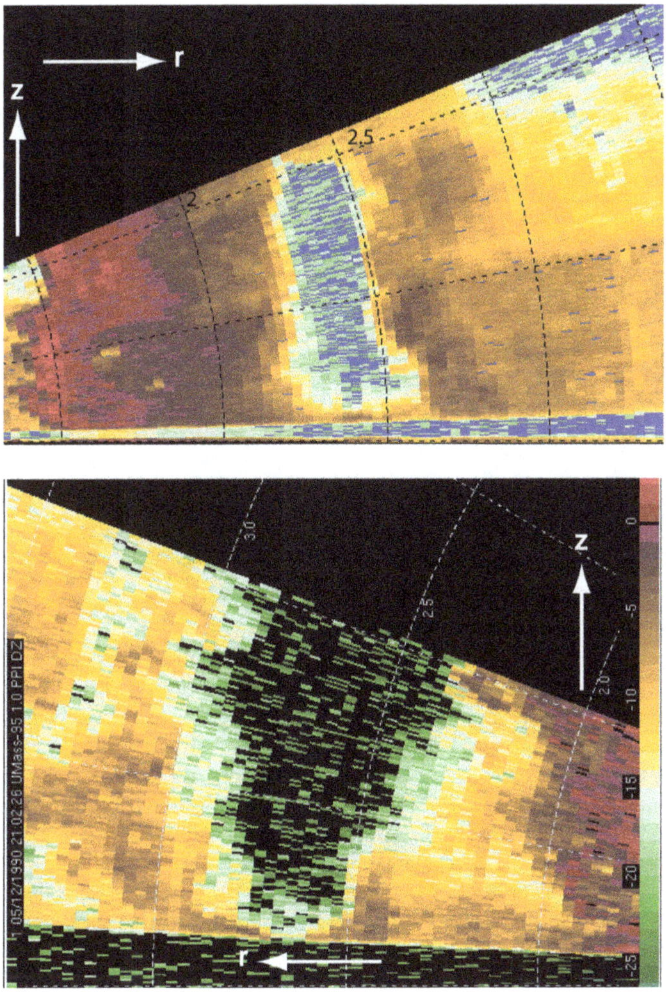

Figure 6.50a. Small-scale vertical structure of weak-echo holes in tornadoes. Vertical cross section of radar reflectivity factor in dBZ_e as detected by the U. Mass. W-band, mobile Doppler radar through the center of a tornado (top) in the Texas Panhandle on May 5, 2002 (range rings shown every 250 m) and (bottom) in south central Kansas on May 12, 2004 (range rings shown every 500 m).

6.6.2 Maximum possible wind speeds in tornadoes

6.6.2.1 Thermodynamic speed limit

The pressure drop (Δp) in the core of a tornado (with respect to the pressure in the environment) may be estimated by integrating the equation of cyclostrophic balance

$$\alpha_0 \, \partial p'/\partial r = v^2(r)/r \tag{6.46}$$

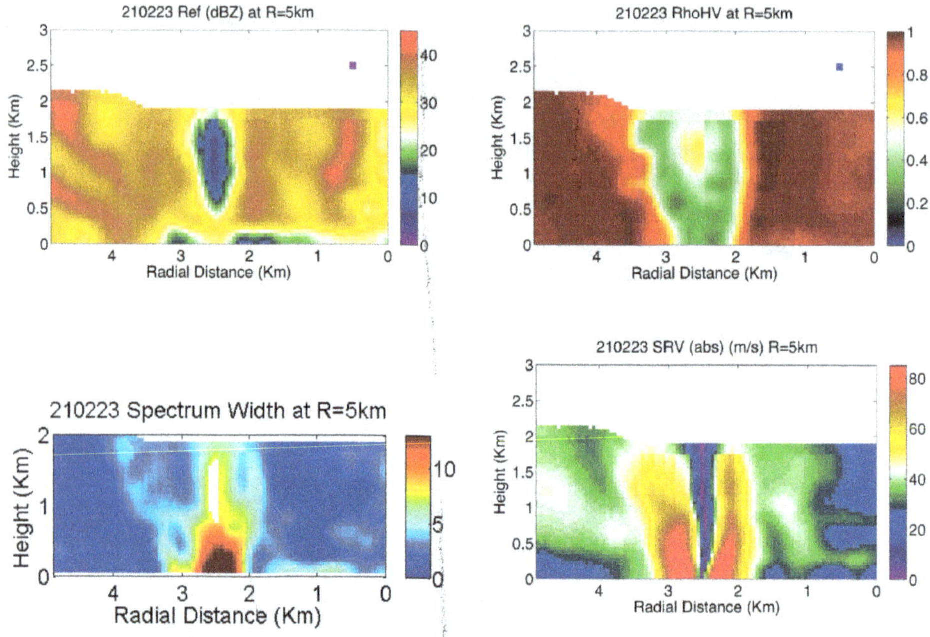

Figure 6.50b. As for Figure 6.50a, but vertical cross section of radar reflectivity in dBZ_e (upper left), co-polar cross-correlation coefficient (ρ_{HV}) when the reflectivity factor is $>10\,dBZ_e$ (upper right), spectrum width (indicative of variance of wind speeds in radar volume; could represent sharp gradients or turbulence or noise, if the signal is relatively weak) in $m\,s^{-1}$ (lower left), and the azimuthal wind speed in $m\,s^{-1}$ (lower right), at constant azimuth, through the center of an EF5 tornado in Oklahoma on May 24, 2011, as detected by RaXPol, a rapid-scan, X-band, mobile Doppler radar. The center of the tornado is at 2.5 km from the origin, which is not colocated with the radar (data processing courtesy of Jana Houser). The WEH closes up near the ground in all cases, probably as a result of frictional inflow. The debris cloud is marked by $\rho_{HV} < 0.8$, the column of which lies inside the region of maximum azimuthal velocities near the ground; the maximum spectrum width, indicative of turbulence, is located at low levels within the radius of maximum azimuthal wind speed. Approximate scale of the radar cross-sectional area is indicated by the purple square in the upper right-hand portions of the top panels.

for an assumed radial profile of azimuthal wind, radially inward in a hydrostatic atmosphere beginning with the environment of the tornado. In a combined Rankine vortex ((6.27) and (6.28))

$$v(r) = \begin{cases} (v_c/r_c)r & \text{for } r \le r_c \qquad (6.47) \\ (r_c v_c)/r & \text{for } r > r_c \qquad (6.48) \end{cases}$$

where r_c and v_c are the core radius and azimuthal velocity at the core radius, respectively. Since v_c is also the maximum azimuthal velocity in the vortex, we re-name it $v_{\max}$. Substituting (6.47) into (6.46) and integrating the result from $r = r_c$

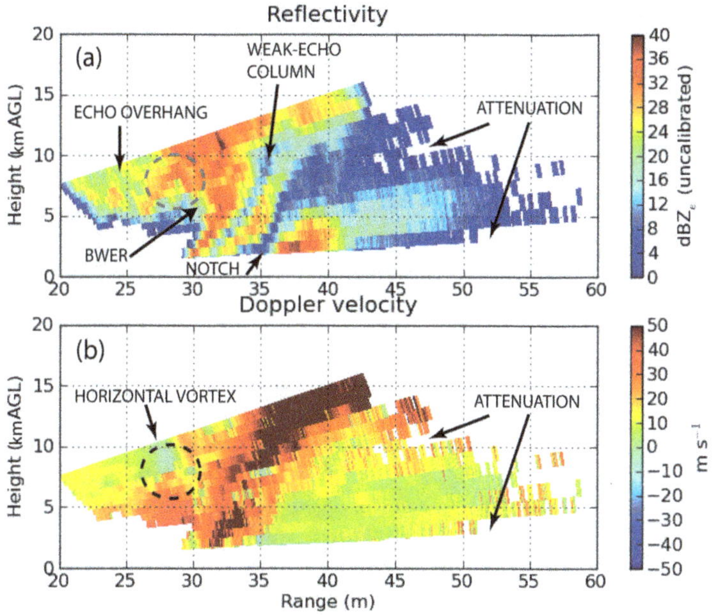

Figure 6.51a. Weak-echo column (WEC) in a tornadic supercell on May 4, 2007 near Greensburg, KS, as detected by the U. Mass. X-Pol, mobile Doppler radar. The WEC extends from low altitude up to at least 12 km ARL. Vertical cross sections through the center of tornado of radar reflectivity factor in dBZ_e (top) and Doppler velocity (bottom). In the top panel, a weak echo column is also seen, but it represents a slice through a notch of precipitation free air that has been advected around the mesocyclone and tornado. A horizontal vortex signature is seen in the echo overhang region, above the bounded weak-echo region (BWER) (courtesy of R. Tanamachi, from Tanamachi *et al.*, 2012).

to $r = 0$, we find that

$$\Delta p_{0\to r_c} = -v_{\max}^2/2\alpha_0 \tag{6.49}$$

where the pressure drop Δp is given here with respect to the pressure at the radius of maximum wind, the core radius. Now, substituting (6.48) into (6.46) and integrating from $r = \infty$ to $r = r_c$, we find that

$$\Delta p_{r_c\to\infty} = -v_{\max}^2/2\alpha_0 \tag{6.50}$$

So, the total pressure drop from the environment of the vortex to its center

$$\Delta p_{\infty\to 0} = v_{\max}^2/\alpha_0 \tag{6.51}$$

For $v_{\max} \sim 100\,\mathrm{m\,s^{-1}}$ and $\alpha_0 \sim 1\,\mathrm{m^3\,kg^{-1}}$, it follows from (6.51) that $\Delta p_{\infty\to 0} \sim 100\,\mathrm{hPa}$. Tim Samaras and his Storm Intercept Group have made measurements using portable instrument packages of pressure drops in tornadoes at the ground of as much as 100 hPa, which is in at least qualitative agreement with theory. Formula (6.51), however, neglects turbulent mixing, asymmetries, and

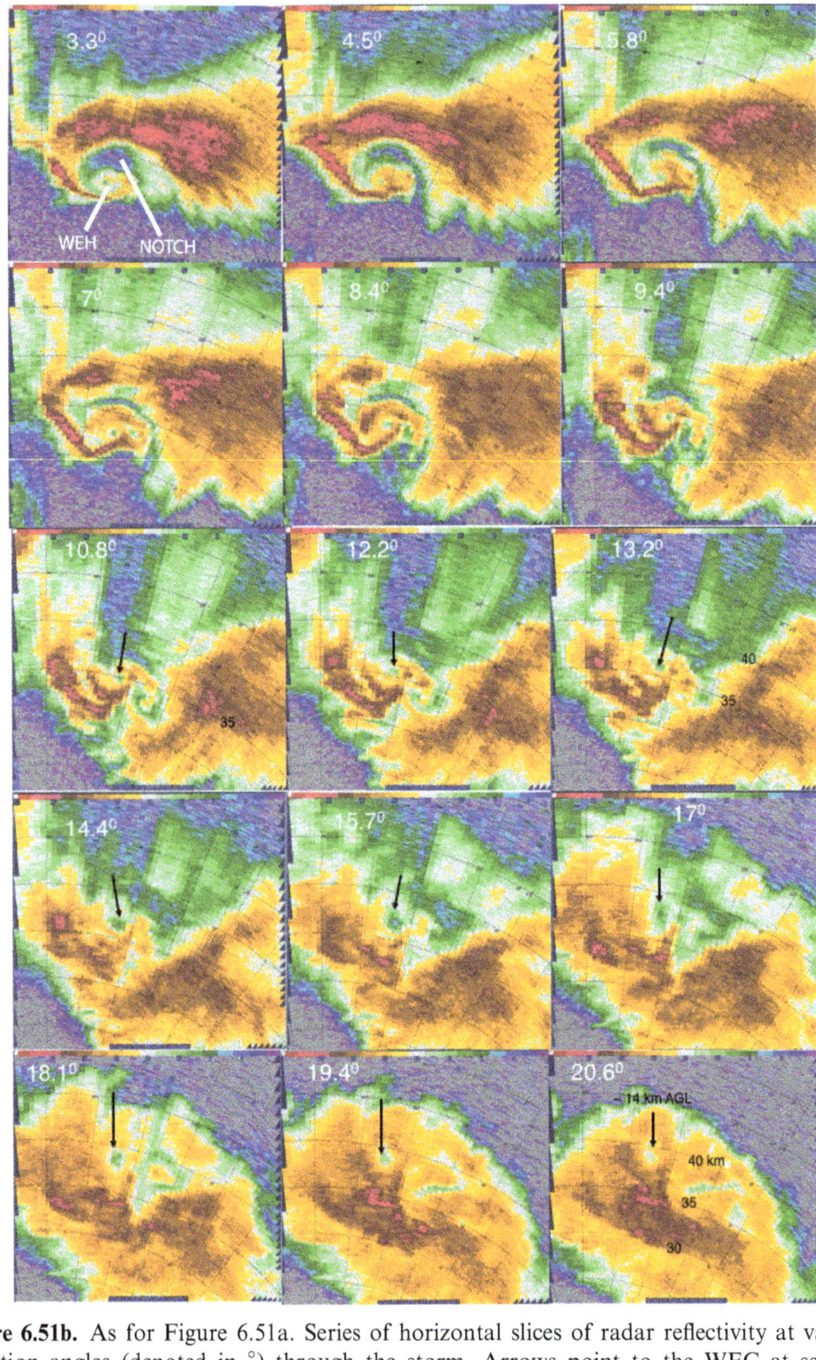

Figure 6.51b. As for Figure 6.51a. Series of horizontal slices of radar reflectivity at various elevation angles (denoted in °) through the storm. Arrows point to the WEC at selected elevation angles to aid visualizing vertical continuity. Range from radar given in km. At the 20.6° elevation-angle scan, the WEC is still evident at 14 km ARL.

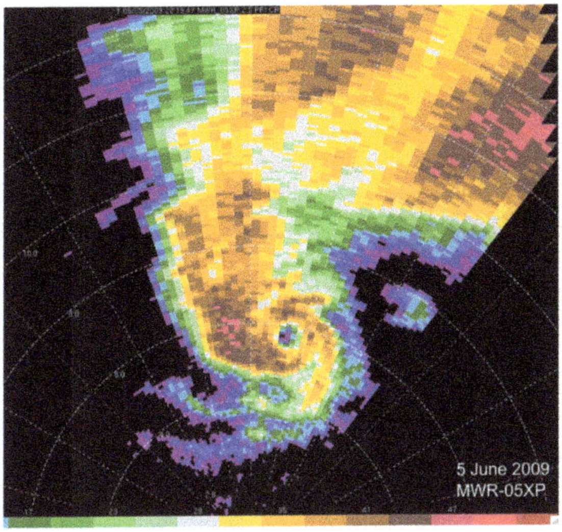

Figure 6.52. Spiral bands of radar reflectivity (dBZ_e) around a tornado marked by a WEH on June 5, 2009 in southeastern Wyoming, as detected by the MWR-05XP, X-band, mobile, phased array Doppler radar from the Center for Interdisciplinary Remotely-Piloted Aircraft Studies (CIRPAS) at the Naval Postgraduate School. Range rings shown every 2 km.

transient phenomena, and is regarded as the simplest (and definitely not the most accurate) one can make. For radial profiles of azimuthal wind in the core that are "smoother" than that of a Rankine vortex, the hydrostatic pressure drop in (6.51) is less (e.g., for a Burgers–Rott vortex the pressure drop is only 59% of that in a Rankine combined vortex).

In nature only part of the pressure deficit in a tornado is a hydrostatic consequence of the warm, buoyant air column above it. The maximum wind speed in a tornado in a hydrostatic atmosphere is referred to as the "thermodynamic speed limit"; the seminal work on this topic was instigated by a scientific report authored by Doug Lilly at NCAR in 1969, which was not published in the refereed literature.

To make the simplest estimate of maximum horizontal wind speeds in tornadoes we apply the hydrostatic approximation to the vertical equation of motion

$$0 = -\alpha_0 \, \partial p'/\partial z + B \tag{6.52}$$

We integrate the hydrostatic equation (6.52) from the surface to the tropopause and find that at $r = 0$ the center of the tornado

$$\int_{p'(z=0)}^{p'(z=z_{\text{trop}})} \alpha_0 \, dp' = \int_0^{z_{\text{trop}}} B \, dz = \text{CIN} + \text{CAPE} \approx \text{CAPE} \approx -\alpha_0 p'(r = 0, z = 0) \tag{6.53}$$

where $p'(z = z_{\text{trop}}) = 0$ (i.e., we assume that the tornado vortex induces no perturbation pressure at the tropopause). The perturbation pressure at the surface at the center of the tornado is identical to the pressure drop at the surface from the environment to the center

$$p'(z = 0) \equiv \Delta p_{0 \to \infty} \tag{6.54}$$

Integrating the cyclostrophic equation (6.46) from the perturbation pressure at the surface at the center of the tornado to the perturbation pressure at the tropopause above the center of the tornado and using (6.47), (6.48), and (6.53) we find that

$$\begin{aligned}\int_{p'(z=0)}^{p'(z=z_{\text{trop}})} \alpha_0\, dp' &= \int_{p'(z=0)}^{p'(z=z_{\text{trop}})} \alpha_0\, \partial p'/\partial r\, dr = \int_{r[p'(z=0)]}^{r[p'(z=z_{\text{trop}})]} v^2/r\, dr \\ &= \int_{r=0}^{r=r_c} \Omega^2 r\, dr + \int_{r=r_c}^{r=\infty} [(v_{\max})^2/\Omega]^2 (1/r^3)\, dr = \text{CAPE}\end{aligned} \tag{6.55}$$

It is seen that $r(p'(z = 0))$ corresponds to $r = 0$ and $r(p'(z = z_{\text{trop}}))$ corresponds to $r = \infty$. It follows that

$$v_{\max}^2 = \text{CAPE} \tag{6.56}$$

so that

$$v_{\max} = (\text{CAPE})^{1/2} \tag{6.57}$$

which is similar to the parcel theory estimate of vertical velocity (3.7), save for the absence of the $\sqrt{2}$ factor. Other more accurate vortex solutions, which also include radial and vertical wind components, may be used to find other expressions for $v_{\max}$ and for the radial pressure distribution.

Above the upper part of the corner region where the vortex is strongest, a dynamic, downward-directed perturbation pressure gradient force develops (cf. (4.48); as in Figure 4.58, but for the tornado—not for the mesocyclone), which acts to induce subsidence. Consider now what happens when the effects of compressibility are taken into account. The dynamically induced subsidence induces adiabatic warming at the center of the tornado and thereby produces a warmer core that hydrostatically results in lower central pressure; this warmer core is manifested as higher CAPE. If the air in the middle of the tornado is unsaturated, then it will descend dry adiabatically rather than moist adiabatically, and CAPE could be substantially increased. Thus, it seems that a dynamically forced downdraft should increase the hydrostatic speed limit (cf. (6.57)). However, it takes work to force a downdraft, which is warmer than its surroundings; if the forcing for the downdraft were to disappear suddenly, the air would be positively buoyant and thus accelerate back upward, a manifestation of static stability.

If a downdraft were to originate at the tropopause and the downdraft were unsaturated, there would be the potential for very substantial warming near the surface. However, the work needed to push an air parcel downward in a colder environment beginning at the tropopause and ending at the surface is enormous. Suppose that the air is saturated, so that it descended at the moist-adiabatic rate (instead of at the greater dry-adiabatic rate): the initial downward vertical velocity at the top of the storm (in the absence of mixing with the environment) would have to be $\geq (2\ \text{CAPE})^{1/2}$, which could be in excess of $50\,\text{m}\,\text{s}^{-1}$, an extremely unlikely occurrence. Robin Tanamachi and collaborators have shown from digital infrared imagery that the lapse rate of temperature on the surface of a tornado condensation funnel is, as would be expected, moist adiabatic, but we do not know what the lapse rate is inside the condensation funnel. Extreme warming has not been observed at the center of tornadoes, though there have been anecdotal accounts of fires going on near tornadoes, which more likely might have been a consequence of the destruction of structures by tornadic winds. Stirling Colgate, in the early 1980s, attempted unsuccessfully to obtain measurements in tornadoes using rockets with sensors launched from an aircraft.

Despite in situ measurements near the ground of temperature in the center of tornadoes being rare (nonexistent aloft), we cannot however *totally* rule out the possibility that subsidence warming might sometimes increase the intensity of tornadoes. In addition to the great amount of work that would be needed to bring air parcels from high up down to low altitude, lateral mixing (entrainment of environmental air) would reduce buoyancy and thus reduce the effects of subsidence warming. Moreover, the time it takes a tornado vortex to develop solid body rotation may be short compared with the time it takes environmental air to descend substantially, even if it did.

When a non-rotating updraft penetrates above the tropopause, a hydrostatic "cold dome" is produced because air parcels become colder than their environment just above the tropopause. Owing to the cold air aloft, there cannot be a hydrostatic pressure deficit underneath the updraft at the ground because CIN above the tropopause should be approximately equal and opposite in sign to CAPE in the troposphere.

A rotating updraft in a tornado, on the other hand, must be accompanied by a pressure deficit at the center. Above the tropopause there is divergence, so that vorticity in the updraft decreases. Because vorticity decreases with height, there should be a downward-directed perturbation pressure gradient force and therefore a dynamically driven downdraft, which may appear as a "crater" in the cloud top. So, we are led to the conclusion that a depression in the height of the anvil region above the updraft in a buoyant cloud that drives a "deep" tornado may belie the hidden updraft below. There have been some observations of collapsing tops in thunderstorm anvils near the time of tornadoes, especially by Ted Fujita. So, while the dynamics of tornadoes depends mostly on what happens near the surface, a look from above via satellite imagery or aircraft flying above a storm may have merit.

Doppler radar estimates of the wind speeds in tornadoes frequently exceed the thermodynamic speed limit (6.57) by a substantial margin, though there are some significant uncertainties regarding the representativeness of radar data and nearby thermodynamic data. We would not expect the thermodynamic speed limit to be realistic, owing to parcel accelerations experienced in the corner region: a tornado is grossly non-hydrostatic. So why even consider the hydrostatic speed limit? The main justification is that it can be thought of as a benchmark. We will explain why shortly.

6.6.2.2 Dynamic effects on the intensity of tornadoes: the swirl ratio

Owing to the stability of the core of a tornado vortex with respect to displacements in the radial direction, *centrifugal wave motion* is possible. Centrifugal waves can be produced in tornadoes when there is an imbalance between the radially inward-directed pressure-gradient force and the radially outward-directed centrifugal force. The resultant restoring force (cf. (6.9)) can be explained as follows: imagine squeezing a vortex of solid body rotation radially inward locally and then letting go; the vortex will then experience a radially outward-directed restoring force and pop back outward and overshoot its equilibrium level, a consequence of the conservation of angular momentum, when there is no friction (Figure 6.53). Above and below, coupled vertical motions and the resultant radial flow will cause the vortex above and below to be squeezed, and vertical wave propagation will follow.

Above the upward jet in an end-wall vortex upward vertical velocity eventually is reduced and may even switch sign (turn into a downdraft) since vorticity decreases with height there and therefore a dynamically induced downdraft above the level at which azimuthal winds are greatest is induced. If rising air in the upward, frictionally induced jet ascends more rapidly than the phase speed of vertically propagating centrifugal waves, then the flow is said to be "supercritical" (with respect to centrifugal waves). Above the jet, vertical velocity becomes less than the phase speed of centrifugal waves, so the flow becomes "subcritical".

The transition from supercritical to subcritical flow can lead to a phenomenon known as "vortex breakdown", which the fluid dynamicist T. Benjamin in a

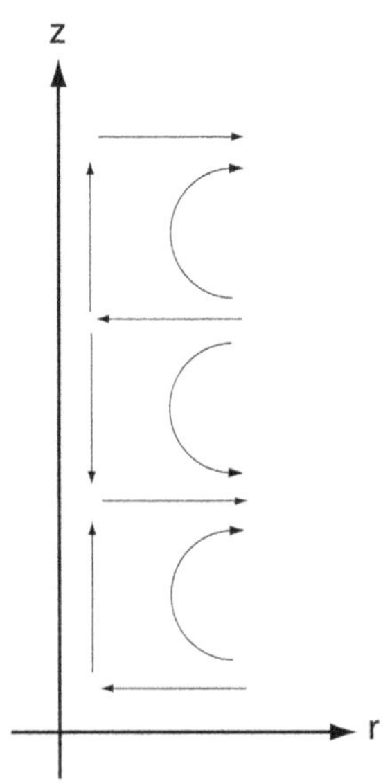

Figure 6.53. Idealized illustration of vertically propagating centrifugal waves in a stable vortex. Streamlines show alternating compression and expansion (based on Shapiro, 2001).

Figure 6.54. An example of vortex breakdown in a laboratory vortex as a transition from a narrow, laminar rotating column below to a wider, turbulent column aloft (from Church *et al.*, 1979).

classic 1962 paper showed is analogous (just turn his setup onto its side by rotating it so that it is horizontally oriented rather than vertically oriented and gravity waves are substituted for centrifugal waves) to the "hydraulic jump" observed in non-rotating, stratified flows when there is a transition from upstream supercritical flow (in this case the flow speed is faster than that of gravity waves) to subcritical flow (the flow speed is slower than that of gravity waves). When there is vortex breakdown, there is a transition from laminar (below) to turbulent (above) flow (Figure 6.54). Air parcels accelerating upward in the jet suddenly encounter much weaker flow and the zone of upward motion becomes wider, much weaker, and turbulent. Rich Rotunno at NCAR was the first to analyze simulations of tornado-like vortices and show that vortex breakdown can occur.

One may interpret vortex breakdown as a consequence of the requirement of a match between the "flow force" or "head" just below the top of the jet and that just above the jet. In other words, there must be a "match" between the vortex imposed from above in the parent storm (the mesocyclone) and the vortex in the corner region below affected by the boundary layer. Downward-propagating centrifugal waves from above cannot continue to propagate downward into the supercritical region, so wave energy is reflected upward and standing waves are produced. Standing centrifugal waves above the jet, in the subcritical regime, must reduce the upstream flow force so that a steady state can be maintained. When the transition between supercritical and subcritical regimes is sharp, the lead centrifugal wave breaks and there are standing waves downstream. When the transition is sharper, even downstream waves break. Because downward-propagating centrifugal waves from the mesocyclone in the parent storm cannot propagate into the supercritical region below, information about the mesocyclone aloft is not "communicated" to the supercritical corner region.

Doppler radars cannot easily document the vortex breakdown phenomenon because it is difficult for radars to detect motions near the surface where vortex breakdown often occurs. Radars cannot detect motions very close to the surface owing to the curvature of the Earth and because the half-power beamwidths of radar antennas are too wide; as a result of the latter there is ground clutter contamination. Vortex breakdown has been seen in nature, especially from airborne platforms that permit a look downward into the corner region of a tornado, which may otherwise be hidden from view at the surface by a debris cloud or a condensation funnel. It perhaps could be verified by observing the spectrum width (a measure of the variation of Doppler wind speeds in a radar volume) and determining if the spectrum width increases with height at the level at which vortex breakdown is expected. It may be, however, that the region of vortex breakdown may not have a high enough density of scatterers to be detected or that scatterers are too small to be detected, or both.

The most important parameter defining idealized vortex behavior in a simulated laboratory vortex, based on many experiments, is the "swirl ratio" (S):

$$S = (R\Gamma)/(2M) \tag{6.58}$$

where R is the radius of the updraft hole; Γ is circulation at the edge of the updraft ($v_0 2\pi R$) (i.e., angular momentum multiplied by 2π); and M is the volume flow rate of the updraft ($w\pi R^2 = 2\pi Rhu_0$), where h is the height of the inflow area, v_0 is the azimuthal wind component at the outer edge of the updraft hole (the edge of the updraft), u_0 is the radial inflow velocity at the edge of the chamber (note that u_0 is positive for radial inflow, the reverse of convention, according to which the radial wind component is negative for radial inflow), and w is the mean vertical velocity in the updraft hole. The swirl ratio is therefore also given by the following:

$$S = R(v_0 2\pi R)/2(w\pi R^2) = v_0/w \tag{6.59}$$

From mass continuity consideration, the flux of mass into the chamber from all sides must equal the flux of mass out of the chamber through the updraft hole, as noted earlier, so that

$$w = (2hu_0)/R \tag{6.60}$$

Then

$$S = (v_0 R)/(2hu_0) \tag{6.61}$$

If the vortex chamber is constructed so that $R/2h = 1$ (i.e., if the depth of the inflow layer is twice the radius of the updraft hole), then

$$S = v_0/u_0 \tag{6.62}$$

The swirl ratio can be thought of as a measure of the relative amount of azimuthal flow compared with the amount of radial flow into the bottom of the vortex or, equivalently, the relative amount of vertical vorticity to (horizontal) convergence.

The physical significance of the swirl ratio is illuminated by considering once again the radial equation of motion for an axisymmetric vortex in a steady-state, inviscid, constant density, incompressible fluid in which $w = 0$ at the bottom surface (6.45)

$$u\,\partial u/\partial r - v^2/r = -\alpha_0\,\partial p'/\partial r \tag{6.63}$$

In this equation, the inertial acceleration term ($u\,\partial u/\partial r$) contains the effect of radial inflow and the centripetal term ($-v^2/r$) contains the effect of swirl, the azimuthal wind component. In a crude, qualitative manner, at a given radius, when the swirl is large compared with the radial flow, the centripetal term dominates; when the swirl is low compared with the radial flow, the inertial acceleration dominates. Also, if the swirl is large, then there is a large drop in perturbation pressure at the center of the vortex: If the pressure drop is large enough, the resultant downward-directed perturbation pressure gradient force can reverse the frictionally induced central updraft to induce a downdraft. The larger the relative amount of aziumthal flow (circulation) to the updraft, the greater the effect of the downward-directed perturbation pressure gradient force.

Rich Rotunno in the late 1970s and early 1980s, building on work done by Davies-Jones, illuminated why the structure of a vortex changes as the swirl ratio ranges from small to large. His work was motivated in part by the finding that the size of the core of the vortex in vortex chambers and the nature of the secondary circulation are determined by the swirl ratio (Figure 6.55). For the approximate wind field

$$u \approx (r/R)u_0 \tag{6.64}$$

$$v \approx (R/r)v_0 \tag{6.65}$$

which depicts potential flow expected outside the core and radial convergence ($\partial u/\partial r < 0$) as air decelerates as it enters the corner region. It follows then from (6.63) that

$$u_0^2/(R^2 r^3)[(r/R)^4 - S^2] = -\alpha_0\,\partial p'/\partial r \tag{6.66}$$

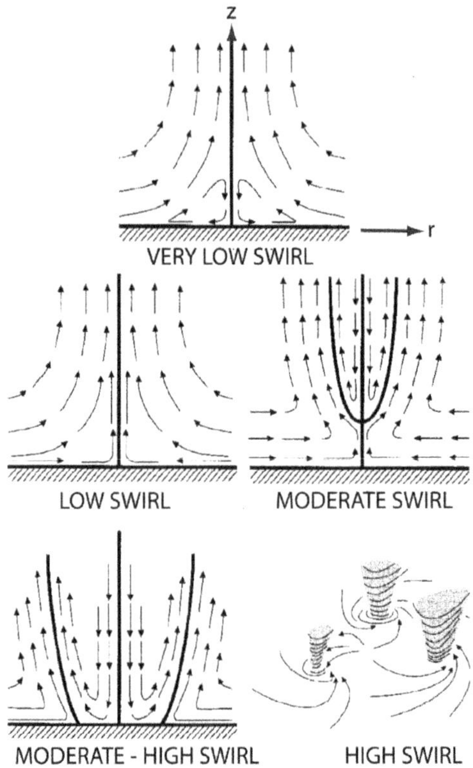

Figure 6.55. Idealized illustration of the relationship between swirl ratio and vortex structure in a tornado simulator. Vertical cross sections of the flow are shown. For very low swirl ratio there is boundary-layer separation and there is no corner flow and no tornado; for low swirl ratio, there is a laminar, one-cell vortex; for moderate swirl ratio, a laminar end-wall vortex "erupts" into a turbulent, wider, two-cell vortex above the level of vortex breakdown; for moderate–high swirl ratio, the downdraft reaches the surface and there is a wide, turbulent, two-cell vortex, with an annular corner region; for high swirl ratio, there are smaller, secondary, satellite vortices rotating about a common axis. The advectives "low", moderate", "high", etc. are relative descriptors—not absolute (based on Davies-Jones, 1986; Davies-Jones *et al.*, 2001; Wakimoto and Liu, 1998).

where the swirl ratio—defined as in (6.62)—is $S = v_0/u_0$. Then

$$-\partial p'/\partial r \sim (r/R)^4 - S^2 \tag{6.67}$$

At small r, the potential flow profile (6.65) is not valid because there is solid body rotation, so we cannot interpret (6.67) for small radius within the core. However, for relatively large radius where potential flow is a good approximation, we find that for a given swirl ratio the radial pressure gradient force that is acting inward (<0) decreases until at some radius the radial pressure gradient force vanishes and then reverses (acts radially outward) at even larger radius. From (6.67) we see that this radius is $R(S)^{1/2}$, which is outside the domain of the vortex chamber for $S > 1$ (i.e., $r > R$); thus, the radius at which the radial pressure gradient vanishes increases monotonically with swirl ratio. Incoming air must therefore decelerate and turn upward at a large distance from the axis of rotation when S is high: surfaces of constant angular momentum are *not* allowed to converge to small radius, so that azimuthal wind speeds are modest when S is large. The physical interpretation of this result, which may be applied to the real atmosphere, is that as an air parcel having a specified angular momentum ($rv = \Gamma$) is forced radially inward, v increases and r decreases, so that for (6.63) expressed as

$$v^2/r - u\,\partial u/\partial r = \alpha_0\,\partial p'/\partial r \tag{6.68}$$

the centrifugal acceleration (v^2/r) increases on the LHS of (6.68). At high r in the inertial region, however, v is small, so that v^2/r is very small and

$$-u\,\partial u/\partial r \approx \alpha_0\,\partial p'/\partial r \tag{6.69}$$

Since $u < 0$ (radial inflow) and $\partial u/\partial r < 0$ (radial inflow increases with increasing radius) $-u\,\partial u/\partial r < 0$ at high $r, \partial p'/\partial r < 0$. The radially directed pressure gradient ($\partial p'/\partial r < 0$) at large radius is therefore adverse, but becomes less negative (adverse) and eventually vanishes as r decreases because the centrifugal term in (6.68) increases and opposes the inertial term ($-u\,\partial u/\partial r$) and eventually overwhelms it.

In addition, if the vortex is confined to low levels, then there is a downward-directed dynamic perturbation pressure gradient force. At small radius there is therefore sinking motion (Figure 6.55), so that there is a *two-cell* vortex, like that in the Sullivan vortex: upward motion at some radius near or beyond where the inertial term and the centrifugal terms cancel each other out, and downward motion at the center and at the far radius beyond where there is rising motion (Figure 6.55). *So, at high swirl ratio, air ascends at some radius and descends both in the center and far from the center owing to (1) the adverse radial pressure gradient resulting from the dominance of the inertial term at high radius; (2) the downward-directed dynamic perturbation pressure gradient resulting from the decrease of vorticity with height; and (3) mass conservation.*

If the updraft is very strong, then there is a relative minimum in pressure associated with the updraft—consistent with (4.18) integrated from the surface up to some level and neglecting B. In this case, not only might a condensation funnel form near the center of the vortex where at a given altitude the pressure is a minimum, but it might also form near the updraft, leading to a double-wall condensation funnel, which is sometimes observed (Figure 6.56). Rotunno has noted that double-pressure minima occur more often with relatively large swirl ratios. This feature may, however, be an artifact, created as a consequence of the limited model domain and the explanation may lie elsewhere.

When the swirl ratio is high, angular momentum does not converge beyond some radius, so that a ring (annulus) of relatively large (cyclonic) shear vorticity forms (inside this radius the air is stagnant and/or stagnant air from aloft is advected downward); this ring of vorticity may be thought of as a curved vortex

Figure 6.56. (Top) Double-walled condensation funnel in a tornado in southwestern Nebraska on June 10, 2004. There is an inner condensation funnel and an outer sheath, which is opaque when looked at side on. (Bottom) Hollow condensation ring in a waterspout in the Florida Keys, as seen from a NOAA helicopter on August 28, 1993. The condensation funnel appears translucent, except at the right and left edges, where the integrated amount of condensate along the line of sight is deeper (photographs by the author).

sheet (Figures 6.57). This annulus of strong shear vorticity (perhaps augmented by the annulus of convergence that accompanies the rising branch of vertical circulation) is barotropically unstable as a vortex sheet and breaks down into multiple, subtornado-scale vortices (suction vortices). First, two small-scale, "secondary" vortices form; with increasing swirl ratio, three to six vortices occur. Secondary, satellite vortices are also seen in some dust devils (Figure 6.58). Wind speeds in secondary vortices can be extremely high, but only very locally and only for very short periods of time.

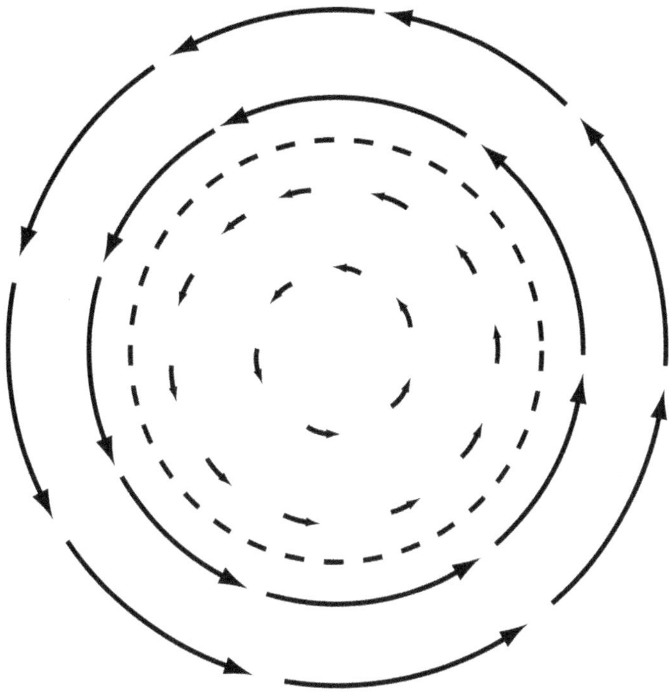

Figure 6.57. Idealized depiction of a vortex sheet, which is located near the dashed line, where the azimuthal wind speed suddenly jumps to a higher value at larger radius. Curved lines with arrows represent the horizontal component of the flow around the two-cell vortex; the length of the arrows is proportional to the azimuthal wind speed.

Vortex sheet instability may affect not only the strongly horizontally sheared region in two-cell vortices, but also vortices produced along shear lines associated with surface boundaries such as fronts, outflow boundaries, the dryline, and the sea breeze/land breeze front.

If the swirl ratio is then decreased, there is hysteresis, such that the transition in the number of secondary vortices decreases, but at higher critical swirl ratios. The transition from a one-cell to a two-cell vortex occurs at swirl ratios of ~0.5–0.7 in a laboratory simulator, depending upon the Reynolds number. Typical swirl ratios in laboratory vortices range from ~0.1–1.5; in a multiple-vortex tornado, swirl ratios of ~2–6 have been estimated from analyses of ground-based, mobile, Doppler radar data; in the parent mesocyclone of a tornado, swirl ratios of ~0.7–8 have been estimated from airborne Doppler radar analyses and have increased rapidly with time when a tornado was forming.

When a vertically oriented vortex line is introduced into a larger scale barotropic vortex, the vortex line moves along with the flow and becomes contorted. For a larger scale cyclonic vortex, there is frictionally induced vorticity in the radially inward direction as a result of the increase in azimuthal wind with

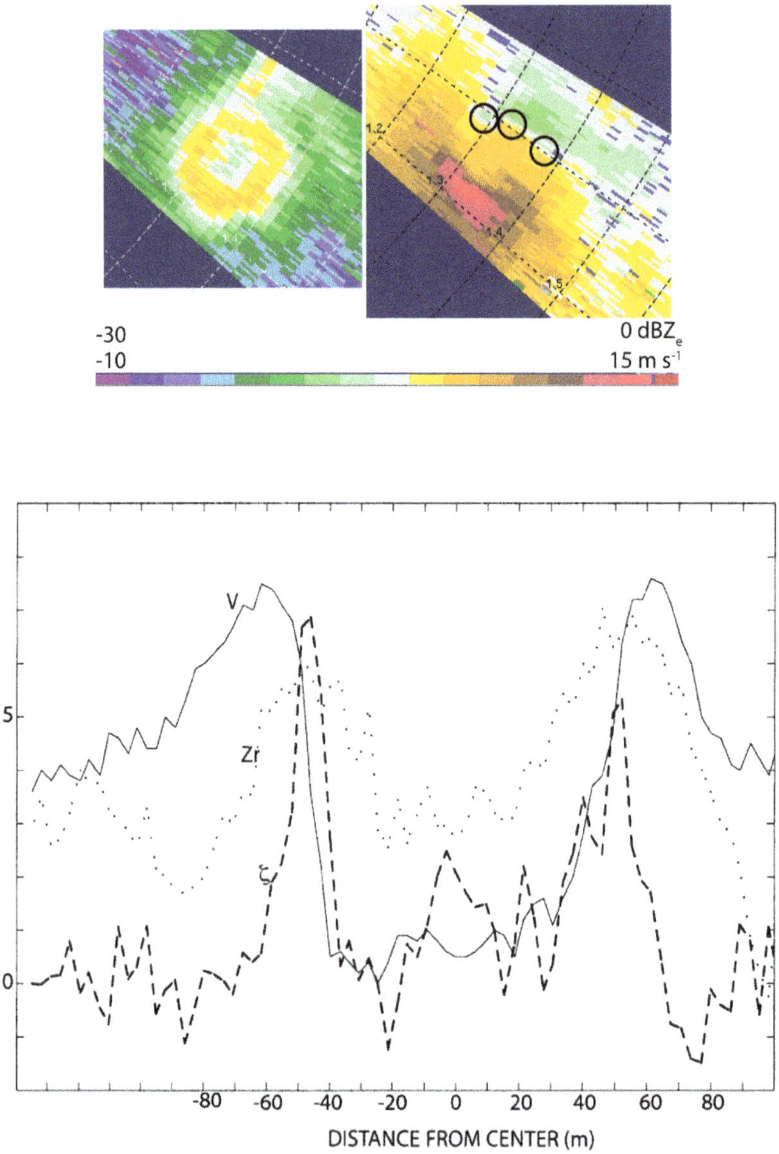

Figure 6.58. Secondary satellite vortices in a dust devil in northwest Texas on May 25, 1999, as depicted by the U. Mass. W-band, mobile Doppler radar. (Top left) Radar reflectivity factor color-coded below; (top right) Doppler velocity color-coded below; cyclonic shear vortex signatures indicated by circles. Range rings shown every 100 m. (Bottom) Approximate vortex-relative azimuthal wind (V, solid line, m s^{-1}), vorticity (ζ, dashed line, $\times 10$ s^{-1}), and relative equivalent radar reflectivity factor (dotted line) $Z_r = (Z_e - Z_{\text{noise}})/C$, where Z_e is the equivalent radar reflectivity factor (dBZ), and C is a dimensionless compression factor chosen subjectively. The vorticity at the center is estimated as the average of the vorticities computed just to the right and left of the center (from Bluestein *et al.*, 2004a).

height and in the clockwise direction as a result of the increase in radial inflow with height near the ground (Figure 6.29, top right panel). The vortex line initially is therefore vertical above the boundary layer, but leans inward and in the clockwise direction with height near the surface. It then moves more quickly in the counterclockwise direction with height owing to increase in cyclonic flow with height. The net effect is that the vortex line curls with height in the clockwise direction and near the surface, forming a helix about the central axis. Secondary vortices are located near the radius of maximum wind (RMW) and propagate more slowly than the mean flow (approximately half the speed, based on observations in a vortex chamber) so that they *retrograde* with respect to the mean flow.

When the swirl ratio is very low, from (6.67) it is seen that

$$-\partial p'/\partial r \sim (r/R)^4 \tag{6.70}$$

In other words, since the RHS of (6.70) is >0 and increases with r, the radial pressure gradient force acts radially outward, and from (6.69) we find that when v^2/r can be neglected (for low v and/or high r) inertial acceleration is radially outward. In this case, radial inflow decreases inward and is forced upward at the very far radius and a strong vortex cannot occur at low levels (Figure 6.55). Boundary-layer flow is forced to "separate" at high radius beyond the core radius, beyond the corner region.

So, we have described what happens at the two extremes, when there is a high swirl ratio and a very low swirl ratio. What happens when there is a "low swirl ratio", a swirl ratio intermediate between that of "very low swirl ratio" and "high swirl ratio"? At low swirl ratio the radius at which the radial pressure gradient force reverses from radially outward to radially inward occurs at small radius—cf. the discussion immediately following (6.45). If it occurs near the core radius, the air will be accelerated inward into the corner region and then it decelerates and turns upward there. In this case, there is rising motion near the axis of rotation and sinking motion beyond the core radius. The resultant circulation is that of a "one-cell vortex".

The reader is reminded that the behavior of an idealized tornado-like vortex in a vortex simulator is summarized in Figure 6.55. Idealized force diagrams at various swirl ratios are illustrated for summary purposes in Figure 6.59. At very low swirl ratio, there is boundary-layer separation at large radius and an intense low-level vortex does not occur.

At low swirl ratio, a one-cell vortex forms, in which there is rising motion along the central axis of the tornado and sinking motion far from the central axis. In the corner region, rapidly rising air along the central axis weakens and encounters the central downdraft and vortex breakdown may occur. Above the level of vortex breakdown, there is a two-cell vortex, in which there is sinking motion along the central axis, rising motion outside of the central axis, and sinking motion far from the central axis. At some higher swirl ratio, the level of vortex breakdown lowers to just above the surface and is referred to as a "drowned vortex jump".

As the swirl ratio is increased some more, the width of the core of the vortex increases, owing to the increase in centrifugal force inhibiting the radial pressure gradient force and forcing air upward at an even greater radius. The vortex is now a relatively wide, two-cell vortex, and the radial profile of azimuthal wind is such that there is an annulus of strong shear vorticity flanked by a downdraft at smaller radii and an updraft at higher radii.

In nature, the swirl ratio is probably controlled by the updraft magnitude in the convective cloud above, which is related to both buoyancy in the cloud and dynamic vertical pressure gradients, and to the dimensions (depth and width) of the inflow layer. A change in updraft intensity while a convective cloud grows or decays or a change in the nature of surface roughness characteristics or a change in how much vorticity is produced in the parent storm or a change in the depth of the moist boundary layer feeding the updraft in the parent storm may change the swirl ratio. It is difficult to relate the swirl ratio defined for a vortex simulator to the swirl ratio in nature because critical swirl transitions depend on the Reynolds number, which is different in the real atmosphere, and because there is uncertainty as to how to interpret the parameters defined in the simulator based on measurements in the real atmosphere; nevertheless, it has been attempted with some claimed success in mesocyclones and tornadoes.

We will now relate the swirl ratio to the maximum intensity of tornadoes. Lewellen *et al.* in 2000 argued that the conventional definition of swirl ratio in simulators (or in their "virtual" representation) is deficient because "... other physical parameters also affect the structure of the central vortex corner flow, so that flows that share the same large-scale swirl ratio can produce different corner flow structures." Lewellen *et al.* therefore defined a swirl ratio *for the corner flow region only*: this parameter describes the ratio of azimuthal velocities in the core region to radial inflow velocities (cf. (6.62)) in the tornado boundary layer only at the outer boundaries of the corner region. The more conventional swirl ratio includes radial inflow at a much larger radius; since radial inflow outside the core radius in the inertial layer is accelerated, it depends on what radius one chooses to measure inflow. In addition, the value used for the azimuthal wind is not necessarily the one at the location of the radius of maximum wind or at the core radius or at some radius that marks some particular characteristic of the flow field; moreover, the azimuthal wind varies with height in the boundary layer. The swirl ratio for the corner region is defined as

$$S_c = r_c\Gamma_\infty/M \tag{6.71}$$

where r_c is the core radius (Γ_∞/v_c); Γ_∞ is a measure of the angular momentum outside the core (i.e., at infinity) and above the boundary layer; and M is a measure of the mass flux flowing into the corner region from the boundary layer. M must be chosen for some conserved quantity such that it does not matter what the radius is in the boundary layer or what the height is in the core region at which we measure it. Lewellen *et al.* therefore define "depleted angular momentum" as

$$\Gamma_d = \Gamma_\infty - \Gamma \tag{6.72}$$

$$Du/Dt = \underbrace{u\,\partial u/\partial r}_{\text{ACCEL}} = \underbrace{\Gamma^2/r^3}_{\text{CEN}} - \underbrace{\alpha_0\,\partial p/\partial r}_{\text{PGF}}$$

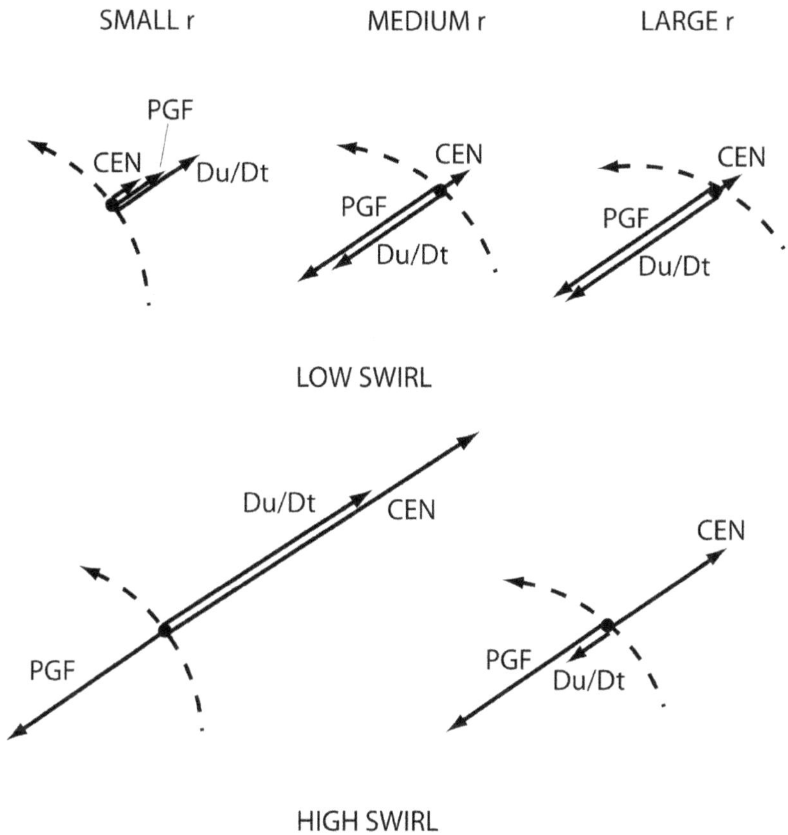

Figure 6.59. Idealized illustrations of force diagrams (vectors) for various swirl ratios and radii, for steady-state, frictionless flow at a level surface (where $w = 0$), for a potential vortex characterized by angular momentum Γ. The curved dashed streamlines indicate the motion of the air around the center of a cyclonic vortex. The governing equation of motion in the radial direction is given at the top. The net acceleration is indicated by ACCEL, centrifugal acceleration by CEN, and pressure gradient force by PGF. For low swirl (ratio), at large r, CEN is very small, and the radial acceleration is determined by the radially inward-directed PGF. As r decreases, CEN increases, but is still relatively small. At small r, the air parcel must decelerate before reaching $r = 0$, so the PGF reverses (especially if the swirl is very low and CEN very low); as the parcel decelerates it turns upward before reaching $r = 0$. For high swirl (ratio), at large r CEN detracts from the inward-directed PGF, so the parcel accelerates radially inward more slowly. At smaller r, CEN increases so much that it eventually overwhelms the inward-directed PGF and the parcel decelerates, finally coming to a halt and turning upward at some relatively large value of r (much larger than that for the low-swirl case). A downward-directed PGF (not shown here) at $r = 0$ drives a central downdraft.

Depleted angular momentum is useful because its flux into the corner region from the boundary layer is nearly equal to its flux out of the corner region into the core (not shown here), whereas angular momentum flux itself is not because it varies significantly within the corner region (also not shown here). So, instead of the mass flux, we will use the total depleted angular momentum flux through the corner region

$$\Upsilon = -2\pi r_1 \int_0^{z_1} u\Gamma_d \, dz \tag{6.73}$$

where z_1 is the height of the top of the friction layer; and r_1 is just outside the corner flow. To make sure that the mass flux M in the denominator of (6.71) has units of $(\mathrm{m\,s^{-1}})\,\mathrm{m}^2$, we scale Υ by Γ_∞ so that $M = \Upsilon/\Gamma_\infty$. It follows that the corner flow swirl ratio in (6.73) may be expressed with the aid of (6.33) as

$$S_c = r_c\Gamma_\infty^2/\Upsilon \equiv (\Gamma_\infty/r_c)/[\Upsilon/(\Gamma_\infty r_c^2)] = v_c/U \tag{6.74}$$

where U is a measure of the component of the wind flowing into the corner region given by $\Upsilon/(\Gamma_\infty r_c^2)$. It is nice that the dimensions of the hole and inflow depth in the vortex chamber do not appear in (6.74), so that is more easily applied to the real atmosphere. The reader is reminded that (6.74) is still an empirical parameter and that it is not the only way to characterize flow.

Another way to interpret the corner flow swirl ratio is, using (6.33), to express the numerator of (6.74) as

$$r_c\Gamma_\infty^2 = v_c r_c^2 \Gamma_\infty \sim w_c r_c^2 \Gamma_\infty \tag{6.75}$$

where vertical velocity upward from the corner region into the core region w_c is proportional to the azimuthal velocity in the core (the stronger the vertical jet in an end-wall vortex, the higher the azimuthal velocity). We interpret (6.33) as expressing the relationship between r_c, the closest to the center of the axis of rotation fluid may converge when the fluid is characterized by environmental angular momentum Γ_∞ and by v_c, the azimuthal wind speed at the core of the vortex imposed from above. The flux of angular momentum upward into the core is given by (6.75), while the denominator of (6.74) is the depleted angular momentum flux into the corner region: thus, the *corner flow swirl ratio is a measure of the ability of the converging boundary-layer flow to supply the core of the vortex with upward-moving fluid depleted of angular momentum subject to the constraints of the size of the core radius r_c and the azimuthal wind speed in the core v_c.*

When the corner flow swirl ratio $S_c \sim S_c^*$, the "critical corner flow swirl ratio", there is an end-wall vortex with a strong upward jet that ends with vortex breakdown just above the surface. This configuration results in the maximum azimuthal wind speed possible as close to the ground as possible. Brian Fiedler showed that the depleted angular momentum flux in the boundary layer of a potential vortex characterized by a high Reynolds number is

$$\gamma = 2.6\,\delta\,\Gamma_\infty^2 \tag{6.76}$$

where δ is the depth of the boundary layer in the limit as $r \rightarrow 0$. From (6.74) it is then seen that the corner flow swirl ratio

$$S_c = r_c/(2.6\,\delta) \tag{6.77}$$

The corner flow swirl ratio can therefore also be regarded as being a measure of the relative size of the core radius in the parent vortex above to the thickness of the boundary-layer inflow region below. Fiedler and Rotunno showed that the azimuthal (swirl) velocity in a supercritical endwall vortex

$$v_1 = 0.5\Gamma_\infty/\delta \tag{6.78}$$

while downstream (i.e., above) from the point of vortex breakdown the swirl is

$$v_2 \approx 0.25\Gamma_\infty/\delta \tag{6.79}$$

If the swirl beyond the vortex breakdown level is equal to the swirl above in the parent vortex ($v_2 = v_c$), then from (6.33) it follows that

$$r_c = 4\,\delta \tag{6.80}$$

Then from (6.77) it is seen that the critical corner flow swirl ratio $S_c^* = 1.5$. (If the flow rate across the vortex breakdown point is not exactly conserved, S_c^* may be slightly less.) This value of S_c is in accord with a series of trial-and-error numerical (LES) experiments performed by Dave Lewellen *et al.* To get the strongest vortex, the depth of the boundary layer must be "matched" to the core radius of the vortex imposed by the parent storm above. For larger values of S_c, a central downdraft reaches the surface, the core radius near the surface increases, and maximum wind speeds are less; for smaller values of S_c, vortex breakdown occurs higher up and maximum wind speeds are higher up (there may be a central downdraft, but it is elevated, above the height at which vortex breakdown occurs). As S_c is reduced even more, eventually there is no vortex breakdown at all. At the critical corner flow swirl ratio, the thermodynamic speed limit is augmented by a factor of ~2. *Thus, to get the strongest tornado possible for the parent vortex produced by the parent supercell above the boundary layer, there must be a "match" between the total depleted angular momentum flux into the corner region at the core radius and angular momentum outside the core. When there is a "match", the radially inward-accelerating air from the inertial layer penetrates the smallest radius possible, as close to the surface as possible.*

If a vortex is characterized by an S_c that is less than S_c^*, then the intensity of the vortex may be enhanced if the total depleted angular momentum flux is decreased; if S_c is greater than S_c^*, then the intensity of the vortex may be enhanced if the total depleted angular momentum flux is increased. It has therefore been suggested that changes in the characteristics of boundary-layer inflow away from the vortex can determine whether or not a parent mesocyclone can be intensified and decreased in scale to tornado intensity and size.

Suppose that initially there is a mesocyclone at low levels that has a relatively high (traditional) swirl ratio and a low corner flow swirl ratio. If boundary-layer inflow at large radius is suddenly cut off, then the corner flow swirl ratio is increased, so that it may now be closer to its critical value. Consider (6.45) again. If the influx of angular momentum is cut off at the large radius, then $-v^2/r$ in the corner region decreases in magnitude. Therefore, the radially inward acceleration $u\,\partial u/\partial r$ must be greater in magnitude when the magnitude of $-v^2/r$ is decreased. *Thus, a reduction in the inward flux of angular momentum allows for greater penetration of rings of air inward and thus the potential for an increase in azimuthal velocity as a potential vortex is brought radially inward farther.* The increase in intensity of a vortex as the influx of angular momentum into the corner region is impeded is known as "corner flow collapse". It has been suggested that corner flow collapse may be triggered when the rear-flank downdraft in a supercell wraps around a mesocyclone at the surface and acts as a barrier to radially inward-flowing air. This hypothesis differs from the one that explains vortex intensification due to enhanced production of surface vorticity when convergence is produced as the RFD wraps around the mesocyclone or as higher angular momentum from aloft is advected downward.

We conclude that the intensity of a tornado is determined both by the intensity of the parent mesocyclone that is produced by storm-scale processes and by boundary-layer processes that allow for greatest radially inward penetration and greatest enhancement resulting from surface friction. The downward-directed perturbation pressure gradient force associated with the vertical decrease in vorticity is a dynamic feedback that limits the amount of frictional enhancement of vortex intensity.

If a vortex is non-steady there may be local "transients" that are characterized by even higher wind speeds than possible through corner flow collapse. If the vortex is asymmetric because, for example, it is translating, then the drag of the air on the surface is greater on the side in which the translational motion vector is added to the azimuthal wind field and less on the opposite side. Therefore, S_c may be decreased in a "medium-swirl" vortex on one side toward S_c^*. Thus, a vortex may experience corner flow collapse if it translates along, but the same vortex may remain relatively weak if it does not translate along. In this case, the reference frame of the parent supercell is important. Furthermore, a vortex that translates along and encounters changes in surface roughness may or may not undergo corner flow collapse depending on how S_c is changed.

If the thermodynamic speed limit is based on faulty physics and almost always exceeded as much as a factor of 2, or maybe more, why refer to it all? The reason is that it might be useful as a forecast tool in that the *maximum possible wind speed* may be predicted based on a forecast of CAPE: take the square root of CAPE and multiply it by 2.

6.7 ECONOMIC AND SOCIETAL IMPACTS

We now leave the realm of science and enter the non-scientific realm of how society reacts to tornadoes as a weather problem. It is argued that in the U. S. taxpayers contributing to government-funded research projects should have access to studies on the return on their investment to tornado research. To be able to formulate an estimate of the return on investments in research, one needs to estimate the reduction in loss of life, injuries, damage, etc. Fatalities occur most often in mobile homes. Most fatalities, injuries, and damage come from a small fraction of all tornadoes. Tornado fatalities in the U. S. have fallen from about 4 per 250,000 people from 1900 to 1930, to about 1 per 250,000 people from the mid-1980s until now, in spite of an increase in population, especially outside urban areas. In 2011, however, the number rose dramatically, owing to an unusually large number of outbreaks. It is thought that our improved understanding of tornadoes, improved forecasts, and nowcasts, in part as a result of the national network of Doppler radars (NEXRAD) in the U. S. and knowledge passed to forecasters from researchers ("technology transfer"), have contributed to the lower fatality rate. In addition, heightened media awareness is also probably responsible for improved dissemination of warnings.

Warning lead times have increased over the years, but there is a tradeoff between the probability of detection and the false alarm rate. If warnings were issued for all possible tornado-producing storms, fewer tornadoes would be missed, but the false alarm rate would increase and the public would be less likely to respond to warnings; if fewer warnings were issued for tornado-producing storms so as to reduce the false alarm rate, some tornadoes would be missed and the public's confidence in warnings would also decrease.

In addition to direct tornado damage, which in the U. S. equals about $1 billion per year (inflation-adjusted to 2007), estimated measures such as the "value of a statistical life" (VSL) and the "value of a statistical injury" (VSI) have been devised by economists. The VSL has been estimated to be in the range of $1 million to $10 million or greater. The VSL has been estimated based on factors such as wages lost, among other things. Structures that can survive even the most powerful tornadoes can be built, but at a cost. Efforts by the National Weather Service to reduce the fatality rate in tornadoes have made tornado shelters less cost effective. Until recently, the overall cost of responding to warnings (and disrupting normal activities) was greater than the costs inflicted by tornadoes.

6.8 UNRESOLVED PROBLEMS AND CHALLENGES FOR FUTURE RESEARCH, WITH SUGGESTIONS FOR IMPROVED MEASUREMENT CAPABILITIES

Much of the theory of tornado boundary layers and tornado structure involves axisymmetric, steady-state flow in an incompressible atmosphere that is homogeneous. In supercells the flow is far from axisymmetric and homogeneous; the real atmosphere is compressible and since tornadoes form near the forward-flank downdraft and the rear-flank downdraft, there can be significant horizontal variations in temperature and static stability. Mesocyclones in supercells are far from steady state. A theory needs to be developed that accounts for the full complexity of real atmospheric conditions.

Tornado formation in which the mesocylone forms first aloft as a result of tilting of environmental vorticity and later a mesocyclone forms at low levels as a result of baroclinic generation or advection of horizontal vorticity from the environment or both, followed by tilting and stretching, is most likely if the mesocyclone aloft and mesocyclone at low levels are superimposed. This situation resembles that of synoptic-scale, mid-latitude cyclogenesis, because both involve the effects of the superposition of high-level and low-level vortices, albeit of vastly different scales (in the case of synoptic-scale flow, ~1,000 km; in the case of mesocyclones, ~1–10 km). How the positions of mid-level and low-level mesocyclones change with time and are influenced by storm-scale processes needs to be understood.

It would be nice to analyze tornadogenesis as an instability process as has been done for extratropical cyclones. Tornadogenesis, however, is highly nonlinear and involves precipitation microphysics, which also contributes to the nonlinearity; it is apparently not as amenable to simple analysis and certainly not linear analysis. Nevertheless, it is hoped that a definitive set of numerical simulations under controlled conditions can be produced some day that will determine what ranges of parameters are necessary conditions for tornadogenesis.

A wish list for improved measurement capabilities includes

a. Rapid-scan, mobile Doppler radars with very narrow beams (~0.1–0.25° half-power beamwidths) and volumetric update times ~10 s. It is likely that to attain very narrow beams on mobile platforms the operating wavelength will have to be short. To scan rapidly, it is likely that electronic scanning will be necessary.

b. Probes to measure thermodynamic and moisture variables mounted on unmanned aircraft and helicopters.

c. Doppler radars and eye-safe Doppler lidars mounted on unmanned aircraft and helicopters. The radars and lidars will have to be small and lightweight, so they will probably emit low power.

d. Storm-penetrating aircraft that can make in situ measurements of thermodynamic, moisture, and cloud microphysics variables.

e. Downward-looking Doppler radars with polarimetric capability that can fly over severe convective storms on manned or unmanned aircraft.

f. Development of spaced antennas or other technologies that allow the construction of mobile Doppler radar systems with effective half-power beamwidths ~10, so that the two-dimensional wind field could be attained (over an entire convective storm). It would be desirable to do this rapidly, about every minute or less. It would also be desirable to have polarimetric capability. There is more discussion on future observational capabilities in Section 7.3.

6.9 GENERAL MONOGRAPHS AND BOOKS

The reader is referred to p. 24 for a list of relevant general monographs and books.

6.10 REFERENCES AND BIBLIOGRAPHY

Agee, E. M., J. T. Snow, and P. R. Clare (1976) Multiple vortex features in the tornado cyclone and the occurrence of tornado families. *Mon. Wea. Rev.*, **104**, 552–563.

Alexander, C. R. (2010) A mobile radar based climatology of supercell tornado structures and dynamics. Ph.D. thesis, University of Oklahoma, 229 pp.

Alexander, C. R. and J. Wurman (2005) The 30 May 1998 Spencer, South Dakota, storm, Part I: The structural evolution and environment of the tornadoes. *Mon. Wea. Rev.*, **133**, 72–96.

Alexander, C. R. and J. Wurman (2006) Mobile radar based climatology of tornado structure and dynamics. *23rd Conference on Severe Local Storms, November 6–10, St. Louis, MO*, American Meteorological Society, Boston (extended abstract not available).

Baker, G. L. and C. R. Church (1979) Measurements of core radii and peak velocities in modeled atmospheric vortices. *J. Atmos. Sci.*, **36**, 2413–2424.

Barcilon, A. I. (1967) Vortex decay above a stationary boundary. *J. Fluid Mech.*, **27**, 155–175.

Barcilon, A. and P. G. Drazin (1972) Dust devil formation. *Geophys. Fluid Dyn.*, **4**, 147–158.

Beck, V. and N. Dotzek (2010) Reconstruction of near-surface tornado wind fields from forest damage. *J. Appl. Meteor. Climatol.*, **49**, 1517–1537.

Bedard, A. J., Jr. (2005) Low-frequency atmospheric acoustic energy associated with vortices produced by thunderstorms. *Mon. Wea. Rev.*, **133**, 241–263.

Benjamin, T. B. (1962) Theory of the vortex breakdown phenomenon. *J. Fluid Mech.*, **14**, 593–629.

Bluestein, H. B. (1985) The formation of a "landspout" in a "broken-line" squall line in Oklahoma. Preprints, *14th Conference on Severe Local Storms, Indianapolis*, American Meteorological Society, Boston, pp. 267–270.

Bluestein, H. B. (1992) *Synoptic-Dynamic Meteorology in Midlatitudes, Vol. I: Principles of Kinematics and Dynamics*, Oxford University Press, New York, 431 pp.

Bluestein, H. B. (1994) High-based funnel clouds in the Southern Plains. *Mon. Wea. Rev.*, **122**, 2631–2638.

Bluestein, H. B. (2000) A tornadic supercell over elevated, complex terrain: The Divide, Colorado storm of 12 July 1996. *Mon. Wea. Rev.*, **128**, 795–809.

Bluestein, H. (2005) A review of ground-based, mobile, W-band, Doppler-radar observations of tornadoes and dust devils. *Dyn. of Atmospheres and Oceans*, **40**, 163–188.

Bluestein, H. B. (2007) Advances in applications of the physics of fluids to severe weather systems. *Reports on Progress in Physics*, **70**, 1259–1323.

Bluestein, H. B. (2009) The formation and early evolution of the Greensburg, Kansas supercell on 4 May 2007. *Wea. Forecasting*, **24**, 899–920.

Bluestein, H. B. and A. L. Pazmany (2000) Observations of tornadoes and other convective phenomena with a mobile, 3-mm wavelength, Doppler radar: The spring 1999 field experiment. *Bull. Amer. Meteor. Soc.*, **81**, 2939–2951.

Bluestein, H. B., W. P. Unruh, J. LaDue, H. Stein, and D. Speheger (1993) Doppler radar wind spectra of supercell tornadoes. *Mon. Wea. Rev.*, **121**, 2200–2221.

Bluestein, H. B., C. C. Weiss, and A. L. Pazmany (2003a) Mobile Doppler radar observations of a tornado in a supercell near Bassett, Nebraska, on 5 June 1999, Part I: Tornadogenesis. *Mon. Wea. Rev.*, **131**, 2954–2967.

Bluestein, H. B., W.-C. Lee, M. Bell, C. C. Weiss, and A. L. Pazmany (2003b) Mobile Doppler radar observations of a tornado in a supercell near Bassett, Nebraska, on 5 June 1999, Part II: Tornado-vortex structure. *Mon. Wea. Rev.*, **131**, 2968–2984.

Bluestein, H. B., C. C . Weiss, and A. L. Pazmany (2004a) Doppler radar observations of dust devils in Texas. *Mon. Wea. Rev.*, **132**, 209–224.

Bluestein, H. B., C. C. Weiss, and A. L. Pazmany (2004b) The vertical structure of a tornado near Happy, Texas, on 5 May 2002: High-resolution, mobile, W-band Doppler radar observations. *Mon. Wea. Rev.*, **132**, 2325–2337.

Bluestein, H. B., C. C. Weiss, M. M. French, E. M. Holthaus, and R. L. Tanamachi (2007a) The structure of tornadoes near Attica, Kansas, on 12 May 2004: High-resolution, mobile, Doppler radar observations. *Mon. Wea. Rev.*, **135**, 475–506.

Bluestein, H. B., M. M. French, S. Frasier, K. Hardwick, F. Junyent, and A. L. Pazmany (2007b) Close-range observations of tornadoes in supercells made a dual-polarization, X-band, mobile Doppler radar. *Mon. Wea. Rev.*, **135**, 1522–1543.

Bluestein, H. B., M. M. French, I. PopStefanija, R. T. Bluth, and J. B. Knorr (2010) A mobile, phased-array Doppler radar for the study of severe convective storms: The MWR-05XP. *Bull. Amer. Meteor. Soc.*, **91**, 579–600.

Bödewadt, V. T. (1940) Die Drehströmung über festem Grunde. *Z. Angew. Math. Mech.*, **20**, 241–253 [in German].

Bosart, L. F., A. Seimon, K. D. LaPenta, and M. J. Dickinson (2006) Supercell tornadogenesis over complex terrain: The Great Barrington, Massachusetts, tornado on 29 May 1995. *Wea. Forecasting*, **21**, 897–922.

Brantley, J. Q. (1957) Some weather observations with a continuous-wave Doppler radar. *Proceedings 6th Weather Radar Conference, Cambridge, MA*, pp. 297–306.

Brooks, H. E. and C. A. Doswell, III (2001) Some aspects of the international climatology of tornadoes by damage. *Atmos. Res.*, **56**, 119–201.

Brooks, H. E., J. W. Lee, and J. P. Craven (2003a) The spatial distribution of severe thunderstorm and tornado environments from global reanalysis data. *Atmos. Res.*, **67/68**, 73–94.

Brooks, H. E., C. A. Doswell, III, and M. P. Kay (2003b) Climatological estimates of local daily tornado probability for the United States. *Wea. Forecasting*, **18**, 626–640.

Brown, R. A. and V. T. Wood (2012) The tornadic vortex signature: An update. *Wea. Forecasting*, **27**, 525–530.

Brown, R. A., L. R. Lemon, and D. W. Burgess (1978) Tornado detection by pulsed Doppler radar. *Mon. Wea. Rev.*, **106**, 29–38.

Brown, J. M. and K. R. Knupp (1980) The Iowa cyclonic–anticyclonic tornado pair and its parent thunderstorm. *Mon. Wea. Rev.*, **108**, 1626–1646.

Bryan, G. and L. Orf (2011) The effects of low-level shear on the intensity and longevity of low-level mesocyclones in simulated supercell thunderstorms. *Sixth European Conference on Severe Storms, Mallorca, Spain* (*http://www.essl.org/ECSS/2011/programme/presentations/126.pdf*).

Bunkers, M. J. and J. W. Stoppkotte (2007) Documentation of a rare tornadic left-moving supercell. *Electronic J. Severe Storms Meteor.*, **2**, 1–22.

Burgers, J. M. (1948) A mathematical model illustrating the theory of turbulence. *Adv. Appl. Mech.*, **1**, 197–199.

Burgess, D. W. (1976) Anticyclonic tornado. *Weatherwise*, **29**, 167.

Burgess, D. W., V. T. Wood, and R. A. Brown (1982) Mesocyclone evolution statistics. Preprints, *12th Conference on Severe Local Storms, San Antonio, TX*, American Meteorological Society, Boston, pp. 422–424.

Burrgraf, O., K. Stewartson, and R. Belcher (1971) Boundary layer induced by a potential vortex. *Phys. Fluids*, **14**, 1821–1833.

Buynitzky, E. (1915) A tentative classification for meteorological literature. *Mon. Wea. Rev.*, **43**, 362–364.

Byko, Z., P. Markowski, and Y. Richardson (2009) Descending reflectivity cores in supercell thunderstorms observed by mobile radars and in a high-resolution numerical simulation. *Wea. Forecasting*, **24**, 155–186.

Carbone, R. E. (1983) A severe frontal rainband, Part II: Tornado parent vortex circulation. *J. Atmos. Sci.*, **40**, 2639–2654.

Carroll, J. J. and J. A. Ryan (1970) Atmospheric vorticity and dust devil rotation. *J. Geophys. Res.*, **75**, 5179–5184.

Church, C. R. and J. T. Snow (1985) Measurements of axial pressures in tornado-like vortices. *J. Atmos. Sci.*, **42**, 576–582.

Church, C. R. and J. T. Snow (1993) Laboratory models of tornadoes. In: C. R. Church (Ed.), *The Tornado: Its Structure, Dynamics, Prediction, and Hazards*, American Geophysical Union, Washington, D.C., pp. 277–295.

Church, C. R., J. T. Snow, and E. M. Agee (1977) Tornado vortex simulation at Purdue University. *Bull. Amer. Meteor. Soc.*, **58**, 900–908.

Church, C. R., J. T. Snow, G. L. Baker, and E. M. Agee (1979) Characteristics of tornado-like vortices as function of swirl ratio: A laboratory investigation. *J. Atmos. Sci.*, **36**, 1755–1776.

Church, C. R., J. T. Snow, and J. Dessens (1980) Intense atmospheric vortices associated with a 1000 MW fire. *Bull. Amer. Meteor. Soc.*, **61**, 682–694.

Colgate, S. (1982) Small rocket tornado probe. Preprints, *12th Conference on Severe Local Storms, San Antonio, TX*, American Meteorological Society, Boston, pp. 396–400.

Cooley, J. R. (1978) Cold air funnel clouds. *Mon. Wea. Rev.*, **106**, 1368–1372.

Davies-Jones, R. P. (1973) The dependence of core radius on swirl ratio in a tornado simulator. *J. Atmos. Sci.*, **30**, 1427–1430.

Davies-Jones, R. P. (1982) Observational and theoretical aspects of tornadogenesis. In: L. Bengtsson and J. Lighthill (Eds.), *Intense Atmospheric Vortices*, Springer–Verlag, New York, pp. 175–189.

Davies-Jones, R. P. (1984) Streamwise vorticity: The origin of updraft rotation in supercell storms. *J. Atmos. Sci.*, **41**, 2991–3006.

Davies-Jones, R. P. (1986) Tornado dynamics. In: E. Kessler (Ed.), *Thunderstorm Morphology and Dynamics*, University of Oklahoma Press, Norman, OK, pp. 197–236.

Davies-Jones, R. P. (2000) A Lagrangian model for baroclinic genesis of mesoscale vortices, Part I: Theory. *J. Atmos. Sci.*, **57**, 715–736.

Davies-Jones, R. P. (2006) Tornadogenesis in supercell storms: What we know and what we don't know. Extended abstracts, *23rd Conference on Severe Local Storms, November 6–10, St. Louis, MO*, American Meteorological Society, Boston (CD-ROM 2.2).

Davies-Jones, R. (2008) Can a descending rain curtain in a supercell instigate tornadogenesis barotropically? *J. Atmos. Sci.*, **65**, 2469–2497.

Davies-Jones, R. P. and H. E. Brooks (1993) Mesocyclogenesis from a theoretical perspective. In: C. R. Church (Ed.), *The Tornado: Its Structure, Dynamics, Prediction, and Hazards*, American Geophysical Union, Washington, D.C., pp. 105–114.

Davies-Jones, R. P., D. W. Burgess, L. R. Lemon, and D. Purcell (1978) Interpretation of surface marks and debris patterns from the 24 May 1973 Union City, Oklahoma tornado. *Mon. Wea. Rev.*, **106**, 12–21.

Davies-Jones, R. P., R. J. Trapp, and H. B. Bluestein (2001) Tornadoes. In: C. Doswell, III (Ed.), *Severe Convective Storms*, AMS Monogr. Vol. 28, Issue 50, American Meteorological Society, Boston, pp. 167–221.

Dessens, J. (1972) Influence of ground roughness on tornadoes: A laboratory simulation. *J. Appl. Meteor.*, **11**, 72–75.

Dessens, J. and J. T. Snow (1989) Tornadoes in France. *Wea. Forecasting*, **4**, 110–132.

Doswell. C. A., III (1980) Synoptic-scale environments associated with High Plains severe thunderstorms. *Bull. Amer. Meteor. Soc.*, **61**, 1388–1400.

Doswell, C. A., III and J. S. Evans (2003) Proximity sounding analysis for derechos and supercells: An assessment of similarities and differences. *Atmos. Res.*, **67/68**, 117–133.

Dowell, D. C. and H. B. Bluestein (2002) The 8 June 1995 McLean, Texas, storm, Part II: Cyclic tornado formation, maintenance, and dissipation. *Mon. Wea. Rev.*, **130**, 2649–2670.

Dowell, D. C., F. Zhang, L. J. Wicker, C. Snyder, and N. A. Crook (2004) Wind and temperature retrievals in the 17 May 1981 Arcadia, Oklahoma, supercell: Ensemble Kalman filter experiments. *Mon. Wea. Rev.*, **132**, 1982–2005.

Dowell, D. C., C. R. Alexander, J. M. Wurman, and L. J. Wicker (2005) Centrifuging of hydrometeors and debris in tornadoes: Radar-reflectivity patterns and wind-measurement errors. *Mon. Wea. Rev.*, **133**, 1501–1524.

Faller, A. J. (1963) An experimental study of the instability of the laminar Ekman layer. *J. Fluid Mech.*, **15**, 560–576.

Fiedler, B. H. (1998) Wind-speed limits in numerically simulated tornadoes with suction vortices. *Quart. J. Roy. Meteor. Soc.*, **124**, 2377–2392.

Fiedler, B. H. (1994) The thermodynamic speed limit and its violation in axisymmetric numerical simulations of tornado-like vortices. *Atmos. Ocean*, **32**, 335–359.

Fiedler, B. B. (1995) On modeling tornadoes in isolation from the parent storm. *Atmos. Ocean*, **33**, 501–512.

Fiedler, B. (2009) Suction vortices and spiral breakdown in numerical simulations of tornado-like vortices. *Atmos. Sci. Letters*, **10**, 109–114.

Fiedler, B. H. and G. S. Garfield (2010) Axisymmetric vortex simulations with various turbulence models. *Computational Fluid Dynamics Letters*, **2**, 112–121.

Fiedler, B. H. and K. M. Kanak (2001) Rayleigh–Bénard convection as a tool for studying dust devils. *Atmos. Sci. Letters*, **2**, 104–113.

Fiedler, B. H. and R. Rotunno (1986) A theory for the maximum windspeeds in tornado-like vortices. *J. Atmos. Sci.*, **43**, 2328–2340.

Forbes, G. S. and R. M. Wakimoto (1983) A concentrated outbreak of tornadoes, downbursts and microbursts, and implications regarding vortex classification. *Mon. Wea. Rev.*, **111**, 220–235.

Fujita, T. T. (1974) Jumbo tornado outbreak of 3 April 1974. *Weatherwise*, **27**, 116–126.

Fujita, T. T. (1975) New evidence from April 3–4, 1974 tornadoes. Preprints, *Ninth Conference on Severe Local Storms, Norman, OK*, American Meteorological Society, Boston, pp. 248–255.

Fujita, T. T. (1981) Tornadoes and downbursts in the context of generalized planetary scales. *J. Atmos. Sci.*, **38**, 1511–1534.

Fujita, T. T. (1989) The Teton–Yellowstone tornado of 21 July 1987. *Mon. Wea. Rev.*, **117**, 1913–1940.

Fujita, T. T., D. L. Bradbury, and C. F. Van Thullenar (1970) Palm Sunday tornadoes of April 11, 1965. *Mon. Wea. Rev.*, **98**, 29–69.

Fujita, T. T., G. S. Forbes, and T. A. Umenhofer (1976) Close-up view of 20 March 1976 tornadoes: Sinking cloud tops to suction vortices. *Weatherwise*, **29**, 116–131, 145.

Gall, R. L. (1983) A linear analysis of multiple vortex phenomena in simulated tornadoes. *J. Atmos. Sci.*, **40**, 2010–2024.

Gaudet, B. J., W. R. Cotton, and M. T. Montgomery (2006) Low-level mesocyclonic concentration by nonaxisymmetric transport, Part II: Vorticity dynamics. *J. Atmos. Sci.*, **63**, 1134–1150.

Gentry, R. C. (1983) Genesis of tornadoes associated with hurricanes. *Mon. Wea. Rev.*, **111**, 1793–1805.

Golden, J. H. (1971) Waterspouts and tornadoes over south Florida. *Mon. Wea. Rev.*, **99**, 146–154.

Golden, J. H. (1974) The life cycle of Florida Keys' waterspouts, I. *J. Appl. Meteor.*, **13**, 676–692.

Golden, J. H. (1974) Scale-interaction implications for the waterspout life cycle, II. *J. Appl. Meteor.*, **13**, 693–709.

Golden, J. and H. B. Bluestein (1994) The NOAA–National Geographic Society Waterspout Expedition (1993). *Bull. Amer. Meteor. Soc.*, **75**, 2281–2288.

Golden, J. H. and D. Purcell (1978a) Life cycle of the Union City, Oklahoma tornado and comparison with waterspouts. *Mon. Wea. Rev.*, **106**, 3–11.

Golden, J. H. and D. Purcell (1978b) Airflow characteristics around the Union City tornado. *Mon. Wea. Rev.*, **106**, 22–28.

Grams, J. S., R. L. Thompson, D. V. Snively, J. A. Prentice, G. M. Hodges, and L. J. Reames (2012) A climatology and comparison of parameters for significant tornado events in the United States. *Wea. Forecasting*, **27**, 106–123.

Grasso, L. D. and W. R. Cotton (1995) Numerical simulation of a tornado vortex. *J. Atmos. Sci.*, **52**, 1192–1203.

Hall, M. G. (1961) A theory for the core of a leading-edge vortex. *J. Fluid Mech.*, **11**, 209–228.

Hanstrum, B. N., G. A. Mills, A. Watson, J. P. Monteverdi, and C. A. Doswell, III (2002) The cool-season tornadoes of California and southern Australia. *Wea. Forecasting*, **17**, 705–722.

Hoecker, W. H., Jr. (1960) Wind speed and air flow patterns in the Dallas tornado of April 2, 1957. *Mon. Wea. Rev.*, **88**, 167–180.

Hoecker, W. H., Jr. (1961) Three-dimensional pressure pattern of the Dallas tornado and some resultant implications. *Mon. Wea. Rev.*, **89**, 533–542.

Houston, A. L. and R. B. Wilhelmson (2007) Observational analysis of the 27 May 1997 central Texas tornadic event, Part II: Tornadoes. *Mon. Wea. Rev.*, **135**, 727–735.

Howard, L. N. and A. S. Gupta (1962) On the hydrodynamic and hydromagnetic stability of swirling flows. *J. Fluid. Mech.*, **14**, 463–476.

Howells, P. C., R. Rotunno, and R. K. Smith (1988) A comparative study of atmospheric and laboratory analogue numerical tornado-vortex models. *Quart. J. Roy. Meteor. Soc.*, **114**, 801–822.

Kanak, K. M. (2005) Numerical simulation of dust devil-scale vortices. *Quart. J. Roy. Meteor. Soc.*, **131**, 1271–1292.

Kanak, K. M., D. K. Lilly, and J. T. Snow (2000) The formation of vertical vortices in the convective boundary layer. *Quart. J. Roy. Meteor. Soc.*, **126A**, 2789–2810.

Karstens, C. D., T. M. Samaras, B. D. Lee, W. A. Gallus, Jr., and C. A. Finley (2010) Near-ground pressure and wind measurements in tornadoes. *Mon. Wea. Rev.*, **138**, 2570–2588.

Kuo, H. L. (1971) Axisymmetric flows in the boundary layer of a maintained vortex. *J. Atmos. Sci.*, **28**, 20–41.

Kumjian, M. R. and B. H. Fiedler (2007) The efficacy of subsidence warming in the core of numerically simulated tornado-like vortices. Paper 2.1, *16th Conference on Atmospheric Oceanic Fluid Dynamics, Santa Fe, NM, June 25–29*, American Meteorological Society, Boston.

Lee, B. D. and R. B. Wilhelmson (1997) The numerical simulation of nonsupercell tornado-genesis, Part II: Evolution of a family of tornadoes along a weak outflow boundary. *J. Atmos. Sci.*, **54**, 2387–2415.

Lee, W.-C. and J. Wurman (2005) Diagnosed three-dimensional axisymmetric structure of the Mulhall tornado on 3 May 1999. *J. Atmos. Sci.*, **62**, 2373–2393.

Lee, J. J., T. M. Samaras, and C. R. Young (2004) Pressure measurements at the ground in an F-4 tornado. Extended abstracts, *22nd Conference on Severe Local Storms, October 4–8, Hyannis, MA*, American Meteorological Society, Boston (CD-ROM 15.3).

Leibovich, S. (1978) The structure of vortex breakdown. *Ann. Rev. Fluid Mech.*, **10**, 221–246.

Lemon, L. R., D. W. Burgess, and R. A. Brown (1978) Tornadic storm airflow and morphology derived from single-Doppler radar measurements. *Mon. Wea. Rev.*, **106**, 48–61.

Leslie, L. M. (1971) The development of concentrated vortices: A numerical study. *J. Fluid Mech.*, **48**, 1–21.

Leverson, V. H., P. C. Sinclair, and J. H. Golden (1977) Waterspout wind, temperature, and pressure structure deduced from aircraft measurements. *Mon. Wea. Rev.*, **105**, 725–733.

Lewellen, W. S. (1976) Theoretical models of the tornado vortex. *Proceedings of Symposium on Tornadoes: Assessment of Knowledge and Implications for Man, Lubbock, TX*, Texas Tech. University, pp. 107–143.

Lewellen, D. C. and W. S. Lewellen (2007a) Near-surface intensification of tornado vortices. *J. Atmos. Sci.*, **64**, 2176–2194.

Lewellen, D. C. and W. S. Lewellen (2007b) Near-surface vortex intensification through corner flow collapse. *J. Atmos. Sci.*, **64**, 2195–2209.

Lewellen, W. S., D. C. Lewellen, and R. I. Sykes (1997) Large-eddy simulation of a tornado's interaction with the surface. *J. Atmos. Sci.*, **54**, 581–605.

Lewellen, D. C., W. S. Lewellen, and J. Xia (2000) The influence of a local swirl ratio on tornado intensification near the surface. *J. Atmos. Sci.*, **57**, 527–544.

Lilly, D. K. (1969) *Tornado Dynamics*, NCAR Manuscript 69-117, National Center for Atmospheric Research, Boulder, CO.

Ludlam, F. H. (1963) Severe local storms: A review. In: D. Atlas (Ed.), *Severe Local Storms*, AMS Monogr. Vol. 27, Issue 5, American Meteorological Society, Boston, pp. 1–30.

Magsig, M. A. and J. T. Snow (1998) Long-distance debris transport by tornadic thunderstorms, Part I: The 7 May 1995 supercell thunderstorm. *Mon. Wea. Rev.*, **126**, 1430–1449.

Mahale, V. N., J. A. Brotzge, and H. B. Bluestein (2012) An analysis of vortices embedded within a quasi-linear convective system using X-band polarimetric radar. *Wea. Forecasting* (in press).

Markowski, P. M. (2002) Mobile mesonet observations on 3 May 1999. *Wea. Forecasting*, **17**, 430–444.

Markowski, P. M., J. M. Straka, and E. N. Rasmussen (2002) Direct surface thermodynamic observations within the rear-flank downdrafts of nontornadic and tornadic supercells. *Mon. Wea. Rev.*, **130**, 1692–1721.

Markowski, P. M., J. M. Straka, and E. N. Rasmussen (2003) Tornadogenesis resulting from the transport of circulation by a downdraft: Idealized numerical simulations. *J. Atmos. Sci.*, **60**, 795–823.

Marquis, J., Y. Richardson, J. Wurman, and P. Markowski (2008) Single- and dual-Doppler analysis of a tornadic vortex and surrounding storm-scale flow in the Crowell, Texas, supercell of 30 April 2000. *Mon. Wea. Rev.*, **136**, 5017–5043.

Marquis, J., Y. Richardson, P. Markowski, D. Dowell, and J. Wurman (2012) Tornado maintenance investigated with high-resolution dual-Doppler and EnKF analysis. *Mon. Wea. Rev.*, **140**, 3–27.

Marshall, T. P. (2004) The enhanced Fujita (EF) scale. Preprints, *22nd Conference on Severe Local Storms, Hyannnis, MA*, American Meteorological Society, Boston (CD-ROM 3B.2).

Maxworthy, T. (1973) Vorticity source for large scale dust devils and other comments on naturally occurring vortices. *J. Atmos. Sci.*, **30**, 1717–1720.

Monteverdi, J. P. and J. Quadros (1994) Convective and rotational parameters associated with three tornado episodes in northern and central California. *Wea. Forecasting*, **9**, 285–300.

Monteverdi, J. P., W. Blier, G. Stumpf, W. Pi, and K. Anderson (2001) First WSR-88D documentation of an anticyclonic supercell with anticyclonic tornadoes: The Sunnyvale–Los Altos, California, tornadoes of 4 May 1998. *Mon. Wea. Rev.*, **129**, 2805–2814.

Monteverdi, J. P., R. Edwards, G. A. Stumpf, and D. Gudgel (2006) An analysis of the 7 July 2004 Rockwell Pass, CA tornado: Highest elevation tornado documented in the U. S. Extended abstracts, *23rd Conference on Severe Local Storms, November 6–10, St. Louis, MO*, American Meteorological Society, Boston (CD-ROM P12.2).

Morton, B. R. (1966) Geophysical vortices. In: *Progress in Aeronautical Sciences*, Vol. 7, Pergamon Press, Oxford, U. K., pp. 145–193.

Niino, A. T. and H. Niino (2005) Genesis and structure of a major tornado in a numerically-simulated supercell storm: Importance of vertical vorticity in a gust front. *Scientific Online Letters on the Atmosphere,* **1**, 5–8.

Nolan, D. S. (2005) A new scaling for tornado-like vortices. *J. Atmos. Sci.*, **62**, 2639–2645.

Nolan, D. S. and B. F. Farrell (1999) The structure and dynamics of tornado-like vortices. *J. Atmos. Sci.*, 2908–2936.

Novlan, D. J. and W. M. Gray (1974) Hurricane-spawned tornadoes. *Mon. Wea. Rev.*, **102**, 476–488.

Orf, L., M. S. Gilmore, R. B. Wilhelmson, J. M. Straka, and E. N. Rasmussen (2006) The role of hook echo microbursts in simulated tornadic supercells, Part I: Association with counter-rotating vortices and tornadogenesis. *23rd Conference on Severe Local Storms, 6–10 November, St. Louis, MO*, American Meteorological Society, Boston (extended abstract not available)

Pauley, R. L. and J. T. Snow (1988) On the kinematics and dynamics of the 18 June 1986 Minneapolis tornado. *Mon. Wea. Rev.*, **116**, 2731–2736.

Pauley, R. L., C. R. Church, and J. T. Snow (1982) Measurements of maximum surface pressure deficits in modeled atmospheric vortices. *J. Atmos. Sci.*, **39**, 369–377.

Rankine, W. J. M. (1882) *A Manual of Applied Physics*, 10th Edn., Charles Griffin & Co., London, 663 pp.

Rasmussen, E. N. and J. M. Straka (2007) Evolution of low-level angular momentum in the 2 June 1995 Dimmitt, Texas, tornado cyclone. *J. Atmos. Sci.*, **64**, 1365–1378.

Rasmussen, E. N., J. M. Straka, M. S. Gilmore, and R. Davies-Jones (2006) A preliminary survey of rear-flank descending reflectivity cores in supercell storms. *Wea. Forecasting*, **21**, 923–938.

Renno, N. O. (2008) A thermodynamically general theory for convective vortices. *Tellus*, **60A**, 688–699.

Rott, N. (1958) On the viscous core of a line vortex. *Z. Math. Phys.*, **9b**, 543–553.

Rotunno, R. (1977) Numerical simulation of a laboratory vortex. *J. Atmos. Sci.*, **34**, 1942–1956.

Rotunno, R. (1979) A study in tornado-like vortex dynamics. *J. Atmos. Sci.*, **36**, 140–155.

Rotunno, R. (1980) Vorticity dynamics of a convective swirling boundary layer. *J. Fluid Mech.*, **97**, 623–640.

Rotunno, R. (1984) An investigation of a three-dimensional asymmetric vortex. *J. Atmos. Sci.*, **41**, 283–298.

Rotunno, R. (1986) Tornadoes and tornadogenesis. In: P. Ray (Ed.), *Mesoscale Meteorology and Forecasting*, American Meteorological Society, Boston, pp. 414–436.

Rotunno, R. (2012) The fluid dynamics of tornadoes. *Ann. Rev. Fluid Mech.* (in review).

Rotunno, R. and J. B. Klemp (1985) On the rotation and propagation of simulated supercell thunderstorms. *J. Atmos. Sci.*, **42**, 271–292.

Shabbott, C. J. and P. M. Markowski (2006) Surface in situ observations within the outflow of forward-flank downdrafts of supercell thunderstorms. *Mon. Wea. Rev.*, **134**, 1422–1441.

Shapiro, A. (2001) A centrifugal wave solution of the Euler and Navier–Stokes equations. *Z. angew. Math. Phys.*, **52**, 913–923.

Sinclair, P. C. (1969) General characteristics of dust devils. *J. Appl. Meteor.*, **8**, 32–45.

Sinclair, P. C. (1973) The lower structure of dust devils. *J. Atmos. Sci.*, **30**, 1599–1619.

Smith, R. L. and D. W. Holmes (1961) Use of Doppler radar in meteorological observations. *Mon. Wea. Rev.*, **89**, 1–7.

Smith, R. K. and L. M. Leslie (1978) Tornadogenesis. *Quart. J. Roy. Meteor. Soc.*, **104**, 189–199.

Snellman, L. W. (1982) Impact of AFOS on operational forecasting. Preprints, *Ninth Conference on Weather Forecasting and Analysis, Seattle, WA.*, American Meteorological Society, Boston, pp. 13–16.

Snow, J. T. (1982) A review of recent advances in tornado vortex dynamics. *Rev. Geophys. Space Phys.*, **20**, 953–964.

Snow, J. T. (1984) On the formation of particle sheaths in columnar vortices. *J. Atmos. Sci.*, **41**, 2477–2491.

Snow, J. T. and D. E. Lund (1997) Considerations in exploring laboratory tornadolike vortices with a laser Doppler velocimeter. *J. Atmos. Oceanic Technol.*, **14**, 412–426.

Snow, J. T. and R. L. Pauley (1984) On the thermodynamic method for estimating maximum tornado windspeeds. *J. Clim. Appl. Meteor.*, **23**, 1465–1468.

Snow, J. T., A. L. Wyatt, A. K. McCarthy, and E. K. Bishop (1995) Fallout of debris from tornadic thunderstorms: An historical perspective and two examples from VORTEX. *Bull. Amer. Meteor. Soc.*, **76**, 1777–1790.

Spratt, S. M., D. W. Sharp, P. Welsh, A. Sandrik, F. Alsheimer, and C. Paxton (1997) A WSR-88D assessment of tropical cyclone outer rainband tornadoes. *Wea. Forecasting*, **12**, 479–501.

Starr, V. P. (1968) *Physics of Negative Viscosity Phenomena*, McGraw-Hill, New York, 256 pp.

Sullivan, R. D. (1959) A two-cell vortex solution of the Navier–Stokes equations. *J. Aerospace Sci.*, **26**, 767–768.

Tanamachi, R. L., H. B. Bluestein, S. S. Moore, and R. P. Madding (2006) Infrared thermal imagery of cloud base in tornadic supercells. *J. Atmos. and Oceanic Tech.*, **23**, 1445–1461.

Tanamachi, R. L., H. B. Bluestein, W.-C. Lee, M. Bell, and A. Pazmany (2007) Ground-based velocity track display (GBVTD) analysis of W-band Doppler radar data in a tornado near Stockton, Kansas on 15 May 1999. *Mon. Wea. Rev.*, **135**, 783–800.

Tanamachi, R. L., H. B. Bluestein, J. B. Houser, S. J. Frasier, and K. M. Hardwick (2012) Mobile, X-band, polarimetric Doppler radar observations of the 4 May 2007 Greensburg, Kansas tornadic supercell. *Mon. Wea. Rev.*, **140**, 2103–2125.

Trapp, R. J. (2000) A clarification of vortex breakdown and tornadogenesis. *Mon. Wea. Rev.*, **128**, 888–895.

Trapp, R. J. and R. Davies-Jones (1997) Tornadogenesis with and without a dynamic pipe effect. *J. Atmos. Sci.*, **54**, 113–133.

Trapp, R. J. and B. H. Fiedler (1995) Tornado-like vortexgenesis in a simplified numerical model. *J. Atmos. Sci.*, **52**, 3757–3778.

Trapp, R. J., E. D. Mitchell, G. A. Tipton, D. W. Effertz, A. I. Watson, D. L. Andra, Jr., and M. A. Magsig (1999) Descending and nondescending tornadic vortex signatures detected by WSR-88Ds. *Wea. Forecasting*, **14**, 625–639.

Trapp, R. J., S. A. Tessendorf, E. S. Godfrey, and H. E. Brooks (2005) Tornadoes from squall lines and bow echoes, Part I: Climatological distribution. *Wea. Forecasting*, **20**, 23–34.

Wakimoto, R. M. and N. T. Atkins (1996) Observations on the origin of rotation: The Newcastle tornado during VORTEX 94. *Mon. Wea. Rev.*, **124**, 384–407.

Wakimoto, R. M. and C. Liu (1998) The Garden City, Kansas, storm during VORTEX 95, Part II: The wall cloud and tornado. *Mon. Wea. Rev.*, **126**, 393–408.

Wakimoto, R. M. and B. Martner (1992) Observations of a Colorado tornado, Part II: Combined photogrammetric and Doppler radar analysis. *Mon. Wea. Rev.*, **120**, 522–543.

Wakimoto, R. M. and J. W. Wilson (1989) Non-supercell tornadoes. *Mon. Wea. Rev.*, **117**, 1113–1140.

Wakimoto, R. M., N. T. Atkins, and J. Wurman (2011) The LaGrange tornado during VORTEX2, Part I: Photogrammetric analysis of the tornado combined with single-Doppler radar data. *Mon. Wea. Rev.*, **139**, 2233–2258.

Walko, R. L. (1988) Plausibility of substantial dry adiabatic subsidence in a tornado core. *J. Atmos. Sci.*, **45**, 2251–2267.

Walko, R. L. (1993) Tornado spin-up beneath a convective cell: Required basic structure of the near-field boundary layer winds. In: R. Church (Ed.), *The Tornado: Its Structure, Dynamics, Prediction, and Hazards*, American Geophysical Union, Washington, D.C., pp. 89–95.

Ward, N. B. (1972) The exploration of certain features of tornado dynamics using a laboratory model. *J. Atmos. Sci.*, **29**, 1194–1204.

Wegener, A. (1917) *Wind- und Wasserhosen in Europa (Tornadoes in Europe)*, Verlag Friedrich Vieweg & Sohn, Braunschweig, Germany, 301 pp. (*http://www.essl.org/pdf/Wegener1917/*) [in German].

Wicker, L. J. (1996) The role of near surface wind shear on low-level mesocyclone generation and tornadoes. Preprints, *18th Conference on Severe Local Storms, San Francisco, CA*, American Meteorological Society, Boston, pp. 115–119.

Wicker, L. J. and R. B. Wilhelmson (1995) Simulation and analysis of tornado development and decay within a three-dimensional supercell thunderstorm. *J. Atmos. Sci.*, **52**, 2675–2703.

Wilson, T. and R. Rotunno (1986) Numerical simulation of a laminar end-wall vortex and boundary layer. *Phys. Fluids*, **29**, 3993–4005.

Wurman, J. (2002) The multiple-vortex structure of a tornado. *Wea. Forecasting*, **17**, 473–505.

Wurman, J. and C. R. Alexander (2005) The 30 May 1998 Spencer, South Dakota, Storm, Part II: Comparison of observed damage and radar-derived winds in the tornadoes. *Mon. Wea. Rev.*, **133**, 97–119.

Wurman, J. and S. Gill (2000) Finescale radar observations of the Dimmitt, Texas (2 June 1995), tornado. *Mon. Wea. Rev.*, **128**, 2135–2164.

Wurman, J., C. Alexander, P. Robinson, and Y. Richardson (2007a) Low-level winds in tornadoes and potential catastrophic tornado impacts in urban areas. *Bull. Amer. Meteor. Soc.*, **88**, 31–46.

Wurman, J., Y. Richardson, C. Alexander, S. Weygandt, and P. F. Zhang (2007b) Dual-Doppler and single-Doppler analysis of a tornadic storm undergoing mergers and repeated tornadogenesis. *Mon. Wea. Rev.*, **135**, 736–758.

Wurman, J., Y. Richardson, C. Alexander, S. Weygandt, and P. F. Zhang (2007c) Dual-Doppler analysis of winds and vorticity budget terms near a tornado. *Mon. Wea. Rev.*, **135**, 2392–2405.

Xia, J., W. S. Lewellen, and D. C. Lewellen (2003) Influence of Mach number on tornado corner flow dynamics. *J. Atmos. Sci.*, **60**, 2820–2825.

Ziegler, C. L., E. N. Rasmussen, T. R. Shepherd, A. I. Watson, and J. M. Straka (2001) Evolution of low-level rotation in the 29 May 1994 Newcastle–Graham, Texas, storm complex during VORTEX. *Mon. Wea. Rev.*, **129**, 1339–1368.

7

Forecasting and future work

"An intelligence which, for a given instant, could know all the forces by which nature is animated, and the respective situation of the beings who compose it, if, moreover, it was sufficiently vast to submit these data to analysis, if it could embrace in the same formula the movements of the greatest bodies in the universe as well as those of the lightest atom—nothing would be uncertain for it, and the future, like the past, would be present to its eyes."

Pierre-Simon Laplace—*A Philosophical Essay on Probabilities*

7.1 SHORT-RANGE FORECASTING

7.1.1 Ingredients-based forecasting

In the early days of severe weather forecasting in the U. S. in the 1950s and 1960s, synoptic conditions associated with severe convection in the Great Plains and to the east were identified. For example, it was noted that a strong low-level southerly jet transporting moisture northward from the Gulf of Mexico surmounted by a more westerly jet aloft were synoptic conditions that seemed to permit the development of severe convective storms. This forecasting technique is one of pattern recognition based on synoptic features. A good example of pattern recognition forecasting is the use of the idealized synoptic patterns ("Types A–F") discussed in the widely used Technical Report 200 for the U. S. Air Weather Service, by Col. Robert C. Miller, in 1972 and revised in 1975. Since local climatology, orography, and topography vary significantly across the globe, it is not possible to formulate a general set of "forecasting rules" that will work everywhere: For example, an easterly wind on the high plains of Colorado, which is indicative of low-level upslope flow, transporting relatively high-dewpoint air westward, reducing CIN, and increasing vertical shear when the winds aloft have a westerly component, is an indicator of possible severe weather. In eastern Massachusetts, however, an easterly wind transports relatively cool air off the Atlantic inland, decreasing CAPE and increasing CIN, although vertical shear might be increased if the winds aloft have a westerly component. The forecasting rule that upslope, easterly flow in eastern Colorado is often associated with severe weather does not work in eastern Massachusetts. On the other hand, *the basic physics and thermodynamics of severe convection are independent of location.*

Forecasters of severe convection therefore often use what is called an "ingredients-based" methodology. The "ingredients" are the necessary, but not sufficient, physical conditions for the occurrence of severe convection. The ingredients are not to be confused with diagnostics, which are the measured quantities that can be used to determine if the ingredients are present.

The main ingredients for supercells are adequate moisture for storm initiation, CAPE, vertical shear, and mesoscale ascent or surface heating or both. Without adequate moisture, there will be no condensation and thus no clouds at all. The minimum moisture content needed depends upon the vertical profile of temperature; surface dewpoints of 50°F or higher are usually needed, though severe convection above elevated terrain can occur with surface dewpoints as low as the 40–49°F range. CAPE is necessary for the release of conditional instability in deep cumulus and cumulonimbus clouds. What constitutes the minimum CAPE (for air parcels originating in the well-mixed boundary layer) needed is not clear, but observational studies show that a minimum of 500–1,000 $\mathrm{j\,kg^{-1}}$ is usually required (Figure 7.1); in some locations (such as Europe or the West Coast of the U. S. during the winter) tornadoes can occur when CAPE is very low. Deep-layer shear is probably the most important requirement for supercells (Figure 7.2) when the other two ingredients for cumulus convection in general are present. Vertical shear in the lowest 6 km of at least 20 $\mathrm{m\,s^{-1}}$ is usually necessary (Figure 7.2) for typical hodographs in the Great Plains of the U. S., though some variations occur depending on the depth of the layer used and the shape of the hodograph. Some forecasters have found the product of vertical shear and CAPE or other combinations of vertical shear and CAPE to be useful. Mesoscale ascent or surface heating or both are necessary to trigger cumulus convection. One cannot set a threshold for the intensity of the ascent or the net amount of surface heating needed because they depend on the nature of the vertical profile of temperature and moisture.

It is not known precisely how to distinguish tornadic from non-tornadic supercells yet, but strong low-level shear/storm-relative environmental helicity (Figures 7.3, 7.4), and especially low-level shear normal to the shear above, is suspected as a possible ingredient. "Significant" tornadoes (those inflicting F2 or greater damage) tend to occur when the surface–1 km shear exceeds a threshold value and the mixed-layer LCL height is not too high (Figure 7.4).

The ingredients needed for non-supercell tornadoes—or large hail or strong straight-line winds at the surface—depend on the nature of the parent storm. Isolated convective cells and mesoscale convective systems can produce severe weather via different physical processes. The reader is encouraged to reread

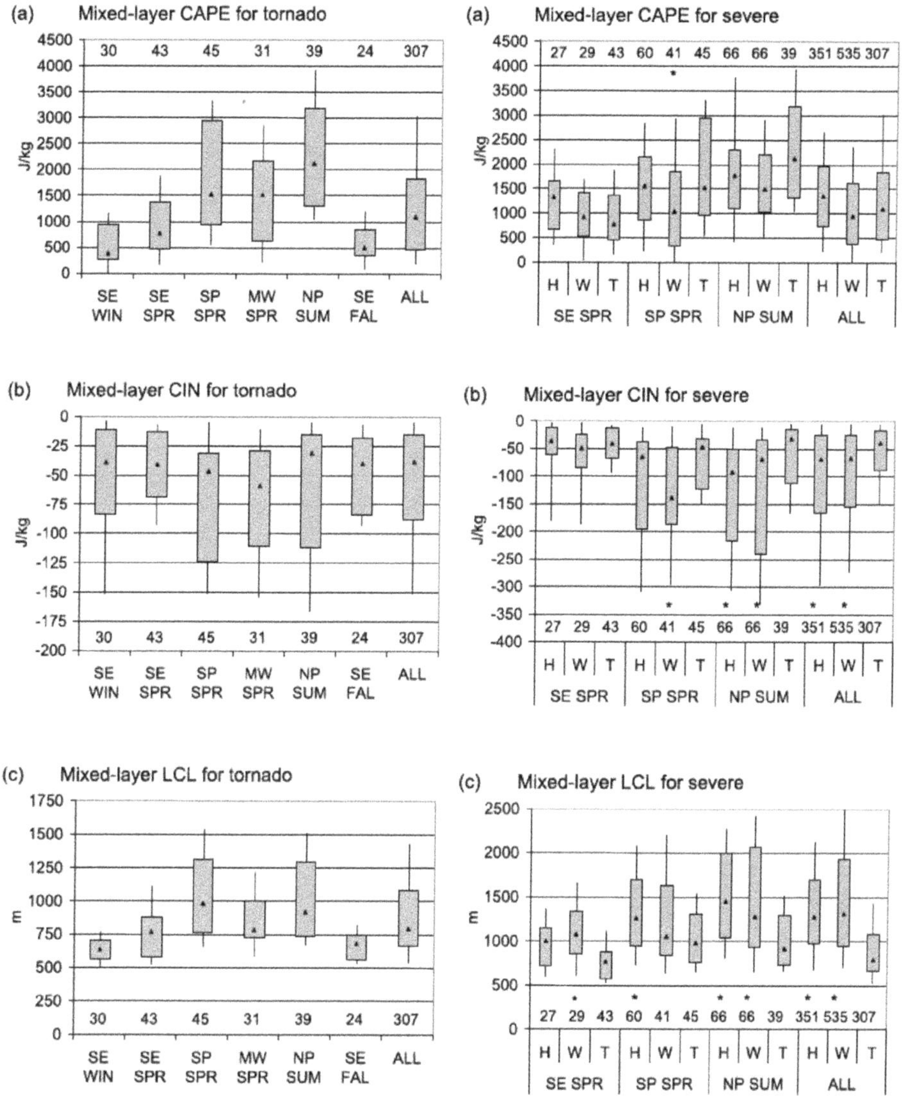

Figure 7.1. Box-and-whiskers plots of (a, left) mixed-layer CAPE for all tornadoes, regardless of whether they were in supercells or ordinary cells or MCSs, (a, right) mixed-layer CAPE for all severe events. (H), (W), and (T) refer to hail, wind, and tornado events, respectively. (b, left) mixed-layer CIN for all tornadoes, (b, right) mixed-layer CIN for all severe events, (c, left) mixed-layer LCL for all tornadoes, and (c, right) mixed-layer LCL for all severe events, in the southeast (SE), Midwest (MW), southern plains (SP), and northern plains (NP) of the U. S. during the spring (SPR), summer (SUM), and fall (FAL) and for the entire sample (ALL), for the contiguous U. S. from 2000 to 2008. The shaded box covers the 25th–75th percentiles, the whiskers extend to the 10th and 90th percentiles, and the median values are marked by a triangle within each shaded box. The number of events is provided for each category along the abscissa (from Grams *et al.*, 2012).

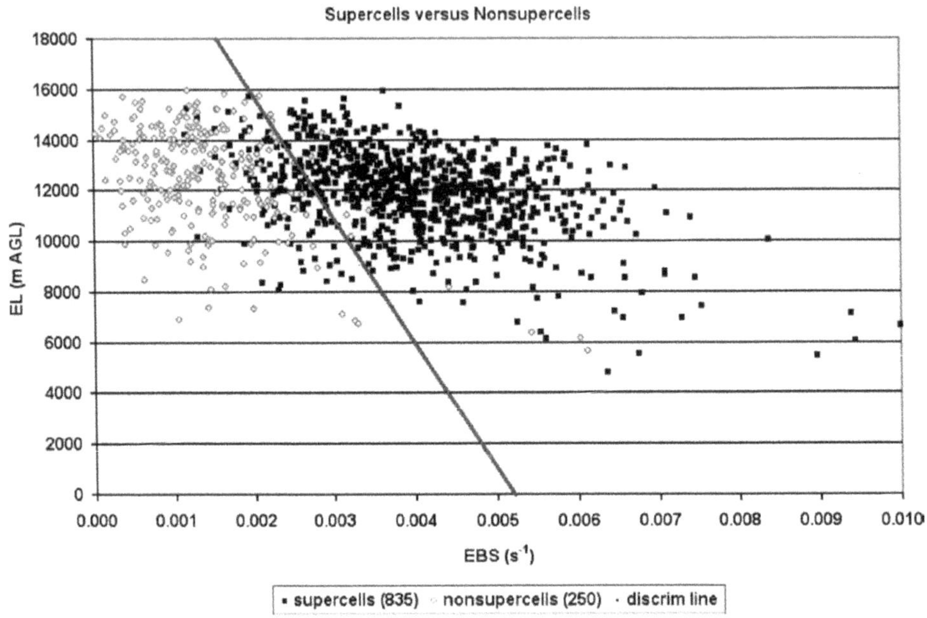

Figure 7.2. Scatterplot of equilibrium level (EL) height vs. EBS ("effective bulk shear," which is the effective bulk wind difference in the lower half of the storm, similar to bulk shear in the lowest 6 km; "bulk" refers to an incremental measure that accounts for the length and curvature of the hodograph) for supercells (solid black squares) and non-supercells (open gray squares). A linear discriminant function is plotted (solid gray line) that separates the supercells (right of the line) from the non-supercells (left of the line) in 89% of all cases. A crude measure of the discriminant would be for an average EL of approximately 12 km, the EBS separating supercells from non-supercells is approximately $0.003\,\mathrm{s}^{-1} \times 6\,\mathrm{km} = 18\,\mathrm{m\,s}^{-1}$ (from Thompson *et al.*, 2007).

Chapters 3–6, which detail the physics of these phenomena, to find out what the ingredients for them are or might be.

Forecasting on time scales of days in advance is made possible by numerical models that do not explicitly represent cumulus convection, but do represent the broad synoptic-scale and some mesoscale features that produce the ingredients needed for severe convection. The presence or absence of the ingredients are assessed through diagnostics such as the water vapor mixing ratio (e.g., in a well-mixed boundary layer), the vertical profile of temperature and dewpoint, the vertical profile of winds, and vertical velocity either represented explicitly in primitive equation models or computed from a quasi-geostrophic ω equation applied to the filtered geopotential height field or from objectively analyzed actual radiosonde data.

In addition to evaluating the parameters necessary to determine whether the necessary "ingredients" are available, pattern recognition (an example of which was given at the beginning of this section) and climatology are also used by fore-

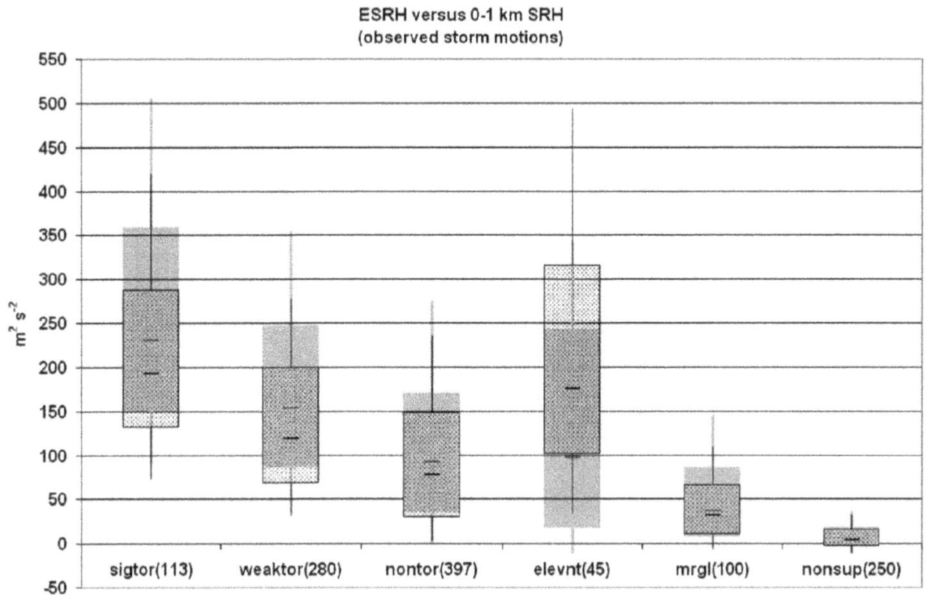

Figure 7.3. Box-and-whisker overlay plots of effective storm-relative helicity (ESRH) when CAPE $\geq 100\,\mathrm{j\,kg^{-1}}$ and CIN $\geq -250\,\mathrm{j\,kg^{-1}}$ (solid gray box and thick gray whiskers) and 0–1 km SRH (dotted black box and think black whiskers) for discrete convective storms that are significantly tornadic (sigtor), weaker tornadic (weaktor), non-tornadic (nontor), elevated non-tornadic (elevnt), marginal supercells (mrgl), and non-supercells (nonsup). Sounding sample sizes are shown in parentheses (from Thompson *et al.*, 2007).

casters. Pattern recognition makes use of often-observed parameters associated with severe convection, even when their physical relevance is not completely understood. Climatology is used to alert forecasters for spatial and temporal biases toward certain ingredients.

7.1.2 Model-based forecasting

The model-based forecasting methodology is in sharp contrast to the ingredients-based methodology. Through this method, cumulus convection is explicitly represented in non-hydrostatic cloud models run over regional domains and possibly over even larger domains. It is beyond the scope of this text to detail the numerical schemes, initialization procedures, etc. of all, or even just a few, of the models. For the latter, important issues include deciding how to assimilate Doppler radar wind data as well as wind, moisture, pressure, and temperature data from observational networks. The model needs time to "adjust" to the perturbations inflicted by incorporating new data.

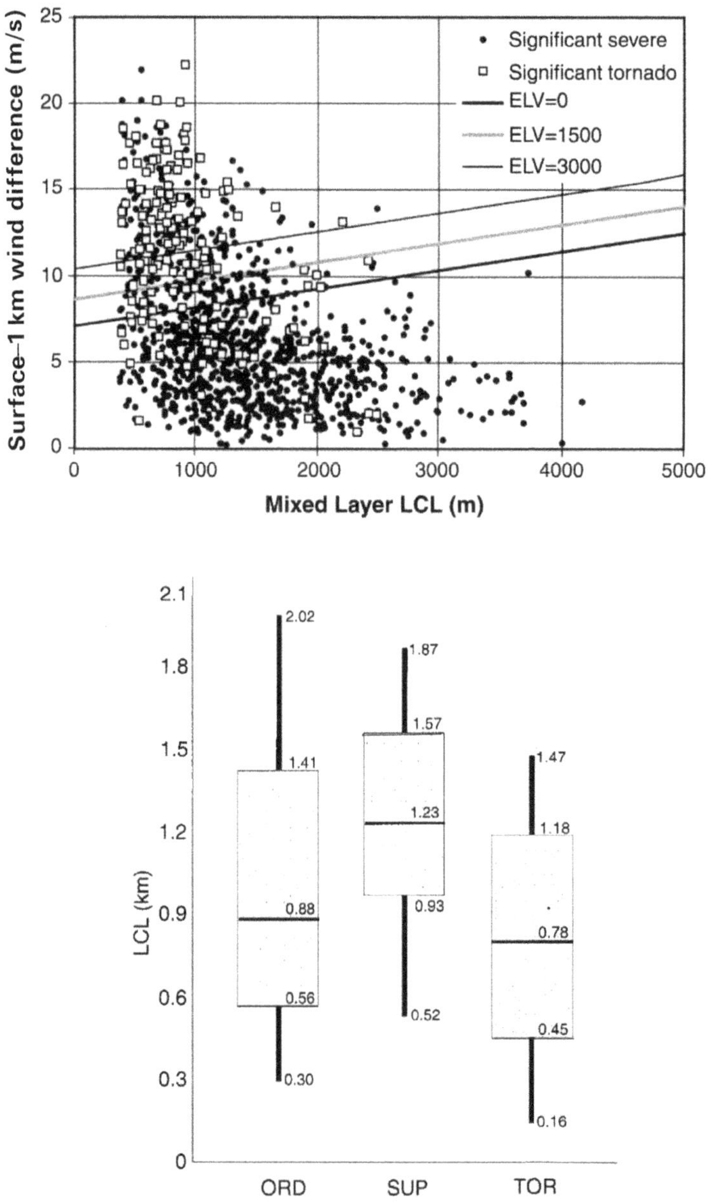

Figure 7.4. Forecast parameters for distinguishing between tornado and non-tornado convective storm environments as a function of elevation (ELV) in m MSL. Distribution of a measure of surface–1 km vertical wind shear and mixed-layer LCL (from Brooks *et al.*, 2003). (Bottom) Box-and-whisker plots of LCL for soundings associated with supercells having spawned "significant" (F2 or greater) tornadoes (TOR), supercells not having spawned significant tornadoes (SUP), and non-supercell thunderstorms not having produced any severe weather (i.e., large hail, strong straight-line winds, or tornadoes) (ORD) (from Rasmussen and Blanchard, 1998).

The excessive reliance on computer-generated model runs (at least on the large scale) without critical analysis has led to what has been called "meteorological cancer" by Len Snellman of the National Weather Service. Forecasters must not simply use computer-generated forecasts as the truth and act merely as communicators, but must critically assess how accurately they might be and recognize the uncertainty of forecasts.

Recently, owing to the relatively extreme uncertainty of explicit convection forecasts resulting from nonlinear advection and other nonlinear processes, "ensemble forecasts" are produced by running a model many times, each time for slightly different initial conditions and model microphysics, to give the forecaster a feeling for the range of possible realizations in the atmosphere. The best forecast averaged over the long run may be the ensemble mean, but a look at all the forecasts gives one a feeling for what low-probability events, but possibly of very high impact, are nevertheless possible. Exactly how to implement ensembles and how many members should be used, etc. are part science and part art.

Warnings for severe weather events are issued by the National Weather Service in the U. S. mainly on the basis of detection of severe weather by Doppler radar observations and public reports. There is, however, a movement now to attempt to issue warnings at least in part based on the output from ensembles of short-range, high-resolution, numerical model runs. This procedure is called "warn on forecast" and, if successful, has the potential to extend warning lead times to allow people to take action to lessen the effects of severe weather by allowing them to get out of harm's way and protect some structures in advance.

7.1.3 Evaluations of forecast skill

It is believed that a combination of ingredients-based and model-based forecasting is necessary for the most accurate severe convection forecasts. The human forecaster plays a vital role in assessing the diagnostic quantities to determine whether the proper ingredients are in place and in interpreting the model output and quantifying uncertainty. He/she can make decisions in which erroneous observations are discovered to have made it into the forecasting process and how much to weight the impact of model forecasts vs. observations. Variables such as precipitation totals, maximum wind speed, etc. may be verified at gridpoints and measures of error used. It is beyond the scope of this text to explore the different measures currently used. I recall once that Doug Lilly, in response to Bruce Morton, a fluid dynamicist, who noted that while a certain phenomenon might seem to be *theoretically impossible*, Lilly noted that it *is in fact* sometimes observed. When a model produces a forecast, which is refuted by simple observations, we must go with the observations.

Forecast verification for events in convective storms (e.g., of tornadoes, large hail, strong winds) may be accomplished via various ad hoc quantities such as probability of detection (POD), which is the percentage of events forecast (# forecast events/sum of # of forecast and # of unforecast events), the false alarm ratio (FAR), which is a measure of false alarms (# of unforecast events/

sum of # of both forecast and unforecast events), the critical success ratio (CSI), which is a measure of the relative number of forecast events to the sum of all the forecast, unforecast, and non-events, and the average lead time, which is the duration of time between when the forecast is issued and when the forecast event begins, if it does.

One may also compute the average value of the differences between forecast variables (precipitation amount, largest hail size, strongest wind speed). These quantities, however, may be misleading indicators of forecast skill. Suppose that a forecast is perfect, save for an error in location. In other words, the aforementioned empirical measures of forecast error would be poor, but if the forecast were shifted in space by a small distance, they could be significantly reduced. Then it is the phase of the forecast that is in error. It is somewhat of an art to devise indicators of forecast accuracy.

Ed Lorenz suggested in the 1960s that errors in numerical forecasts should grow more rapidly as the initial estimate of atmospheric variables is improved and smaller and smaller scales are resolved. The predictability of convection may not be much greater than the time it takes air parcels to enter a storm and exit it, which is the advective time scale. If so, then there could be a theoretical limit that is relatively short. Errors on very small scales are amplified by nonlinear terms and show up on resolved larger scales. On the other hand, some convective systems like supercells persist for time periods longer than the advective time scale, which could result in extended predictability. This extended predictability could be caused by a reduced effect of nonlinear advection terms. It was noted earlier, for example, how some nonlinear advection terms are small when there is low helicity.

7.2 FORECASTING AND CLIMATE CHANGE

In the last decade, when temperatures averaged globally have been near record highs (at least with respect to the last century or so), and in the aftermath of a spring with record tornado activity in the U. S. (2011), the topic of how climate change will affect the climatology of severe convective storms has become of greater interest to the public, in part as a result of enhanced coverage by the media. In particular, global climate models have been run which have included the physical process of enhanced radiative forcing due to increased greenhouse gases. Two of the ingredients associated with supercells, CAPE and vertical shear, can be measured from the very long-range forecasts from these models to determine whether these ingredients will be made more or less conducive to supercell formation. It is recognized that these climate models do not explicitly represent convection and that many processes are parameterized using methods that have varying degrees of accuracy, some not very accurate.

It is thought that low-level moisture will increase as low-level temperature increases, which, if the temperature aloft does not warm appreciably, will result in greater CAPE. However, it is also thought that the overall pole-to-equator temperature gradient will decrease which, according to the thermal wind relation, will lead to a decrease in deep-layer vertical shear. Since supercells are probably more sensitive to the amount of vertical shear in the atmosphere than CAPE, it might be that the likelihood of severe weather associated with supercells will decrease, while the likelihood of heavy rain from mesoscale convective systems will increase. The models, however, do not tell us anything about the change, if any, in the likelihood that convective storms will actually be triggered and it is possible that there could be fewer storms even if CAPE is higher. Furthermore, it is possible that future distributions of CAPE and shear could have significant regional differences. Moreover, while the long-range mean values of CAPE and shear may change, it is possible that there could be increased variability, such that there could be more instances when CAPE and shear deviate significantly from the long-term mean: there could be major outbreaks in spite of an unfavorable environment in the mean.

Using a technique known as "dynamical downscaling", Jeff Trapp at Purdue University and collaborators have undertaken an exercise in which a large-scale model is run for short time periods and then a non-hydrostatic, fine-grid cloud model is run to produce forecasts given larger scale forecasts as initial conditions. Using such a procedure, one can assess regional changes in severe convection. Currently, efforts have been undertaken to look at non-hydrostatic model runs based on decades of reanalysis data to find statistics on the relation between actual severe weather events and model output.

7.3 FUTURE RESEARCH

There are several areas of research that are evolving now and several that are just emerging. The current areas of evolving research include advancing our ability to probe severe convective storms by remote sensing with Doppler radars, especially those dealing with rapid-scan, polarimetric radar technology. The development of electronic scanning using phased array technology, imaging radar technology, and spaced antenna technology, should have the potential for improving our ability to map the wind and hydrometeor fields in severe convective storms and tornadoes on short time scales with high spatial resolution. If these advances could be adapted for high-frequency radars, then tornadoes could be much better observed when attenuation is not too great. In the far future, it would be advantageous to mount these radars, when made small and lightweight, on mobile, airborne platforms, such as helicopters in particular, which can hover safely near tornadoes and not be restricted to road networks necessary for ground-based vehicles, and do not require longer update times as do aircraft, which must fly by at some minimum speed necessary to keep the aircraft airborne. Developing airborne radars likely requires solid state, low-power transmitters, which will have to make use of pulse compression techniques to enhance their sensitivity. NASA has already begun this area of technology development with their radar on the Global Hawk UAV, and Gerry Heymsfield and colleagues have used it to collect data in hurricanes over the ocean. Research groups at OU (Oklahoma University), NCAR, U. Mass., and Colorado State University, among others are currently pursuing advances in radar technology.

Pulsed Doppler lidars, which can estimate winds in clear air from aerosol movements or molecular movements, should be improved so that their range resolution can be decreased to be closer to their azimuthal resolution, which is currently only ~20 cm; range resolution now is ~50 m, but should be reduced to ~10 m or less. Pulsed Doppler lidars can complement Doppler radars, which work best when there are hydrometeors or insects (in clear air), but lidars cannot penetrate very far into precipitation or clouds.

In situ probes will be improved to increase the chances of successful measurements. With the upcoming armored A-10 storm-penetrating aircraft in the U. S., we should be able to obtain in situ measurements of hydrometeor type, thermodynamic measurements, and vertical velocity measurements in severe convective storms. Robotic helicopters could be used to make thermodynamic measurements in and around convective storms, but not in regions of large hail or high winds. An advantage of robotic helicopters over UAVs is that they can hover: they do not need to keep moving, like sharks, to "stay alive" in the air.

Swarms of small, lightweight, expendable probes that drift with the wind, in part enabled by nanotechnology, may be used to provide detailed thermodynamic measurements and wind measurements where radars or lidars are not appropriate. These probes, an example of which were first introduced in the 1996 movie *Twister* as science fiction, might someday be realized as science fact. They could be released by airborne platforms such as manned aircraft, UAVs, or robotic helicopters.

As computers evolve to be faster and contain even more memory, we will be able to simulate both tornadoes and their parent convective storms with sufficient spatial resolution to resolve both. Better representation of microphysical processes will also be possible. Improved short-term forecasts should be possible at very high spatial resolution, with the advent of improved cloud microphysics, and can be made for many ensemble members.

On the theoretical side, and probably in conjunction with faster computers, it may be possible to improve our understanding of boundary-layer physics and to analyze tornadogenesis as an instability problem. There are currently many researchers involved in improving radar technology, many making use of new instrumentation, and many doing numerical experiments and assimilating data into numerical models but, arguably, fewer doing more basic theoretical work.

7.4 GENERAL MONOGRAPHS AND BOOKS

The reader is referred to p. 24 for a list of relevant general monographs and books.

7.5 REFERENCES AND BIBLIOGRAPHY

Brooks, H. E., C. A. Doswell, III, and R. A. Maddox (1992) On the use of mesoscale and cloud-scale models in operational forecasting. *Wea. Forecasting*, **7**, 120–132.

Brooks, H. E., C. A. Doswell, III, and M. P. Kay (2003) Climatological estimates of local daily tornado probability for the United States. *Wea. Forecasting*, **18**, 626–640.

Brooks, H. E., J. W. Lee, and J. P. Craven (2003) The spatial distribution of severe thunderstorm and tornado environments from global reanalysis data. *Atmos. Res.*, **67/68**, 73–94.

Doswell, C. A., III (1987) The distinction between large-scale and mesoscale contribution to severe convection: A case study example. *Wea. Forecasting*, **2**, 3–16.

Grams, J. S., R. L. Thompson, D. V. Snively, J. A. Prentice, G. M. Hodges, and L. J. Reames (2012) A climatology and comparison of parameters for significant tornado events in the United States. *Wea. Forecasting*, **27**, 106–123.

Heymsfield, G. M., S. Guimond, L. Tian, L. Li, M. McLinden, M. Perrine, and A. E. Reynolds (2012) Results and lessons learned from the first flights of the HIWRAP on the Global Hawk during GRIP. *30th Conference on Hurricanes and Tropical Meteorology, Ponte Vedra Beach, FL*, American Meteorological Society, Boston, 12A.2.

Houston, A. L., R. L. Thompson, and R. Edwards (2008) The optimal bulk wind differential depth and the utility of the upper-tropospheric storm-relative flow for forecasting supercells. *Wea. Forecasting*, **23**, 825–837.

Johns, R. H. and C. A. Doswell, III (1992) Severe local storms forecasting. *Wea. Forecasting*, **7**, 588–612.

Lorenz, E. N. (1969) The predictability of a flow which possesses many scales of motion. *Tellus*, **21**, 289–307.

Miller, R. C. ([1972], 1975) *Notes on Analysis and Severe-Storm Forecasting Procedures of the Air Force Global Weather Central*, Tech. Report 200 (Revised), Air Weather Service, U. S. Air Force, Offutt Air Force Base, NE, 190 pp.

Moller, A. R. (2001) Severe local storms forecasting. In: C. Doswell, III (Ed.), *Severe Convective Storms*, AMS Monogr. 28, no. 50, American Meteorological Society, Boston, pp. 433–480.

Rasmussen, E. N. and D. O. Blanchard (1998) A baseline climatology of sounding-derived supercell and tornado forecast parameters. *Wea. Forecasting*, **13**, 1148–1164.

Rosenfeld, D. and T. L. Bell (2011) Why do tornados and hailstorms rest on weekends? *J. Geophys. Res.*, **116**, D20211, 14 pp.

Snellman, L. W. (1977) Operational forecasting using automated guidance. *Bull. Amer. Meteor. Soc.*, **58**, 1036–1044.

Stensrud, D. J., J. V. Cortinas, Jr., and H. E. Brooks (1997) Discriminating between tornadic and nontornadic thunderstorms using mesoscale model output. *Wea. Forecasting*, **12**, 613–632.

Stensrud, D. J., M. Xue, L. J. Wicker, K. E. Kelleher, M. P. Foster, J. T. Schaefer, R. S. Schneider, S. G. Benjamin, S. S. Weygandt, J. T. Ferree *et al.* (2009) Convective-scale warn on forecast: A vision for 2020. *Bull. Amer. Meteor. Soc.*, **90**, 1487–1499.

Thompson, R. L., C. M. Mead, and R. Edwards (2007) Effective storm-relative helicity and bulk shear in supercell thunderstorm environments. *Wea. Forecasting*, **22**, 102–115.

Trapp, J. R., N. S. Diffenbaugh, H. E. Brooks, M. F. Baldwin, E. R. Baldwin, E. D. Robinson, and J. S. Pal (2007) Changes in severe thunderstorm environment frequency during the 21st century caused by anthropogenically enhanced global radiative forcing. *Proc. National Acad. of Sciences*, **104**, 19719–19723.

Zhang, F., C. Snyder, and R. Rotunno (2003) Effects of moist convection on mesoscale predictability. *J. Atmos. Sci.*, **60**, 1173–1185.

Appendix

Doppler radar analysis techniques

Dual- (or multiple) Doppler analysis is possible when two or more Doppler radars scan the same volume at the same time from two different viewing angles. The equations relating the Doppler velocities—V_{r_i}, where i refers to the first (1), second (2), ..., nth (n) radar—to the three wind components (u, v, and w) in, for example, Cartesian coordinates are, as originally formulated by Larry Armijo in the late 1960s, are as follows (Figure A.1):

$$V_{r_1} = 1/R_1[xu + yv + z(w + W_t)] \tag{A.1}$$

$$V_{r_2} = 1/R_2[(x - x_2)u + yv + z(w + W_t)] \tag{A.2}$$

$$\vdots$$

$$V_{r_n} = 1/R_n[(x - x_n)u + (y - y_n)v + z(w + W_t)] \tag{A.3}$$

where the first radar is at the origin, the second is at $(x_2, 0)$, and the nth is at (x_n, y_n); R_n is the range of the nth radar from the target volume; and W_t is the terminal fall speed of scatterers in the volume. It is assumed that targets move along with the wind, but also have a component of terminal fall velocity. In the case of sharply curved flow and relatively massive hydrometeors or other targets, the centrifugal force may also make targets deviate from air motion.

Much attention must be given to how one interpolates radar data to grids not native to the data collection mode (most data from fixed radars are collected in tilted planes in cylindrical coordinates), such as Cartesian coordinates (e.g., as above). Gaussian weighting (or interpolating) functions are commonly used, though in some instances others may be more appropriate (e.g., polynomial functions and functions whose radii of influence are dependent upon direction). Specifying the proper spatial filter (e.g, via parameters in the Gaussian weighting function) is necessary to extract the maximum detail from the data, given the

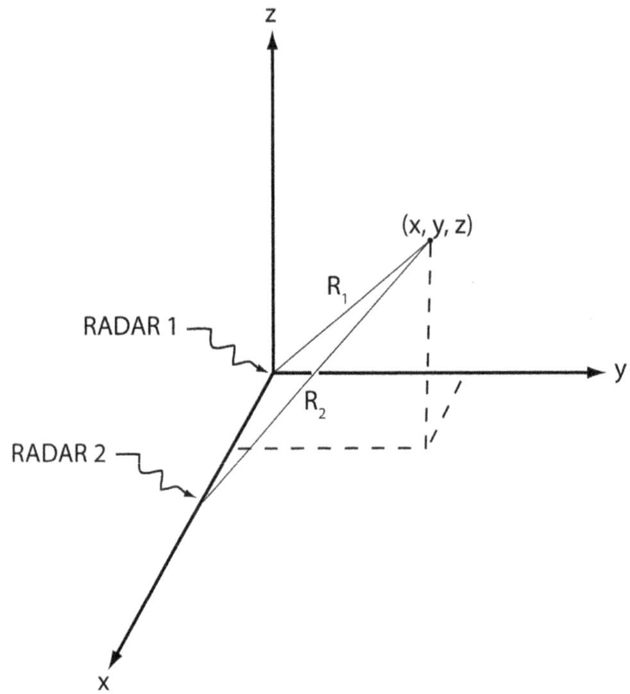

Figure A.1. Geometry relating the positions of two radars $(0, 0, 0)$ and $(x_1, 0, 0)$ and a point in Cartesian space (x, y, z) to the distances from each radar to the target (R_1 and R_2), the three components of the wind (u, v, w), and the terminal fall velocity (W_t).

spatial sampling rate and the scale of the phenomenon one is trying to resolve. Multiple passes (successive corrections of the analysis interpolated to the data points) of the objective analysis scheme steepens the spatial frequency response of the spatial filter so that there is less damping at resolved spatial scales, but the same amount of suppression of high-frequency noise. It is somewhat of an art to determine precisely how to adapt an "objective analysis" technique to a given set of radar data collected.

Horizontal velocity error variances in the x and y-directions σ_u and σ_v at low-elevation angles are related to error variances of the mean Doppler velocity from each radar σ_1 and σ_2 by the following:

$$(\sigma_u^2 + \sigma_v^2)/(\sigma_1^2 + \sigma_2^2) = \csc^2 \beta \tag{A.4}$$

where β is the between-beam angle and the Doppler velocities from each radar are uncorrelated with each other. Based on experience, for dual-Doppler analyses of acceptable quality, the between-beam angles from adjacent radars (Figure A.2) should be at least 30°, but less than 150°, though studies have been conducted with slightly narrower or wider between-beam angles, respectively, with some success. It is apparent that if the between-beam angle is 180°, there is no informa-

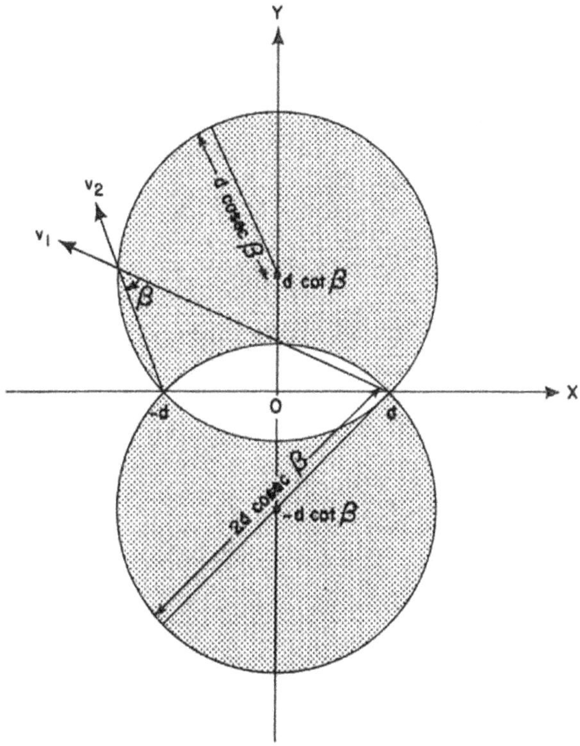

Figure A.2. Area (stippling) for which the between-beam angle β lies between β and $180° - \beta$. In this figure the two radars are located at $x = \pm d$ and $y = 0$ (not at $x = 0$ and $x = x_2$, and $y = 0$ as in Figure A.1) (from Davies-Jones, 1979).

tion about the component of the wind normal to the "baseline", which is defined by the line connecting the locations of two adjacent radars. Similarly, if the two beams are nearly parallel to each other, so that the between-beam angle approaches 0°, there is also no information about the component of the wind normal to the radar beams. The optimum situation is one in which the radar volume is positioned such that the between-beam angle is 90°. For targets very close to the baseline, the between-beam angle approaches 180°; for targets far from and normal to the baseline, the between-beam angle approaches 0°. It is clear that the target must be located in certain restricted areas with respect to the radars. For fixed site radars, one can only hope that a target will pass through the optimum region. The length of the baseline is chosen so that target scatterers lie within the acceptable between-beam angle to minimize errors and to ensure that the spatial resolution of the radar volume is sufficient to map out the wind field for spatial scales that must be resolved.

During field operations using mobile radars, one can position the radars so that the target will pass through the optimum region, but often this cannot be done when the road network is inadequate. Airborne radars do not have this restriction, but they cannot fly too close to storms for safety, and do not sense motions close to the ground well, owing to ground clutter contamination. In addition, it takes a relatively long time (~5 min) for an aircraft to fly by a storm. Unless two aircraft are available to do simultaneous scans, one aircraft typically passes by a storm and the radar antenna scans alternately in the fore and aft directions at an angle to the aircraft (Figure A.3). Since overlapping beams are not valid for the exact time, the analysis of the wind field from Doppler radar wind measurements is called "pseudo dual-Doppler analysis.

For a network of just two Doppler radars, since we have two equations (in terms of two measured quantities V_{r_1} and V_{r_2}) in four unknown variables (u, v, w, W_t), we need two additional, independent equations. Typically, an equation of continuity is used in conjunction with a kinematic lower-boundary condition. The equation of continuity that is usually used is the incompressible form (relates spatial derivatives of u, v, and w with respect to each other—the Boussinesq form) or one modified to include vertical variations in density (the anelastic form); these equations are discussed in Chapter 2.

To complicate matters further, the radar measures the Doppler velocity of scatterers such as raindrops and hailstones, which have terminal fall speeds relative to the air. It is customary to approximate fall speeds based on a formula that relates reflectivity to fall speed. Such a method is expected to work better at longer wavelengths such as S-band (10 cm) than at shorter wavelengths such as C-band (5 cm) or X-band (3 cm), because it is more likely at shorter wavelength that the hydrometeors detected will be in the Mie range, rather than in the Rayleigh range. For example, at X-band, large hailstones are well into the Mie range (for hailstones 2–3 cm or larger in diameter), and resonance effects lead to a non-monotonic relationship between the radar reflectivity factor and hail size (Figure A.4). In the future, estimates of fall speed may be refined using polarimetric measurements and fuzzy logic.

For a network of three Doppler radars, there are three equations related to network geometry, in four unknowns. For a network of four Doppler radars, there are four equations related to geometry, in four unknowns, so in principle one could solve for u, v, w, and W_t exactly without resort to any other equations such as continuity or vorticity. In practice, however, the more radar members there are in a network, the less likely it is that a storm will be positioned optimally. Moreover, no matter how many radars there are, none of them will resolve much of vertical velocity, which is nearly perpendicular to all beams except for mid to high levels when the convective storm is very close to the radars (i.e., within ~10–15 km).

To solve for u, v, and w in practice, in a dual-Doppler network, one typically iteratively finds an exact solution to (A.1), (A.2), and the continuity equation, or uses variational analysis to find the best "fit" to the observations, subject to boundary conditions (at the surface and possibly aloft) and "weak" constraints such as continuity or a vorticity equation or both. To find an exact solution, one can make a first guess, for example, by neglecting w and making use of a reflectivity-based relationship for terminal fall velocity W_t, and solving only (A.1) and

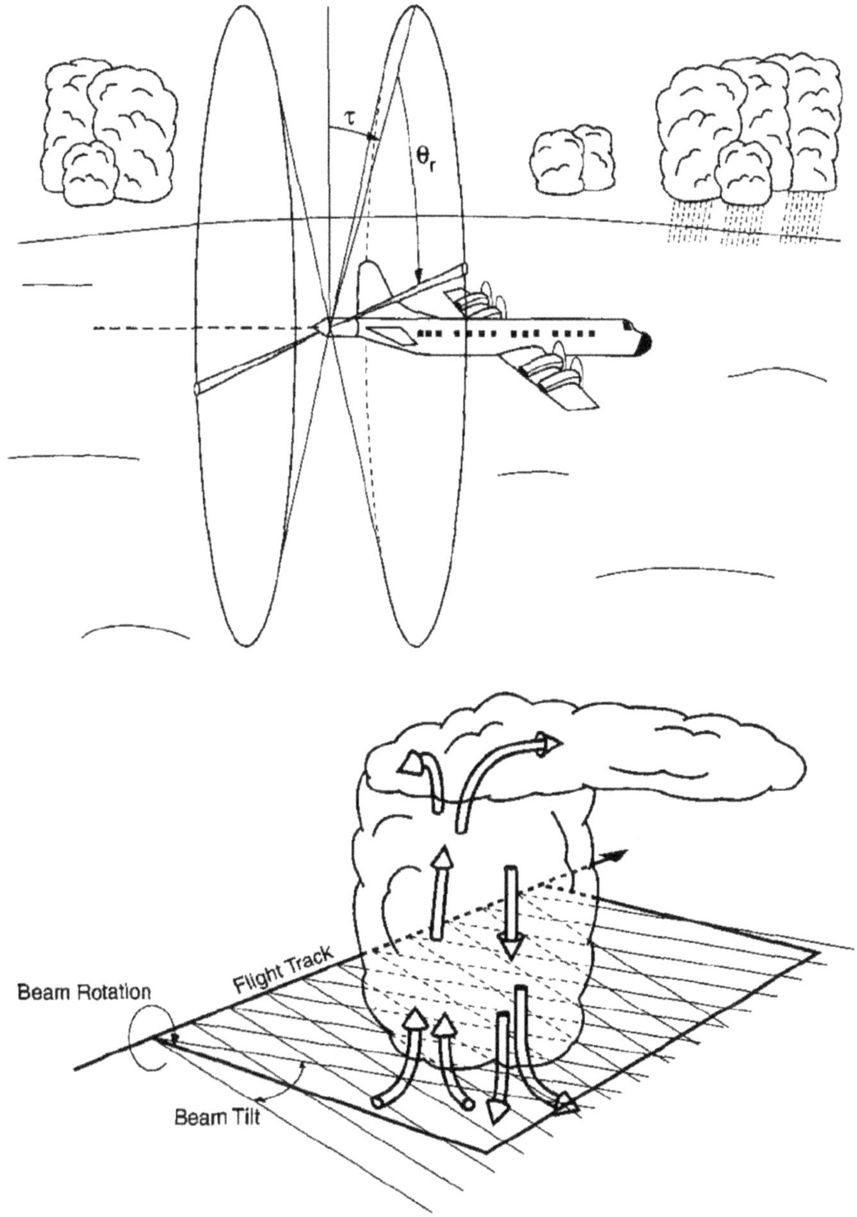

Figure A.3. FAST: scanning technique used by the airborne radar ELDORA and illustration of the pseudo dual-Doppler synthesis of the wind field for beams that trace out cones alternately in the fore and aft directions. (Top) Scanning illustrated at a fixed time; the antennas and radome rotate as a unit about an axis parallel to the longitudinal axis of the aircraft. The fore and aft angles are $\pm\tau$ and the angle measured in the clockwise direction about the longitudinal axis of the aircraft is θ_r. (Bottom) Idealized example of beam intersections in a flight by the convective storm (from Hildebrand *et al.*, 1996).

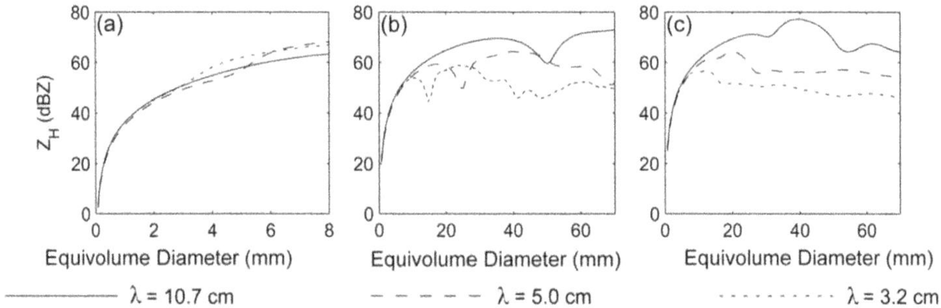

Figure A.4. Comparison of the variation of horizontally polarized radar reflectivity factor (Z_H) for hydrometeor type and size at the S-band (solid line), C-band (long-dashed line), and X-band (short-dashed line): (a) for rain; (b) for "dry hail" with 0% fractional water; (c) for "wet hail" with 10% fractional liquid water content. Z_H was calculated for rain at 10°C, a mean canting angle and a canting angle standard deviation of 0°, and a liquid water content of $10\,\mathrm{g\,m^{-3}}$; for dry hail with ice content of $2\,\mathrm{g\,m^{-3}}$, 0° mean canting angle, and 60° canting angle standard deviation; for wet hail with conditions the same as those for dry hail, but with a canting angle standard deviation of 55.2°. Note how, for both dry and wet hail, the relationship between Z_H and size is not monotonic for equivolume diameters in excess of 10–15 mm, well into the Mie scattering range especially at the X and C-bands (from Snyder *et al.*, 2010).

(A.2) for u and v. Vertical velocity w is then computed kinematically using a lower-boundary condition (such as $w = 0$ at $z = 0$) and then u and v are solved again using (A.1) and (A.2), and so on until the solutions for u, v, and w converge. (It is also possible to impart an upper-boundary condition also, if data are available high enough up in the storm so that one can assume, for example, that $w = 0$ at the top; if so, then an adjustment scheme is employed to redistribute the error in horizontal divergence with height. Or, for airborne radars, one may integrate downward.)

The variational method, pioneered by Yoshi Sasaki in the 1960s, makes use of a cost function J, which is minimized. The cost function is arbitrary and may be composed of a number of components, so that, for example

$$J = J_1 + J_2 + \cdots + J_n \tag{A.5}$$

The first cost function J_1 represents the square of the difference between interpolated Doppler (radial) wind speeds and actual, measured Doppler wind speeds at each radar at each gridpoint, weighted by the square of the distance from the radar to account for beam spreading; the second cost function is the square of the continuity equation set to zero, which acts as a "weak" continuity constraint on the three-dimensional wind field (i.e., the wind field that minimizes the residual left over in the least square sense when applying the continuity equation, so that the equation of continuity is not satisfied exactly; it is therefore termed a "weak" constraint); other cost functions may also be used, such as the square of the time-dependent vorticity equation set to zero, which acts as a weak constraint and the square of the first or second gradient of the three-dimensional wind field components, the latter representing the smoothness of the solution. The cost functions are minimized by setting the first derivatives of the total cost function to zero and then solving for u, v, and w. When the wind components are estimated from observational wind data only (e.g., radar data), with no use of a numerical model, the analysis technique is broadly referred to as 3DVAR, or three-dimensional variational analysis.

There are many sources of error in dual/multiple Doppler analysis, especially in the vertical component of the wind (a component of which is only partly resolved except at close range and high altitude), and in any other horizontal wind component for which the radar beam measures only a small component. In addition to instrument error and problems with the non-uniformity of scatterers and air motions within radar volumes, measurements made at low altitude, where boundary-layer variability is likely to be great, are difficult owing to ground clutter; thus, use of the continuity equation near the surface may suffer from a lack of good observations. Perhaps just as significant is that it is very unlikely that radar volumes are sampled at exactly the same time, and it is unlikely that radar volumes are of exactly the same volume in space or same size. To account for the non-simultaneity of radar observations horizontal advection schemes have been used, but if the wind field is rapidly evolving there are errors introduced due to time evolution. The use of rapid-scan (by "rapid-scan" we mean that the volumetric update time is small compared with the advective time scale) radars can reduce errors due to advection and evolution. Dual/multiple Doppler analyses also require data to be interpolated to a grid, thus introducing some filtering and degradation of the intrinsic spatial resolution of the radar. As noted earlier, when air motions are sharply curved scatterers may be centrifuged outward radially so that air motion is not identical to scatterer motion. Finally, there may be substantial errors introduced when the estimate of terminal fall speed is poor, for example, when there is large hail having very high terminal fall speeds ($\sim 35\,\mathrm{m\,s^{-1}}$ or greater), but the terminal fall speed for large raindrops is assumed ($\sim 10\,\mathrm{m\,s^{-1}}$). This terminal fall speed error may be particularly serious when the target storm is at close range and the elevation angle is high, so that vertical motions are well resolved by the radar. Such a situation is likely when a mobile Doppler radar is probing a tornadic supercell at close range, especially near a tornado. Dual/multiple Doppler analyses must therefore be viewed with caution when making quantitative measurements, even though there are a number of techniques employed to mitigate some of the aforementioned problems; the reader is referred to the current literature to find state-of-the-art solutions. Since commonly used quantities derived from the Doppler wind field—such as the three-dimensional vorticity vector, circulation, and trajectories—are very sensitive to the three-dimensional wind field, these derived quantities must be viewed with caution.

To get around the need for two or more Doppler radars, *spaced antenna* techniques are being tested so that the component of the wind parallel to the radar beam is used in the conventional way, but the component of the wind normal to the beam is estimated from a pair of closely spaced antennas. Such a technique requires a relatively long dwell time and averaging in space, so that the spatial and temporal resolutions are degraded. However, one does not need to worry about having the target storm appear only in a very limited volume.

It is also possible to estimate the wind field using only one Doppler radar by making use of the tracking of reflectivity elements. The equation for the conservation of radar reflectivity is

$$\partial R/\partial t = -u\,\partial R/\partial x - v\,\partial R/\partial y - w\,\partial R/\partial z + (\text{source and sink terms}) \quad \text{(A.6)}$$

If we ignore the source and sink terms and if the local time derivative term and advection terms are known from the radar reflectivity field, then (A.1), (A.6), and the equation of continuity are three independent equations in three unknowns. Again, the time-dependent vorticity equation or mass continuity may be used as constraints to solve for the three-dimensional wind field, and terminal fall velocity must be accounted for.

The two-dimensional wind field has also been estimated using TREC (tracking radar echoes by correlation), in which the horizontal wind direction and speed are estimated from the highest correlation between the radar reflectivity field for two successive scans of a limited area, one scan lagged with respect to the other. This procedure and the previous one fail when vertical motions are significant, when condensation or evaporation/sublimation is significant, when the radar reflectivity field is very noisy, or when the radar reflectivity field is perfectly uniform.

A retrieval technique for estimating thermodynamic variables based on the wind field derived from Doppler radar wind data is discussed in the main body of the text in Section 2.5.2. This technique may be improved upon by using a thermodynamic equation as a weak constraint. The retrieval of thermodynamic fields from wind data is sensitive to the vertical velocity field and its time derivative, which are not estimated very well.

More recently, researchers have incorporated Doppler radar data into numerical models using data assimilation techniques to estimate thermodynamic variables, which are not measured directly by Doppler radars, and to obtain estimates of the three-dimensional wind field when multiple-Doppler analyses are not possible. It has been found that when data from radars at different locations are assimilated, the accuracy of retrieved variables is increased. If a good fit of observational data to model "data" (i.e., to numerically simulated variables) can be obtained, then one can do diagnostic studies of the model data, which contain thermodynamic information in addition to the wind field. How to assimilate data into models is an area of current research and it is somewhat of an art to determine the best way to process the data. Ideally, one would expect the model, if left alone, to make a relatively "good" prediction of future events in order for one to trust retrieved variables enough to make diagnostic computations based on them, relevant to testing hypotheses.

When a best fit in the least square sense (i.e., through a variational analysis, involving cost functions) is found between wind data and their temporal evolution, the technique is referred to as 4DVAR, or four-dimensional variational analysis. An alternative to 4DVAR is the ensemble Kalman filter (EnKF) approach, in which statistics are employed to find the "most likely" estimate based on a numerical forecast and observational data. The statistics needed are the spatial and multivariate covariances of forecast errors. These covariances are estimated from an ensemble of forecasts, each of which is based on similar, but slightly different initial conditions or model physics. The reader is referred to papers listed in the reference list for more details on 3VAR, 4DVAR, and the EnKF techniques.

GENERAL MONOGRAPHS AND BOOKS

The reader is referred to p. 24 for a list of relevant general monographs and books.

REFERENCES AND BIBLIOGRAPHY

Armijo, L. (1969) A theory for the determination of wind and precipitation velocities with Doppler radars. *J. Atmos. Sci.*, **26**, 570–573.

Davies-Jones, R. P. (1979) Dual-Doppler radar coverage area as a function of measurement accuracy and spatial resolution. *J. Appl. Meteor.*, **18**, 1229–1233.

Doviak, R. J., P. S. Ray, R. G. Strauch, and L. J. Miller (1976) Error estimation in wind fields derived from dual-Doppler radar measurement. *J. Appl. Meteor.*, **15**, 868–878.

Dowell, D. C., F. Zhang, L. J. Wicker, C. Snyder, and N. A. Crook (2004) Wind and temperature retrievals in the 17 May 1981 Arcadia, Oklahoma, supercell: Ensemble Kalman filter experiments. *Mon. Wea. Rev.*, **122**, 1982–2005.

Gal-Chen, T. (1982) Errors in fixed and moving frame of references: Applications for conventional and Dopper radar analysis. *J. Atmos. Sci.*, **39**, 2279–2300.

Gao, J., M. Xue, A. Shapiro, and K. K. Droegemeier (1999) A variational method for the analysis of three-dimensional wind fields from two Doppler radars. *Mon. Wea. Rev.*, **127**, 2128–2142.

Hildebrand, P. H., W.-C. Lee, C. A. Walther, C. Frush, M. Randall, E. Loew, R. Neitzel, R. Parsons, J. Testud, F. Baudin, and A. LeCornec (1996) The ELDORA/ASTRAIA airborne Doppler weather radar: High-resolution observations from TOGA COARE. *Bull. Amer. Meteor. Soc.*, **77**, 213–232.

Houtekamer, P. L. and H. L. Mitchell (1998) Data assimilation using an ensemble Kalman filter technique. *Mon. Wea. Rev.*, **126**, 796–811.

Majcen, M., P. Markowski, Y. Richardson, D. Dowell, and J. Wurman (2008) Multipass objective analyses of Doppler radar data. *J. Atmos. Oceanic Technol.*, **25**, 1845–1858.

Rhinehart, R. E. (1979) *Internal Storm Motions from Single Non-Doppler Weather Radar*, Tech. Note NCAR/TN-146+STR, National Center for Atmospheric Research, Boulder, CO.

Sasaki, Y. (1970) Some basic formalisms in numerical variational analysis. *Mon. Wea. Rev.*, **98**, 875–883.

Shapiro, A., C. K. Potvin, and J. Gao (2009) Use of a vertical vorticity equation in variational dual-Doppler wind analysis. *J. Atmos. Oceanic Technol.*, **26**, 2089–2106.

Shapiro, A., K. M. Willingham, and C. K. Potvin (2010) Spatially variable advection correction of radar data, Part I: Theoretical considerations. *J. Atmos. Sci.*, **67**, 3445–3456.

Snyder, J. C., H. B. Bluestein, G. Zhang, and S. J. Frasier (2010) Attenuation correction of high-resolution, X-band, dual-polarized mobile radar measurements in severe convective storms. *J. Atmos. Oceanic Technol.*, **27**, 1979–2001.

Sun, J. and N. A. Crook (1994) Wind and thermodynamic retrieval from single-Doppler measurements of a gust front observed during Phoenix II. *Mon. Wea. Rev.*, **122**, 1075–1091.

Trapp, R. J. and C. A. Doswell, III (2000) Radar data objective analysis. *J. Atmos. Oceanic Technol.*, **17**, 105–120.

Tuttle, J. D. and G. B. Foote (1990) Determination of the single boundary layer airflow from a single Doppler radar. *J. Atmos. Oceanic Technol.*, **7**, 218–232.

Weygandt, S. S., A. Shapiro, and K. K. Droegemeier (2002) Retrieval of model initial fields from single-Doppler observations of a supercell thunderstorm, Part I: Single-Doppler velocity retrieval. *Mon. Wea. Rev.*, **130**, 433–453.

Index

www.ingramcontent.com/pod-product-compliance
Lightning Source LLC
Chambersburg PA
CBHW080943211125
35560CB00037B/702
* 9 7 8 3 6 4 2 0 5 3 8 0 1 *